George N. Tsoudis

THE SHIPBROKER'S WORKING KNOWLEDGE

akakia

Published in England by AKAKIA Publications, 2021

George N. Tsoudis

THE SHIPBROKER'S WORKING KNOWLEDGE

ISBN: 978-1-912935-45-1

Cover Image:
Source: shutterstock.com / Copyright: Sorapop Udomsri / File No: 195546176
Source: shutterstock.com / Copyright: zebra0209 / File No: 67518703
Mixed and Designed by AKAKIA Publications

akakia

46 Warberry Road, London, N22 7TQ
T. 0044 02082457849

www.akakia.net
publications@akakia.net

2021, London, UK

CONTENTS

PART I

OVERVIEW OF THE DRY CARGO SECTOR

Carriage of goods from one place to another by sea is the most common and the most economical method of transportation.

Billions of tons of dry goods are transported by sea around the world each year and thousands of ship-fixtures are concluded and successfully executed, under the terms of the governing Charter Party.
The entire maritime transaction, which should be inhered in the working knowledge of any shipbroker who is involved in the dry-cargo chartering field, can be grouped into seven short steps as follows:

STEP ONE.
Sale of the cargo.

To enable an international sales contract (international trade transaction) and marine transportation to take place in the dry cargo sector (by the international sea-going ships), the majority of the following contractual parties should be involved:
-the Shippers/Suppliers and/or Sellers, who are the Producers and/or Exporters of the intended goods for export.
-the Freight Forwarders, who will take over and perform on behalf of the Shippers/Sellers and/or the Consignees/ Buyers all the required/necessary arrangements for the completion of the transaction, including the documentation procedures.
-the Charterers (who could also be the Shippers/Sellers, or the Buyers, or the Freight Forwarders of the cargo, or the Operators), who will Charter (fix) a ship from the open shipping market through their shipbrokers, for the transportation

of the goods from one place to the other (from the port of shipment to the port of discharge).
-the Ship-Owners (they have the possession and legal rights to the ship), or the Managers (they are the party hired by the Owners to 'run' the ship according to applicable International rules), or the Disponent Owners (they have hired a ship for a Time Charter trip out or for a short/ long period), who will provide a ship in the right position, that is seaworthy, suitable for the shipment and Charterers' needs, for the carriage of the intended goods by sea.
-the Buyers and/or Consignees, who are the Importers and/or final Receivers of the carried goods for import.

When the parties involved of a sales contract finally reach an agreement for an international marine transportation, the Buyers' or Consignees' bank will issue a document, namely a 'Letter of Credit' that will guarantee the payment of goods under specific conditions to the bank of Shippers/Sellers and the documentation part of the sales process should be deemed to have been concluded.

The majority of the sales contract terms are defined under the International Commercial Terms (INCOTERMS), which, with its standard terms, assigns and allocates the duties/obligations of each involved party (Sellers and Buyers), the cost, as well as the financial and commercial risks and responsibilities for any commercial transaction by sea among countries.
The International Chamber of Commerce created and published the first Incoterms in 1936, which has been accepted and adopted by all countries/legal authorities and has been updated 6 times since then.
There are thirteen different terms of sale, as embodied in Incoterms, with each term determining the variety of transactions.
These terms are grouped into four categories, according to their first letter:

1. Ex Works (EXW)
2. Free Carrier (FCA)
   Free Alongside Ship (FAS)
   Free On Board (FOB)
3. Cost and Freight (C+F or CFR)
   Carriage Paid To (CPT)
   Cost, Insurance and Freight (CIF)
   Carriage and Insurance Paid to (CIP)
4. Delivered At Frontier (DAF)
   Delivered Duty Paid (DDP)
   Delivered Ex Ship (DES)
   Delivered Ex Quay (DEQ)
   Delivered Duty Unpaid (DDU).

For the dry cargo sector, the following three main terms of sale apply:

A. Free On Board (FOB).
The selling price of the cargo and its terms binds the Sellers/Shippers under their risks/costs to:
- carry the cargo from their mines/farms/factories etc. and present it as per the sales contract to the named loading port/place within the agreed dates for shipment.
- load the ship (free in/ no cost to Carrier), as the ship will be presented by the Buyers.
- arrange for the customs formalities and clearance of the cargo for export.

The selling price of the cargo and its terms binds the Buyers/Consignees under their risks to:
- enter into a contract of carriage (fix/present a ship) with the Ship-Owners or Disponent Owners and pay for all costs of transportation by sea (freight-demurrage etc.) to the place of destination, as will be nominated by them.
- discharge the ship under their cost (free out/ no cost to Carrier) and carry it to its destination.
- arrange and pay for the cargo insurance.
- arrange and pay for the customs formalities and clearance of the cargo for import and
- take the risk of loss of or any damage to the cargo that is being shipped (has passed the ship's rails)/after discharge.

B. Cost and Freight (C+F).
The selling price of the cargo and its terms binds the Sellers/Shippers under their risk to:
- carry the cargo from their mines/farms/factories etc. and present it as per the sales contract to the named loading port/place within the agreed dates for shipment.
- enter into a contract of carriage (fix/present a ship) with the Ship-Owners or Disponent Owners and pay for all costs of transportation by sea (freight-demurrage etc.) to the agreed with the buyers destination and sales contract.
- load and discharge the ship under their cost (free in and out/ no cost to Carrier).
- arrange and pay for the customs formalities and clearance of the cargo for export.

The selling price of the cargo and its terms, binds the Buyers/Consignees under

their risk to:
- arrange and pay for the cargo insurance.
- arrange and pay for the customs formalities and clearance of the cargo for import and
- take the risk of loss of or any damage to the cargo that is being shipped (has passed the ship's rails) /after discharge.

C. Cost Insurance and Freight (CIF).
The selling price of the cargo and its terms binds the Sellers/Shippers under their risk to:
- carry the cargo from their mines/farms/factories etc. and present it as per the sales contract to the named loading port/place within the agreed dates for shipment.
- enter into a contract of carriage (fix/present a ship) with the Ship-Owners or Disponent Owners and pay for all costs of the transportation by sea (freight-demurrage etc.) to the agreed destination, that is nominated by the buyers and sales contract.
- load and discharge the ship under their cost (free in and out/ no cost to Carrier).
- arrange and pay for the cargo insurance.
- arrange and pay for the customs formalities and clearance of the cargo for export.

The selling price of the cargo and its terms, binds the Buyers/Consignees under their risk to:
- arrange and pay for the customs formalities and clearance of the cargo for import and
- take the risk of loss of or any damage to the cargo that is being shipped (has passed the ship's rails) /after discharge.

STEP TWO.
The cargo and the ship.

The cargo:
For carriage by sea, dry cargo is the good of any grade, type, quality and of any quantity, which is extracted or manufactured or grown or produced from the mines, factories, farms located inland or ashore and moved/carried in any form, close to the port, for the purpose of being loaded into the cargo ship's spaces, for transportation to another foreign or within the same country's port/s.

All solid bulk cargoes (solid are all cargoes other than liquid or gas) and especial-

ly dangerous solid bulk cargoes should comply with the
. International Maritime Solid Bulk Cargoes Code (IMSBC Code).
The cargoes listed in the IMSBC Code have an official name (Bulk Cargo Shipping Name-BCSN) and the sea transport documentations issued for the cargo must be identified with the BCSN sign.

The dry bulk cargoes are classified by the IMSBC Code into three groups:
Group A:
Solid bulk cargoes which may liquefy during the sea voyage, resulting in cargo shift and possible ship capsizing, are listed in this group. Such cargoes are nickel ore/mill scale/iron ore fines/ pyrites etc. Special care must be taken by the Master/Owners prior to commencement of the loading, in the areas where the cargo is exposed to rain without proper protection. The moisture content of the cargo must be measured/determined by the recognised international methods, which in any case must not be in excess of their Transportable Moisture Limit (TML). However, dry cargoes are not liable to liquefy.
Group B:
Solid bulk cargoes, which may pose a chemical hazard and could expose the ship and crew to danger, are listed in this group. These cargoes are dangerous, because when they come in contact with air or by interaction with water, they can come to be self-heating or emit flammable gases in dangerous quantities during the voyage or turn into radioactive materials or even contain toxic substances etc.
Dangerous goods are classified according to their particular characteristics (4.1/ 4.2/ 4.3/ 5.1/ 6.1/ 7/8/9), while certain others, when carried in bulk, are characterized as 'Materials Hazardous only in bulk' (MHB). Shipments will only be executed on specific ships that can provide the Appendix 'B' certificate, as will be issued by a classification society (ship's class) and can provide a CO2 fire-fighting system in their cargo holds.
Group C:
Solid bulk cargoes, which are simply not liable to liquefy and do not pose chemical hazards are listed in this group.

However, certain mineral cargoes are listed in both Group A and Group B, i.e., pose chemical hazards, and may liquefy during the sea voyage.

For all cargoes, whether they have been classified as dangerous or not, the Shippers are obliged before loading to provide the Master of the ship with a cargo declaration, as outlined in the IMSBC and the BCSN, as well a valid certificate with the information of the physical and chemical properties and the required conditions for the carriage.

Shippers and Stevedores will load and discharge the cargo strictly as per the agreed ship's loading sequence/stowage plan/method of cargo operations and as per provisions of SOLAS chapter VI, regulation 7.3 (loading, unloading and stowage of solid bulk cargoes) and the Code of Practice for the Safe Loading and Unloading of Bulk Carriers (BLU Code).

The cargo can be transported:

. As full cargo.

Even if the ship's holds and/or deck cargo spaces have not been filled with cargo, the Owners are not allowed by the terms of the Charter Party to search the market and load any other 'completion cargo' in the cargo holds or on deck.

. As part of the cargo (part cargo).

If, the ship's holds and/or deck cargo spaces have not been filled with cargo, the Owners have the right, under the terms of the Charter Party, to search the market for a 'completion cargo' (until it reaches the maximum deadweight which the ship can provide), mainly en route to its final destination. During the negotiations of any 'part cargo', both Charterers and Owners could negotiate whether the cargo intended to be loaded is to be 'first in-last out' or 'last in-first out' etc. from the cargo holds.

This chosen method of transportation will be incorporated into the Charter Party.

The ship:

A commercial ship for dry cargo is a 'steel lady' that is seaworthy, navigable and was built and designed for the carriage by sea, from one place to another, of any grade, type, quality, and form of goods for transportation.

Dry cargo ships are designed in different types and sizes, depending on the global demand for marine cargo transportation. The world population is constantly increasing, resulting in increased needs for transportation and imposing the need to build new larger ship sizes that should comply with the environmental protection and in addition to provide improved fuel efficiency. This is also supported by the construction of new modern ports/berths/facilities, which can accommodate these large sizes and can provide means of very fast loading and discharging.

The size of the ship is identified by the

. deadweight all told (abbreviated as 'dwat'),

which is the maximum deadweight which the ship can provide, as designed by the shipyard based on transactions within zones of summer, at the designated scantling draft after calculating the following complete weights:

- the cargo,
- the fuel including lubricants,
- the fresh water,
- the ballast water,
- the ship's crew and their effects,
- the stores, spares, provisions, equipment, inventories etc. on board the ship.

When the ship navigates within winter or tropical zones,
- the draft will fluctuate, from reducing or increasing the deadweight.

The
. deadweight cargo capacity (abbreviated as 'dwcc'),
is the total (but not standard) weight of the cargo that the ship can accommodate, when loading by virtue of the maximum permitted weight, net of current fuels on board the ship, fresh and/or ballast water and all other weights as stated above.

The main types of commercial cargo ships are:
A. Dry cargo ships,
which are designed to carry dry cargoes in any form, in bulk or in bags/palletized etc. such as grains/minerals/cement/ projects/steels etc.
These ships can provide one or two decks and some of them even three decks in part of their hold/s. They can be fitted with cargo gear (cranes, derricks etc.) on their decks for self-loading / discharging or they can be gearless and proceed only to terminals that provide shore loading/ discharging means.
B. Tanker ships,
which are designed to carry liquid cargoes such as oils, water, fuels etc.
C. LNG-LPG ships,
which are designed to carry Liquefied Natural Gas)/LPG - carry Liquefied Petroleum Gas
D. Container ships,
which are designed to carry cargoes that are stowed in containers. The containers are iron boxes with a standard size of 20' or 40' feet. The size and capacity of the container ship is defined by how many TEU (Twenty feet Equivalent Units) it can carry. They can provide very fast loading and discharging (subject to the facilities provided by the container terminal) making the ship's contact with the ports very short.
E. Reefer ships,
which are designed with refrigerated holds that can carry fresh or frozen types of cargo such as bananas, citrus, meat, fish etc. and perishable goods in general.
F. Ro-ro (roll on-roll off) ships,

which are designed to carry cargoes that can be usually loaded and discharged using their own means (machines), with wheels that roll on and roll out of the ship's ramps such as cars/trailers/ vehicles etc. These ships do not have cargo hatches.
G. Livestock or Cattle carriers,
specialized ships, which are designed to carry only live animals.

The Dry cargo ships are divided into two main categories:

1- the bulk carriers/singledeckers (abbreviated as 'bc/sid' or 'bc/sd')
are designed to carry homogeneous bulk (loose) cargoes and can only provide one deck.
This type of ship is segregated as follows, depending on the dwat and render services:
A- 'coasters'/small sized ships,
up to about dwat 10,000 mtns.
'Coasters' (say up to 2-2,500 dwat) and small ships, because of their size, usually trade in specific areas, such as the Mediterranean Sea including Black Sea/ Caribbean Sea/ Continent including Baltic/ Arabian Gulf including India /Southeast Asia including China etc. The shallow draft they can provide, their short overall length, their low height (they can navigate under bridges) and their ability to trade in confined waters (tidal ports/rivers etc.), make these ships workable for operations, at ports/places where the largest sea-going ships cannot approach. They will trade along the coast and have the advantage of transporting small quantities of cargo and assisting in the development of the economy between the countries of the region they serve, while also assisting the small inland factories/ producers/mines and agricultural farms in exporting their products.
B- 'handy' sized ships,
ranging in size from about dwat 10,000 mtns to about dwat 40,000 mtns.
The 'handy' ship is the base of the pyramid and the most circulated ship in the maritime sector. It can carry any type of dry cargo and can approach many ports around the world. Due to its convenient size, it has access to almost all terminals, which can either provide shore facilities for loading and discharging (many of them are gearless) or are restricted and the ship will operate with the provided own means.
The 'handy' ships can be grouped as follows:
-the small 'handysize' (from about 10 to about dwat 25,000mtns),
-the large 'handysize' (from about 25 to about dwat 40,000 mtns) .
The ships which are equipped with stanchions and can carry logs are called 'loggers'.
The ships which can approach and trade within the Great Lakes and pass through

the seaway locks are called 'lakers'.
The ships that are designed to carry woodchips are called 'woodchip carriers'.
C- 'handymax' sized ships,
ranging in size from about dwat 40,000 mtns to about dwat 59,000 mtns.
For many brokers, these ships are the most favorable size and the most profitable investment. The majority of them can provide cargo gear and can carry any grade/ type of dry cargo.
The 'handymax' ships can be grouped as follows:
- the normal 'handymax' size ranging from about dwat 40,000 to about dwat 50,000 mtns. They can trade around the world, but a number of them are gearless or provide insufficient gear.
- the new modern 'handymax' is called the 'supramax', and ranges from about dwat 50,000 to about dwat 59,000 mtns. The original 'supramax' is about dwat 52,000 mtns with 5 holds/ hatches that can provide 30 mtns of gear capacity (cranes) with a grain capacity similar to a small 'Panamax' ship. These types of ships can offer huge services, while many of them can also be fitted with grabs.
A new ship, designed about 10 mtrs longer than a 'supramax', called 'Ultramax'. Its size is about dwat 60-64,000 mtns and can provide cranes, improved fuel efficiency and much better grain or bale cubics than an old 'Panamax' of similar size.
D- 'panamax' sized ships,
ranging in size from about dwat 60,000 mtns to about dwat 85,000 mtns.
The oldest of the maritime community argue that if an Owner has not owned a 'Panamax' size ship, he has not reached the top.
The name of the ship itself 'Panama Maximum' indicates that the ship has been constructed to pass the Panama Canal locks (was opened in 1914-two sets of locks), according to the maximum passage width dimensions (beam), while maximizing the cargo capacity in accordance with the prevailing tropical draft.
The Panama Canal can provide maximum a l.o.a 294 mtrs- beam 32,31 mtrs-draft 12,0 mtrs in tropical fresh water and air draft of 57 mtrs. The normal transit capacity of the Panama Canal is approximately 40 ships per day.
The expansion of the Panama Canal with the third set of locks opened in June 2016, larger than the original two opened in 1914, allow the transit of larger ships.
The Canal expansion increased the allowed dimensions to loa 366 mtrs-beam 49 mtrs and draft 15,20 mtrs and the 'New Panamax' or 'Neo Panamax' ship, capable of handling more cargo, is the next big investment.
The 'Panamax' ships carry mostly any type of grains/coal/minerals etc., always in bulk form.
A few of them accommodate cranes and can provide services to specific transactions.

The 'Panamax' ships can be grouped as follows:
- 'Panamax' is the normal size of Panamaxes ranging from about dwat 60,000 to about dwat 70,000 mtns. However, nowadays, new building of this size is limited, due to the increased global transport demand. In the 70's this size was the 'king', and in the 80's it was a typical panamax.
-Large 'Panamax' including the 'Large Modern Economic' (abbreviated as 'LME') ships that extend over dwat 70,000 mtns can provide a large cargo capacity and improved fuel efficiency. In the 90's and 2000's this size was a typical panamax.
'Kamsarmax' is a new type of ship and is nowadays the most favorable for the market needs. It can provide dwat about 82,000 mtns to a maximum length over all (loa) of 229 mtrs, which allows it to load at Kamsar port in Guinea, maximizing its cargo capacity in accordance with the prevailing restrictions. The port of Kamsar is one of the world's biggest ports, exporting bulk bauxite.
E- 'Cape' sized ships,
ranging in size from about dwat 85,000 mtns and over.
The name of the ship itself states that the ship is constructed so that it can only pass through either Cape Horn or the Cape of Good Hope and is located at the shipping's apex (on the top of the shipping pyramid).
The 'Capes' can trade only between terminals especially constructed to accommodate these types of ships, providing deep waters and special means for loading/discharging for the ships' quickest despatch.
The 'Capes' are characterized as 'dirty' when carrying coal/minerals etc., or 'clean' when carrying only grains.
Ships with a size ranging from about dwat 85,000 mtns to about dwat 100,000 mtns, are called 'Post Panamax', i.e., bigger than a Panamax and cannot cross the Panama Canal. However, the ships called 'Neo Panamax' or 'New Panamax' of this size or even larger can cross the new Panama Canal locks.
The 'Oil Bulk Ore' (abbreviated as 'OBO') is the 'Cape' which can carry liquid and dry cargo in separated holds at the same time or not. The additional maintenance required and the difficulties arising when loading dry cargo after the discharge of liquid cargo, as well as the instability of the ship, lead the owners to use the ships either as tankers or as bulk carriers.
'Mini Capes' characterize ships ranging in size from about dwat 100,000 mtns to about dwat 120,000 mtns.
'Cape' ships over dwat 200,000 mtns are called 'Very Large Bulk Carriers'.
'Dunkirk Max' is a 'Cape' with a dwat about 175,000 mtns, a maximum length over all 289 mtrs/ beam 45 mtrs, which can trade in Dunkirk port, maximizing its cargo capacity in accordance with the prevailing restrictions there.
'Newcastle Max' is a 'Cape' with about 200,000 dwat, maximum beam 47 metres, which can load from Newcastle in Australia, maximizing its cargo capacity in accordance with the prevailing port restrictions there.

2- the general cargo ships,

which can actually provide more than one deck, therefore dividing the cargo space ('multipurpose'/'tween-decker'/'triple-decker' ships etc.) and are designed to carry homogeneous or inhomogeneous general cargoes (break-bulk cargoes), in the same or different grade /type of cargoes, whether in bags/ slings/ palletized/boxes/barrels/bundles or in any other form of manufacture. General cargo ships are equipped with their own means for the loading and discharging operation ('geared' ships).

The 'tween-decker' ships were famous before the advent of the container ships, which eventually replaced them. In this category the most famous standard designed ship was the 'Liberty' type ship, built in the USA during World War II, in order to replace the lost merchant ships and to cover the need for transportation of all kinds of war supplies. These ships provided about 9,000 mtns cargo capacity in five holds plus their deck space for the lashing of war materials. After the end of the war, the ships were purchased by various Owners (mainly by Greeks) and entered the shipping market. Another general cargo ship was the 'SD 14' (British type-Shelter Deck, 14,000 mtns dwat) actually designed in the middle of the 1960's to replace the 'Liberties'. At that time, it was the ideal size with a low building cost. There was also the 'Freedom' type ship (Japanese type) of about dwat 15,100 mtns equipped with derricks, as well as the 'Santa Fe' type (Spanish type) and the 'Mark 2' (Germania type).

The 'multipurpose' ships (abbreviated as 'mpp') are modern general cargo ships that do not have a standard design and size. They can carry any type/grade of general cargo that cannot be fitted into containers. They can carry standard containers, as well as, reefer containers, large or even heavy machinery, forest products, steel, heavy lift cargoes, project cargoes, bulk cargoes, dangerous goods etc. and most of them provide wide hatches, collapsed tweens and heavy lift cranes for loading or discharging purposes.

Ship's class:

The ship, from the time that the order for its construction is placed, should be followed/monitored by a classification society, which will initially survey/ supervise the construction and set the standards of the build. When it is delivered to the Owners for trade and throughout its 'life', the classification society will certify its operating condition and maintenance.

Every five years, the ship is under a class commitment to pass a 'Special Survey' (abbreviated as 'SS'), where all ship's parts/equipment/areas including the thickness of the plate, the hatch covers, holds, machineries etc. should be examined and where necessary to be repaired and/or replaced.

Approximately every two and a half years, the ship is under the commitment to

enter Dry-Dock (abbreviated as 'DD'). The ship should be removed from the sea in a dock that the Owners have previously chosen, for its exterior parts to be examined and repaired and/or painted accordingly.
For older ships, both surveys are expensive and along with the fact that the ship will remain off-hire for a long time, a cost that may not be able to be recovered by the existing market, often leads the Owners to sell the ship for 'scrap'.

The major classification societies are:
-UK- Lloyd's Register of Shipping (Abbreviated as 'LR').
-FRANCE- Bureau Veritas (Abbreviated as 'BV').
-NORWAY- Det Norske Veritas (abbreviated as 'DNV').
-USA- American Bureau of Shipping (abbreviated as 'ABS').
-JAPAN -Nippon Kaiji Kyokai (abbreviated as 'NK').
-GREECE -Hellenic Register of Shipping (abbreviated as 'HRS').
-GERMANY- Germanischer Lloyd (abbreviated as 'GL').
-ITALY - Registro Italiano Navale (abbreviated as 'RINA').
-CHINA- China Classification Society (abbreviated as 'CCS').
-INDIA -Indian Register of Shipping (abbreviated as 'IRS').
-POLAND- Polski Rejestr Statkow (abbreviated as 'PRS').
-SOUTH KOREA- Korean Register of Shipping (abbreviated as 'KRS').

Ship's flag:
The ship will be controlled and regularly inspected by the 'flag state', under the national flag of which the ship is registered. The ship will operate under the law of its flag, which will inspect/ certify the equipment/crew and will issue the safety and pollution prevention documents.
Ships hoisting a 'flag of convenience' have reduced or low operating and tax/registration costs, as well as the ability to employ cheap crew/officers, but should, however, avoid trading worldwide and especially in countries where the International Transport Workers' Federation (abbreviated as 'ITF') is strict.
ITF is a union that
. protects and assists seafarers worldwide
and represents their interests and rights, covering all crew nationalities, irrespective of ship's flag/ownership.

Ship's insurance coverage:
The ship must be fully covered for Marine Risks/liabilities to third parties, since the delivery to its Owners and before it begins to trade, through their insurance brokers, who are in contact with a number of insurance companies.

The Marine Insurance mainly includes three types of coverage:

-P&I Club insurance (protection and indemnity),
-H&M insurance (hull and machinery),
-War Risk insurance.

The P&I Club (Protection + Indemnity) provides cover to their members (not only to ship-Owners, but also to Charterers or Operators), for a wide range of liabilities to third parties, such as
. cargo claims, oil pollution, wreck removal, injury, or sickness of crew etc.
The International Group of P&I Clubs with its thirteen members (Groups) covers more than the 90 percent of the world's trading ships.
The thirteen Groups are:
-American Steamship Owners Mutual Protection and Indemnity Association, Inc.
-Assuranceforeningen Skuld.
-The Britannia Steam Ship Insurance Association Limited.
-The Japan Ship Owners' Mutual Protection & Indemnity Association.
-The London Steam-Ship Owners' Mutual Insurance Association Limited.
-The Swedish Club.
-The Steamship Mutual Underwriting Association (Bermuda) Limited.
-The North of England Protecting & Indemnity Association Limited.
-The Standard Club Ltd.
-Gard P&I (Bermuda) Ltd.
-The Ship owners' Mutual Protection & Indemnity Association (Luxembourg).
-United Kingdom Mutual Steam Ship Assurance Association (Bermuda) Limited.
-The West of England Ship Owners Mutual Insurance Association (Luxembourg).

The H&M insurance will cover the ship from
. marine perils/its total loss (actual or constructive) and for damage caused to the ship's structure and machinery and those that are connected therewith, either to collision or to any other way that the ship can be damaged, such as a storm/crew negligence etc. and will contribute to the General Average and pay for the salvage expenses, but always provided that the ship is trading within the International Navigating Limits (I.N.L.). If the ship trades outside of the I.N.L. a written acceptance/confirmation should be obtained by its insurance company. In any case excludes the coverage in case of damage or losses caused due to wars/ civil wars/ riots/ revolutions/terrorist acts/piracy etc. However, the Owners or Managers will arrange for the ship's coverage for all these risks, through the recognised
. War Risk insurance companies.
A basic war risk annual premium is payable by the Owners to them, provided that the ship is not trading within high-risk areas.

If the ship enters ports, places, countries, zones, or areas, where there are wars or war like operations or piracy etc. and are generally considered as a 'high-risk areas', as listed from time to time by the War Risk insurance companies, an extra cover then is required, and certain additional premiums (Extra War Risk Insurance) are in effect and will be charged on the ship.

Ship's inspection at the ports of contact for operation:
During the operations in the port, the ship may be subject to inspection by the inspectors
. of the Port State Control (abbreviated as 'PSC').
The PSC are regional organizations covering the entire coastal world as follows:
-The Europe and the north Atlantic is covered by the Paris MOU.
-The Asia and the Pacific Ocean from the Tokyo MOU.
-The Caribbean Sea from the Caribbean MOU.
-The Mediterranean Sea from the Mediterranean MOU.
-The Indian ocean from the Indian Ocean MOU.
-The Latin America from the Acuerdo De Vina Del Mar.
-The West and Central Africa from the Abuja MOU.
-The Black Sea from the Black Sea MOU.
-The Arab States of the Gulf from the GCC MOU.

The PSC inspectors will check/inspect/ensure:
-the ship's compliance with the International conventions (MLC- Maritime Labour Convention / SOLAS -Safety of Life at Sea / STCW – International Convention on Standards of Training).
- the ship's condition (if there is lack of maintenance and if affecting its seaworthiness).
- the ship's documents, (if they are valid and up to date. All ship's certificates according to SOLAS must be originals with certain exceptions).
- the crew's certifications for inadequate certifications (safe manning with qualified officers and crew).
- if the equipment is working properly.
- the operation of lifeboats (life-saving equipment).
- the fire safety on board.
- the cleanliness in the engine room etc.
and generally, if the ISM procedures are followed by the Master/officers/crew, as well as, ensuring whether the ship meets and maintains the safety and the security standards and in an operational condition in compliance with the international maritime conventions/law.
The well-trained officers and crew and their familiarization with the company's safety management system (SMS) is of the obligation of the Master/company.

Officers'/crew's failure to comply with the plans for shipboard operations will result in possible ISM failure, with huge consequences to the ship.
If the ship fails to pass the PSC inspection, it may be detained in the port and its sailing prohibited. The ship, depending on the nature of the deficiency/ies, must comply with any repairs and/or recommendations made by the Authorities, rectifying them, as the case may be, before departure or until the ship reaches the next port or within a certain time.
The most important codes to action taken are:
- Code 15: rectify deficiencies by next port.
- Code 16: rectify deficiencies within 14 days.
- Code 50: rectify deficiencies within 30 days.
- Code 17: rectify deficiencies prior to departure.
- Code 30: detained ship.

If the ship was detained (code 30), it means that the deficiencies were serious, and the ship was considered by the PSC inspectors as unsafe to trade and sail from the port. The ship will remain
. detained,
until all the deficiencies are rectified, either at the port of occurrence, or otherwise, and, if due to lack of repair yards or competent shore equipment etc., the ship under special conditions and provided it has no cargo on-board, may be granted permission to sail to a nearby port/yard, which can provide all the required means to rectify the deficiencies.
The ships under its flag, are categorised as being in a
. black or grey or white list.
Obviously, the ships which are black, or grey listed constitute a 'red flag' for the inspectors.
It is a fact that a ship is profitable for the Owners only when it is working properly and without delays, hence the ship's Management should take all the necessary measures, to ensure that the ship's hull condition, maintenance, equipment, and crew to be to the PSC inspectors' full satisfaction, with adequate manning and in such an operational condition, as to be always in compliance with the international maritime conventions. Normally, even if the ship is detained, it is not usually for a long time and the loading or discharging operation/program is not disrupted.

Ship's supply with Marine fuels:
The Owners on Voyage Charter and the Charterers on Time Charter remain responsible for the quality of the fuel as ordered and supplied on the ship. The fuel supplied must be in accordance with the latest ISO standards, of a stable and homogeneous nature and of such specifications and grades suitable for burning in vessel's engine as well as to comply with the various regulations within the ar-

ea of trading and especially on the level of sulphur content.
Fuel used in marine transportations is regulated by the IMO's (International Maritime Organization) Marine Environment Protection Committee (MEPC70). Since 1st January 2020, all ships operating globally (always outside of the designated emission control areas/ European and American ECA with less than 0,10% of sulphur as well as regions/ports implemented their own sulphur emission limits) comply with the IMO MARPOL new regulations as addressed in Regulation 14 of the Annex VI, and burn in their engine marine fuel with a reduced sulphur from the previous limit of 3,50% to the current limit of 0,50%. This restriction is expected to reduce the sulphur oxides emitted from the ships mainly in the coastal areas.
Some Owners, in order to be a step ahead, and avoid paying the more expensive low sulphur (0,50%), they selected to install the Exhaust Gas Cleaning Systems (EGCS). The scrubbers allow the ships to keep burning heavy sulphur fuel oil (3,5%) as remove the Sulphur Oxides (Sox) and Nitrogen Oxides (NOx) from the exhaust gasses generated. As the wastewater generated from scrubbers can be released into the sea, some countries have imposed bans on ships with scrubbers and prohibit the discharge within their port limits.
From each bunkering operation samples must be taken from the ship's manifold, for fuel quality analysis purposes.

The workable ship for a particular 'cargo order':
Charterers will select a ship from the market that is the most workable for the needs of the shipment (singledecker or multipurpose or ship, with box-holds or not, geared, or gearless etc.), taking into consideration various parameters (port restrictions/means of loading or discharging/type and form of cargo etc.) and what method of fixing (Voyage or Time Charter) is the most profitable for them. However, and most of the times, the ship's description/flag/ownership etc. is subject to the Letter of Credit terms and/or to the Shippers'/Receivers' approval. Sometimes, the method of fixing is also under discussion if the Owners are only Voyage or Time Charter minded.
The selected and subsequent fixed ship will proceed for loading from the Shippers' mines, or from the place where Shippers will present the cargo for delivery to the Carriers, towards the Consignees' mines or place of discharge from where the Consignees will be receiving their cargo.

The ownership and the type/size of the ship will be selected depending on:
-the type/grade of cargo, if it is wheat or cement, forest products, project cargo, steel etc. and if it contains dangerous/flammable/harmed goods etc., in which case a class approved ship for loading and transportation of such cargoes will be required.

-the total available quantity of cargo and if a 'coaster'/'handy'/'panamax'/'cape' ship etc. will be required for loading
-the cargo form if it is in a bulk/bags/pallets/bundles etc. condition. in which case, as the case may be, a straight bulk carrier or multi-purpose, tween-decker ship etc. will be required for loading.
-the availability of facilities at both ends, i.e., if shore cranes, evacuators or other forms of loading or discharging equipment are available or not, in which case a geared or gearless ship will be fixed.
-the intended place/port for loading and/or discharging, as an Ice class ship, River ship, Laker, Kamsarmax or other type of ship will be required for loading.
-the specific terms of the Letter of Credit, or restrictions from cargo insurance companies for the maximum acceptable age of the ship which will be fixed, indicating if an over aged or not ship will be required for loading.
-the ship's flag/nationality of crew/ownership/previous trade etc. in relation to political constrains and possible boycott in the port of loading or discharging.
-the port restrictions in effect in the port of load and/or discharge that will determine the size of the ship.

The legitimate employment instructions to the ship:
The Master of the ship will be receiving legitimate employment instructions and in accordance with the terms of the Charter Party
. either directly by the 'Head Owners' and provided that he is executing a Voyage Charter,
. or by the 'Managers' and provided that he is executing a Voyage Charter,
. or by the 'Disponent Owners'/ 'Operators', who have taken (fixed) the ship on Charter (as Time Charterers) for a specified short/long period or even for a trip out.

The Owners/Managers will provide safe manning, an efficient crew, updated charts and if on a Voyage Charter, also sufficient bunkers for the execution of the voyage and generally produce a seaworthy ship for the satisfactory prosecution of the intended contract of carriage or Time Charter period. Valid and up-to-date original certificates should be in the ship's possession, complying with the law of its class/flag etc. and with all international and local regulations in the ports that will be contacted. The ship should be kept in perfect condition and maintained the same throughout the duration of the voyage or Time Charter period, for its obligations under the terms of the Charter Party to be fulfilled.

Dry cargo ships can perform as:
-Tramp ships: operating in a non-standard schedule of worldwide trade and searching the open market for the best offer. The ship will trade for any voyage between different countries and with different cargoes, in accordance with the

terms of the issued Charter Party and of the issued Congen Bills of Lading, or otherwise depending on the cargo carriage and Charter Party terms.
-Liner ships: operating under a standard schedule/program between fixed countries and within a certain period of time, under the issued Conline Bills of Lading. Liner ships have berthing priority upon arrival in port.

STEP THREE.

Methods of chartering a ship.

When the cargo intended for the carriage is sold and the Letter of Credit is in order (open), the next step should be for the Charterers to search in the open market, through their brokers, for the charter of a suitable and in the right position ship.

A cargo ship can be chartered based on the following three methods:
1. on Voyage Charter.
2. on Time Charter.
3. on Bareboat Charter.

1. Voyage Charter.
The Voyage Charter is the most common method of transporting goods by sea and the agreement/ fixture/Charter Party is a contract of carriage between the parties involved (Owners and Charterers).
The ship remains in control and maintenance of the Owners or Managers and for an agreed 'freight rate' as compensation that is multiplied 'per metric ton' or for certain cargoes 'per cubic metre' of the quantity of cargo loaded onboard or an agreed ' lumpsum' amount as 'freight', the Owners will execute the Voyage Charter with the provided and subsequently fixed named ship.

The Owners are obliged in accordance with the terms of the issued Charter Party:
- to transport the agreed grade/type and quantity of cargo (the cargo is known),
- from the identified by name port or range of loading, to the identified by name port or range of discharge,
and
- to produce a seaworthy ship and ensure that the ship can safely trade for the satisfactory execution of the intended voyage.

The Charterers are obliged in accordance with the terms of the Charter Party
- to supply the fixed quantity/type/grade of cargo,
- at the agreed port/ designated berth for operations,

- with the cargo in sound condition,
and
- with the ship completing the operations in the ports within the lay-time agreed, or otherwise the Owners will be compensated by demurrage.

The Owners bear the cost of the ship's daily fuel consumption and all costs incurred at sea or in port, whether ordinary or unexpected.
The Master will be receiving trading instructions only by the Owners/Managers and NOT/NEVER by the Charterers.
The ship will operate in the ports according to the facilities and the capacity that every specific port can provide, which is usually in line with the agreed Charter Party loading and discharging daily rate (speed). However, the lay-time will cease to count if interrupted for reasons, such as bad weather, strikes, force majeure and many others, even if they are beyond the fault of the Owners or ship, unless the ship is already on demurrage.

2. Time Charter.
On Time Charter,
. the ship remains on the management and maintenance of the Owners / Managers, but the Charterers, by paying an agreed daily or pro rata 'hire rate' to the Owners,
. will put the fixed named ship, under their legitimate instructions and directions, as employment regards (the Master will be receiving trading instructions only by the Charterers).
Will trade and direct the ship to ranges, areas, ports, places, berths as specified by them, but always in compliance with the terms and conditions of the governing Time Charter Party.
They will bear the cost of the daily fuel consumption and all costs incurred at sea or at ports, whether ordinary or unexpected, excluding the costs incurred by the Owners' husbandry matters or due to their fault or ship's defect. The Charterers have the right to utilize the cargo spaces throughout the ship, and load and discharge any permitted by the Charter Party type/grade and quantity of cargo, as full or part cargo, but always in compliance with the ship's stress and trim.
The ship will remain 'on-hire', even if the operations in ports will be interrupted, due to reasons of force majeure, bad weather, strikes etc. or other reasons beyond the Owners' or ship's fault/ defect.
The Owners are responsible for providing a seaworthy ship, with valid and up-to-date original certificates on-board, complying with all international and local regulations in the ports that will be contacted. The ship should be maintained in good condition throughout the duration of the Charter, for its obligations under the terms of the Time Charter Party to be fulfilled.

3. Bareboat Charter.
The bareboat charter could be considered as a type of Time Charter.
The
- management is transferred to Charterers, who, by paying an agreed daily 'hire rate' to the Owners, will put under their control/orders and directions the ship 'bare' of crew/stores/fuels etc.
Charterers are responsible for the right Management/trading and maintenance of the ship and will provide the ship with efficient officers /crew/stores/ provisions/fuels etc. and keep it throughout the duration of the Charter in such a condition, as to comply with safety and health standards / regulations and in general to trade in compliance with the international maritime conventions. The bareboat Charter is usually fixed for a very long Charter period and Charterers frequently have the option to purchase the ship from the Charter Party at the end of the Charter.

STEP FOUR.
Shipbrokers-the intermediation party.

The chartering brokers are the main part, which should be involved between the two contracting parties, Charterers and Owners, for the successful execution of the shipping leg (transport by sea) and are in fact a valuable resource in the shipping field.

After receiving the 'cargo order' (the biz) from the Charterers' office, they will enter and search the market for a workable ship and proceed with the commencement of the negotiations of the terms, where, with their experience and professionalism/integrity, they should assist all parties to conclude the fixture, ensuring a smooth shipping operation.

Shipbroking is an activity that could be expanded across the field of the shipping business, as it could be the intermediation party, not only for the dry-cargo ships, but also for tankers, reefers, containers, livestock carriers etc. In addition, for the ship's marine insurance, sale, and purchase (S+P brokers), organization and/or representation of shipyards for repairs worldwide, arrangements for the supply of fuel etc.
Their successful work will be compensated by a
. 'brokerage commission',
which varies depending on the services provided.

In a dry cargo fixture,
they can act alone, by serving both contracting parties, the Charterers and Own-

ers, or otherwise the shipbrokers that serve only the Charterers (working directly with the Charterers) are called
. 'Charterers-brokers',
while those that serve only the ship-Owners (working directly with the Owners) are called
. 'Owners-brokers'.
Both will enter the market with the available cargoes for fixture or the position of the unfixed /open ships, searching for suitable ships or cargoes respectively, aiming to genuinely protect the interests of their clients.
The Owners are the only party liable for the payment of the nominal brokerage commission of 1,25 percent of the gross freight/dead-freight/demurrage paid by the Charterers on Voyage Charter or earned 'hire' and possible agreed ballast bonus on Time Charter, to each broker involved in the cargo fixture. The Charterers put an 'address commission' in every cargo fixture, but the charged amount varies from Charterer to Charterer, depending on the method of fixing a ship. The charged address commission traditionally covers the Charterers' expenses but is nowadays charged for the reduction of the due 'freight'/ demurrage or 'hire' and 'ballast bonus'. The involvement of many brokers is not preferred by the Owners, as the total commission will not be kept to a minimum and will greatly affect the amount of 'freight' or earned 'hire' and therefore the daily income and profit of the ship.

Dry cargo shipbrokers have the following basic tasks:
- to get an overall view and idea of all types of ships and to board ships at every given opportunity during loading and/or discharging operations or while in repair yards, in order to gain personal experience about the method of operations in ports, the ship's structure etc. For the shipbrokers, the continuous learning is a key toll to success.
- at any time and by any cost they are obligated to be increasingly aware of anything concerning the shipping field, giving free advice to their clients.
- to circulate the available cargoes/ships in the shipping market, searching for interested parties.
- upon interest, to obtain all possible information and convey the same to their clients, as the case maybe, regarding the cargo (total available quantity, stowage factor etc.), the position of the ship and its itinerary, its Owners' and Charterers' financial standing etc. To obtain information concerning the main port restrictions (drafts etc.) by the Charterers, also 'freight ideas' and any other specific/useful information that they could be asked to gather, in order to assist the Owners/Operators in calculating the initial voyage estimation and to encourage both parties to enter firm negotiations.
- to have a close cooperation with the Owners' operations and Technical

department, as to the ship's suitability in loading the intended quantity/grade/type of cargo.
- upon request, to inform the interested parties about the direction of the market and the existing competition in the area where the cargo or the ship will be available.
- to follow and take personal action and influence during the progressive negotiations on main terms and to conclude the fixture after having both parties successfully agree on the Charter Party details.
- the broker serving the Charterers, or the sole broker involved in the fixture to issue the Charter Party on behalf of their clients.
- to forward the freight or hire invoices, the Time Sheets etc. to the interested parties, to assist the Owners in collecting/tracing the freight/hires etc. and to generally keep the Post Fixing (which is the entire process which takes place after fixing) up to high standards.
- in the case of a dispute arising between the Owners and Charterers and before the case is decided to be referred to arbitration, costing money and time to all parties involved, to try and resolve the dispute by proposing a compromise.
- to remain involved in the fixture, from the time of commencement and maintain close follow up with post fixtures operations, in order to ensure the smooth execution of the terms and conditions of the governing Charter Party and to generally:

. be alert and ready to travel/offer services to their clients even if they are based in a different time zone/trading zone (America-Asia-Europe) 24 hours a day, 7 days a week.

The shipbrokers do not have the right
. to conceal the truth, falsify documents, misrepresent facts, or alter the information received and/or alter the messages passing through them.
In case of violation of any of the above they will be liable and responsible for any consequences that may arise therefrom.

A 'cargo order', or the position of the ship could be placed/circulated on the market by the chartering brokers, in accordance with the following ways:
a. as 'exclusive'.
Cargoes or positions entered the market by a broker mentioning that he acts as an 'exclusive' broker, or that they are 'exclusive in our hands', indicates that this broker is the only source, from which someone can fix the particular circulated cargo or ship.
b. as 'semi-exclusive'.
The quotes as 'semi-exclusive' indicate that these two different brokers are the only source from which someone can fix the particular circulated cargo or ship.

c. as 'pool brokers'.
Represents a certain number of brokers and that if anyone has an interest in the circulated biz or ship, then they should choose the most active broker from the 'pool' and contact him for further development.
d. as 'direct'.
This indicates that the brokers work directly with the circulated in the market clients. The Owners and/or Charterers are freely open to the market and are agreeable in giving the 'cargo order' or the position of the open ship to anyone requesting the same. However, when quoted as 'direct/ close', pay attention to interested parties, as apart from 'direct', they are also 'close' to the Owners or Charterers, having good business relationships with them and there should be a preference in getting feedback from them, as they know better than others how the ship or biz can be easily fixed.
In a word they have
. 'go'
with them.

The Chartering brokers working on the following basis:
a. as competitive brokers.
Brokers which are indeed 'snipers' and work with freely based cargoes/ships and their success is the speed with which they
. 'first inform their clients of the cargo or ship.
and upon gaining their interest to commit the cargo or ship ahead of any other broker.
They are always alert, searching for new clients to expand their activities.
b. as exclusive brokers.
They work only on an exclusive basis of tonnages or cargoes and depending on the volume of work they have, do not care about other cargoes or ships on the market.
c. as 'home' or 'house brokers'.
They have a desk at the Owners' or Charterers' offices, looking to cover only the company's fleet or Charterers' circulated cargo orders.

STEP FIVE.

Negotiations- Charter Party-Post fixture operations.

When the Charterers (holding the right cargo) and the Owners or Operators (who control the ship in the right position), have agreed to work in one of the three aforementioned methods of fixing a ship and after their shipbrokers have completed the preliminary work and the usual questions and/or information has been replied/exchanged, the firm negotiations will commence and begin making progress, with both parties trying to conclude the agreement.

The amount of voyage 'freight' or the 'hire rate' that will be paid to the Owners of the ship as compensation for the dry marine transportation should essentially be determined by the:
- global or the local prevailing conditions,
- volume of the 'cargo orders' in local areas,
- volume of the available ships globally or locally,
and
- the provided characteristics of the ship.

The fixture will be concluded in two parts:
the negotiations and agreement on Main Terms and then the negotiations and agreement on the provided Charter Party proforma with its additional clauses (details).

a. the Main Terms
The Owners' initial offer with its included important terms and the Charterers either
. clean acceptance,
or
. counter-offer to 'accept/except' on the basis of the Owners' terms,
or
. the Charterers declined reply to the Owners' terms and the presentation of their own terms, which in fact represents the negotiations or the commencement of the negotiations of the Main Terms.
The Main Terms are certain basic terms that should be agreed upon between the Owners and Charterers, before entering the discussion of the Charter Party terms. There are no standard forms, which could include standard recognised/accepted terms by all interested parties and the time that is required to reach a successful agreement varies, depending on the selected method that is needed to fix a ship and the gap on the negotiated terms. In fact, each voyage has its own peculiarities by the addition of terms for the needs of the shipment, some of which are often modified/updated, in accordance with the Arbitration awards that provide additional protection for both sides.
The Owners' initial 'offer' will be presented to the Charterers for acceptance/response within a specified period, after which it cannot be revoked or asked to be modified by the Owners.
The Charterers' reply to the Owners' offer, as well as all subsequent counter replies, always within the specified period as will be stated from both sides, is ex-

pressed as
. 'counter offers'.
The exchanged 'counter offers' are
. on an 'Accept/Except' basis,
where both sides simply reply with the terms that they want to amend/alter. All other terms are considered 'acceptable' and any reference back to them for re-negotiation should be considered as
. 'back-trade',
which could easily result in the termination of the negotiations.

b. the Charter Party terms (Details).
Following the Main Terms' agreement, the parties that are involved, depending on the law, which is applicable to negotiations, or in accordance with the shipping ethics, through the negotiation of the Charter Party proforma, will be obliged or required to develop further, with its included standard and additional terms and conditions to be referred to as 'details'.
On Voyage Charter the Charterers and sometimes on Time Charter the Owners, will produce their Charter Party pro forma (usually a previously executed and based on a standard type of Charter Party) to the Owners or Charterers respectively, for their consideration and for any likely comments.
The type of the Charter Party that will be in use on a Voyage Charter, should be known in the market from the time of the cargo order's circulation in it, otherwise it should be the latest from the beginning of the negotiations on the Main Terms.

There are various types of Charter Parties (standard forms) that have been adopted or issued or recommended by various maritime organizations, covering the chosen method of marine carriage. There are also private Charter Party forms, as issued by various large Charterers' houses. Some of the Charter Parties are the most commonly used in the shipping industry, as it is an advantage that these forms and their standard printed terms, issued for the transportation of specific cargoes, are known to the contracted parties, making it easier for them to eventually get the fixture.
These are the Gencon Charter Party adopted for general cargoes, cement etc., the Norgrain and the Synacomex for grain cargoes, the Sugar Charter Party for sugar cargo, the Phosphate Charter Party (Africanphos) for the phosphates cargo, the Ferticon/Fertivoy for the fertilizers cargo etc.
The most common Time Charter Party is the New York Produce Exchange Charter Party (NYPE) that has been amended from time to time, with the most used form being the one that was amended in 1946.

The Charter Party consists of two parts:
1. the Main Body
The Main Body of the Charter Party is a standard wording that was adopted or issued or recommended by various shipping organizations and amended from time to time. These terms set the limit of obligations and responsibilities of each contracting party. However, all printed terms are negotiable, and the Charterers and/or Owners make their own deletions/amendments, taking into consideration their interests and in accordance with the prevailing conditions and circumstances.
2. the Rider or Additional Clauses
The Rider or Additional clauses cover any omissions to the Main Body or add new regulations, as well as other specific/special terms that apply to the trip/s and/ or to protect the Owners' and /or Charterers' interests. Many important clauses/terms allocate liabilities and responsibilities for specific risks for each contracting party, trying to make each case "crystal clear', without doubts or room for any disagreement, which might otherwise lead to extra costs and lengthy litigations.

For the terms of the Charter Party nothing is taken for granted, fix it with crystal clear terms, so as to prevent the contracting parties from referring to arbitration.

As almost always happens, every attempt to conclude a fixture within specific dates (lay-can) for shipment should be subject to cargo readiness ('subject to enough merchandise' or as abbreviated 'stem') and subject to Shippers' and/or Receivers' approval of the ship's characteristics.
As it is commonly called, after the completion of the negotiations, the ship will be
. on 'subs'.
The 'Subjects' or 'subs' for short, are the Charterers duly kept reservations to conclude the fixture. They will present the ship, as the case may be, to Shippers/ Sellers and/or to Consignees/ Buyers for approval of the ship's characteristics and will get the readiness of the cargo for loading.
The Owners will be advised in writing, within the mutual agreed specified time, of whether the fixture is concluded, and the ship is
. 'fully fixed'
or if it has otherwise
. 'failed on subs' and it is free to look for other biz.

When all negotiations have been completed and the subjects are lifted, the fixture is concluded, and the Charter Party is issued.

The Charter Party is the document
. of proof of the conclusion of a fixture,
for a specific Voyage Charter or Time Charter trip/ period and is the legal link between the contractual parties (Owners and Charterers), requiring the fulfilment of the mutually agreed terms /conditions/exceptions, under the law in force.
The importance of the Charter Party is such, that it should only contain all the agreed written terms/ conditions and exceptions. Verbal agreements have no value or legal status and cannot be inserted into the Charter Party or if they are inserted without written proof, they can easily be denied.
The Charter Party will be issued at the Charterers' place of business, or at the place, where the Charterers have their Chartering department. The Charter Party date is the date when the full conclusion of business (fixture) is achieved and 'hands given'.
The issue of the Charter Party is one of the brokers' involved duties (the Charterers' broker). Two original Charter Parties will be issued (and some working copies if it is requested), which should be exchanged between the Owners and Charterers after they have all signed.
Every writing abbreviation that has been used during the negotiations, including those on rider/ additional clauses, will not exist when the Charter Party is due to be issued. Whole words must be entered, in this case to minimize disputes regarding any misunderstanding due to abbreviations.

Once a ship is fixed, the post-fixing/operations department at the Owners/ Charterers/ Brokers offices will become active. The liaison between the parties involved on various issues will ensure a smooth voyage and execution of the terms of the Charter Party. The post- fixture persons will handle a full operation 24 hours a day, 7 days a week, track all events, assist, and keep posted the parties throughout the Charter period, mitigate risks, and generally should be fully dedicated to all operational matters mainly as follows:
-establish communication with all parties involved (owners/charterers/brokers /attending agents at the ports of contact etc.).
-forward to the parties involved the messages received during the execution of the Charter. They do not have the right to conceal the truth, misrepresent the facts, alter the messages passing through them.
- for Time Chartered ships will issue hire invoices / keeping records of the payments/issuing the final hire invoice/ ensure timely payments/ handle Charter Party disputes.
- for Voyage Chartered ships will issue freight invoice /make the laytime calculations/ handle demurrage claims / Charter Party disputes.
-monitor and report the loading/discharging progress.
-negotiate the ship's port expenses/track the expenses involved/ finalizing port

disbursement account with the attending agents.
-issue the original Charter Party,
as well as many other tasks.

STEP SIX.

The shipping leg at sea/port/berth (marine transportation).

The ship, after the conclusion of the fixture, if on Voyage Charter will sail to the port of loading where it will arrive within the agreed lay-days and before the cancelling date to load the fixed cargo, or, if on Time Charter, will sail to the delivery point or as otherwise directed by the Charterers after its delivery.
The Lay-days/Cancelling (abbreviated as 'lay-can' or as 'l/c')
. is the range of dates,
when the cargo is expected to be ready for export, or delivery of the ship to Charterers' services and within which the ship must be presented for loading or for delivery purposes.
- 'Lay-days' is actually the first day when the ship can tender the N.O.R. in load port if on Voyage Charter or the delivery notice to the Charterers if on Time Charter
and the
- 'Cancelling' date is the last date when the N.O.R, or the delivery notice could be tendered until 24:00 hrs., unless otherwise stated in the Charter Party.

Master will receive clear sailing/voyage instructions from the Owners' operations department if on Voyage Charter or from Charterers' operations department if on Time Charter, and obviously the ship will be required to perform and fulfil its obligations, under the agreed terms of the issued Charter Party.

At the loading port:
A. The Master, for a number of days (as stipulated in the Charter Party) prior to the arrival of the ship at the port of shipment, either it is fixed on Voyage or Time Charter, should begin to tender specific notices of its estimated time of arrival (abbreviated as 'e.t.a.'). The initial 'approximate' and during the last days before the arrival the 'definite' notices should be sent to the attending Agents and to the parties specified in accordance with the terms of the Charter Party.
The ship's notices and relevant communications are necessary, so that the parties involved to be ready for loading and the Shippers have time to carry the cargo from the mines/farms etc. near the port for temporary storage there.
B. Before the ship enters the territorial waters of the relevant country, the Master should be requested by the Agents to provide the crew list and report on their health status (Maritime Declaration of Health). He should also provide cer-

tain basic certificates and declare the last 10 ports of call for the 'free pratique' and ISPS purposes.

C. When the ship approaches the commercial limits of the port, the Master must be in touch with the Port Authorities/port control and the Agents, in order to receive the appropriate instructions as to the ship's next movement. He will be directed to the load berth, or if the port or berth is congested or closed for any reason, he will receive instructions for the ship to anchor in a safe place of anchorage, usually within the limits of the commerce of the port, to wait its turn for safe berthing. Lacking safe anchorage, the ship can be instructed to remain 'drifting' (in the open sea and outside the limits of the port), until further instructions.

D. Upon arrival, the Notice of Readiness should be tendered to the attending Agents and to specified parties.

E. If the berth is available, or when the berth becomes available, the ship will move to the first or sole pilot station, for the port pilot to embark. A number of tugs, depending on the size of the ship, should be hired to assist in its safe berthing.

F. When the ship is 'all fast' at berth and before anyone else, the attending Agents along with the Port Authorities, Customs and Immigration will board for identification and grant the 'free pratique' to the ship. The pratique is the granted permission of the contacted country, allowing the ship to remain there for the purpose of its call. The Port Authorities will check that all corresponding certificates are in compliance with the health regulations and will obtain the Master's assurance/ confirmation that the crew is healthy and complies with the vaccination and sanitary regulations at the port of contact.

G. The Notice of Readiness in a hard copy in writing will be delivered to the Agents.

H. After the completion of the Port Authorities' work, the Charterers' and/or Shippers' holds inspector/s-surveyor/s will follow and should inspect and ensure the cleanliness and readiness of the cargo holds/ hatch covers to receive the intended cargo. In the case of rejection, the crew or shore workers should re-clean and then apply for re-inspection. If the holds are too dirty from the previous cargo or if rust/rust scales prevail on them, then the Port Authorities can instruct the ship to vacate the berth and clean the cargo holds in the anchorage or to shift to another berth that is not in use for operation (waiting berth). In all cases (except if the ship is on Time Charter and the 'intermediate holds cleaning' clause is in effect), lost time and extraordinary costs incurred, are for the Owners' sole account.

I. Before the commencement of loading, the ship should carry out the initial draft survey. The draft survey shows the displacement of the ship before the commencement of loading, which allows the determination of the quantity loaded after completion of loading. Shippers will conduct their own draft survey or do so jointly with the ship or can even rely on the indicators fitted on the loading means (shore figures).

J. If the ship is on Time Charter, the on-hire surveyor should attend to ascertain the quantity of fuel on board and the ship's condition for previous stevedoring damage that could later be transferred to the current Charterers.

K. After all these procedures have been completed/successfully passed and the ship's hatches are open, together with the readiness of the means of loading/ loading system (ship's cranes/derricks or shore cranes/scale etc.), the loading operation will commence, usually from the ballast hold, allocated amidships (for bigger ships).

The loading process should be in accordance with the stowage plan and loading sequence, as delivered by the Master of the ship, before the commencement of loading, to the leader of the stevedores (called the 'foreman'), who is the person responsible and the supervisor of the loading process.

L. Prior to the commencement of loading and depending on the type and grade of the cargo and how it is stored (open stockpiles etc.), or how it is carried to the port (with rusty or uncovered wagons, barges with open hatches, uncovered trucks etc.), the cargo should be inspected by the Master of the ship along with the ship's P&I Club's surveyors. Their task is to determine whether the cargo is in apparent good order and condition and can be loaded onto the ship. They will check if the bulk cargo is clean and dry, or for general cargoes, if there are torn bags/broken pallets etc. for loading, or if the steel is rusty etc. When it is not to their entire satisfaction and in apparent good order and condition, or not in accordance with the prevailing standards for this particular cargo, they should require the cargo which is rejected and not allowed for loading to be substituted with cargo that is acceptable (sound cargo). For any type of dangerous cargoes (iron ore fines, concentrates, nickel ore etc.) that could liquefy during the sea voyage, a laboratory analysis is necessary to be performed before and/or during the loading operation, which should indicate the content moisture and Transportable Moisture Limit (abbreviated TML).

For all cargoes, either classified as dangerous or not, the Shippers are responsible before loading for providing the Master of the ship with a cargo declaration, as set out in the IMSBC and the BCSN, as well as a valid certificate with the information on the physical and chemical properties and the conditions required for the carriage.

The ship's officers/crew should monitor the loading process at all times and be ready to stop the loading if the loading sequence, as provided by the Master, is not followed, or if damaged/wet cargo is loaded or is going to be loaded, or if the men working the cargo cranes are not competent and/or inexperienced, thereby leading to damage to the ship's holds/hatches.

M. During the loading and before departure, the ship may be subject to inspection by the inspectors of the Port State Control (PSC). The inspection could also take place at the discharge port. They should check/ inspect/ensure the ship's condition/documents/equipment/lifeboats etc. and in general whether the ISM procedures are followed by the Master/crew. They will ensure whether the ship is safe and secure/operates in compliance with the international maritime conventions. In case of failure, the ship could be detained, and its sailing prohibited. The deficiencies must be rectified before departure or before approaching the next port or within a specific period of time, as should be stipulated by the PSC inspectors.

N. The International Transport Workers' Federation (abbreviated as 'ITF', a union that protects and assists seafarers worldwide and represents their interests and their rights) could also visit the ship, irrespective of flag/Ownership or the nationality of the crew. They will check whether the ship is navigating in accordance with the terms and conditions of the ITF, or in accordance with terms/ conditions that are in any case acceptable by them. The ITF could also visit the ship at the discharge port.

O. The ship could be supplied with fuel oil and/or fresh water and/or lubricant oils (luboils) and/or to change crew and/or to receive provisions or spares etc. that are necessary to continue its employment.

P. After the completion of the loading operation, the final draft survey report will be issued. The draft surveyors and ship's officers will measure the displacement from the ship's draft marks and having already performed the initial measure of displacement, can determine the weight of the cargo loaded.

Q. The Agents will provide the Master with the following documents:

1-cargo manifest:

a document which will be issued for the customs' use and contains:

- the name of the ship,
- its loading origin and destination,
- the description of the cargo and the cargo quantity loaded onboard,
- the Shippers and Consignees and if any remarks.

2- mate's receipts:

This document is a 'receipt' showing and defining goods that have been received and their apparent condition. The Bill of Lading should be issued in strict conformity with the mate's receipts (as should be stipulated in the Charter party), regarding the quantity, description, condition of cargo and the date of shipment and if there are any Master's remarks.

. The Mate's Receipts do not show evidence of the contract of carriage and cannot replace the Bills of Lading.

The Mate's Receipt contains:

- the name of the Shippers
- the Consignee and the Notify party,
- the ship's name,

- the load/discharge port,
- the description of the cargo and its quantity,
- the date of shipment,
- which holds or hatches have been loaded,
- if there are any Master's remarks.

The Consignee/Notify party and the port of discharge are simply mentioned as an indication and can be changed when the Bill of Lading is issued.

3-Bill of Lading:

The Bill of Lading is a legal document showing evidence of the contract between the Shipper and the Carrier, under which specific terms the shipment is made from one place to another, as well as the receipt of goods.

The Bill of Lading should be issued from the Shippers or Agents on the Carriers' behalf.

After signing, it is the link between the Owners and the Buyers/Consignees of the cargo, who are to receive their cargo upon the ship's arrival at the discharge port and after their presentation of one originally issued Bill of Lading.

It contains:

- the Shippers' name,
- the Consignees and the Notifying party,
- the ship's name,
- the load port and the discharge port,
- the cargo description and the weight,
- the place and date of issue,
- who will sign the Bill of Lading,
- the number of the original issued Bill of Lading,
- the conditions of carriage,
- the Master's remarks (if any).

The Master is the only party authorised to sign the issued original Bill/s of Lading. However, if the issue would delay the departure, then it should be signed by an authorised party as should be clearly stated in the Charter Party, after departure. The Master will deliver a letter of authorization to the authorised party, assigning the conditions under which the Bills of Lading should be signed on his behalf.

4-Statement of Facts (abbreviated as 'S.O.F.):

It is an important document that records all the events from the time of the ship's arrival, until the last pilot disembarks from the ship and it sails to open sea towards its next destination.

The importance of this document is that the lay-time computation (the time spent for loading or discharging) for a party to claim for demurrage/despatch, must be based on it.

The same should be signed by the attending Agents (also sometimes by the

Shippers or Receivers) and the Master of the ship, who in any case has the right to insert reservations/objections on the events mentioned.
The issued S.O.F. ought to contain:
- the ship's name,
- when it arrived and when it sailed,
- when the Notice of Readiness tendered and when it was accepted,
- when the ship berthed/unberthed, when it passed free pratique and holds inspections,
- when the draft surveys were carried out and the lost time,
- when the ship commenced the operation and when it completed it,
- when any stoppages incurred and the reason (rain, stevedores' meals, cranes' breakdown etc.),
- the international/national or public holidays applicable during the stay of the ship in the port which prevented or not the loading or discharging process,
- the quantity loaded or discharged,
- the Master's or Shippers' remarks.

In addition, the Agents will issue the cargo stowage plan which shows the distribution of the cargo quantity loaded in each hold.
R. Upon the completion of all works and when the ship is ready to sail, the Port Authorities will issue the 'port clearance' (a document that allows the sailing of the ship) and the Agents will hand it over to Master.
S. The pilot will embark, the ship will unmoor and with the tugs assistance it will sail to the last pilot station for the pilot to disembark and 'full away' to the open sea, where with the utmost dispatch it will navigate to the port of discharge.

<u>At the discharging port:</u>
A. The Master, upon departure, will tender the e.t.a. of the ship, together with the cargo quantity onboard and the expected arrival draft to the attending Agents and to the parties specified in accordance with the terms of the Charter Party. Thereafter, the 'approximate' and during the last days before the arrival the 'definite' notices should be sent to the attending Agents and to the parties specified in accordance with the terms of the Charter Party.
The notices and relevant communications are necessary, so that the parties involved can be ready to receive the cargo and present in a timely manner the means for cargo carriage from the port to their warehouses etc. (by lorries/wagons etc.)
B. Before the ship enters the territorial waters of the relevant country, the Master should be requested by the Agents to provide the crew list and report on their health status (Maritime Declaration of Health). He should also provide certain of the ship's certificates and declare the last 10 ports of call for the 'free pratique' and ISPS purposes.

C. When the ship approaches the commercial limits of the port, the Master should be in touch with the Port Authorities/port control and Agents, in order to receive the appropriate instructions as to the ship's next movement. The ship should be directed to the discharge berth or if the port/ berth is congested, the Master will receive instructions for the ship to anchor in a safe place of anchorage, usually within the limits of the commerce of the port, to wait its turn for berthing. Lacking safe anchorage, the ship should be instructed to remain 'drifting' (in the open sea and outside the limits of the port) until further instructions.
D. On arrival the Notice of Readiness will be tendered to the attending Agents and to specified parties.
E. If the berth is available or when the berth becomes available, the ship will move to the first or sole pilot station, for the port pilot to embark. Tugs will be hired to assist in the ship's safe berthing.
F. When the ship is 'all fast' at the berth and before anyone else, the attending Agents along with the Port Authorities/Customs/ Immigration will board the ship for identification and to grant the 'free pratique'.
G. The Notice of Readiness in a hard copy in writing will be delivered to the Agents.
H. Before the commencement of the discharging operation the attending Agents or the Consignees' representative should deliver to the Master of the ship one original Bill of Lading, duly endorsed by the Consignees, or (subject to Charter Party terms) a Letter Of Indemnity for delivery of the cargo without presentation of the original Bill of Lading.
I. The initial draft survey will be carried out.
J. The ship, having its hatches open and after the means of discharge are ready, will commence the discharging operation in accordance with the stowage plan and discharging sequence, as will be delivered by the Master of the ship to the Foreman.
K. During discharging and before departure, the ship could be inspected by the inspectors of the Port State Control (PSC) and additionally the ITF might board the ship.
L. The ship could be supplied with fuel and/or fresh water and/or luboils and/or change crew and/or receive provisions/spares etc. that are required to continue its employment.
M. After the completion of discharge, the final draft survey should be carried out to determine the discharged quantity. The sailing could be blocked by the Receivers/Port Authorities, in the case of heavy cargo shortage (draft survey figures much less than the Bill Of Lading figures). Usually, the ship's P&I Club will issue a Letter of Undertaking (L.O.U.), confirming that it will cover the cost of shortage, following the court's decision.
N. If the ship is on Time Charter and the redelivery should take place in this port,

the off-hire surveyors will board the ship to determine the quantity of fuel remaining on board at the time of redelivery and the condition of the ship for stevedoring damage.
O. The Port Authorities will issue the 'port clearance' and the Agents will hand it over to the Master.
P. The Statement of Facts should be signed accordingly.
Q. The pilot will embark, the ship will unmoor and with the tugs assistance it will sail to the last pilot station for the pilot to disembark and 'full away' to the open sea for its next employment.

STEP SEVEN.

Voyage file.

On the ship's full away from the discharge port, the shipping leg at sea/port /berth has been completed/executed, but the file in the Owners'/ brokers'/ Charterers' office will remain open for a period of time.

The file should be considered closed:
-if all economic obligations have been settled and the final hire statement or the final freight statement including the demurrage statement has been issued and paid to the Owners accordingly and when the commissions have been paid by the Owners to the brokers involved in the fixture.
-if no any claim, from any party involved has been presented.
However, Charter Party disputes or minor cargo claims for shortage/damage etc. could be presented much later, but usually, if there is a dispute/claim, it will be raised sooner rather than later.
For some years (and according to the rules of the ISM) the parties involved should keep the files safe and ready to re-open, whenever it is requested.

The presentation of the aforementioned/undermentioned contents-Disclaimer notice:
The presentation of the aforementioned/undermentioned contents is without any liability and responsibility, as well as the recommendations/comments/ information as will be provided here below. The contents, as well as the recommendations /comments/ information contained in this book may not be accurate at times and presented without any responsibility and liability for the use of the same, and for any errors or omissions. The recommended full clauses can be altered/ amended in accordance with the prevailing negotiations/conditions and it is up to the Owners' or Charterers' or any other parties' choice on whether they will adopt the same or not.

INTERNATIONAL TRADE TRANSACTION BY SEA

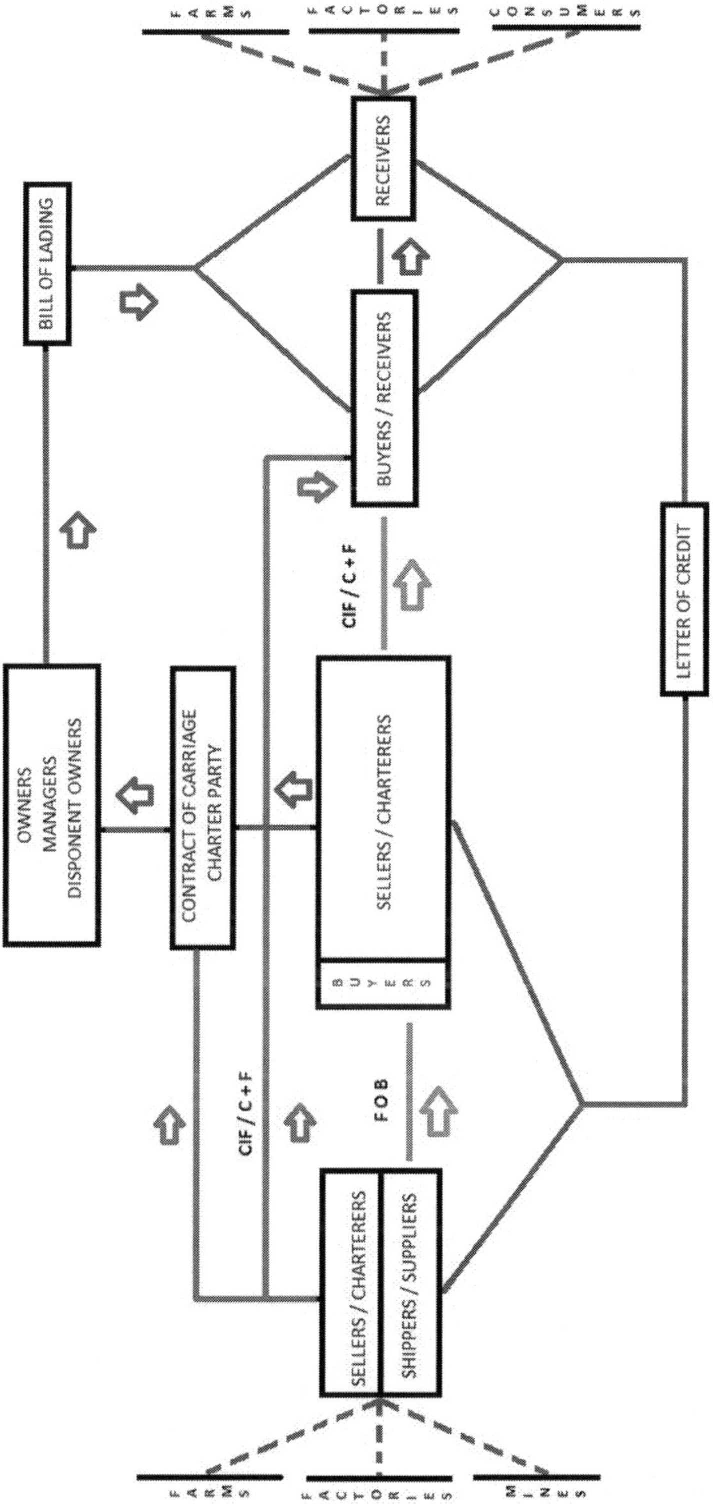

N O T I C E   O F   R E A D I N E S S

PORT OF: ZHENJIANG

DATE : 10/DEC./2011

TO: [redacted]

Cc: [redacted]

CC: [redacted]

CC: [redacted]

DEAR SIR, (S)

M/V [redacted]

THIS IS TO ADVISE THAT THE ABOVE VESSEL UNDER MY COMMAND ARRIVED AT CJK P/S AT 0200 UTC OR 1000 LT HOURS ON THE 10 DEC. 2011 AND IS READY IN ALL RESPECTS TO DISCHARGE HER CARGO AS PER TERMS /CONDITIONS / PROVISIONS / EXCEPTIONS OF THE RELEVANT CHARTER PARTY DATED 08 NOV 2011.

YOURS FAITHFULLY

[redacted]

MASTER (STAMP)

TENDERED: 0200 UTC OR 1000 LT HOURS DATE: 10 DEC. 2011

ACCEPTED: UTC OR LT HOURS DATE: 10 DEC. 2011

IN ACCORDANCE WITH TERMS AND CONDITIONS OF THE RELEVANT CHARTER PARTY DATED 08 NOV 2011

[redacted]

AGENT (STAMP)

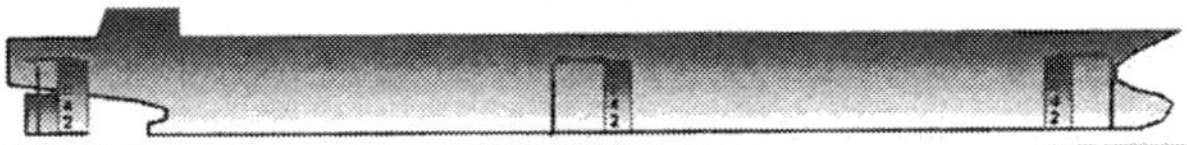

Aft	9.500	0.000	midship	1.700	Forward
LBP	217.000	205.800	LBDM	32.500	Beam

Load Port	SANTOS, BRAZIL	Cargo	MAIZE	S.F.	45.5000 f3/Mt		
Discharge Port	TOMAKOMAI, JAPAN	Date	12-12-2012 1400LT	Date	21-12-2012 1830LT		
Voyage #	4L/2012						

		INITIAL Survey			FINAL Survey		
		Port		Starboard	Port		Starboard
1 Draft @ Draft Marks	Fwd	11.920		11.890	4.350		4.260
	Mid	11.890		11.850	5.850		5.770
	Aft	11.820		11.780	7.400		7.350
		Fwd	Mid	Aft	Fwd	Mid	Aft
	Mean	11.905	11.870	11.800	4.305	5.810	7.375
	Trim	-0.105	by Head		3.070	by Stern	
2 Draft @ Perpendicular	Draft Correction	0.0009		-0.0048	-0.0254		0.1417
F Corr'n 0.00826	Draft Corrected	11.9059		11.7952	4.2796		7.5167
A Corr'n 0.04616	Trim Corrected	-0.111	by Head		3.237	by Stern	
3 Deflection	Midship	11.870			5.810		
	Fore & Aft Mean	11.851			5.898		
	Deflection	-0.019	Sagging		0.088	Hogging	
	Listing	0.040	Port		0.080	Port	
	Heel	-0.071	Degrees		-0.141	Degrees	
4 Quarter Mean Draft		11.8651			5.8320		
5 Hydrostatic Particulars							
	a. Displacement	71,214.566			33,186.600		
	b. TPC	66.46			60.36		
	c. LCF	0.210			-8.445		
	d. MTC 1 + 0.5	1075.9			811.7		
	e. MTC 2 - 0.5	1032.4			791.2		
	f. MTC Difference	43.5			20.5		
6 Trim Correction							
	a. ((trim x 100 x lcf x tpc) / lbp)	-0.715			-760.368		
	b. ((trim x trim x 50 x mtc diff.) / lbp)	0.124			49.407		
	c. A & B Difference	-0.591			-710.962		
	d. Displacement Corrected for Trim	71,213.975			32,475.638		
7 Dock Density							
	a. Observed Density	1.0280	tons/m3		1.0270	tons/m3	
	b. Density Correction	208.431			63.367		
	c. Displacement Corrected for Density	71,422.406			32,539.005		
8 Deductibles							
	a. Fuel	630.000			609.960		
	b. Diesel	59.000			43.930		
	c. Lubricants						
	d. Ballast	236.096			20,577.565		
	e. Fresh Water	56.994			144.732		
	f. Others (Slop tank port)	85.914			85.914		
	g. Constant	199.904			199.904		
	TOTAL	1,267.908			21,662.005		
9 Summary							
	a. Actual Displacement	71,422.406			32,539.005		
	b. Total Deduction	1,267.908			21,662.005		
	c. Lightship	10,877.000			10,877.000		
CARGO WEIGHT / Constants		59,277.498			0.000		

Cargo as per B/L 59,255.630

Remarks:

Officer In-Charge

Page | 1

STATEMENT OF FACTS

Vessel's Name	[redacted]
IMO Number	
Flag	
Owners	
Charters	
Agents	Cargill Agricola S/A
Port of Loading	Santos, Brazil

Free Pratique Granted	Monday, August 04, 2014	15hrs 08min
Vessel Arrived on Roads	Sunday, August 03, 2014	22hrs 48min
Dropped Anchor	Sunday, August 03, 2014	23hrs 30min
NOR Tendered and Acknowledge Receipt	Thursday, August 07, 2014	00hrs 01min
Having Anchor Aweigh	Thursday, August 07, 2014	13hrs 25min
Pilot on Board for Berthing	Thursday, August 07, 2014	13hrs 36min
First Line Ashore	Thursday, August 07, 2014	14hrs 32min
All Fast at TEG Terminal	Thursday, August 07, 2014	15hrs 00min
Cleared by Customs	Thursday, August 07, 2014	16hrs 00min
Holds Insp. by Superinspect obo Terminal	Thursday, August 07, 2014	15hrs 10min
Cmtd Hlds Insp. by Superinspect	Thursday, August 07, 2014	15hrs 45min
Loading Commenced at TEG Terminal	Thursday, August 07, 2014	16hrs 04min
Loading Completed at TEG Terminal	Sunday, August 10, 2014	22hrs 10min
Fumigation Commenced	Sunday, August 10, 2014	23hrs 00min
Fumigation Completed	Sunday, August 10, 2014	23hrs 59min
Pilot on Board	Monday, August 11, 2014	02hrs 30min
Vessel Sailed	Monday, August 11, 2014	02hrs 50min
ETA at Malaysia	Thursday, September 18, 2014	If agw wp

Bunkers ROB on Arrival	IFO:	279.10	MT	DO:	8.40	MT	FW:	130.00	MT
Bunkers ROB on Berthing	IFO:	275.79	MT	DO:	7.56	MT	FW:	120.00	MT
Drafts on Berthing	FWD	4.29	MTRS	AFT	7.05	MTRS			
Bunkers ROB on Sailing	IFO:	1,146.0	MT	DO:	42.189	MT	FW:	225	MT
Drafts on Sailing	FWD	13.20	MTRS	AFT	13.20	MTRS			
Bunkers Replenishment	IFO:	879.91	MT	DO:	39.964	MT	FW:	100	MT

Shippers	Quantity	Cargo	Terminal
[redacted]	15.000.000 Kgs	Maize	TEG Terminal
	53.462.000 Kgs	Maize	TEG Terminal
TOTAL	**68.462.000** Kgs	Maize	TEG Terminal

HOLIDAYS OCCURRED WHILE SHIP STAYED AT SANTOS

NIL

RAIN PERIODS RECORDED AT TERMINAL WHILE SHIP WAS ON SANTOS ROADS

Monday, August 04, 2014	2145/0000
Tuesday, August 05, 2014	0000/0713

*** FENAMAR and FONASBA Quality Standard Approved ***
65, Bento Pedro da Costa Av. – Guarujá city, SP/Brazil – Zip 11472 000
Tel: + 55 (13) 33081800 – Fax: + 55 (11) 28452183
UK 94075693 Carg G – email: santosagentvops@cargill.com

Page | 2

GENERAL REMARK
Vessel tendered her Nor on August 07th at 00:01hrs, meanwhile the pre-inspection were booked to be carried out on same date from 09:45 until 11:15 hrs Lt. Being all holds approved.

PORT LOG

TEG Terminal

Date	Gangs	Remarks
Thursday, August 07, 2014	0001	Notice of Readiness Tendered. Time to Count as per Terms, Conditions and Exemptions of the Governing Charter-Party
	1336/1500	Berthing maneuvering
	1500/1545	Holds inspection / approved
	1545/1604	Preparation ashore
	1604	Commenced loading operation
	1604/1900	Loading into h3
	1900/1914	Shifting gang from h3 to h5
	1914/2159	Loading into h5
	2159/2212	Shifting gang from h5 to h2
	2212/2400	Loading into h2
Friday, August 08, 2014	0000/0058	Loading into h2
	0058/0114	Shifting gang from h2 to h6
	0114/0406	Loading into h6
	0406/0414	Shifting gang from h6 to h4
	0414/0542	Loading into h4
	0542/2400	Rain
Saturday, August 09, 2014	0000/1351	Rain
	1351/1511	Loading into h4
	1511/1527	Shifting gang from h4 to h1
	1527/2043	Loading into h1
	2043/2124	Warping vessel by rope to ship loader reach h7
	2124/2134	Shifting gang from h1 to h7
	2134/2400	Loading into h7
Sunday, August 10, 2014	0000/0108	Loading into h7
	0108/0117	Loading stopped owing to Stevedores passage to board
	0117/0344	Loading into h7
	0344/0351	Shifting gang from h7 to h5
	0351/0645	Loading into h5
	0645/0700	Loading stopped owing to Stevedores passage ashore
	0700/0716	Loading into h5
	0716/0723	Shifting gang from h5 to h3
	0723/0730	Loading stopped owing to Stevedores passage to board
	0730/0743	Defect ashore
	0743/1040	Loading into h3
	1040/1043	Loading stopped owing to Port operator passage to board
	1043/1050	Awaiting fully open h3
	1050/1108	Loading into h3
	1108/1119	Shifting gang from h3 to h6
	1119/1314	Loading into h6
	1314/1321	Trimming
	1321/1338	Loading into h6
	1338/1350	Trimming

****** FENAMAR and FONASBA Quality Standard Approved ******
65, Bento Pedro da Costa Av. – Guarujá city, SP/Brazil – Zip 11472 000
Tel: + 55 (13) 33081800 – Fax: + 55 (11) 28452183
UK 94075693 Carg G – email: santosagencyops@cargill.com

1350/1425	Loading into h6
1425/1440	Shifting gang from h6 to h2
1440/1633	Loading into h2
1633/1644	Shifting gang from h2 to h4
1644/2021	Loading into h4
2021/2120	Intermediary draft survey by Terminal / Chief Mate
2120/2127	Shifting gang from h4 to h6
2127/2135	Loading into h6
2135/2150	Shifting gang from h6 to h2
2150/2210	Loading into h2
2210	Completed loading operations

GENERAL REMARK
After completion, vessel stayed idle waiting high tide until August 11, 2014 at 02:30 hrs, when POB was confirmed.

Please Note Bunkers Replenishment:

Barge Alongside	Thursday, August 07, 2014	21 45
Started Pumping	Thursday, August 07, 2014	23 40
Finished Pumping	Friday, August 08, 2014	02 50
Cast Off	Friday, August 08, 2014	03 53
Quantity Supplied	879.910 mt	
Barge Alongside	Thursday, July 07, 2011	15 55
Started Pumping	Thursday, August 07, 2014	17 32
Finished Pumping	Thursday, July 07, 2011	18 00
Cast Off	Thursday, August 07, 2014	18 35
Quantity Supplied	39.964 mt	

FOR AND ON BEHALF OF MESSRS

WEIGHT, MEASURE, QUALITY, QUANTITY, CONDITION, CONTENTS AND VALUE UNKNOWN
NO DELAY ON SHIP'S ACCOUNT

Master M/V

MONROVIA

TEG - Terminal Exportador do Guarujá Ltda
- As Port Operators Only -

*** FENAMAR and FONASBA Quality Standard Approved ***
65, Bento Pedro da Costa Av. – Guarujá city, SP/Brazil – Zip 11472 000
Tel: + 55 (13) 33081800 – Fax: + 55 (11) 28452183
UK 94075693 Carg G – email: santosagencyops@cargill.com

TEG

SANTOS, BRAZIL

OFFICIAL

MATE'S CARGO RECEIPT

DATE Aug 07th, 2014

THIS CERTIFIES THAT HAS BEEN RECEIVED IN APPARENT GOOD ORDER AND CONDITION

ON BOARD THE VESSEL: MV [redacted]

FROM: [redacted]

QUANTITY SAID TO BE: 13,291.250 METRIC TONS

COMMODITY SAID TO BE: BRAZILIAN MAIZE

PACKING IN BULK

HOLDS:

H2:	3,283.175	MT
H3:	5,003.510	MT
H5:	5,004.565	MT

WEIGHT, MEASURE, QUALITY, QUANTITY, CONDITION, CONTENTS AND VALUE UNKNOWN

DESTINATION: PORT KLANG AND/OR BUTTERWORTH, MALAYSIA

SIGNATURE: [redacted]

TITLE: MASTER

CARGONAVE

CARGO MANIFEST

SHEET: 1 Of 1

NATIONALITY	NAME OF SHIP	VOY.No.	MASTER NAME	FROM	TO
LIBERIAN	M/V [illegible]		[illegible]	PARANAGUA, BRAZIL	NANTONG PORT

SHIPPER	CONSIGNEE	NOTIFY PARTY	MARKS AND NUMBERS	DESCRIPTION OF GOODS	WEIGHT	FREIGHT
[illegible]	To Order	[illegible]	Packing: In Bulk	Brazilian Soybeans, Crop 2012	40.565,825	PAYABLE AS PER C/P
				Grand Total	**40.565,825**	

PARANAGUA, JUNE 15TH, 2012
AGENCIA MARITIMA CARGONAVE LTDA

BIMCO

CONGENBILL 2007
BILL OF LADING
To be used with charter parties
Page 1

Shipper [redacted]	**Bill of Lading No.** **TFN/MY-01**	**Reference No.**
Consignee TO ORDER	**Vessel** M/V [redacted]	
Notify [redacted]	**Port of loading** **SANTOS, BRAZIL** **Port of discharge** PORT KLANG AND/OR BUTTERWORTH, MALAYSIA	

Shipper's description of goods — **Gross Weight**

BRAZILIAN MAIZE
PACKING IN BULK

CLEAN ON BOARD
FREIGHT PREPAID

29,400.000 METRIC TONS

ORIGINAL

(of which on deck at Shipper's risk; the Carrier not being responsible for loss or damage howsoever arising)

Freight payable as per CHARTER-PARTY dated:
AS PER RELEVANT C/P

FREIGHT ADVANCE
Received on account of freight:

SHIPPED at the Port of Loading in apparent good order and condition on the Vessel for carriage to the Port of Discharge or so near thereto as the Vessel may safely get the goods specified above.
Weight, measure, quality, quantity, condition, contents and value unknown.
IN WITNESS whereof the Master or Agent of the said vessel has signed the number of Bills of Lading indicated below all of this tenor and date, any one of which being accomplished the others shall be void.
FOR CONDITIONS OF CARRIAGE SEE OVERLEAF.

Date shipped on board	Place and date of issue	Number of original Bills of Lading
	SANTOS, BRAZIL, AUG 10th, 2014	3 – (THREE)

Signature
(i) .... Master
Master's name and signature
Or
....
AS AGENT FOR AND ON BEHALF OF THE MASTER [redacted]
.... **as Agent for the Owner**
s name and signature
.... **Owner**
tion (iii) filled in, state Owner's name above

CONGENBILL 2007
BILL OF LADING
To be used with charter parties
Page 2

Conditions of Carriage

(1) All terms and conditions, liberties and exceptions of the Charter Party, dated as overleaf, including the Law and Arbitration Clause/Dispute Resolution Clause, are herewith incorporated.

(2) **General Paramount Clause**
The International Convention for the Unification of Certain Rules of Law relating to Bills of Lading signed at Brussels on 25 August 1924 ("the Hague Rules") as amended by the Protocol signed at Brussels on 23 February 1968 ("the Hague-Visby Rules") and as enacted in the country of shipment shall apply to this Contract. When the Hague-Visby Rules are not enacted in the country of shipment, the corresponding legislation of the country of destination shall apply, irrespective of whether such legislation may only regulate outbound shipments.

When there is no enactment of the Hague-Visby Rules in either the country of shipment or in the country of destination, the Hague-Visby Rules shall apply to this Contract save where the Hague Rules as enacted in the country of shipment or if no such enactment is in place, the Hague Rules as enacted in the country of destination apply compulsorily to this Contract.

The Protocol signed at Brussels on 21 December 1979 ("the SDR Protocol 1979") shall apply where the Hague-Visby Rules apply, whether mandatorily or by this Contract.

The Carrier shall in no case be responsible for loss of or damage to cargo arising prior to loading, after discharging, or while the cargo is in the charge of another carrier, or with respect to deck cargo and live animals.

(3) **General Average**
General Average shall be adjusted, stated and settled according to York-Antwerp Rules 1994 in London unless another place is agreed in the Charter Party.

Cargo's contribution to General Average shall be paid to the Carrier even when such average is the result of a fault, neglect or error of the Master, Pilot or Crew.

(4) **New Jason Clause**
In the event of accident, danger, damage or disaster before or after the commencement of the voyage, resulting from any cause whatsoever, whether due to negligence or not, for which, or for the consequence of which, the Carrier is not responsible, by statute, contract or otherwise, the cargo, shippers, consignees or the owners of the cargo shall contribute with the Carrier in General Average to the payment of any sacrifices, losses or expenses of a General Average nature that may be made or incurred and shall pay salvage and especial charges incurred in respect of the cargo. If a salving vessel is owned or operated by the Carrier, salvage shall be paid for as fully as if the said salving vessel or vessels belonged to strangers. Such deposit as the Carrier, or his agents, may deem sufficient to cover the estimated contribution of the goods and any salvage and special charges thereon shall, if required, be made by the cargo, shippers, consignees or owners of the goods to the Carrier before delivery.

(5) **Both-to-Blame Collision Clause**
If the vessel comes into collision with another vessel as a result of the negligence of the other vessel and any act, neglect or default of the Master, Mariner, Pilot or the servants of the Carrier in the navigation or in the management of the vessel, the owners of the cargo carried hereunder will indemnify the carrier against all loss or liability to the other or non-carrying vessel or her owners in so far as such loss or liability represents loss of, or damage to, or any claim whatsoever of the owners of said cargo, paid or payable by the other or non-carrying vessel or her owners to the owners of said cargo and set-off, recouped or recovered by the other or non-carrying vessel or her owners as part of their claim against the carrying Vessel or the Carrier.
The foregoing provisions shall also apply where the owners, operators or those in charge of any vessel or vessels or objects other than, or in addition to, the colliding vessels or objects are at fault in respect of a collision or contact.

For particulars of cargo, freight, destination, etc., see overleaf.

THE LADY

AND
THE HUNTERS

==

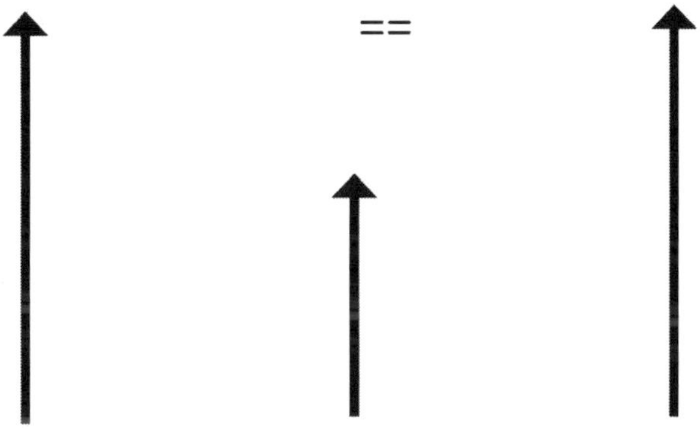

The owners / The managers / The banks /The master / The officers / The crew / The office staff
The provisions, tickets, transportations etc / The class / The flag / The yards / The surveyors
The inspectors / The repair shops / The trucks, wagons, trailers etc / The mines / The farms
The factories / The consumers / The charterers / The operators / The forwarders / The traders
The shippers /The buyers / The receivers / The ship-brokers/ The s+p brokers / The yard's brokers
The insusrance brokers / The fuel oil brokers / The spare parts suppliers / The customs
The immigration / The bunker suppliers / The agents / The labourers / The stevedores
The port authorities / The pilots / The tugs / The linesmen / The launches / The port state control
The port expenses / The taxes / The dues / The adjusters / The solicitors / The tribunals
The p+I club insurance / The h+m insurance / The war risk insurance / The canals-straits
The cargo insurance / The pirates / The marine security /The boycott / The ITF / The salvage

PART II

CHAPTER I

FIXING ON VOYAGE CHARTER

This is one of the three methods of fixing a ship, which applies to all types and sizes in the dry cargo sector.

The majority of global fixtures are made on Voyage Charter. It is a 'rule' for all the tramp and liner coasters/smaller sizes, as well as, for most of the handy sized ships. The number of cargo fixtures that are based on Voyage Charter, for large capacity ships will increase, when the shipping market has a downward trend, and their rates are under continued pressure.
Charterers and/or Operators can fix on a 'back to back' basis and avoid the risk of 'hiring' a ship (i.e. fixing on Time Charter). They will not suffer consequences from the cost of time lost due to adverse weather conditions and/or rough seas and/or for other reasons that are attributed not only to the prevailing congestion at the ports and/or strikes and/or force majeure etc., but also to cargo shortage claims and/or increased fuel prices and/or increased port expenses and dues and/or unforeseen conditions and circumstances, which may arise during the execution of the Charter etc. Additionally, the Charterers are not obliged to have the ship in a condition, after completion of the discharging operation, as clean as, when it entered for loading.

When fixing on Voyage Charter:
. the type/grade/quantity of cargo and the loading and discharging port/s or range/s are known to the Owners.
The Owners will execute the agreed voyage and will be compensated based on

a. a fixed price as 'freight rate'.
After completion of the loading operation, the agreed/fixed 'freight rate' will be multiplied with the intaken Bill/s of Lading cargo quantity, measured in metric tons, resulting in the amount due as 'freight' to the Owners, i.e., the Owners will be compensated (usually in US Dollars) 'per metric ton' or for certain cargoes 'per cubic metre', of the quantity of cargo loaded onboard.
b. a 'lumpsum' amount as 'freight'.
The Owners will be compensated with a 'lumpsum' agreed/fixed amount, which is payable by the Charterers, regardless of the quantity of cargo loaded. However, all designated cargo spaces are available to Charterers, who have the right to load the ship up to its full 'cargo capacity', when loading by virtue of the maximum permitted weight, or of the maximum available grain or bale capacity.

Generally, the ship based on an agreed 'freight rate' or on a 'lumpsum' amount as 'freight', will load the agreed grade/type/form of cargo, at the agreed quantity or until 'full and complete' all the cargo holds, as the case may be, from the known port/s or range/s as its loading area, to the known port/s or range/s as its discharging area.
The Owners should ensure that the ship can load and safely carry the cargo intended to be loaded, with the close cooperation of the Chartering department and the Operations and Technical departments of the Owners' office.

Fixing on Voyage Charter:
a. with named ports, implies Owners' obligations and responsibilities to check/ ascertain before fixing, the ability/suitability of the ship to approach/ enter the specified (named) ports and berth/un-berth/depart in a seaworthy and safe stress/ trim condition and comply with the port restrictions/ regulations and safely load and discharge the contractual cargo quantity.
b. with named ranges, the burden lies on the Charterers to nominate safe port/s, safe berth/s and safe anchorage/s without restrictions and to ensure, in accordance with the Charter Party terms, the 'safe' approach/ loading/discharging/sailing of the fixed ship.

The ship remains:
. in control of the Owners/Managers and maintenance,
who will organize and maintain safe manning on board, provide an efficient officers/crew, updated charts and sufficient quantity of fuel and qualities that are required for the completion of the voyage and generally present a seaworthy ship for the satisfactory execution of the intended voyage. Valid and up to date original certificates should be in the ship's possession, complying with the law of its class /flag etc. and with all international and local regulations in the ports

that will be contacted. The ship should be kept in a perfect condition and maintained the same for the entire duration of the voyage, for its obligations under the terms of the Charter Party to be fulfilled.

Owners undertake to pay for the costs, which include (unless otherwise agreed on by the Charter Party marked as ';'):

-the crew's wages and all related expenses, such as transportation/repatriation etc. and additional agency fees or costs resulting therefrom.
-the provisions/stores/spare parts/dry-dock and special survey or other commitments to the ship's class/flag etc. and generally all costs required to be spent for the ship's maintenance, including all additional Agency fees or costs resulting therefrom.
-the H&M (Hull and Machinery), P&I Club (Protection and indemnity).
-the basic War Risk insurance and the additional War Risk insurance (;), due to the ship's call to ports which are designated as dangerous or likely to be dangerous.
-the supply of fuel required for the execution of the voyage at sea and in ports.
-the supply of lubricants required for the execution of the voyage at sea and in ports.
-the passage through the canals/straits.
-the communications/victualing/entertainment.
-the consumption of fresh water for domestic use and cleaning of cargo holds.
-the consular fees/taxes associated with the ship's flag/crew/ownership etc.
-the imposed fines/penalties and/or security that is required, as a result of the Master's/ Officers'/crew's illegal actions.
-the ordinary port expenses in all ports of contact (tugs/pilots/light port dues/ agency fees etc. (;)
-the extra insurances (additional premiums) as will be charged by the cargo insurances, due to the ship's age/flag/ownership etc. (;)
-the possible despatch money to the Charterers, because of the earlier completion of loading and/or discharging. (;)
-the address commissions and the nominal percentage (rate) of 1,25 percent to each broker involved in the ship's fixture, which is calculated on the gross amount of the due 'freight' and possible demurrage incurred at all ports of contact.
-the office expenses (staff wages/insurance/rental costs etc.).
-the installments on loans, obtained from the banks, for the ship's purchase
All above costs should be covered by the amount obtained as 'freight'/ demurrage.

Owners/Ship will bear the cost and will suffer consequences for any loss of time:

-at sea, due to a swell/stormy sea, reduced speed due to adverse currents or for any other reason attributed to the ship's fault/defect at port.
-due to adverse weather conditions, or because of icy conditions, while ashore or in the waiting place/anchorage at the load/discharge port/s.
-due to reasons of force majeure (which should be specifically stated in the Charter Party), such as the act of god, war, hostilities, rebellion, mobilization, revolution, insurrection, acts of public enemies, civil commotions, sabotage, acts of government, fires, floods, earthquakes, explosions or other catastrophes, epidemics, quarantines, strikes or lockouts, embargos, railway accidents etc.
-if the ship is not, at any time during the voyage, in a seaworthy and safe stress /trim condition and in possession of all valid and up-to-date original certificates on board, complying with safety and health regulations and/or requirements at the countries that the ship intends to trade.
-due to out-of-date charts.
-in case the ship is not fitted for passing through the canals/straits and also the delays and time required for passing through them.
-due to the deviation from its route, for purposes of refueling or due to failure or damage of machinery, collision, stranding, fire, or accidental damage to the ship, dry-docking or periodical survey, sickness or accident of the Master/Officers/Crew, disembarkation of stowaways and for any other reason pertaining to the Owners' matters or fault.
-if the ship is captured or seized or arrested in the port or while passing the canals/straits etc., attributed to the Owners'/ship's fault/defect, or due to its current or previous trade.
-due to smuggling from the Master/Officers/Crew.
-because of the ship's detention for quarantine, due to the Master's/ Officers'/ Crew's pestilence, epidemics, illness or due to failure to pass 'free pratique' or for any other reason attributed to the ship's/crew's/Owners' fault.
-due to breakdown, partial or otherwise, or non-conformity or disablement or malfunction, in one or any/all winches/cranes or any other part of the ship's equipment used for the loading/ discharging of the cargo, that prevents the full operation of the ship.
-due to insufficient power to operate all the winches/cranes at the same time.
-due to unclean holds from the previous voyage and/or rust scale and/or loose rust in holds and generally, not being ready to load or discharge the cargo in every respect.
-if the ship is detained as unseaworthy by the Port State Control inspection, or by the International Transport Workers' Federation (ITF).
-due to the crew's strike and/or deficiency of officers/crew or stores.

-due to detention for oil or air pollution.
-due to supply of improper quality of fuel.

Duties imposed on the Owners/Master, who undertake to fulfil all the following obligations and are responsible for:

-ensuring that the ship is fit and can actually load, safely carry and discharge the fixed type/ grade/quantity of cargo.
-loading and discharging the ship in strict conformity with the international agreed provisions for safe loading, discharging, shipment and stowage of solid bulk cargoes (SOLAS 74) and the Code of Practice for the safe loading and unloading of bulk carriers (BLU Code) and always keeping it in a seaworthy and safe stress/trim condition.
-damage/shortage to the cargo due to the Owners'/ship's fault/defect.
-the prosecution of the voyage with the utmost dispatch and the avoidance of delays in delivery of the cargo to the Consignees.
-the ship being in possession, throughout the duration of the Voyage Charter, of all necessary valid equipment, updated charts and original certificates that comply with safety and health regulations, national and international regulations and all current requirements applicable to all intended ports for contact or crossing of the canals/straits and various straits.
-the ship to be presented for loading having its holds clean, swept, dried, free of cargo residues from the previous voyage, free of rust scales and loose rust and in all respects ready to load.
-the ship to be in such condition, as to pass the Port State Control inspections without deficiencies, which could cause significant delays in sailing, with severe consequences.
-arranging and taking preventative measures for the safety of the cargo/crew/ ship.
-providing an efficient crew and keeping the safe manning in accordance with the requirements of the flag.
-checking and ensuring that the ship can trade at the intended ports, taking into consideration the ship's characteristics, flag, class, previous trade, ownership etc. and that no port restrictions are in effect.
-obtaining all the necessary information concerning the intended cargo for loading, enabling the Master to ensure that the cargo is suitable for the ship and can be stowed/trimmed and secured as per the ship's stress and can be loaded/ transported/discharged in accordance with the local and international regulations.
- the consequences/delays/loss/expenses that can arise when they have provided any misleading information on the ship's characteristics and on the terms agreed in this Charter Party.

-providing enough fuel quantity for the safe execution of the voyage.

Charterers undertake to pay for (unless otherwise agreed on by the Charter Party marked as ';'):

-the amount due as 'freight'.
-the amount due as demurrage, possibly.
-the amount due as dead-freight, possibly, due to insufficient presentation of the agreed and requested by Master cargo quantity or due to inaccurate/ incorrect declaration of the cargo stowage factor.
-the loading/discharging costs (free in/out terms to Carrier) which should have been paid by the party responsible (according to incoterms that include the stevedoring expenses, hiring of shore means, trucks etc.).
-all taxes/dues calculated/levied on cargo.
-possible stevedore damage during the operations in ports. (;)

Duties imposed on the Charterers, who undertake to fulfil the following obligations and are responsible for (unless otherwise agreed on by the Charter Party marked as ';'):

-presenting the agreed type/ grade/form and quantity of cargo for loading
-the presentation of the cargo in a sound condition.
-making the payment/s correctly and timely, in terms of the Charter Party.
-the trim/extra trim (;) /stowage of the cargo in the ship's holds.
-the actual stowage factor of the cargo to remain as declared.
-the proper stowage and operation by competent and experienced stevedores, without cause/infliction of damage to the ship's cranes/hull/equipment.
-providing before loading (the same should be incorporated in the Charter Party, otherwise it might fall on to the Owners to provide the necessary dunnage materials) and removing after discharge, all dunnage materials placed amongst the cargo, for the proper stowage / separation of cargo and the lashing/ unlashing/securing of cargo, under their own time/risk/expenses.
-providing grabs that are compatible with the ship's gear, as described by the Owners.

The 'cargo orders', for Voyage Charter will enter the market in accordance with the following modes:
1. Single voyage (one-way voyage):

The ship will execute
. a single voyage (one-way voyage), from the specified/named port/s of loading

to specified/ named port/s of discharge, strictly in accordance with the terms of the issued Charter Party.
Before completion of discharge, the Owners will enter the market again with the ship's actual position and its expected dates open and looking for the next employment.

The interested and subsequent fixed ship must be a
. 'named' ship
and under the terms of the Charter Party, the Owners do not have the right to substitute it with another or to sell it (and change name/flag/class etc.) during the execution of the voyage without the Charterers' consent.
The Charterers have the right to refuse the commencement of loading, if the Owners, in breach of the Charter Party terms, have presented any other ship, even one with better characteristics, than the fixed one.

The cargo can be transported:
- As full cargo.
Even if the ship's holds and/or deck cargo spaces have not been filled with cargo, the Owners are not allowed by the terms of the Charter Party to search the market and load any other 'completion cargo' in cargo holds or on deck.
- As part of the cargo (part cargo).
If, the ship's holds and/or deck cargo spaces have not been filled with cargo, the Owners have the right, under the terms of the Charter Party, to search the market for a 'completion cargo' (until it reaches the maximum deadweight which the ship can provide), mainly en route to its final destination. During the negotiations of any 'part cargo', both Charterers and Owners could negotiate whether the cargo intended to be loaded is to be 'first in-last out' or 'last in-first out' etc. from the cargo holds.
This chosen method of transportation will be incorporated into the Charter Party.

As an example, for a single voyage (a cargo order as entered the market):
-Rouen France /Bari Italy
-3,000 mtns upto 6,000 mtns bulk wheat about 44'
-Next week/onwards
-3,000 mtns shex uu/3,000 mtns sshex eiu
-Synacomex 90 with 5 pct ttl commission
-SID-Max 25 years – gearless ok-no part cargo

One-way voyage will be executed:
-from Rouen France as load port.

-to Bari Italy as discharge port.
-the cargo quantity ranging from 3,000 up to 6,000 mtns.
-the type of cargo is wheat in bulk.
-the loading dates are from the next week and onwards from the time the cargo entered the market i.e., ships that are in a position to load immediately (available for spot or even prompt loading) are not workable.
-the loading rate is 3,000 mtns shex uu (Sundays/holidays are excluded as working period Unless Used) for any size loaded.
-the discharging rate is 3,000 mtns sshex eiu (Saturdays/Sundays/holidays are excluded Even If Used) for any size loaded.
The ship which should be workable for this particular biz has been indicated (sid-max 25 years-gearless) as are the dates which are available for loading. No other cargo is allowed to be loaded. After execution of the voyage and completion of discharge and since Charterers have fulfilled their financial obligations to the Owners and no cargo claims have been raised/presented, there is no further commitment to the Owners for this particular biz. Nevertheless, for some years (according to the rules of the ISM) the parties involved should keep the files safe and ready for re-opening, whenever the same is requested.

2. Consecutive Voyages in direct continuation:

The ship will execute:
- a number of consecutive voyages, as mutually agreed upon.
- in direct continuation.
- from a named load port to a named discharge port.
- with cargo quantities in equal lots.
- with a fixed 'freight rate', which is in effect for all voyages and is payable 'per metric ton' of cargo loaded onboard.
- under the terms of a uniquely issued Charter Party, which covers all voyages.

The ship, after the completion of each voyage will proceed directly to the load port for the execution of the next voyage, i.e., every voyage starts being calculated in direct continuation and on the basis of the previous one's completion.

The ship has no right to perform any other intermediate voyage and will remain committed to the contract, until the execution of the entire number of the agreed voyages.

Charterers or Shippers and Receivers should take care to ensure the availability of the loading and discharging berths respectively, upon the ship's arrival. The consecutive voyages are usually profitable, when the cargo is moved to or from private berth/s or Shippers'/ Receivers' own mines, enabling them to control the ship's schedule, thus preventing the consequences of delays in berthing.

The whole cargo quantity will be transported
. as sole cargo (no part cargo).

Usually, the interested and subsequently fixed ship should be a
. 'named' ship and,
under the terms of the Charter Party, the Owners do not have the right to substitute it with another or to sell it during the execution of the voyages, without the Charterers' consent.

As an example (a cargo order as entered the market):
-6 to 10 consecutive voyages in direct continuation
-each 4500/5000 mtns bagged cement
-1 spsb Volos Greece/1sp 1-2 sb/s Alexandria Egypt
-commencing ppt onwards
-2000/1200 mtns sshex bends
-Gencon 5 ttl

As per circulation, consecutive voyages will be executed:
-from 1 safe port/berth in Greece as loading area (from Volos port),
-to 1 safe port- 1/2 safe berth/s in Egypt as discharging area (to Alexandria port),
-with cargo cement in bags.
-2,000 mtns as loading rate/1,200 mtns as discharging rate.
-sshex (Saturdays/Sundays/holidays are excluded-no working period) at both ports.
The number of voyages will vary, between a minimum of six up to a maximum of ten. The exact number will mutually be agreed upon, during the progress of negotiations.

The long distance between the load and discharge port/s increases the costs, making a deal in consecutive voyages obviously unattractive/unprofitable. However, such a type of cargo biz could be fixed on a c.o.a. basis.

3. Contract Of Affreightment (abbreviated as 'C.O.A.'):

Fixing on a c.o.a. basis is when:
-large or small cargo quantities,
-are to be shipped from one specific place/port to another (irrespective of the distance between them),

-in various or equal lots,
-within a certain period of time (but not in direct continuation),
-with the same or various ships,
-under the terms of a unique issued Charter Party.

The 'freight rate' is fixed once and is in effect/valid for all shipments, but it is not the norm. The Owners and/or Operators could take precautionary measures in the Charter Party, for certain unforeseen parameters, such as rising fuel prices or increased port expenses or imposed new port taxes etc. that could arise during the execution of the c.o.a. The 'freight rate' should be re-adjusted proportionally.

The whole cargo quantity can be transported
. as sole or with the option of part cargo, as will be agreed upon.
The Owners or Operators, who would be interested in fixing on a c.o.a. basis, will be contractually required to present ship/s:
a. with the minimum characteristics required for the shipment and have the capacity to accommodate the intended quantity of each shipment/lot,
b. to transport the whole fixed cargo quantity within the prescribed period of time.
The contract should be fixed on the basis of:
-ships, which are 'named' (nominated by the Owners or Operators) and whose characteristics are obviously known before the conclusion of the fixture. The Owners/Operators will only use ships, which are trading under their own Management or Time Charter Party. However, they could retain the right, under the terms of the Charter Party, to substitute the nominated ships with other sister ships or with ships that have similar or better characteristics.
- ships, which are 'to be nominated' or 'to be advised' (abbreviated as 'tbn' or 'tba' respectively). However, the nominated ships, along with their trading characteristics will be disclosed, upon tendering certain mutually agreed notices, before the first day of the dates declared by the Shippers for loading.
Charterers have the right to reject a nomination, in the case that there are reasonable grounds to do so and Owners/Operators are to nominate a substitute ship that should comply with the needs of the shipment and the agreed Charter Party terms. If the Owners/Operators fail to meet their contractual obligations and fail to nominate a suitable ship within the agreed time frame, the Charterers, under the terms of the Charter Party, might have the right to fix a substitute ship from the market and then claim damages from the Owners/Operators.
The party, which will fix on a c.o.a. basis has the following options for the execution of the contract:
1. to fix the whole contract with others on a 'back-to-back' basis, or partly, earn-

ing from the difference of the 'freight rate'.
2. to fix through the open market on a voyage-by-voyage basis:
a- either on Voyage Charter, paying a 'freight rate' per metric ton of intaken Bill of Lading cargo quantity,
b- or on a Time Charter Trip Out, paying a daily 'hire rate' to Owners.
c- or to fix a ship on a long or short Time Charter period, so that after completion of discharge to search for cargoes from the open market, to reposition the ship near to the c.o.a.'s loading area and be ready for the execution of the next shipment.

As an example:
-C.O.A. commencing January 2... thru Dec 2...
-Min 700,000 mtns/ Max 1,100.000 mtns Bauxites in bulk
-By lots of 40,000 or 45,000 mtns 5 pct moloo each
-Tonnage to be geared or gearless single deck max 30 years old
-1 sb New Amsterdam, Guyana/1 sb Dneprobugsky, Black Sea
-8,000 mtns pwwd shinc/10,000 mtns pwwd shinc
-chabe
-3,75 ttl here

Owners and/or Operators will offer:
-for a total cargo quantity of a minimum 700,000 to a max 1,100.000 mtns bulk bauxite.
-in lots of 40,000 mtns or 45,000 mtns with a tolerance of 5 pct more or less, at the Owners' option.
-the transport of the total quantity should be executed from January 2... until December 2...
-the frequency of trips that are going to be performed should be discussed and agreed upon during the negotiations.
-the workable tonnages should be singledeckers, geared or gearless, not, however, exceeding the age of 30 years.

However, before the presentation of any offer, the Owners or Operators should imperatively check:
- the prevailing available draft/salinity and further port restrictions that could prevail at all ports, allowing them to calculate the right size of the intended ship for loading.
- the approximate port expenses in all ports and the cost for passing the straits (in this case the Bosporus straits).
- if any taxes/dues could be in effect for freight/ship/flag/ownership in all ports of contact.

- the existing market's fuel prices.
- the existing political situation in all ports and at the places the ships will cross (straits).
- the prevailing market and the expectations during the execution of the c.o.a.

A fixture on Voyage Charter in the dry cargo sector is characterized by the following parties involved and terms, which along with certain others will apply and most of which could affect the 'freight rate':

1. Owners, Managers and Charterers.

- Owners:

they are the party, who have the possession and legal rights to the ship.

-Managers:

they are the party hired by the Owners to 'run' the ship, according to applicable International rules.

The Owners/Managers will present the seaworthy and, in all respects, ready for the shipment and Charterers' needs, ship, which should be in the right position for the carriage of the goods by sea.

Charterers should avoid chartering those who are known in the market as being unreliable,

- either because they do not fulfill their obligations under the terms of the Charter Party,
- or because of a ship's poor state of maintenance, which could lead to its detention in the ports of contact.

-Charterers:

they are the party, who will Charter (fix) a ship from the shipping market to execute the particular shipment/s.

Those who:

- have very little activity and are not well established in the market,
- are not considered as 'first class',
- have provided negative experiences to certain Owners in the past,

may not be approved, as no one should take the risk of fixing with someone who might step down from full or partial fulfillment of their contractual obligations.

However, a fixture with Charterers with little activity or who are unknown in the market could be saved if:

a. Charterers nominate

. as 'guarantors'

another well-known or better-known reputable shipping company in the shipping community, which could be fully approved by the Owners.

A letter of guarantee, based on the Owners' wording, should be taken direct-

ly from the guarantors, before the final agreement is reached, confirming that they guarantee the Charterers' performance/payments. The guarantors should be requested to countersign the original Charter Party, showing that, apart from the guarantees that are provided to the company, they also admit that they are also well aware of the existing Charter Party terms.
b. Charterers provide
. a revolving bank guarantee,
as will be issued by a first-class bank, which roughly covers the 'freight' amount for the cargo which is expected to be loaded and an approximate amount for possible demurrage that can be incurred in the ports of contact.

The Master will be receiving legitimate employment instructions
. from the Owners and/or Managers and not/never by Charterers.

2. The ship-The description of the ship.

THE SHIP:
A commercial ship for dry cargo is a 'steel lady' that is seaworthy, navigable and was built and designed for the carriage by sea, from one place to another, of any grade, type, quality, and form of goods for transportation.

Dry cargo ships are designed in different types and sizes depending on the global demand for marine cargo transportation. The size of the ship is identified by the deadweight all told (abbreviated as 'dwat'), which is the maximum deadweight, which the ship can provide, as designed by the shipyard based on transactions within zones of summer, at the designated scantling draft, after calculating the following complete weights:
- the cargo,
- the fuels including lubricants,
- the fresh water,
- the ballast water,
- the ship's crew and their effects,
- the stores, spares, provisions, equipment, inventories etc. on board the ship

When the ship navigates within winter or tropical zones, the draft will fluctuate from reducing or increasing the deadweight.

Some Charterers, due to their contractual commitments with Shippers and/or Receivers to load a minimum cargo quantity, could request from the Owners to provide and guarantee, before the fixture, the deadweight cargo capacity (abbreviated as 'dwcc'), which the ship can offer for this biz.
. The 'cargo capacity' (abbreviated as 'cc')

is the maximum cargo, which the ship can load at the port of loading (when loading by virtue of the maximum permitted weight), which is subject to:
- quantity of fuel on board the ship at the time of loading, or until the next intended fuel supply,
- quantity of fresh water on board the ship,
and
- total weight of the constants.

The fixture may be lost, if the ship has been previously heavily bunkered, or the standard weights (constants) are heavy, resulting its 'cc' not to be enough to accommodate the required minimum quantity of cargo.

As the types of ships are designed to serve different requirements/cargo orders, the Charterers will select the ship from the market, in accordance with the needs of the shipment, depending on:
- the cargo if it is wheat, cement, forest products, project cargo, steel etc.
- the type of cargo and if it contains dangerous/inflammable/harmful goods etc. in which case a ship, which is approved/classed by a classification society for the loading and transport of such type of cargoes will be required to be fixed.
- the total available quantity of cargo and if a coaster/handy/panamax/cape sized ship etc. will be required for loading.
- the cargo form; if it is in the form of bulk/bags/pallets/bundles etc., in that case and as the case may be, a bulk carrier (with hoppered holds or box type) or multi-purpose, tween-decker ship etc. will be required for loading.
- the availability of facilities in all ports of contact, i.e., if shore cranes, evacuators or other forms of loading or discharging equipment are available or not, in which case a geared or gearless ship will be required to be fixed.
- the intended place/port for loading and/or discharging, in which case an Ice class ship, or a River ship or a Laker or a Kamsarmax or another type of ship will be required for loading.
- the specific terms of the Letter of Credit or restrictions from Insurance companies for the maximum acceptable age of the ship which will execute the voyage, in which case an over aged or not will be required for loading.
- the ship's flag/nationality of the crew/ownership/previous trade etc. in relation to political constrains and a possible boycott in the port of loading or discharging.
- the port restrictions that are in effect at the port of loading and/or discharging that will determine the size of the ship.

The description, as will be provided by the Owners must be an accurate representation of the characteristics of the ship, which can only be modified with the usual and acceptable shipping interpretation of the word 'about'. The Owners should be kept responsible and liable for the consequences, delays, loss, expenses that could arise when providing any

. misleading information on the ship's characteristics.

Throughout the duration of the Voyage Charter, the ship must be in possession of all necessary valid equipment and certificates that verify the size and type of the ship, its seaworthiness, its compliance with the health and safety regulations, national and international regulations and all current and local requirements, at all the agreed ports/regions that the ship will contact, including valid certificates for passing the Panama and/or Suez canals and several other straits. The Master, before the ship's arrival at the port, will report to the Port Authorities through the ship's attending Agents, that the ship complies with the regulations on health and that the crew is healthy, without infectious diseases or symptoms of infectious diseases. The Port Authorities will allow the ship to approach the port and will either drop the anchor in the anchorage, awaiting its turn for berthing, or it will be directed to the berth. Owners will be responsible and should have arranged for the Master, officers, and crew of the ship to comply with vaccination and sanitary regulations in the ports of call. The Master, Officers and Crew of the ship must hold vaccination certificates against certain serious diseases (yellow fever, cholera, smallpox etc.).

The Port Authorities (along with the Customs and Immigration) will board the ship before anyone else and will check that all corresponding certificates are in compliance with the health regulations and will obtain the Master's assurance/confirmation that the crew is healthy and complies with the vaccination and sanitary regulations at the port of contact.

If all is found in order, they will grant clearance, allowing the ship to stay at the port/berth and to commence its operational activities, with the crew members free to disembark from the ship. The granted health clearance is commonly known as

. 'free pratique'

and is often directly related with the tendering of a Charter Party valid Notice Of Readiness. Before the free pratique is granted to the ship, operational activities are not allowed, nor is any other person except for the pilot, Customs, Immigration, attending Agents and the Port Authorities allowed boarding the ship. In the case of failure, the Port Authorities will deny the commencement of any operation within the port of call and will put the ship in quarantine or will detain it, resulting in its shifting to a waiting place (at berth or anchorage), until it complies with the health regulations.

The ships that trade, either on Time Charter, or on Voyage Charter must carry on board, along with certain others, the following original certificates:

1. Certificate of registry.

This is a certificate that shows where the ship is registered. The certificate of registry, inter alia, contains the ship's identity, its main dimensions, engine data, the name of the Owners' company along with their address, etc.

2. Certificate of classification (Class certificate).

It is a certificate that is in effect for 5 years and is issued by a classification society. It will be issued after the inspection of the ship's hull / machinery and equipment, according to the rules and regulations of the inspected firm.

3. International load line certificate.

It is a certificate which is valid for 5 years, issued under the provisions of the international convention on load lines, by a classification society on behalf of the ship's registry and defines the allowed summer/ winter/tropical ship's draft.

4. Cargo ship safety equipment certificate.

It is a certificate which is valid for 5 years, issued in accordance with the provisions of the International Convention for the safety of life at sea, by a classification society on behalf of the registry, which certifies that the ship complies with the requirements of the Convention with regard to fire safety systems, its life-saving appliances and the equipment of the lifeboats etc.

5. Cargo ship safety construction certificate.

It is a certificate which is valid for 5 years, issued under the provisions of the International Convention for the safety of life at sea, by a classification society on behalf of the registry of the ship, which certifies that the ship has been inspected and that the structure, machinery, and equipment was found to be satisfactory and the ship has complied with the relevant requirements of the Convention.

6. International oil pollution prevention certificate.

It is a certificate which is valid for 5 years, issued under the provisions of the International Convention for the prevention of pollution from ships, by a classification society on behalf of the registry of the ship, which certifies that the ship has been inspected and that its structure, equipment, systems, fittings, arrangements, and materials on board as well as their condition is satisfactory and that the ship complies with the applicable requirements of the convention.

7. Cargo ship safety Radio certificate.

It is a certificate which is valid for 5 years, issued under the provisions of the International Convention for the safety of life at sea, by a classification society on behalf of ship's registry, which certifies that the ship complies with the requirements of the convention as regarding radio installations.

8. P&I Club cover entry certificate.

It is a certificate which is valid for 1 year, issued by the ship's insurances (Protection and Indemnity Club), covering cargo claims, pollution claims, wreck removal, dock damage, seamen/crew risks etc.

9. International tonnage certificate.

It is a certificate issued by the ship's flag, under the provisions of the International Convention on Tonnage Measurement of Ships 1969, without an expiry date (unless its flag is changed in which case re-issuance is required), which certifies the ship's tonnages (gross tonnage/net tonnage).

10. Minimum Safe Manning document.

It is a document issued by the flag of the ship, under the provisions of the Merchant Shipping Law and regulations of the International Convention for the safety of life at sea, stating that the ship is safely manned, carrying no less than the number and grades/capacities of personnel as defined in this document.

11. International air pollution prevention certificate.

It is a certificate which is valid for 5 years, issued under the provisions of the Protocol of 197, by a classification society on behalf of ship's registry, which certifies that the ship has been inspected and that its equipment, systems, fittings, arrangements, and materials fully comply with the Convention's applicable requirements.

12. International sewage pollution prevention certificate.

It is a certificate which is valid for 5 years, issued under the provisions of the International Convention for the prevention of pollution from ships, by a classification society on behalf of the ship's registry, which certifies that the ship is equipped with a Sewage Treatment Plant and a Holding Tank.

13. Document of compliance.

It is a certificate which is valid for 5 years, issued under the provisions of the International Convention for the safety of life at sea, by the ship's flag or a classification society, which certifies that the Company's safety management system has been audited and that it complies with the requirements of the International Management Code for the safe operation of ships and for pollution prevention (ISM Code).

14. Safety management certificate.

It is a certificate which is valid for 5 years, issued under the provisions of the International Convention for the safety of life at sea, by the ship's flag or a classification society, which certifies that the Ship's safety management system has been audited and that it complies with the requirements of the International Management Code for the safe operation of ships and for pollution prevention(ISM Code), following verification that the Document of Compliance for the company is applicable to this type of ship.

15. Deratting exemption certificate.

It is a certificate issued after the inspection of Health Authorities, mainly for rodent infestation, and is valid for six months.

16. International Ship Security Certificate.

It is a certificate which is valid for 5 years, issued under the provisions of the International Code For The Security of Ships and of port facilities (ISPS), by a clas-

sification which certifies that the security system and any associated equipment on board has been verified and it is in all respects satisfactory and that the ship complies with the applicable requirements of the Convention and ISPS Code.
17. Panama Canal tonnage's certificate and Suez Canal certificate.
These certificates are required to be submitted to the Port Authorities on arrival in the canals and will determine the cost of the passage.
18. Ship security plan.
It is a certificate which contributes to the ship's protection and provides measures for prevention of any acts against the ship/cargo and its crew.
19. Continuous Synopsis Record.
It is a certificate issued for the history of the ship that contains its name, flag, class, registered Owners, IMO number, port of registry etc.
20. Document of compliance for the carriage of Solid bulk cargoes.
It is a document which is valid for 5 years and the expired date is always harmonized with the Class period. The document covers the cargoes as listed in the current edition of the BC Code.

The Owners are the sole party responsible for checking the ship's ability and suitability to approach, berth, load, discharge, sail to/from the specified/named ports and pay for the ordinary charged port expenses. Under such circumstances and in comparison, with the ship's Time Charter description, many of the characteristics will be omitted, such as the consumption, fuel specifications, gt/nt etc. as no one that is fixing the ship particularly cares about it.
However, when an agreement is reached on the basis of
. free port disbursements ('free d/a')
of the ship (Charterers will cover the ordinary ship's port expenses in any amount), or when ranges and not ports have been nominated, or general cargoes, logs, steel etc., or any other special cargo is to be loaded, then the holds' dimensions, hatch dimensions, tank top strength, gt/nt, depth moulded etc. should be provided by the Owners to Charterers.

The basic description applicable on Voyage Charter should include at least:
-the ship's name and ex name/s – the type - when it was built -the flag.
-the Class -the Call sign-the IMO number-the P&I Club.
-the summer dwat and at which summer draft.
-the l.o.a. (length over all) /the beam.
-the total grain capacity /the total bale capacity.
-the number of holds and number of hatches.
-the 'cargo gear' description for geared ships.

The Owners, during the progress of negotiations could be asked to guarantee certain customary but important terms, which ensure that the ship trades in accordance with the International rules and regulations and that the ship complies with the needs of the shipment.

However, brokers and Owners should be suspicious, when confirming these terms and seek explanations and/or clarifications and have their reservations where necessary, otherwise,

. *a simple and customary term, which could easily be confirmed by all, could incur additional obligations and responsibilities that could lead to losses.*

Some of these terms could be:

a- that the ship:

-is self-trimming bulk carrier (abbreviated as 'STBC').

-is steel floored throughout.

-has engine/bridge aft.

-has holds/hatches/tank tops clear and unobstructed without centreline bulkhead, walkways, centrebeams and free of all/any obstructions which might incur difficult access to the cargo and free use of grabs/bobcat/shovel operations during loading/trimming/discharging operations.

-is suitable for grab discharge and no cargo is to be loaded in deep tanks, sidetanks, wing tanks and bunker spaces or any other confined spaces unreachable to the means of loading and discharging.

Comments:

A Self-trimming ship is one that, by virtue of the design of its cargo holds, does not need any extra cargo trimming, when loading full bulk cargo, during or after completion of the loading operation.

However, this is not in effect for all types of bulk cargoes (such as pig iron, cement, gypsum, aggregates etc.), but only for free-flowing grain cargoes.

Owners should explicitly state, when confirming that the ship is self-trimming, that (as per Maritime Organizations recommendation) the ship is

. *'STBC when loading of free-flowing grain cargoes only'.*

The presentation by the Charterers or Receivers, of unreasonably large grabs, which could delay the discharging operation due to holds configuration, could give rise to claims against the Owners, because of the incorrect description that the ship is suitable for grab discharge. Owners should clarify and state, as a minimum act to protect the ship, *that* the ship is suitable for a normal grab discharge operation, with normal size/type grabs and as far as a standard bulk carrier/singledecker can be.

The Owners should also be requested to guarantee that the intended cargo will not be loaded in places inaccessible to loading and/or discharging shore means and that it is a bulk carrier/ singledecker without obstructions in its cargo holds.

b- that throughout this Charter period, the hatch covers of the ship will be water-tight and will be kept in top condition.
Comments:
The hatch-covers are on top of the cargo holds and cover them, in order to prevent any ingress of water or humidity, therefore the Owners are always obliged to keep the hatch covers
. watertight and without any leakages which might damage or affect the quality of the carrying cargo.
For expensive or sensitive cargoes, such as alumina/sugar/steels etc., the Charterers could ask to be entitled, before the commencement of the loading operation, to carry out a hose or chalk test, so as to ensure its water tightness, and/or for an ultrasonic test on all hatch covers, in order to ensure the thickness of the steel. In the case of defect, the Owners must repair it immediately, under their own time and expenses.
If the ship fails the second inspection, the Charterers then may become entitled, under the terms of the Charter Party, to cancel the Charter Party.

c- that the ship will have holds strengthened and classed for carriage of cargo of ...... and will be capable of loading the contractual cargo quantity in all or in alternative holds, as will be required by the Charterers.
Comments:
Many Charterers, when loading large sized ships, prefer loading the whole cargo quantity for carriage using the minimum possible holds, thus enabling the discharging operation to proceed faster, avoiding the shifting of the shore cranes in the berth and delays that may be incurred thereof.
Such a method of loading is also sometimes preferred by the Owners, when dirty cargoes are going to be loaded, in which case some of the holds will remain clean and ready for the next shipment.
This method of loading is in effect:
-for class approved ships allowing for alternative loading.
-for cargoes with a very low stowage factor.
-for ships, which provide sufficient tank top strength.

d- that the ship will proceed via the most direct route and with the utmost despatch directly to the discharge port/s and will not deviate for any purpose.
Comments:
The changing of route to the discharge port for whatever purpose, might be considered as a violation of the Charter Party terms. However, the Owners, through the progress of the negotiations could apply to have the right to deviate, under the Charter Party terms:

-for bunkering purposes en route to discharge port (the Bimco's P&I bunker deviation clause could be in effect).
-in the case of emergency (accidents or illness of crew etc.).
-to avoid adverse weather conditions (typhoons etc.).
-for saving lives or to assist a ship in distress (the salvage attribution will be for the Owners' sole benefit).

e- that the ship will not be dry-docked during the currency of this Charter Party, except in case of emergency.
Comments:
It is not logical and quite impossible for a ship to ask to enter dry-dock during the execution of the Voyage Charter, except of course in case of emergency. Charterers will try to exclude any such possibilities through this request.

f- that the ship will not change ownership and/or class or be sold during the Voyage Charter, without the Charterers' written consent.
Comments:
The changing of class or ownership during the fixed voyage could incur huge delays/extra expenses and unpredictable consequences.

g- that the ship is fully P&I Club covered (as member of the International Group of P&I Clubs) and has been covered for ship's oil pollution risks and wreck removal risks and will remain so throughout the duration of this Charter.
h- that the ship's H&M insurance will be fully maintained and will not be changed.
Comments:
Owners should guarantee that the ship is P&I Club covered by a Club, which is a member of the International Group of P&I Clubs and additionally that it is covered for H&M risks. Charterers could request for the relevant certificates of coverage to be faxed to them before the fixture's conclusion.

i- that the ship is fully classed as Lloyd's 100A1 or equivalent with a classification society that is a full member of the International Association of Classification Societies (IACS) throughout the duration of the Charter.
Comments:
Owners should guarantee that the ship is classed by a classification society that is a member of the IACS. Charterers could request the relevant certificate of class coverage to be faxed to them before the fixture's conclusion.

j- that the ship has no outstanding recommendations, restrictions or any other requirements of any nature in respect to flag state, port state control or classification inspections that may affect this voyage and the Charter Party.

Comments:
Owners should guarantee that the ship is free to trade without outstanding recommendations of any nature, which could result in the ship's detention in the ports, with huge consequences for both sides.

k- that the ship will comply with all applicable laws, regulations and requirements in force at or applicable to load and discharge ports and is in all respects eligible for trading to the ports, places or countries specified in the Charter and has all valid and up-to-date certificates and equipment and other documents on board that are required for the particular voyage.
Comments:
Owners should guarantee that the ship is in all respects ready to execute the Voyage and it is in possession of all necessary valid equipment and certificates that verify its size and type, its seaworthiness, its compliance with the health and safety regulations, national and international regulations and all current /local requirements at all agreed ports/regions that the ship will contact for operation.

l- that the ship is suitable for entering and berthing at the specified/named ports and loading / discharging the minimum contractual cargo quantity.
Comments:
Owners should guarantee that all port restrictions have been checked and that the ship will load and discharge the minimum contractual quantity of cargo with-out problem.

m- that the ship can open/close hatches with its own means.
Comments:
In any other case it could cause delays and complications with the Shippers/Receivers or Port Authorities.

n- that the ship and the company is fully in accordance with the ISPS/ISM requirements.
Comments:
Owners should guarantee that the ship and the company will comply with the requirements of the International Safety Management (ISM) Code and with the International Ship and Port Facility Security (ISPS) Code.

o- that the ship has not carried mammalian meat, bone meal, fishmeal, asbestos cullet, or nuts during the past 12 months.
Comments:
In the case of loading grains/agricultural products, Charterers could request the ship to not have been loaded with certain specific cargoes during the last year.

p- that the ship will not and has not been blacklisted/boycotted or arrested due to ship's flag/ownership/crew/employment/age/past trading or for any other reason, from local authorities and or unions in all countries designated under this Charter Party.

Comments:

The ship's flag/ownership/crew's nationality or ships blacklisted due to their previous trades may be cause for the ship to be boycotted or detained in the port of contact. Charterers seek to protect their interests and not to face such problems when they fix the ship.

q- that the ship is free of all encumbrances, arrests, and maritime liens.

r- that the ship has no .... (countries which may affect the voyage) interests and is not blacklisted by the Arab league countries.

s- that that the ship is ITF compliant or bona fide trade union agreement or equivalent and shall remain so throughout the duration of this Charter and the Owners will remain responsible for all consequences and/or delays if the ship is sailing under a flag of 'convenience'.

Comments:

Charterers look to ensure that the ship is free to trade in the ports as specified by them.

t- that the description of the ship, as it is provided by the Owners, is an accurate representation of the characteristics of the ship which can only be modified with the usual and acceptable shipping interpretation of the word 'about'. This clause shall override any inclusion of the words 'without guarantee' in the ship's description. The Owners are responsible and liable for the consequences/ delays/ losses/ expenses that might arise when providing any misleading information regarding the ship's characteristics and the terms agreed in this Charter Party.

Comments:

The description of the ship must be an accurate representation of the characteristics of the ship and Owners are liable and responsible for any misleading information provided, regarding the ship and the terms agreed. Besides, this term covers the Charterers, in case they have overlooked a description of the ship as 'without guarantee'.

u- that the ship provides adequate mooring lines and tends them properly, whilst in the loading or discharging port.

Comments:

Ports regulations should require the ship to provide adequate and in good condition mooring lines for the ship's safety, while it is operating in the ports.

v- that the ship will give to Charterers, free of expenses, the full use of its lighting on deck and in cargo compartments, for the cargo operation during the night.
Comments:
The ship should allow and provide to Shippers and Receivers, the full use of its lighting on deck and in cargo holds, for an unconditional operation, even during the night.

w- that the ship will, throughout the duration of the Charter Party, have on board and make available to the Charterers or their Agents or to the appointed surveyors, the capacity plan, the deadweight and displacement scales, the hydrostatic curves and tables of displacement, the calibration scales for all ship's tanks and trimming correction tables, and all sounding tables, and all sounding tubes shall be in good maintenance condition and free from impediments. Plimsoll marks amidships and draft marks on the 'port side' (on the left side of the ship) and 'starboard side' (on the right side of the ship), bow, amid ship and stern shall be always clearly cut/marked on shell platting and readable.
Comments:
A correct and accurate draft survey could be contacted only when the above-mentioned conditions have been met by the ship.

x- that the ship's cranes and grabs are tested/certified to work simultaneously and continuously with lifts of minimum capacity as described and being able to serve all holds/hatches unconditionally without stoppages.

y- that the cargo ship's gear and all other equipment will be kept in good working order/services and in compliance with the regulations of the countries which the ship will be in contact with and that the ship is in possession at all times of valid and up-to-date gear certificates of efficiency in compliance with such regulations.
Comments:
Owners must guarantee that the ship's cargo gear will work unconditionally and without stoppages.

3. Ports, ranges, berths, anchorages, and port restrictions/expenses.

I. Ports, ranges.

Every fixture should be concluded only after having agreed on one of the following terms that identify by name the ports/ranges that the ship will load and dis-

charge from:
a- identified by name safe port/s for loading and/or discharging purposes.
In this case, the port/s of loading and discharging will be disclosed/named by the Charterers upon circulating the 'cargo order' in the market.
For example:
the ship will load from Santos port, Brazil and will discharge in Piraeus port, Greece.
The named port makes the Owners sole responsible to check/ascertain the ship's ability/suitability to approach, berth, shift, load, discharge and depart in accordance with the prevailing regulations and restrictions that exist there.
b- identified by name a number of safe ports (it is mainly in effect for discharge), out of which the Charterers will confirm, usually en route to ports and within a certain time frame, the port that will be in use.
For example:
the ship will discharge in Egypt, in one port out of Alexandria or Damietta or Abukir at the Charterers' option.
The Owners remain to have the sole responsibility to check/ascertain the ship's ability/suitability to approach, berth, shift and depart from the nominated ports/ berths and load and discharge in accordance with the prevailing regulations and restrictions there.
C- identified by name ranges at the same or different countries in vicinity.
'Range' is a specific area in a country or a specific area consisting of more than one country in vicinity, where the loading or discharging operation will take place.
For example:
the ship will discharge in one port within the Black Sea.
The port is unknown to the Owners and the nomination lies on the Charterers, who should clearly disclose the port well before the ship reaches the range (the exact time of nomination is subject to the agreed Charter Party terms), in order for the ship to choose the correct route towards the port of operation.
Charterers' failure to nominate the port within the agreed period
. will possibly result in their liability for the Owners' losses.
Ranges, implies obligations to the Charterers, to nominate port/s, berth/s, with the capacity to accept the characteristics of the ship and approach, berth, shift and sail safely, without exposing the ship to danger of any nature, after having loaded and/or discharged the contractual cargo quantity. The Charter Party should explicitly incorporate that the ship will trade between 'safe ports' and 'safe berths'.
The Owners, after nomination
. are obliged to accept and proceed to the port/s
and the Charterers, after nomination, cannot revoke the order, unless the Own-

ers have agreed and compensated for the deviation incurred and the value of bunkers consumed.

If two ports are to be used within the range, the Charterers will nominate them

. in geographical rotation,

as will be stipulated in the Charter Party, otherwise, they have the right to choose the rotation at their discretion and/or needs.

The Owners might have taken the risk of unknown port expenses, port dues and/ or taxes, but they are usually sufficiently experienced, when trading through these fixed countries. In addition, it could be supported by checking the respective costs in certain major ports, which are located within this range or alternatively, they could agree with the Charterers to a 'freight rate' based on 'free port expenses', i.e. the ordinary port expenses which will be charged on the ship's account, will be paid/ covered by the Charterers to the attending Agents, in any amount, except for extra costs incurred outside the normal costs, due to ship's/Owners' matters and/or fault. Obviously, the 'freight rate' will be adjusted downwards.

On every occasion on Voyage Charters, the load and discharge ports or the fixed ranges are well known to the Owners. The distance between all the ports of operations will be the first point of calculation, where the Owners make the initial Voyage Estimation. Obviously, the longer distance between the ports means extra fuel costs, additional running expenses and generally a longer voyage duration rapidly increases the 'freight rate'.

The berth/s and anchorage/s designated to accommodate the ship for the loading or discharging operation within the port/s, should provide all the features so as to be considered 'safe', without the ship being in danger and/or risk of any kind. This provision should be explicitly incorporated in the Charter Party, as a contract term. The ports, berths, anchorages should provide competent pilots, tugs, lighthouses for safe berthing and except for a normal tide or swell, the ship must not to be exposed to strong winds and rough seas and the port must generally be organized, accepting the size of the particular ship. However, the Charterers will not be kept liable for damage to the ship caused due to Master's/ crew's negligence and/or fault, during the ship's shifting, or in berth.

The classic definition of a safe port/berth that is in effect for the maritime sector is as follows:

"A port will not be safe unless, in the relevant period of time, a particular ship can reach it, use it and return from it without, in the absence of some abnormal occurrence, being exposed to danger which cannot be avoided by good navigation and seamanship".

The Owners should ask from the Charterers to ensure that the fixed ports, berths, anchorages will remain:

. *always afloat*

or for tidal ports

. *always afloat at all times of the tide*

It declares that the ship will not touch the bottom due to tidal changes, at any time during the cargo operation at the ports. The Owners are covered for possible damage to the ship's hull, in case of possible grounding.

The provision of

. *always accessible,*

declares that the ship will have direct access to the berth upon arrival at the port, but provided it is in all respects an 'arrived ship'. This term could also be valid for 'berth' Charter Parties i.e., when the Notice of Readiness can be tendered only at berth. In all cases the Charterers run the risk of receiving a claim by the Owners for damages for detention for the period of delay, in case of a violation of this accepted term. If the Charter contains this term, the Owners transfer the risk of delays to Charterers not only for the berthing but also for the departure delays when cargo operations are completed. If Charterers wish to limit their warranty, they should insist on negotiating the term

. *'always reachable on arrival only',*

or

. *'always accessible in relation to arrival only and not to departure'.*

The above two provisions 'always afloat, always accessible' could be expressed during the negotiations as the four 'a', i.e., 'aaaa'.

II. Berths, anchorages.

The number of berths or anchorages, which will be used in order for the ship to complete its cargo operation at the ports, could be mutually agreed as follows: a- one loading and/or discharging safe berth (1 safe berth or as abbreviated '1sb')

The ship will use one safe berth at each port of contact.

If the ship, for any reason, is instructed by the Charterers to shift to another berth for completion or by the Port Authorities to vacate the berth for a while and return later to complete, then

. all shifting expenses incurred (tugs/pilots etc.) and time lost will be of the Charterers sole account.

The Owners might be entitled to seek compensation for any additional costs set, including extra fuel consumed during the shifting to another berth, as well as crew's services in overtime etc., which is a basis for the Charterers' violation of the Charter Party terms.

b- one or two loading and/or discharging safe berth/s (1/2 safe berth/s or as abbreviated '1/2 sb/s')

Charterers have the option to use one or two berths in order to complete the operation at each port of contact. The shifting expenses and the time lost that is needed to change berth are negotiable, as to for whose account it should be. However, during the negotiations, the shifting costs are usually agreed to be on behalf of the Owners, but the time required for shifting will continue to count as lay-time or time on demurrage.
For protection reasons, the Charterers could retain the option in the Charter Party to use more than one berth, even if the port/s does not naturally provide a second berth.
c- one loading and/or discharging safe anchorage (1 safe anchorage or as abbreviated (1sa') or one/two safe anchorages (abbreviated as '1/2 sa').
The ship will operate in the anchorage/s and will load or discharge from or into cargo barges respectively, by using its own means (gear) or by a floating crane situated between the ship and the cargo barge.
The prevailing port expenses and the possible shifting expenses from one anchorage to another are obviously limited.
Nevertheless, the operation in the anchorage is risky, as bad weather/rough seas may result in significant delays to the ship's completion and its exposure to risk of damage to its hull. The Charter Party should incorporate
. specific terms for the protection of the ship, crew, and cargo.
It should be emphasized that the loading and discharging operation could also take place as:
a. transshipment or transhipment.
Transshipment takes place when the ship loads the goods directly from another ship that for various reasons cannot execute or cannot complete the trip or generally cannot reach the port of destination to discharge the quantity of cargo already on board. The ship will 'double bank' to the other ship in the berth or anchorage and will load either by using its own means, or with a floating crane situated between them.
b. lightering or lighterage/top off.
Lighterage takes place in a safe anchorage, when a large ship, in order to complete its loading from a restricted berth (partial cargo already loaded) or to reduce its draft before approaching the restricted discharge berth, will load (top off) from or discharge into smaller ships (lighters or barges etc.) respectively, as will be presented by the Charterers at their cost.
When discharging, the Charter Party should explicitly state
. that the delivery over side (over the ship's rails) into the other ship/barges etc. will constitute 'right and true' delivery of the cargo under the relative Bills of Lading and the Owner, Master and Ship will have no responsibility or direct or indirect liability for any claim that is made by Consignees, or by any other party for shortage or admixture or damage or loss of cargo.

The ship's gear or a floating crane will be used for the operation.
Charterers should provide sufficient fenders, as well as, securing and mooring equipment, to the Master's satisfaction, who should keep the right under the terms of the Charter Party to instruct the lighters/barges etc. to move away from the ship, whenever he considers that the operation is unsafe and could cause damage to the ship's hull.
It should be a success for the Owners, if the lay-time continues to count during the prevailing rough seas, while the ship operates in the anchorage.
The location of the load and/or discharge port could require the search for ships, which can break the International Navigating Limits (I.N.L.) and are Class approved for trading in ice ports or are within war or war-like zones. The extra costs, as will be imposed by the insurance companies, will be included in the 'freight rate' or will be paid separately by the Charterers, but on presentation of the relevant original vouchers.

III. Port restrictions/expenses.

The Owners should be aware of all information with which the Charterers for a voyage have updated the market regarding the port restrictions and/or port expenses, dues, taxes etc.
Each port, operates at
. different costs,
in accordance with the local tariffs and/or Agents competition and of the existing ports' Labour unions.
However, any information that is provided by the Charterers as to prevailing port restrictions/ port expenses/dues etc., remains as information and contains no guarantees, unless the same is agreed to be incorporated in the Charter Party, as a contract term. The Owners should rely on it, sending an offer, but will remain the party responsible for ascertaining /ensuring the same, through the attending Agents at the named ports and always before the conclusion of the fixture.
The available draft and the salinity of the sea or river at the port of the operation, will determine the loadable quantity of cargo and the resulting 'freight' amount, which should be due for payment to Owners. The berth's maximum permissible length over all (abbreviated as 'l.o.a.' or 'loa'), the maximum permissible beam and the distance from the water line to the top of the hatch coaming (abbreviated as 'wltohc'), are some of the other major port restrictions, which could make the ship unworkable, therefore,
. *everything should be checked before fixing.*
If, because of lack of time or time differences between the countries indeed prevents the Owners from checking the port restrictions and/or existing port

expenses or because of the ship's schedule it loads and/or discharges in private berths/ mines that charge different port disbursements account than the commercial berths, it may be required by the Charterers

. to expressly guarantee for the prevailing port restrictions there

and/or

. to pay for any amount of the charged port expenses, which should obviously result in a reduced 'freight rate'.

This process is called as 'free disbursement account' or as 'free port expenses' for the Owners, or 'free d/a' for short. All should be incorporated in the Charter Party as contract terms.

4. Cargo and its stowage in cargo holds or on deck/stowage factor -Seaworthy and safe stress/trim condition of the ship- Different grades/Cargo separation in the same hold.

THE CARGO:

For carriage by sea, dry cargo is the good of any grade/type/quality and of any quantity which is extracted or manufactured or grown or produced from the mines/factories/farms located inland or ashore and moved/carried, in any form, close to the port, for the purpose of being loaded into the cargo ship's spaces, for transportation to another foreign or within the same country's port/s.

The type/grade of cargo and the form of its transportation will determine which type of ship (logger, multipurpose, singledecker etc.) is the most workable for accommodating the intended cargo quantity.

The ships, which are designed to accommodate and are approved by a classification society for the transport of dangerous/ inflammable goods and generally non-harmless goods and/or those that can provide heavy lifts, usually achieve a better 'freight rate' than others.

Provided that no port restrictions are in effect, the basic cargo quantity and its actual stowage factor, in combination with the provided tolerance on quantity will determine the

. size of the ship,

which is workable for the execution of the particular voyage.

I. Cargo quantity.

The cargo quantity and the type/grade that can be loaded and carried safely by the ship should be ensured by the relevant departments of the Owners' office.

The Charterers must provide the fixed quantity/type/grade of cargo at the agreed port/designated berth, for loading into the presented fixed/named ship and in sound condition. If the Charterers are unable to perform or present different cargo for loading and the ship/fixture is cancelled, the Owners are entitled to claim for damages for detention and consequential losses. However, the Owners should search the market for alternatives without delay, in order to mitigate their losses.

The cargo quantity for transportation form part of the negotiations on the Main Terms, which could be expressed as follows:
a- with a tolerance (adjustment upwards or downwards) on the basic fixed cargo quantity.
This is the most common method for determining the maximum and the minimum cargo quantity available for transport.
The cargo will be described as:
. a basic quantity,
but should remain in the Owners' choice (option), sometimes 5 or 10 percent, for upward or downward adjustment, to the basic agreed quantity of cargo. The 'option' is expressed as:
. More Or Less at Owners' Option (abbreviated as 'moloo') during the negotiation of the Main Terms.
However, when loading steel cargoes or other special cargoes, the adjustment could be for the Charterers' option, ranges of 1-2 percent to the basic quantity of cargo (More Or Less at Charterers' Option /abbreviated as 'molchopt').
For example, a quantity of cargo that provides tolerance could enter the market as follows:
-15,000 mtns bulk cement with 10 pct moloo.
The basic quantity is 15,000 mtns and provides a tolerance of 10pct. The Owners have the right to load from a minimum of 13,500 mtns to maximum 16,500 mtns.
The key feature on the expected quantity of cargo for loading, is distinctive at the port of loading (provided no port restrictions at discharge port). That is since the cargo ship's capacity/exact loading cargo quantity is subject to:
- actual quantity of fuel on arrival,
- the actual cargo stowage factor in the ship's holds,
- the stress of the ship,
- the actual depth of the port/berth and its salinity.

The upward or downward adjustment:
- gives Owners the privilege to load any quantity of cargo within the allowed edges,
- brings the ship to a seaworthy and safe stress/trim condition,
- protects them from the receipt of claims for any non-laden quantity of cargo.

The 'freight' should be agreed to be paid 'per metric ton' on the intaken Bill of Lading cargo quantity.

b- on a minimum/maximum basis (abbreviated as 'm/m').
The Ship/Master will load the cargo quantity:
- on a minimum/maximum basis, as will be mutually agreed upon during the negotiations of the Main Terms.
There is not upwards or downwards adjustment and the Owners are liable to receiving claim
. if the ship fails to accommodate the whole cargo quantity, for various reasons
. or if, after the loading of the contracted cargo quantity is completed, the ship is overloaded or is not within the permitted safe stress/trim condition to execute the voyage, resulting in the prohibition of its sailing.

For cargoes with a high stowage factor that will obviously 'full up' the cargo spaces, the maximum stowage factor
. should be guaranteed by the Charterers,
who should additionally arrange, under their own time/ expenses, for any extra trimming that may be required, in order to complete the holds' edges.
For example, a cargo quantity without tolerance could enter the market as follows:
m/m 15,000 mtns bulk cement stw max 28'
Based on m/m the 'freight' should be agreed to be paid 'per metric ton' on the intaken Bill of Lading cargo quantity.

c- up to full and complete cargo capacity.
The Charterers have the option:
. of loading any cargo quantity until the cargo holds are 'full up',
either by the maximum permitted weight or in accordance with the maximum available grain or bale cubics in cargo holds.
However, and in any case, the ship should sail from the port in a condition of safe stress/trim.
The Charterers might be entitled and have
. *no contractual obligation to 'full up' the ship's holds and eventually load the ship up to their available quantity of cargo and not as per ship's declarable cargo capacity.*
Consequently, the agreement of payment of 'freight' 'per metric ton' of the loaded cargo quantity, is a risk for the Owners. If the ship does not load the cargo quantity calculated for loading, the Owners should have failed on their initially calculated voyage estimation, as no claim for 'dead-freight' can be made.
Even if the Charterers have the right to use all the spaces designated for loading

cargo, they have no right to make any
. conversions or to change the configuration of the ship's holds, in order to fully load the cargo spaces.
The Master, before the commencement of the loading operation, should also request and be well informed about the available cargo quantity. He will be able to prepare, as accurately as he possibly can, a stowage plan, allowing the ship to sail in a safe trim and under the permissible stress.
Based on 'up to full and complete' term, the 'freight' is preferred to be agreed 'on lumpsum' basis, paid regardless of the available quantity of cargo, in which case the Charterers can load any amount of load.

d- full and complete cargo capacity.
All spaces designated for cargo will be available to Charterers who are
. of the contractual obligation to 'full and complete' them,
either by virtue of the maximum permitted weight as declared by the Master on arrival or in accordance with the maximum available grain or bale cubic feet in cargo holds. The cargo quantity should on no occasion exceed what the ship can actually load and carry.
Again, Charterers have no right to make any conversions or to change the configuration of the ship's holds, in order to fully load the cargo spaces.
If Charterers fail or are unable, for any reason, to provide a full load and the freight is payable 'per metric ton' of cargo loaded, the Owners then have the right to recover losses and make a claim
. for dead-freight
for the non-laden cargo quantity and, additionally, the Charterers will bear the cost of any consequences that may arise (un-seaworthiness of the ship etc.).
The 'freight' should be agreed to be paid 'per metric ton' on the intaken Bill of Lading cargo quantity or 'on lumpsum' basis. If 'on lumpsum', the Charterers will ask the Owners to guarantee
. the ship's minimum available cargo capacity.

II. Cargo stowage.

The bulk cargoes have no difficulty in being secured, as the majority of the ports around the world provide appropriate facilities for the loading/trim and discharge of the cargo. The Master should supervise the Shippers/Receivers and Stevedores for loading and discharging the cargo strictly as per the agreed sequence/stowage plan/method of cargo operations and as per the provisions of SOLAS chapter VI, regulation 7.3 (loading, unloading and stowage of solid bulk cargoes) and the Code of Practice for the Safe Loading and Unloading of Bulk Carriers (BLU Code).

Difficulties may arise, when the ship is loading general cargoes or steel/coils etc. and especially when the cargo is going to be discharged in more than one port. Charterers are to provide all dunnage materials placed among the cargo, before the commencement of the loading operation (the same should be incorporated into the Charter Party, otherwise and depending on the law applicable on fixture, it might fall on to the Owners to provide the necessary dunnage for the proper stowage, separation, lashing and securing of the cargo).
The Charterers are not obliged to have the ship in a condition, after completion of discharge, as clean as when it entered for loading, therefore, exceptionally a provision should state that
. the Charterers will remove and discharge, under their own risk and expenses, all such dunnage supplied to the ship, after discharge and the time will continue to count as lay-time or time on demurrage.
The handling of the cargo will be performed by experienced stevedores and will be loaded and discharged and stowed and/or trimmed, lashed, unlashed, secured by the Charterers, free of any risk/expenses and liability whatsoever to the Owners.
However, the Master should supervise and confirm the safe loading and the ship's seaworthy and safe stress condition and ensure that the cargo cannot shift, if, during the sea passage, the ship encounters bad weather.
However, the Owners/ship will likely be held liable for damage resulting from improper stowage, if it is proven that the bad stowage was caused by the significant intervention of the Master and/or Owners.
The requested stowage plan by the Charterers during the negotiations, or before the ship's arrival at the port of loading should be given as a 'preliminary stowage plan' and should state that:
- the cargo quantity, which will be loaded, and the holds' breakdown is subject to
. the actual stowage factor of cargo in the ship's holds.
. the ship's actual trim/stability/strength/stresses (hogging/sagging bending moments/ shear forces etc.).
. the available draft/salinity of the port.
. load line zones limit.

For cargo that is in the form of bags/pallets etc. the wording could be amended accordingly.

III. Stowage factor.

Each ship provides a number of
- *grain cubic feet* (or cubic meters) in each cargo hold, for the measurement of the maximum cargo space, when loading cargoes in bulk condition.

The
- *bale cubic feet* (or cubic meters) are used when loading any other form of cargo except bulk, such as bags/palletized cargoes etc.
Shippers/Charterers will declare the stowage factor, which varies subject to:
-the type/grade (nature) of the cargo,
-the way of loading/port facilities and of the provided trimming,
-the structure of holds,
-the form of the cargo for transport and from the left 'broken spaces' into the cargo holds.
Heavy cargoes (such as bulk cement/minerals/bauxite etc.) with a very low stowage factor are loadable by virtue of the ship's
. maximum permissible weight.
In fact, the cargo spaces will not 'full up'.
On the other hand, light bulk cargoes (such as bulk soya bean meals/sunflower seeds/cotton etc.) with a high or very high stowage factor are loadable by virtue of the ship's
. maximum available grain capacity for bulk cargoes or of the maximum available bale capacity for bagged/palletized cargoes etc. and is expected to 'full up' the cargo spaces.
The loadable cargo quantity is subject to stowage factor, which should adjust the 'freight rate' depending on the quantity of cargo expected to be loaded.
The approximate loadable cargo weight will be calculated as follows:
1. Based on low stowage factor (without draft/other port restrictions to be in effect):
the ship: Dwat about 6,050 mtns.
cargo capacity about 5,500 mtns.
cubic feet about 280,000 cbft.
the cargo: bulk cement with stowage factor about 28' (each metric ton of the cargo loaded will occupy about 28 ft in the cargo hold)
The ship's cubic feet (280,000 cbft) will be divided with the stowage factor of the cargo (28'), resulting in the ship, according to stowage factor, being able to load about 10,000 mtns.
However, the ship will load by virtue of its
. maximum permissible weight, i.e., about 5,500 mtns and no more.
The cargo holds will not 'full up' with cargo, i.e., the cargo will occupy a height of certain meters above the tank top but will not reach the hatchcovers. Apparently, there are empty cargo spaces, and the ship could load additional cargo, but if it does, it will be considered 'overloaded' and its sailing will not be allowed by the Authorities or will result in the ship's possible capsizing in the port of loading.
2. Based on high stowage factor:

the ship: Dwat about 6,050 mtns.
cargo capacity about 5,500 mtns.
cubic feet about 280,000 cbft.
the cargo: bulk sunflower seeds with stowage factor about 80'
The ship's cubic feet (280,000 cbft) will be divided with the stowage factor of the cargo (80'), resulting in the ship, according to stowage factor, being able to load about 3,500 mtns.
In this case the ship will load by virtue of its
. maximum grain capacity,
i.e., about 3,500 mtns.
The cargo holds will 'full up' with cargo, i.e., the cargo is expected to occupy all the cargo spaces up to the hatchcovers and the ship should not be able to load any other additional cargo.
As the stowage factor affects the seaworthiness of the ship and adjusts the 'freight rate', the Owners must be aware of such important information, preferably before sending any firm offer. However, for common bulk or even bagged cargoes that are transported with an almost standard stowage factor, regardless of the place of loading, the Owners can forward the offer, but it should be subject to the actual stowage factor, which should be declared at the latest with the Charterers' first counter offer.

. *Never fix without any reference to the stowage factor of the cargo.*

Charterers usually avoid committing themselves with the exact stowage factor and any such declaration should be qualified by the word 'about', which is in most cases followed by the word 'wog' (without guarantee). The word 'about' provides a margin (could be about 5 percent) of a little more or a little less stowage factor and there should not be any breach of contract terms if the stowage factor lies within this margin.
However, when it is followed by the word 'wog', there is no contractual commitment, and the Charterers are free of obligations as a result of incorrect declaration of the cargo stowage factor and obviously from its consequences (payment of dead-freight etc.).
In all cases, the Agents of the ship at the load port should be the party experienced enough to inform the Owners about the actual cargo stowage factor and of the way of loading and trimming of the cargo in the cargo holds.

IV. Deck Loading.

For the ship to be fully loaded with cargo, the Owners or Charterers may request for part of the cargo to be loaded on deck/hatchcovers after completion

of loading the cargo holds.
This is a usual loading method for multipurpose ships/loggers etc. Obviously, the deck cargoes must be in a form/condition that enables the stevedores to lash/secure them, allowing the ship to navigate in a safe condition.
Under the Charter Party terms, the Charterers must provide any and all requisites, lashing and securing materials and/or any other extra fittings/equipment, necessary for the safe stowing/carriage. Cargo on deck and hatch covers should be loaded at the Charterers'/Shippers' and Receivers' entire risk/ responsibility and expenses, but always strictly in accordance with the ship's deck/hatch covers permissible strength/ stability and the ship's visibility. The time required for the deck and/or hatch covers loading, under the terms of the Charter Party should continue to count as lay-time or time on demurrage.
The Master must supervise the loading, which must be performed to his absolute full satisfaction.
Owners are to be aware that, when loading deck cargo, except in certain specific circumstances (ship's carrying containers/logs etc.), they might be in breach of the contract of carriage and the ship may fall outside the P&I Club's rules.
Nonetheless, the Owners should seek advice from the ship's P&I Club, as the Bill of Lading must be issued and marked strictly as per their recommendations.

-The Charter Party clause for deck and hatch covers loading could be worded as follows:
Charterers have the right to load cargo on deck and/or hatch covers at the Charterers' and/ or Shippers' and/or Receivers' entire risk, responsibility, liability and expenses, but always strictly in accordance with the ship's deck and/or hatch covers permissible strength and with the ship's stability and/or visibility.
It is expressly agreed that:
1. the cargo on deck and/or hatch covers will be loaded, stowed, dunnaged, lashed, carried, secured and unlashed and discharged always in accordance with the international regulations, by shore labourers, at the Charterers' and/or Shippers' and/or Receivers' entire risk, liability, responsibility, and expenses and to the supervision and satisfaction of the Master and/or the ship's P&I Club's surveyor, who will be appointed by the Owners, at the Charterers' cost. The Charterers are to reimburse the Owners for the P&I Club's Surveyor's fees, once the Owners have submitted their final freight invoice, supported with the original vouchers. The time required for the deck and/or hatch covers loading, stowing, dunnaging, lashing, securing, unlashing, and discharging will continue to count as lay-time or time on demurrage.
2. the Charterers and/or Shippers at their own risk, cost with shore labourers will weld padeyes and/or supply, fit and install stanchions and/or sockets, as well as, will supply sufficient lashing, securing, dunnaging materials and/or any other

extra fittings, equipment etc. requisite for the safe stowage, carriage of the cargo on deck and/or hatch covers, to the Master's and/or ship's P&I Club's surveyor's supervision and satisfaction and the time lost for the execution of these operations will continue to count as lay-time or time on demurrage.
Charterers guarantee and will remain responsible for providing the Master of the ship, under their time and expenses and before the ship's sailing from the port, with an original phytosanitary certificate issued by the plant protection organization of the port/s of loading to the plant protection organization of the discharge port/s, certifying that the dunnages and/or other materials have been inspected according to appropriate procedures and fully comply with the regulations and is fully accepted by the plant protection organization of the importing country.
Any gear, stanchions, lashing etc. materials supplied by the Charterers, are to remain at the Charterers' property and the Master should keep a record for all materials that were supplied in this way and should be kept in good condition, but without guarantee and always subject to wear and tear and for any damage which occurred and is connected or attributed to the voyage.
At the discharged port, the welded padeyes and/or the fitted, installed stanchions and/or sockets and/or debris, dunnages and/or any other extra fittings, equipment etc, supplied to the ship and it were necessary for the safe stowage, lashing, dunnaging and carriage of the cargo, will be dismantled, removed and discharged and/ or delivered in the custody of the Charterers' Agents, by shore labourers, before completion and the ship's sailing from the port, at the Charterers' own risk and expenses and the time lost will continue to count as lay-time or time on demurrage.
The ship will not sail from the port, unless all such materials onboard have been removed and discharged from the ship and any delays, expenses, consequences and/or possible fines imposed by the Port Authorities, arising from the Charterers' inappropriate discharge of the materials or inappropriate delivery ashore, will be for the Charterers' account.
3. any directly related expenses resulting there from and incurred thereby will be for the Charterers' account and those covered by the Owners are payable on request, supported with the original vouchers, which are acceptable, even by fax or e-mail.
4. for all cargo loaded on deck and/or hatch covers, separate Bill/s of Lading will be issued, which will be marked strictly as per the Owners' P&I Club's recommendations.
5. Charterers guarantee that they will remain liable and responsible for any consequences, loss, damages, costs, expenses, possible fines Owners may suffer and will indemnify the Owners against third party claims (inclusive of any interest and costs, which may be incurred for defending any such claims or recovering

any such loss), including, but not limited to, cargo claims and/or oil and/or air pollution arising from loading cargo on deck and/or hatch covers. If the ship is detained by any party and generally its sailing is not allowed from the port, the Charterers then undertake the responsibility to release the ship, by taking all the necessary and appropriate measures (issuance of security, etc.), as the case may be or arise and any expenses incurred, including, but not limited to, additional port expenses, shifts, Agency fees etc. will be for the Charterers' account, payable on request and upon presentation of the original vouchers, which are acceptable, even by fax or e-mail. The time lost will continue to count as lay-time or time on demurrage, until the ship is free to sail from the port and the 'port clearance' is issued and delivered to the ship.

6. the Charterers will be fully responsible for any damage caused to the ship's deck, rails, hatch covers, hydraulic motors etc.

For damage which:

a. affects the effectiveness of the ship and the safe operation and/or safety of the crew and/or affect the trading capabilities and/or seaworthiness and/or cargo worthiness and/or is subject to classification requirements, then such damage as will be ascertained by the ship's Class surveyor, who will be appointed by the Owners at the Charterers' cost, is to be repaired by the Charterers at their own expenses, risk, to the Master's and/or Class surveyor's entire satisfaction and the time lost will continue to count as lay-time or time on demurrage until completion of all repairs and the 'port clearance' is issued and delivered to the ship.

Charterers are to reimburse the Owners for the Class Surveyor's fees, once the Owners have submitted their final freight invoice, supported with the original vouchers.

The damage will be repaired at the place of occurrence, if it is possible and there are adequate facilities for permanent repairs, or otherwise, temporary repairs are to be made on the spot or as when detected, at the Charterers' risk and expenses and to the Master's and/or Class surveyor's satisfaction and the time lost will continue to count as lay-time or time on demurrage. The ship will sail or shift to the nearest shipyard, as chosen by the Charterers, with the fuel consumed at sea or during shifting to be for the Charterers' account, which will be settled together with the Owners' final freight statement, in order to complete the repairs.

The time lost during the sailing or shifting to the shipyard and until completion of the works with the damage being repaired to the Master's and/or Class surveyor's entire satisfaction and the 'port clearance' is issued and delivered to the ship or, at the Owners' option, at equidistance from the discharge port, will count as lay-time or time on demurrage.

Shipyard's expenses and/or port expenses, including in and out tugs and/or pi-

lots and/or agency fees etc., will be entirely for the Charterers' account.
b. does not affect the effectiveness of the ship and the safe operation and/or the safety of the crew and/ or the trading capabilities and/or the seaworthiness and/ or cargo worthiness and/or is subject to classification requirements and cannot be repaired by the stevedores, the Owners then undertake to accept the ship's sailing, in accordance with the terms of the Charter Party.
However, the said damage is to be assessed before the ship's departure from the discharge port and the costs to be agreed upon between the Master, Owners, Charterers and the respective survey company and to be paid together with the Owners' final freight statement. The survey company will be mutually appointed by the Owners and Charterers, if considered necessary, at the Charterers' and Owners' equal expenses and will promptly inform the Charterers and Owners with the damage report, which will be issued before the ship sails, with a copy of all correspondence exchanged in this connection. The time lost will continue to count as lay-time or time on demurrage, until the 'port clearance' is issued and delivered to the ship.
Owners have the option of having the damage repaired at their next convenient time or at the ship's next Dry-Docking.
In the case that the Master, Owners, Charterers cannot reach an agreement on the cost with the respective survey company, the Charterers will then undertake to reimburse for the actual cost of the repairs, but not for the time lost, upon presentation of the original bills from the repairers or dockyard, that are consistent with the surveyors' report on the extent of such damage.
c. is small and has been repaired by the crew, the Charterers will reimburse the Owners for the cost on request, but not for the time lost and upon presentation of the Owners' repair statement, which is acceptable, even by fax or e-mail.
However, and in any case:
- the Charterers are responsible for any directly related expenses resulting therefrom and incurred thereby and must pay compensation for any damage caused to the ship and/or restore the ship to its initial condition, under their own time and expenses, whether or not any payment has been made by the respective parties and/or stevedores to them.
- the Charterers will not to be held liable for damage, when the Master has accepted a statement in
writing from the stevedores and/or respective parties that the operations were carried out without any damage, or that the repairs were made to his satisfaction.
If the ship's P&I Club does not supervise the loading operation on deck, then the above term should be amended, and the following wording could be added:
If any dispute for over stowage, lashing, securing of such deck and/or hatch covers cargo arises, then an internationally recognized independent surveyor will

be mutually appointed by the Charterers and Owners, in order to ascertain and confirm that such deck and/or hatch covers cargo has been stowed/lashed/dunnaged and secured for safe carriage at sea. The cost of the independent surveyor and any extra expenses resulting therefrom and incurred thereby and any detention through any of the above will be for the Charterers' account and the time lost will continue to count as lay-time or time on demurrage.

V. Seaworthy, safe stress/trim.

Problems could arise if the ship is obliged to load or discharge from or at a second or more ports/ berths and there is no provision in the Charter Party for whose account the expenses and time required to bring the ship in a seaworthy and safe stress/trim condition from one port/berth to the next should be.
Charterers should inform the Owners/Master, as soon as possible, about their intention to use more than one port/berth and of the intended cargo quantity from/to each port/berth. The Owners/ Master should calculate and arrange for the seaworthiness of the ship before its sailing from the port/berth.
The seaworthiness could be achieved:
-by loading or discharging the cargo holds, according to the ship's safe stresses/trimming,
-by taking sea water ballast and filling some or all the ship's empty tanks,
or
-in the worst case, if the cargo is grain, by filling empty bags with cargo and stowing them on the surface, which is usually covered over by tarpaulins.
The Charter Party should explicitly state that the time required bringing the ship to a seaworthy and safe stress and trim condition and expenses incurred will be entirely for the Charterers' account, otherwise, the lay-time should be deemed to have been ceased from the time of completion of the operation and extra expense incurred will be for the Owners' account.

-The Charter Party clause could be worded as follows:
The final sailing configuration is always subject to the Master's stowage plan and satisfaction, which is in conformity with the ship's loading manual.
The ship should always be left in a seaworthy and safe stress/trim condition and with cargo on board safely stowed during its sailing and/or its lightering and/or its shifting between all ports, berths, anchorages, and wharfs to the Master's full satisfaction.
Upon completion of the loading and/or discharging and/or lightering operation at the first or subsequent port and/or berth/ anchorage/wharf or at any time it is requested by any party (Master, Port Authorities etc.) to bring the ship in a seaworthy and safe stress, trim condition, then all the expenses and/or extra

port expenses incurred will be entirely for the Charterers' account and the time lost will continue to count as lay-time or time on demurrage, even during ballasting and de-ballasting the ship's tanks and until the Master is fully satisfied and the ship is ready to sail or shift.

VI. Different grades/Cargo separation in the same hold.

During negotiations, the Owners could be asked to confirm the loading of two or more different grades of cargo or different type of cargoes in the same hold, with or without artificial separation.
In general, the artificial separation of cargo is difficult and often costly to arrange. The Owners should have strong reservations as to whether, amongst other things, the method of loading and packaging/stowing/separation will be sufficient to prevent
. commingling or admixture
and the puncturing of the cargoes by the grabs during discharge.
Additionally, the risks
. of contamination of the cargoes are imminent,
resulting in damage to the cargo, i.e., one grade loaded in the bottom of the hold and covered with plastic dunnage and/or tarpaulin, which could make the cargo sweat, with the slightest humidity making, for example, the grains develop mould.
Whilst Shippers/Charterers may be possibly able to construct a cargo worthy barrier, thus keeping the cargoes separate and breathable, concerns should remain, however, on whether or not it will work and does not cause either the cargo to sweat or any other unforeseen problems.
The Owners must be aware that even if an artificial separation is made and the Bills of Lading have been claused (marked) or a Charterers' Letter Of Indemnity has been presented, they cannot avoid the liabilities of separating and carrying cargo with such methods and the ship may be subject to the ship's P&I Club's coverage. In respect with issuing a l.o.i. and in the event that the l.o.i. is found to be inadequate, the risk and any loss will rest with the Owners. In every occasion, the l.o.i. must clearly incorporates that the Charterers agree to undertake any and all liabilities and responsibilities arising from separating the cargo artificially in whatever manner and with whatever method and they also undertake the arrangements for the loading and separation at their own time/risk and cost. Furthermore, it must state that they agree to indemnify the Owners for any and all costs and/or liabilities arising out of carrying cargo with this method.
In the case that such a method of loading is inevitable and the fixture for various reasons must be solved and provided that the Owners are confident with the Charterers' financial standing and/or the intended shipment is a regular ship-

ment for them that have always been completed without problems, then there should obviously be a little risk. However, the Owners must take care and be covered by clausing (marking) the Mate's Receipts & original Bills of Lading, always in accordance with the recommendations of the ship's P&I Club.

Owners' initial offer should logically always include that:
-cargo ...... of one type/grade/quality,
or if there is more than one grade/type/quality of the intended cargo to be loaded:
-the Charterers will provide for loading only one type, grade and quality of cargo, otherwise and in the case of more than one type, grade and quality of cargo, the separation will be made according to the ship's natural compartments,
or
if there is more than one grade/type/quality of the intended cargo to be loaded in the same hold/s:
In the event that an artificial separation is required by the Charterers, in addition to the ship's natural compartments, then it is expressly agreed that:
1. Charterers guarantee that it will be made and the cargoes for separation will be loaded, in a maximum of 2 holds at their entire risk, responsibility, liability and expenses and the time required for loading will continue to count as lay-time or time on demurrage.
2. Charterers guarantee that under their own risk and expenses will supply, erect and install all the necessary and required means of separation, as well as dismantle, remove and discharge them at the last discharge port and will rectify, repair any damage caused to the ship and/or conversions made in the ship's holds and restore the ship to its initial condition, always before the completion of discharge and/or the ship's sailing from the port. The time lost for the execution of these operations will continue to count as lay-time or time on demurrage.
3. the separation will be made under the supervision and the full satisfaction of the Master and/ or the ship's P&I Club's surveyor, who will be appointed by the Owners, at the Charterers' cost. The Charterers are to reimburse the Owners for the P&I Club's Surveyor's fees, once the Owners have submitted their final freight invoice, supported with the original vouchers.
4. all the presented Mate's Receipts and original Bill/s of Lading will be marked strictly in accordance with the ship's P&I Club's recommendations.
5. the ship will always trade in a seaworthy and safe stress, trim condition, between berth/s, port/s, under the Charterers' expenses and the time lost, in order the Master to bring the ship in a seaworthy and safe stress, trim condition even during ballasting and de-ballasting the ship's ballast tanks will continue to count as lay-time or time on demurrage.
6. Charterers guarantee that will hold the Owners harmless and will keep them

fully indemnified against all claims whatsoever (inclusive of any interest and costs, which may be incurred for defending any such claims or recovering any such loss), which may arise and be made by any party against them, due to making separation in the ship's cargo holds. Furthermore, all risks and all expenses incurred, and/ or liabilities and responsibilities and/or loss, damages and/or possible fines and/or third parties' claims which may arise from mixture of goods and/or contamination and/or co-mingling and/or cargo damage, as a result of the loading of more than one cargo type, grade and quality in the same hold, howsoever caused, whether this is due to stevedores' negligence or otherwise, and/or consequential losses will be entirely for the Charterers' account.

If the ship is detained by any party and generally its sailing is not allowed from the port, because of mixture of goods and/or contamination and/or co-mingling and/ or cargo damage, as a result of the loading of more than one cargo type, grade, quality in the same hold, the Charterers then undertake the responsibility to release the ship, by taking all the necessary and appropriate measures (issuance of security, etc.), as the case may be, or arise and any expenses incurred, including, but not limited to, additional port expenses, shifts, Agency fees etc., will be entirely for the Charterers' account, payable on request and upon presentation by the Owners of the original vouchers, even by fax or e-mail. The time lost will continue to count as lay-time or time on demurrage, until the ship is free to sail from the port and the 'port clearance' is issued and delivered to the ship.

Usually for small cargo quantities, only one Shipper is in effect. However, for larger quantities different Shippers could provide similar type/grade/quality of cargo for loading in the same hold.

If Charterers are not to provide separated Bills of Lading for each Shipper involved in the loading operation, but with one issued Bill of Lading then it could be agreed that:

-If a cargo hold/s is loaded by different cargo shippers with or without artificial separation, the Charterers guarantee that the cargo will be of the same grade, type and quality and all cargo documents, including the Mate's Receipts and Bills of Lading, will have exactly the same cargo description. The Charterers should indemnify the Owners against all claims, whatsoever, which may arise and be made against them and/or hold the Owners harmless from any liabilities, loss or damages or consequences which may arise from mixture and/or contamination and/or co-mingling of cargoes, because of being loaded from different Shippers in the same hold/s.

5. Loading/discharging cargo expenses-Trimming/stowing expenses.

The ship will execute the voyage either as:
- Free In and Out basis (abbreviated as 'F.I.O.').
All costs that will be incurred for loading or discharging, such as stevedoring fees, hiring cranes and trucks etc., will be covered by the Shippers/Receivers/ Charterers.
- Charterers should remain liable under the terms of the Charter Party if the Shippers/ Receivers are unable to pay these costs.
Owners are to pay only for the ordinary port disbursements at each port of contact, such as tugs/pilots/light dues/mooring/unmooring/Agency fees etc.
or as on
- Liner terms (liner in and/or liner out).
In this case, the Owners will bear the cost and will cover not only the ordinary port expenses/ dues/taxes but also all costs incurred in order to load/discharge the cargo to/from the ship's holds respectively, such as stevedoring fees, rental expenses for shore gear, the transportation of the cargo to/from the ship etc. For each port, the liner cost works as a standard tariff and the Owners should check the same before the fixture is concluded. The ship will load/discharge in accordance with the 'custom of the port' and no lay-time calculations should be made, i.e., no demurrage/ no despatch.
The liner terms are not necessarily applicable for both ends. It could be agreed 'free in' to load, but 'liner out' to discharge or the opposite.
The designated 'liner ships' operate only under a standard schedule/program between specific countries and load/discharge on liner terms within a certain time and have priority in berthing.

The F.I.O. terms, depending on the cargo form, could be expressed in the Charter Party as follows:
a- Free In and Out Trimmed (abbreviated as 'F.i.o.t.').
Besides f.i.o terms and if the Master requests cargo trimming in the ship's holds over and above what is customary and the port can provide, in order for the cargo to fill up the ends in the cargo holds, either in order to accommodate the whole cargo quantity or for stability/stress reasons, the Charterers are to arrange for the same under their time and expenses.
This term is only in effect, when loading bulk cargoes.

b- Free In and Out spout and/or conveyor and/or grabbed trimmed.
The f.i.o terms are still in effect, but the requirement for Charterers to trim the cargo holds is limited up to the trimming which the spout and/or the conveyor and/or the grab can provide.

Extra trimming required by the Master will be arranged by the Owners, under their time and expenses.
This term is only in effect, when loading bulk cargoes.

c- Free In and Out Stowed Lashing/Securing/ Dunnaging/Unlashing (abbreviated as 'F.i.o.s. l/s/d/unl').
The f.i.o terms are still in effect, but Charterers, under their own time/expenses, will also stow/ lash/ secure/ dunnage the cargo in holds in the port of loading and unlash the cargo in the port of discharge.
This term is only in effect when loading general cargoes (cargoes not in bulk form).

D- Free In and Out Stowed and Trimmed (abbreviated as 'F.i.o.s.t.').
The f.i.o terms are still in effect, but Charterers, under their own time/expenses, will also stow, and, if it is required by the Master, will additionally trim the cargo in the ship's holds.
This term is in effect, when the cargoes are in different forms, requiring both, stowing and trimming to be arranged.

6. Means of loading and discharging - Loading and discharging speed.

I. Loading and discharging systems.

The loading and discharging operation will take place either with the ship's gear (fitted with grabs or not), or with the use of shore means/floating cranes.
The shore operation can be carried out with
-shore cranes with grabs, or if dealing with scrap cargo with magnets fitted on cranes, (for loading and discharging purposes),
-evacuators or mobile cranes placed on the ship's deck (for discharging purposes),
-conveyor belts (for loading purposes) and vertical chain conveyors (for discharging purposes),
-silo (for loading purposes),
and many others.
The use of grabs is a mean of loading or discharging, while Bulldozers/Payloaders - assist in discharging the cargo.
Ports designated for bulk cargoes such as coal, ores, pig iron, grains etc., can provide grabs for the loading and/or for discharging operation. The most suitable type of grabs, depending on the grade of cargo, will be accommodated
. either by shore means/floating cranes
. or by the ship's provided gear.

Mainly for gearless ships, the Owners are to confirm that the ship is suitable for normal grab discharge and that the cargo is not to be loaded in places . inaccessible to grabs (deep tanks etc.).
Loading with grabs is much easier compared to discharging. The cargo could be spread into the holds from high above or even from almost the hatch opening during the loading operation. However, difficulties could arise at the discharge port, where grabs must touch the bottom for the completion of discharge or if unreasonably large grabs are used, incurring damage to the ship's structure, giving rise to disputes, as to whether the ship is fully suitable for grab discharge or not.
Bulldozers/payloaders could be used for stowing/trimming purposes at load ports or for removing the cargo from the sides of the ship and carrying it to the places easily accessible to the grabs at the discharge ports.
When bulldozers should be used, the Owners should require the same to be with rubber wheels and their weight, including the weight of the cargo, not to exceed the tank top strength.
However, when steel, scrap, pig iron cargo, etc., is to be loaded, caterpillar tractors should be used.

a. Geared ships.
The ship's cargo gear and all other equipment must be kept in good working order and properly serviced and complying with the regulations of the countries which the ship will be in contact with. Additionally, the Owners must ensure that the ship is always in possession of valid and up-to-date certificates of efficiency, complying with such regulations, or otherwise, the Owners should take immediate corrective measures. The Owners will bear the cost of all consequences, delays and all consequential losses, detention and/or possible fines, damages, costs, and expenses that may arise, due to the ship's failure to comply with the port regulations.
In the case of a breakdown or malfunction or disablement of the ship's cargo gear, or if the provided electric power for working the cargo gear is not continuous or is insufficient, will lead under the terms of the governing Charter Party, the time to cease to count as lay-time or time on demurrage. The Owners will be asked to hire shore cranes in order to continue the operation and until this time, all extra expenses incurred, including the time waiting and/or demurrage charges for trucks, barges and/or rail wagons and/or overtime charges for Shippers', Receivers', Charterers' employees/ servants and all other directly related expenses incurred, including shifts, whether the affect is partial or not, could be charged for the Owners' account.
If the ship accommodates sufficient gear for the intended voyage and it will be in use, a cargo gear clause should be negotiated and incorporated in the Char-

ter Party, in order to stipulate the events in case of a breakdown or malfunction or disablement of the provided cargo gear.

The Charter Party clause could be worded as follows:
Charterers have the right to load and/or discharge by using the ship's cargo gear as described, at no cost and have at their disposal any related equipment as on board the ship.
Owners guarantee that the ship's cargo handling gear, including derricks and/or cranes, and/or winches and/or grabs and/or equipment, have been serviced and will be maintained in good working order, throughout the duration of this Charter Party. All the provided, as per the ship's description, derricks and/or cranes and/or winches can work simultaneously and can serve all cargo holds, during day and/ or night and/or during weekends and holidays, as may be required, free of expenses to the Charterers. The ship's auxiliary engines and/or generators have been serviced and will be maintained in good working order, throughout the duration of this Charter Party and can provide continuous and sufficient power to the ship's cargo gear.
It is expressly agreed that:
1. -Cargo gear certification.
Owners guarantee that, throughout the duration of this Charter, the ship will be in possession of all necessary valid and up-to-date cargo gear certification, in compliance with safety regulations, national and international regulations and all current requirements at all ports of contact.
If the ship is in not possession of valid and up-to-date cargo gear certification, the Owners then will take immediately corrective measures and all consequences and all consequential losses, delays, detention and/or possible fines, damages, costs and expenses of claims and possible shifts will be entirely for the Owners' account. Expenses for Stevedores, longshoremen or other shore labourers' standby, but up to a maximum of one shift, supported with the original vouchers will be for the Owners' account. Time lost for the Owners to obtain any necessary permission or time lost for issuance of the missing and/or expired certificates and/or documents and time lost, costs, including those relating to the inspection of the ship and/or ship's cargo gear by any Authority, and/or ship's class and/or ship's flag etc. will not count as lay-time or time on demurrage.
2. -Cargo gear breakdown and the hire of shore means.
In the case of a breakdown and/or malfunction and/or disablement of the ship's full cargo gear and/or if the provided electric power to work the full cargo gear is not continuous and/or is insufficient, the time lost will not count as lay-time or time on demurrage. However, should a breakdown and/or malfunction and/ or disablement of the ship's full cargo gear occurs and/or if the provided electric power to work the cargo gear is not continuous and/or is insufficient to a par-

ticular hatch/es, the lay-time or time on demurrage then will count pro rata, in relation to the number of hatches that are workable.
In the case of a breakdown and/or malfunction and/or disablement of the ship's full cargo gear or to a particular hatch/es and/or if the provided electric power to work the full cargo gear or to a particular hatch/es is not continuous and/or is insufficient, the Charterers will provide the Owners with a maximum period of 48 hours, in order to rectify the damage and/or failure etc. before ordering for shore means and/or equipment and/or shore engine, and/or for a possible shift to a berth, as will be ordered by the Charterers and/or Port Authorities, which can provide shore means and/or equipment and/or shore engine.
The cost of the possible shift will be for the Owners' account and the same will not be considered as a Charter Party shift. The cost of the shore means and/or equipment and/or shore engine, if they are available, will be for the Owners' account, but should be ordered after the Owners' approval of the cost. Otherwise, and in the case of a disagreement, the Owners should provide and arrange for cheaper shore means and/or equipment and/or shore engine. The lay-time or time on demurrage will start to count again in full or pro-rata, from the time when the ship will be in a full or pro-rata operational condition again, commencing the loading or the discharging operation with the shore means and/or equipment and/or shore engine, as the case may be.
In all cases, any extra expenses incurred, including, stevedores' standby charges, but only up-to one shift, as well as the waiting time and/or demurrage charges for trucks, barges and/or rail wagons and/or overtime charges for Shippers', Receivers', Charterers' employees, servants and all other expenses incurred, including possible shifts, whether their affect is partial or not, will be for the Owners' account, but will be supported with the original vouchers. The Owners will also be responsible for all the consequences, fines, and all liabilities in respect thereof or arising therefrom.
Any actual time lost due to inefficiency in working the ship with shore means and/ or equipment and/or shore engine will be for the Owners' account, even if the ship is on demurrage.
3.- Competent and experienced drivers.
All cargo gear drivers, including grab operators, will be arranged, and be employed by the Charterers at their own time, risk and expenses at all ports of contact. The Charterers guarantee that they will arrange and employ competent and well experienced cargo gear drivers.
The Master has the right at any time to stop the loading or discharging operation if the drivers and/or stevedores are not competent and experienced and cause or could cause damage to the ship's gear and/or equipment. The Charterers, with immediate effect, will substitute the drivers with competent and experienced ones and time lost will count as lay-time or time on demurrage. If any

breakdown and/ or damage and/or stoppages are caused due to bad handling by the drivers and/ or stevedores and/or other Charterers' servants, then any expenses incurred, including, but not limited to, additional port expenses and/or shifts and/or Agency fees etc., and for any delays, damages, loss, consequences will be for the Charterers' account and the time lost will continue to count as lay-time or time on demurrage.
4.- Ambient temperature.
If the ambient temperature falls below 15 degrees Centigrade while the ship is operating its cargo gear, then the ship is under no circumstances to be held responsible for the general low performance of the gear or for any other delays thereof, because of the low temperature and the time lost will continue to count as lay-time or time on demurrage.

-For ships discharging with grabs fitted on its gear and Bulldozers/Payloaders will assist in discharging the cargo, the Owners could add to the above:

Charterers may fit suitable grabs on the ship's cranes, in accordance with the class of the ship and remove the same after the loading and/or discharging operation is completed. Risk and expenses incurred to fit, un-fit and restore the grabs is to be for the Charterers' account and the time lost for the execution of these operations will count as lay-time or time on demurrage. The loading and/ or discharging operation with grabs, is to be entirely for the Charterers' risk and expenses, who are also responsible and liable for any damage caused to the ship and/or cargo and/ or third party. The ship and/or Owners are not to be kept responsible and liable, in the case of improper grabs fitted on the ship's cranes.
No cargo will be loaded in places inaccessible to grab discharge, such as deep tanks etc. However, the ship's holds sides and/or frames are not to be considered as an inaccessible location.
The Charterers will only use rubber-wheeled bulldozers and payloaders and/or tractors in the ship's holds and/or similar equipment, as deemed necessary by the custom of the port. The unit weight of bulldozers and payloaders and/or tractors and/or similar equipment plus cargo weight, will not exceed the ship's tank top strength and its operation will be under the Master's supervision and satisfaction.

-For gearless ships for grains or other bulk cargoes, the clause could be worded as follows:

The ship will be suitable for a normal grab discharge operation, with normal size/ type grabs and as far as a standard bulk carrier/ singledecker can be. No cargo will be loaded in places inaccessible to grab discharge, such as deep tanks etc. The ship's holds sides and/or frames are not to be considered as an inaccessible location. The Charterers will only use rubber-wheeled bulldozers and/or tractors in the ship's holds and/or similar equipment, as deemed necessary by the port's customs. The unit weight of bulldozers and payloaders and/or tractors and/or similar equipment plus cargo weight, will not exceed the ship's tank top strength and its operation will be under the Master's supervision and satisfaction.

b. Evacuators or mobile cranes placed on the ship's deck.
If the discharging operation is to be performed with mobile cranes and/or evacuators placed on the ship's deck, then the Charterers should be informed before they fix the ship as to the allowable weight per square meter of the available spaces on the deck, in order to provide the most functional Mobile cranes/ evacuators. The Owners should take all the necessary precautions, in order to avoid possible damage to the ship, however, Charterers will be kept fully responsible for any damage caused to the ship.

The Charter Party clause could be worded as follows:
Charterers have the right to fit and weld padeyes on deck and/or fit, install and weld any additional equipment and fittings and/or placing mobile cranes and/or evacuators on deck, in order to perform the discharging operation.
All these operations will take place provided that:
1. an independently recognized surveyor will be mutually appointed, but at the Charterers' cost, to take photos and make a report on the condition of the ship's deck, hatch covers, rails etc., before the fitting and welding of padeyes and/or the fitting and welding of any additional equipment and before the commencement of the work for the placement of the mobile crane/s and/or evacuators. After completion of the operations and the removal, discharge of any padeyes and/ or any additional equipment and fittings and/or removal of the mobile crane/s and/or evacuators and when all repairs have successfully been performed, the surveyor will board the ship again and take photos and make a report, ensuring that the ship has been returned to its original condition. The time lost for the surveyor's inspection will continue to count as lay-time or time on demurrage.
2. the ship's Class Surveyor, as will be appointed by the Owners, will attend the ship's full operation in the port and all works which will be carried out, from the time of berthing until sailing. The Charterers are to reimburse the Owners for the Class Surveyor's fees, once the Owners have submitted their final freight invoice, supported with the original vouchers.
3. the overall operations, installations, fittings, welding, placement of any additional equipment and fittings and/or placing mobile crane/s and/or evacuators on deck, as well as, their lashing, securing, and unlashing, dismantling, remov-

al and discharge upon completion of the operations from the ship's deck, will always be carried out at the Charterers' risk and expenses, by shore labourers and to the supervision and entire satisfaction of the Master and/or ship's Class surveyor. The time lost for the execution of these operations will continue to count as lay-time or time on demurrage.
4. the weight of any additional equipment and fittings as well as that of the mobile crane and/or evacuators plus cargo weight, should not exceed the deck's allowable weight per square meter. The Charterers will provide and install the required dunnage underneath so as to evenly distribute the weight, to the Master's and/or ship's Class surveyor's entire satisfaction.
5. the costs and risks for any cutting and/or welding and/or strengthening that is needed to be performed, in order to strengthen the ship's hull for accommodating the placement and/or is needed to be performed in order to secure the additional equipment and fittings and/or the Mobile cranes and/or evacuators on deck and/ or modification of the hatch covers that have been made, will be for the Charterers' account and all works will be carried out by the Charterers' trained and skilled shore labourers, always under the Master's and/or Class surveyor's supervision and to their entire satisfaction at all stages and time lost will always count as lay-time or time on demurrage.
6. if the ship shifts to another berth or as will be directed by the port Authorities, the Charterers then will provide and install all the required lashing and securing materials at their own expenses. The shifting will be performed at the Charterers' risk and the time lost will continue to count as lay-time or time on demurrage.
7. after completion of the discharging operations and the dismantling, removal and discharge of the additional equipment and fittings, as well as that of Mobile crane/s and/or evacuators, the hatch covers are to return to their original condition to the Master's and/or Class surveyor's entire satisfaction, at the Charterers' entire expenses and the time lost will continue to count as lay-time or time on demurrage. A hose test should be carried out to ensure that all hatch covers are watertight and their integrity is verified. Any degradation of the ship's paint work (on hatches and/or coamings and/or deck and/or tanks etc.) must be returned to its original condition, with such paints as are in accordance with the Owners' original specifications and the time lost for the execution of these operations will continue to count as lay-time or time on demurrage.
8. the Charterers guarantee that they will remain liable and responsible for any consequences, loss, damages, costs, expenses, fines Owners may suffer and will indemnify the Owners against third party claims (inclusive of any interest and costs which may be incurred for defending any such claims or recovering any such loss),, including, but not limited to, cargo claims and oil and/or air pollution arising from placing any additional equipment and fittings and/or evacuators and/or mobile cranes on the ship's deck. If the ship is detained by any par-

ty and generally its sailing is not allowed from the port, the Charterers then undertake the responsibility to release the ship, by taking all the necessary and appropriate measures (issuance of security, etc.), as the case may be or arise and any expenses incurred, including, but not limited to, additional port expenses/ shifts/ Agency fees etc. will be for the Charterers' account, payable on request and upon presentation of the original vouchers, which are acceptable, even by fax or e-mail. The time lost will continue to count as lay-time or time on demurrage, until the ship is free to sail from the port and the 'port clearance' is issued and delivered to the ship.

9. the Charterers will be fully responsible for damage caused to the ship's deck and/or hatch covers and/or rails and/or hydraulic motors etc.

For damage which:

a. affects the effectiveness of the ship and the safe operation and/or safety of the crew and/or affect the trading capabilities and/or seaworthiness and/or cargo worthiness and/or is subject to classification requirements, then such damage, as will be ascertained by the ship's Class surveyor, who will be appointed by the Owners at the Charterers' cost, is to be repaired by the Charterers at their own expenses and risk, to the Master's and/or Class surveyor's entire satisfaction and the time lost will continue to count as lay-time or time on demurrage until completion of all repairs and the 'port clearance' is issued and delivered to the ship.

Charterers are to reimburse the Owners for the Class Surveyor's fees, once the Owners have submitted their final freight invoice, supported with the original vouchers.

The damage will be repaired at the place of occurrence, if it is possible, and there are adequate facilities for permanent repairs, or otherwise, temporary repairs is to be made on the spot or as when detected, at the Charterers' risk and expenses and to the Master's and/or Class surveyor's satisfaction and the time lost will continue to count as lay-time or time on demurrage. The ship will sail or shift to the nearest shipyard, as chosen by the Charterers, with the fuel consumed at sea or during shifting to be for the Charterers' account, which will be settled together with the Owners' final freight statement, in order to complete the repairs.

The time lost during the sailing or shifting to the shipyard and until completion of the works with the damage being repaired to the Master's and/or Class surveyor's entire satisfaction and the 'port clearance' is issued and delivered to the ship or, at the Owners' option, at equidistance from the discharge port, will count as lay-time or time on demurrage. Shipyard's expenses and port expenses, including in and out tugs and/or pilots and/or agency fees etc., will be entirely for the Charterers' account.

b. does not affect the effectiveness of the ship and the safe operation and/or the safety of the crew and/or the trading capabilities and/or the seaworthiness and/

or cargo worthiness and/or is subject to classification requirements and cannot be repaired by the stevedores, the Owners then undertake to accept the ship's sailing, in accordance with the terms of the Charter Party.
However, the said damage is to be assessed before the ship's departure from the discharge port and the costs to be agreed upon between the Master/Owners/ Charterers and the respective survey company and to be paid together with the Owners' final freight statement. The survey company will be mutually appointed by the Owners and Charterers, if considered necessary, at the Charterers' and Owners' equal expense and will promptly inform the Charterers and Owners with the damage report, which will be issued before the ship sails, with a copy of all correspondence exchanged in this connection. The time lost will continue to count as lay-time or time on demurrage, until the 'port clearance' is issued and delivered to the ship.
Owners have the option of having the damage repaired at their next convenient time or at the ship's next Dry-Docking.
In the case that the Master, Owners, Charterers cannot reach an agreement on the cost with the respective survey company, the Charterers will then undertake to reimburse for the actual cost of the repairs, but not for the time lost, upon presentation of the original bills from the repairers or dockyard, that are consistent with the surveyors' report on the extent of such damage.
c. is small and has been repaired by the crew, the Charterers will reimburse the Owners for the cost on request, but not for the time lost and upon presentation of the Owners' repair statement, which is acceptable, even by fax or e-mail.
However, and in any case:
- the Charterers are responsible for any directly related expenses resulting therefrom and incurred thereby and must pay compensation for any damage caused to the ship and/or restore the ship to its initial condition, under their own time and expense, whether or not any payment has been made by the respective parties and/or stevedores to them.
- the Charterers will not to be liable for damage when the Master has accepted a statement in writing from the stevedores and/or the respective parties that the operations were carried out without any damage, or that the repairs were made to his satisfaction.

II. Loading and discharging speed.

The Owners' work on the Voyage Estimation is based on the speed, which each particular port can provide for loading and discharging.
The
- distance between the respective ports,
- the speed of loading and discharging,

- the exception periods will mainly determine the approximate total duration of the voyage at sea/ports,
and together
-with the expected loading cargo quantity/port expenses and fuel prices will lead to the required 'freight rate', in order to execute a particular voyage.

The faster loading and/or discharging speed, which could be achieved by the Owners during the negotiations (provided that the same will not affect the achievable 'freight rate')
- will make the ship enter on demurrage earlier.
Instead, the Charterers will try to obtain a slower speed (by paying the same 'freight rate') than that which can be provided by the port, in order to:
- delay the ship's entry on demurrage.
- earn despatch money.
- cover any delays which may be occurred due to unforeseen circumstances.

The loading and discharging speed could be expressed in the Charter Party as follows:
a- with a daily rate.
It is the amount of cargo that can be loaded or discharged daily and is calculated in metric tons; it is the most common method when fixing on Voyage Charter.
- The allowed lay-time for the operation will result when the intaken Bill of Lading cargo quantity is divided with the agreed daily rate for the completion of the operation.
For example:
if the ship intends to load 34,000 mtns and the daily rate for loading is 7,000 mtns, then the allowed lay-time for the operation is 34,000: 7,000 = 4,8571 = 4 days -20 hours-34 minutes

b- with a number of days.
The ship should complete the operation within a number of days as provided by the Charterers/ Charter Party terms.
For example:
-12,000 mtns bulk wheat
2 days loading/4 days discharging
or even
-12,000 mtns bulk wheat
6 total days for loading and discharging

c- with a number of hours.
The ship should complete the operation within a number of hours as provided

by the Charterers/ Charter Party terms.
For example:
-12,000 mtns bulk wheat
12 hours loading/24 hours discharge
or even
-12,000 mtns bulk wheat
96 hours loading/discharging

d- with Customary Quick Despatch (abbreviated as 'CQD').
The ship will operate according to the maximum speed which the port can provide where:
- there is no maximum allowable time for the operation, therefore no lay-time calculation can be made.
- Charterers are required to take all proven and reasonable steps/actions for loading or discharging as soon as possible or they will otherwise be exposed to receiving claims for damages for detention.
If on the ship's arrival at the port, or after the berthing and commencement of the operation, the Charterers or Shippers/Receivers are not able to keep up with the ship's loading rate or arrange it to be taken away in order to keep up with the ship's discharging rate, a detention rate then could be agreed to be in effect for every lost day/pro rata.
However, the Owners cannot claim for detention if:
-the berth is congested, and the Shippers or Receivers cannot load or discharge two or more ships respectively, at the same time,
-due to breakdown of the loading or discharging means.
This term is commonly used for Panamax sizes of cargo and larger and in certain specific countries.
For example:
-60,000 mtns bulk iron ore
20,000 mtns loading/ cqd discharging
Or
-60,000 mtns bulk iron ore
cqd both ends

e- with custom of the port (abbreviated as 'COP') or on Liner terms (Liner in/ Liner out -abbreviated as 'LI/LO').
In both cases, the ship will load and/or discharge in accordance with the Custom of the Port (COP) where:
- no claims for delays in the ship's berthing or delays for completion of the operations in the port/s can be presented by the Owners and obviously no lay-time calculations can be made.

However, the designated liner ships have priority in berthing.

f- as Fast as can load or discharge.
It is the maximum daily rate which the ship can reach, when it is in full working condition, where:
- there are no fixed terms for loading or discharging.
- no lay-time calculations can be made.
The Charterers and the Shippers or Receivers are of the obligation to load or discharge the cargo respectively, as fast as the ship can take or deliver and the same should be incorporated as a contract term in the Charter Party.

g- scale load.
Some specific Charter Parties, which were adopted for the loading of specific bulk cargoes such as fertilizers, phosphates etc. from certain specific ports, provide a printed standard loading rate and a standard rate for demurrage / despatch, in relation to the actual quantity of cargo loaded (Bill of Lading cargo quantity), i.e., the loading rate will be determined by the quantity loaded on board the ship.
For example, and as printed in the Phosphate Charter-Party (Africanphos 1950):
up to 9,999 mtns loading rate 3,000 mtns.
between 10,000 mtns and 14,999 mtns loading rate 3,600 mtns.
between 15,000 mtns and 19,999 mtns loading rate 4,500 mtns.
between 20,000 mtns and 24,999 mtns loading rate 6,000 mtns.
etc.

h- per hatch per day.
Charterers provide a daily cargo quantity which can be loaded per each existing available hatch. The loading rate is calculated by multiplying the daily cargo quantity per ship's hatch, by the total number of the ship's available hatches.
Often, when a ship has a large number of hatches (usually more than 5), Charterers will limit the total maximum daily cargo quantity, not only to protect themselves from paying demurrage, but also because this rate of loading cannot be provided by the port facilities.
For example:
-loading rate 200 mtns per hatch but maximum 1,000 mtns.
If the ship has 4 hatches, then the loading rate is 800 mtns daily/pro rata.
If it has 5 hatches the loading rate is 1000 mtns.
However, even if the ship has 6 hatches the loading rate will remain 1,000 mtns daily.

i- per workable or working hatch per day.

Charterers provide a daily rate per workable or working hatch. The largest quantity of cargo loaded in a hold will be divided by the agreed daily rate per workable or working hatch, resulting in the total allowable number of days for loading or discharging. If the hold provides two hatches, the daily rate will count double.

The contract terms for the loading and discharging may differ, i.e., the daily rate could be in effect for loading but c.o.p. discharge or total hours for loading but total days for discharge etc.

7. Notice of Readiness (abbreviated as 'N.O.R.').

Tendering the Notice of Readiness is one of the Master's main tasks
. as it well-defines/declares the arrival of the ship in the port and its readiness for the purpose of call.
Among the tendered N.O.R. and the commencement of the lay-time there is always a time gap. During this time the Port Authorities/Customs/ Immigration / Surveyors etc., will board the ship and grant the 'free pratique', the holds inspection will take place etc. and also the Charterers/ Shippers/ Receivers/ Stevedores will get the loading or discharging facilities ready for operation.

The N.O.R. will be tendered to the attending Agents and/or to the specified parties, as will be clearly stated in the Charter Party, or as will be requested by the Charterers during the steaming to the port.
There is no requirement/obligation of the Master and/or Owners, to find out the Shippers' and/ or Consignees' premises/contact numbers, in order to tender the NOR, unless otherwise specifically stated in the Charter Party.
The effective N.O.R. should be tendered and re-tendered if necessary:
- within the Charter Party laydays-cancelling dates at load port.
- when arriving at the commercial limits of the port, or at the place as will be instructed by the Port Authorities to wait, or in a berth as specified in the Charter Party.
- when the ship is ready and, in all respects, to operate for the purpose of call.
- within the working days and working hours, as stipulated in the Charter Party.
Violation of these terms could lead to the tendered N.O.R.
- being given prematurely or not in accordance with the terms of the Charter Party.
As a result, the N.O.R. could be considered as invalid, leading to the postponing of the commencement of the lay-time or even leading to the referral to Arbitration, in order to resolve the dispute.

Subject to the Charter Party terms, the N.O.R. could be rejected, either in berth

or at the waiting place if:
a- it fails to pass the 'free pratique' and the inspection by the Port Authorities (invalid certificates etc.)
If the ship has direct access to the berth or Port Authorities boarded at anchorage/ waiting place upon arrival and it fails to pass the 'free pratique' and/or the ship possesses invalid certificates, then the tendered N.O.R. could be considered invalid under the terms of the Charter Party and a new N.O.R. should be tendered, but only when the problem is solved, and the ship is ready for operation.
The provision of 'wibon/wipon/wifpon/wccon' will trigger the commencement of the lay-time for ships delayed in berthing, if, however, it fails to pass the 'free pratique' or possesses invalid certificates upon berthing, then under the terms of the Charter Party, the actual time lost in passing the 'free pratique' or until the ship is in all respects ready to operate will not count as lay-time.
b- the holds/hatches are not ready for loading.
In the case that the ship has direct access to the berth, or the cargo holds inspectors board at the anchorage/waiting place upon arrival and the ship fails to pass the holds/hatches cleanliness inspection, then the N.O.R. could be considered invalid under the terms of the Charter Party and a new N.O.R. should be tendered, but only upon the successful passing of all holds.
The provision of 'wibon/wipon/ wifpon/ wccon' will trigger the commencement of the lay-time for ships delayed in berthing and if it fails to pass holds/hatches inspection upon berthing, then under the terms of the Charter Party, the actual time lost in passing the holds inspection will not count as lay-time.
c- it fails to pass the hatch-covers watertightness inspection.
Charterers could require, before the commencement of the loading operation, to carry out a hose or chalk test on hatchcovers, in order to ensure their water tightness and/or an ultrasonic test to ensure the thickness of the steel. In the case of a defect, it could cause the tendered N.O.R. to be considered invalid under the terms of the Charter Party and should be re-tendered again, after all hatch-covers have been repaired and successfully passed the re- inspection.

A provision should be incorporated in the Charter Party, clearly stating that:
- the 'NOR is deemed to be tendered and accepted from the time of tendering and the lay-time will commence counting in accordance with the terms of the Charter Party.
When the Master has doubts as to its validity, or it has been tendered during the exception periods or outside the limits of the port's working hours,
. a new NOR
should be tendered within the working days/time limits, but will be imperatively marked

. without prejudice to the preceding Notice Of Readiness......',
stating the date and time of the ship's arrival at the port.

There is no standard form for the tendering of the N.O.R. and each Ship-Owning company uses its own wording that is similar in meaning.
Under the Charter Party terms, the following two methods could be used for the tendering:
a. in writing upon berthing. If the ship arrives and 'all fast' without berthing delay, the Master will present the N.O.R. to the attending Agents in a hard copy in writing. The lay-time will commence counting in accordance with the terms of the Charter Party.
b. by cable/telex/e-mail/fax either from the anchorage or waiting place (both within the commercial port limits) or from the area where it will be instructed to wait (open sea-outside of port limits). This method of tendering the N.O.R. is in effect when the port is congested, and the ship's berthing will be delayed. After berthing, the N.O.R. in a hard copy in writing will be given to the attending Agents, stating the date and time of its arrival at the anchorage/waiting place/ outside port limits.

8. Lay-time and when it will commence counting.

I. Lay-time.

- Lay-time is the allowed period of time during which the ship must timely complete the operation at each port of contact and is directly resulted from the division of the Bill of Lading intaken cargo quantity with the daily loading and/or discharging rate at each port, or as specified under the terms of the Charter Party.
The time used in excess of the allowed period, will count as 'demurrage'. The time saved due to earlier completion will count as 'despatch'.
The lay-time will commence to count, when the ship has arrived at the port or berth and it is ready in all respects for the purpose of call and a valid Notice of Readiness is tendered within the specified working days/hours, or otherwise as per the provisions of the Charter Party.

Generally, a ship has arrived and the lay-time commences when:
- it has arrived at the commercial limits of the port, or at the place as will be instructed by the Port Authorities to wait, or in the berth as specified in the Charter Party.
- the ship is ready to operate in all respects.
- a valid NOR is tendered, in accordance with the terms and conditions of the Charter Party.

Failure to grant free pratique or unreadiness of the cargo holds/hatches for loading etc. could lead to the rejection of the N.O.R. and the commencement of the lay-time will be postponed or cease to count, as the case may be, until the failure is rectified.
The lay-time will cease counting, causing significant delays to the ship's sailing, or even entry on demurrage, if the operation is interrupted due to (unless otherwise stated in the Charter Party):
- force majeure reasons, which are usually specified in the Charter Party.
- adverse weather conditions, such as strong winds/rain/fog/high humidity which prevent the operation or could cause damage to the cargo.
- exception periods, as stipulated in the Charter Party.
- faults attributed to the Owners/ship, such as the breakdown of cranes/ ITF or Port State Control detention etc.
- ballasting/de-ballasting of the ship's tanks for trim/stability or other reasons.

II. Commencement of lay-time.

The lay-time will commence counting when all terms/conditions of the Charter Party have been met, in one of the following ways:
a- 08:00 hrs. /14:00 hrs.
This term is fulsomely used on Voyage Charter for all small/handy/handymax sized ships and it is in effect when carrying general/bulk cargoes and mainly when the Gencon Charter Party is in use.
The lay-time will commence counting either
. at 08:00 hrs.
or
. at 14:00 hrs.,
depending on the time that the valid Notice of Readiness (N.O.R.) is tendered to the concerned parties.
As defined by most of the Voyage Charter Parties, the N.O.R. must be tendered within working hours, which is usually determined as from 08:00 hrs. to 17:00 hrs. of weekdays. If the N.O.R. is tendered before noon (from 08:00-to 11:59 hrs.), the lay-time will commence counting
. at 14:00 hrs. of the same day,
but if it is tendered in the afternoon (from 12:00 hrs. - to 16:59 hrs.), the lay-time will commence counting
. at 08:00 hrs. of the next working day.
The working hours in the Charter Party should be agreed as:
'from 08:00 hrs. –up to and including 17:00 hrs.'
And for the tendering of the N.O.R.:
'from 08:00 hrs. -up to and including 12:00 hrs.' and from 12:01 hrs.- up to and

including 17:00 hrs.".

b- 08:00 hrs.' of the next working day.

This term was a standard term in the older Synacomex Charter Party form, which was in effect for grain shipments. The lay-time will commence counting

. at 08:00 hrs.' of the next working day, if the valid N.O.R. is tendered during working hours (which is usually determined as from 08:00 hrs.- up to and including 17:00 hrs.' of weekdays) of the previous day.

This term could cost to the Owners.

For example:

the ship is fixed/based on this particular term and Saturdays/Sundays/Holidays have not been designated as working days for tendering the N.O.R.. The N.O.R. should be tendered within working hours from 08:00- up to and including 17:00 hrs., from Monday to Friday.

However, the ship arrived at the port on Friday at 1800 hrs.

Therefore, the N.O.R. is valid only for the following Monday from 08:00 hrs., resulting in the lay-time to commence counting on next Tuesday at 08:00 hrs.

The ship lost 3 days and 14 hours. However, if the ship commenced the operation before the commencement of the lay-time, the Charter Party agreed provision, which refers to 'unless sooner commenced' will be a success for the Owners, as the actual time or half actual time used (depending on the agreement) before commencement of the lay-time will count as lay-time.

c- Turn time.

It is valid for bulk cargoes, especially for very large quantities of cargo (panamax/ cape sizes). The ship upon arrival will tender the N.O.R. and the turn time will start counting. The lay-time will commence counting

. upon the Turn time expiring.

For example:

if the agreed Turn time is 8 hrs. and the arrival of the ship was at 11:00 hrs., then the

lay-time will commence counting at 19:00 hrs. of the same day.

For large ships, the N.O.R. will usually be agreed to be tendered at any time, during day or night, seven days a week, including holidays. Otherwise, if the turn time expires on an officially recognized closed port holiday, usually and as per the Charter Party terms, the lay-time will commence counting at 08:00 hrs. on the next working day.

d- upon arrival or upon the first resumption of work.

This term is in effect when the ship is fixed to load or discharge in more than one port.

When it is 'upon arrival', the lay-time will commence counting immediately after the ship's arrival in port and tendering of the N.O.R., regardless of the hour or day of arrival (during day/night or weekends/holidays).
When it is 'upon the first resumption of work', the lay-time will commence counting from the first resumption of work in the port, regardless of the hour or day of arrival.

e-Unless sooner commenced.
If the ship has commenced the operation before the commencement of the lay-time, the actual or half actual time (depending on the agreement) that was used before the commencement of the lay-time will count as lay-time.

f-Reversible/non reversible lay-time.
Under the terms of the Charter Party, for the purpose of calculating the demurrage or dispatch incurred, as the case may be, the lay-time should be calculated: . as non-reversible
between the ports that were used.
The lay-time will be calculated in a separately issued Time Sheet, with the resulting demurrage or despatch due for each port of contact.
. *as reversible*
between the ports that were used for loading or discharging or between the loading and discharging ports, i.e., offset the time used at load port with the time used at discharge port. Particularly, for the time used for the completion of loading, either resulting demurrage or despatch will be added along with the time used to complete the discharge.
One Time Sheet will be issued for all ports.
Compulsory reversible lay-time is in effect, when the ship is fixed with a number of days or hours, for the completion of the operation at the ports of contact.
For example:
-12,000 mtns bulk wheat
-Rouen/Piraeus
-5,000 mtns loading/5,000 mtns discharging
The lay-time will count as 'reversible' or 'non reversible' as will be agreed upon.
But when:
-12,000 mtns bulk wheat
-Rouen/Piraeus
-5 ttl days loading/discharging
then 'Reversible' lay-time will be in effect.

g-Lay-time counting during several operations/stoppages in the ports.
Usually and under the terms of the Charter Party, the lay-time ceases to be cal-

culated after the completion of the loading or discharging operation, resulting in all time lost for any reason, not to be counted as lay-time or time on demurrage.
However, the brokers and the Owners and/or Charterers should exercise care, if the lay-time or time on demurrage will count during the time required for several operations to be executed in the ports or during stoppages incurred for various reasons, when working on the Charter Party.
Incorporation in the Charter Party will specify if the loss of time will count during:
1-the first shifting time from the waiting place due to congestion (anchorage or standby berth or pilot station etc.) to the load or discharge berth.
Comments:
Many Charter Parties incorporate the term that this particular shifting time will not count as lay-time, even if the ship is already on demurrage.
2-the waiting for a sufficiently high tide for berthing/on berth or for sailing.
Comments:
The incorporation of 'always afloat at all times of the tide' could lead the Owners to insist that this particular time should count as lay-time or time on demurrage
3- the time waiting for daylight to berth or sail.
Comments:
In many ports, a sailing or berthing or ship maneuver during nighttime is not allowed. Owners should be aware of such restrictions before fixing the ship, as the lost time should not count as lay-time, unless otherwise stated in the Charter Party.
4- the period of time in which tugs/pilots wait for the Port Authorities to grant permission for the ship's transit towards berthing or sailing.
Comments:
It could happen that the pilots board the ship, or the tugs are alongside the ship, but the transit to the berth or from the berth to the open sea is not allowed by the Port Authorities, due to heavy traffic/ prevailing congestion in the port (in/out of the port) or priority to cruise or war ships or for any other reason.
Care should be taken in the Charter Party if these particular delays should count as lay-time or not.
5- the time required for the initial/intermediate/final draft survey.
Comments:
The time that is required for the draft survey which will be carried out in order to determine the cargo quantity on board or discharged could cause significant delays to the ship's sailing and the Owners could suffer from the lost time, which could result in increased port expenses (tugs/pilots working overtime) or even for the ship, because of the delays incurred, to have to wait until daylight,

in order to sail.
The Owners should negotiate a time limit to be in effect for the completion of the draft survey or for the time lost to be shared equally between the Owners and Charterers.

6- the time required for the opening and closing of the cargo hatches.

Comments:

The first opening and the last closing of hatches or during the loading or discharging operation due to adverse weather conditions may cause some delays. Usually, in the Charter Parties only the time required for the first opening and last closing does not count as lay-time.

7- the holidays of all types.

Comments:

There are various types of holidays that could be incorporated into the Charter Party, such as national and/or local holidays, legal/general holidays or Charter Party holidays or Bimco holidays etc.

The lay-time will usually cease to count, even if the ship worked during this period.

However, the 'type of holidays' should be clearly specified in the Charter Party, otherwise,

. the general expression of 'holidays' should have no limitations to it.

8- the time required by the Owners to make certain necessary repairs, thus interrupting or interfering with the normal loading or discharging operations.

Comments:

Such time should not count as lay-time and the Owners should be kept responsible for all consequences and expenses that might arise, such as stevedoring and trucks' stand-by etc.

9- the time required for the trimming/leveling of the cargo in cargo holds or for extra trimming.

Comments:

Lost time could arise because of the need for cargo trimming, in order to fill up the ends in cargo holds. Under the terms of the Charter Party the same could be performed by the shore/loading facilities or by arrangements in case extra trimming is required due to insufficient trimming as provided by the shore means. It should be specified if the time lost will count as lay-time or not.

10- the time required for the shifting of the shore cranes, between the cargo holds.

Comments:

The ship must load and discharge strictly as per the agreed sequence/stowage plan.

The shore cranes could shift back/forwards between cargo holds multiple times during the work which could delay the loading or discharging completion. The lost time actually counts as part of the loading or discharging operation, however and because those shifts are inserted into the issued Statement of Facts, it could be incorporated in the Charter Party that the time required for the shifting of the shore cranes will count as lay-time.

11- the time required for the ship's shifting/warping alongside a quay by the mooring lines.

Comments:

The ship could shift in the berth by using the mooring lines, as could be asked by Charterers or Terminal, in order to make all hatches available for the loading or discharging purposes.

Under the terms of the Charter Party, this type of shifting cannot be considered as 'berth shifting', however, it should clearly state that:

- time lost will count as lay-time
- it will be done by the ship's crew free of expense to the Charterers, provided port regulations permitting, otherwise the Charterers are to employ shore labourers at their own time/expenses.
-if tug assistance is necessary for any reason, either because of weather conditions or to port regulations or as a result of distance, then it would be considered as 'berth shifting'.

12- the time required for the placing/removing/manoeuvring of bulldozers in the ship's holds.

Comments:

The Statement of Facts will include the time required to place or remove the bulldozers into or out of cargo holds respectively, as well as the time required to manoeuvre into the cargo holds. All time lost will count as lay-time and should be incorporated as a contract term in the Charter Party.

13- the time required for the fumigation.

Comments:

After the completion of the loading operation or during loading, the Shippers may require the cargo that is on board to be fumigated, in which case it may result in a significant loss of time, delaying the ship's sailing. Charter party terms should clearly state on account of whom the time lost should be.

14- the interruptions to the loading or discharging operation, beyond the Owners' or ship's fault.

Comments:

Lost time due to Charterers or Charterers' Agents or the terminal operator or Shippers or Receivers or any Authority prohibiting, impeding, or restricting the loading or discharging operation beyond the Owners'/ship's fault, should be stated if counted as lay-time or not.

15- the time waiting for the Charterers or Shippers or Port Authorities to present the cargo documents or the clearance of the ship and the possible shifting for

whose account should be.
Comments:
In some ports the presentation of the cargo documents (Bills Of Lading/ Mate's Receipts/Cargo Manifest etc.) or the issuance of the Statement of Facts or the port clearance is delayed for more than a reasonable amount of time. It could be agreed that the lay-time should recommence counting after the time allowed to present the relevant documents has expired and if the ship shifts to another berth, then the shifting expenses will be for the Charterers' account and the time lost will count as lay-time or time on demurrage.
16- the time required for the shifting of the ship, beyond the number of shiftings allowed by the Charter Party.
Comments:
The ship might be ordered to shift between berths or be instructed to vacate the berth for a while and return later, beyond the number of shifts allowed by the Charter Party. Obviously, the time lost will count as lay-time.
17- the time used on overtime, which is ordered by any party including Port Authorities, except for the Owners.
Comments:
Overtime could be arranged and instructed by the Shippers or by Receivers, in order to speed up the loading or discharging operation, resulting in the ship's earlier departure from the berth. Additionally, the Port Authorities, for safety or other reasons, could order the parties to work on overtime. It should be clearly incorporated in the Charter Party that if any overtime is ordered, except by the Owners:
- the time lost and expenses will be for the account of the party ordering the same and for the Charterers' account.
- if ordered by the Port Authorities or other third parties will be for the Charterers' account.
However,
-crew's/officers' overtime will be for the Owners' account.
18- the operations needed to bring the ship to a seaworthy and safe stress/trim condition.
Comments:
The ship must be left in a seaworthy and safe stress/trim condition to the Master's satisfaction during any shifting which will be carried out between loading and discharging berths, as well as, also between ports.
Under the terms of the Charter Party the time lost will be for the Charterers' account.
19- the interruptions to the loading or discharging operation due to the insufficiency of the ship's ballast/de-ballast pump capacity.
Comments:

If the loading and/or discharging operation is interrupted or extended due to the insufficiency of the ship's ballast/de-ballast pump capacity in relation to loading/ discharging capacity, then there should be a provision about whether or not the lost time should count as lay-time. Many ports and Charter Parties grant some time for this operation to be completed, however, if the delay is the ship's fault, the time should cease to count as lay-time.

20- the unloading of any overloaded cargo quantity.

Comments:

If the quantity of the cargo loaded exceeds the maximum contractual cargo quantity or the ship is overloaded and exceeds the maximum draft, then the time lost to discharge the overloaded cargo will be for the Owners' account.

21- the ship's detention arising due to unpaid port or other expenses.

Comments:

Lost time due to the Owners' failure to provide Agents with sufficient funds for covering the port expenses or other of the Owners' expenses incurred before the ship sails from the port, will be for the Owners' account.

22- the ship's detention due to labour boycott arising due to its flag/ownership/ registry/crew.

Comments:

If the labourers refuse to operate due to the ship's flag/ownership/registry/ crew then the time lost will not count as lay-time or time on demurrage.

23- the time lost due to the holds' ladders not being in a good state of repair and not complying with the safety at work and health act.

Comments:

The holds' ladders must be kept in good condition during the operations in the ports, otherwise, time lost will not count as lay-time or time on demurrage.

24- the ship's detention due to failure of the Port State Control and/or ITF inspection.

Comments:

If the ship is detained during the inspection carried out by the Authorities, the time lost in order to comply with the repairs or recommendations made by the Authorities will not count as lay-time or time on demurrage.

III. The term of 'wibon/wipon/wifpon/wccon'.

The agreement that the lay-time will commence counting:

. *whether in berth or not/whether in port or not/whether in free pratique or not/ whether customs cleared or not'* (abbreviated as 'wibon/wipon/ wifpon/ wccon' or even shorten 'w/w/w/w '),

which is commonly used with Charter Parties that are governed by English Law, is one of the most *basic terms* that should be incorporated in all the Voyage

Charter Parties, otherwise the Owners could sustain significant financial losses. This terms states that if the ship's berthing is prevented due to congestion and the berth being occupied or due to bad weather or tidal conditions or due to Port Authorities' instructions or due to unreadiness of cargo for loading and generally for any reason beyond the Owners'/ship's fault and if they are not responsible and provided that the ship is in all respects ready to operate, then the Master has the right to tender the N.O.R. and the lay-time will commence counting as per the Charter Party terms w/w/w/w.

IV. Lay-time calculation.

Upon completion of the discharging operation and when the ship has sailed from the port, then the shipping leg is finished, and the Owners will further proceed to close the file by claiming for possible balance freight plus demurrage incurred in the ports of contact.

The demurrage or despatch money will be the result of the lay-time calculation, which will be made on

. the Time Sheet form,

and will be based

. on the Statement of Facts (abbreviated as 'S.O.F.'), as will be issued by the attending Agents in each port of contact.

The Time Sheet will be issued separately for each port of contact, unless the lay-time is 'reversible', in which case only one will be issued for all ports. It will be drawn-up by the Owners and will include in detail the actual working time calculated as lay-time and the time that was excluded in accordance with the terms of the Charter Party. The resulting overall lay-time used will determine if the ship was actually delayed from its sailing and a demurrage amount is due, or if it sailed earlier, in which case despatch money is due to the Charterers.

The S.O.F. is an important document which records

. all the events from the time of the ship's arrival, until the last pilot disembarks from the ship and it sails to open sea towards its next destination.

The Time Sheet, the original Statement of Facts and any other necessary document will be presented to the Charterers, together with the Owners' final freight statement, for the Charterers' consideration and /prompt settlement.

9. Exception periods (non-working days).

The exception periods at the ports are the periods as incorporated in the Charter Party, where the ship under regular conditions and in accordance with the custom of the port

. *will not work, resulting in the lay-time to cease counting.*

The exception periods will be determined in accordance with the applicable standard working days to ports which the ship will contact or to the Shippers' or Receivers' private mines. Usually for small /handy ships operating within Europe, Saturdays/Sundays are the standard days of the week that are excluded. Different weekdays are in effect when a ship contacts Muslim countries and Israel. Nevertheless, the very large capacity ships usually operate without exception periods being in effect.

The N.O.R. tendered during the exception period could easily be considered invalid, and must therefore be re-tendered during the resumption of the first normal working period, but always 'without prejudice to the pre-sending N.O.R. '.

The lay-time will count 'per weather working day of 24 consecutive hours', however, when it is raining/snowing or even high humidity prevailing, the conditions, which could affect the quality of the cargo, will result in the interruption of the loading or discharging operation and the actual lost time will count as exception periods.

Strikes in the port or breakdowns of ship's cranes and force majeure periods in general, are periods which will result in the lay-time to cease, unless otherwise has clearly stated in the Charter Party.

When the allowed lay-time for loading or discharging has expired and the ship is actually on demurrage, then

. *no exception periods are in effect,*

and stoppages due to rain/snow/weekends/holidays/force majeure etc. will be calculated as time on demurrage, unless otherwise has clearly stated in the Charter Party. The same is known as

. *'once on demurrage, always on demurrage'.*

The agreement with exception or without exception periods for each individual port could be expressed as follows:

a- Saturdays/Sundays/Holidays Excluded (SSHEX).

The time lost in port during

. Saturdays/Sundays and Holidays does not count as lay-time.

It usually ceases from 17:00 hrs. on Friday until 08:00 hrs. on Monday and on Holidays from 17:00 hrs. of the day preceding a holiday until 08:00 of the next working day.

b- Sundays/Holidays Excluded (SHEX).

The time lost in port during

. Sundays and Holidays does not count as lay-time.

It usually ceases from 12:00 hrs. on Saturday until 08:00 hrs. on Monday and on Holidays from 17:00 hrs. of the day preceding a holiday until 08:00 of the next working day.

c- Saturdays/Sundays/Holidays Included (SSHINC).

No exception periods are in effect and the lay-time will count seven days a week

as well as during holidays.
SUPER Holidays could be agreed to be excluded, such as the 25th of December, 1st of January and Easter day.
d- Thursdays/Fridays/Holidays Excluded (TFHEX).
This term reflects Arab countries and Muslim countries in general, where weekend holidays differ from those applicable to the Western World. This term is the equivalent to the term of SSHEX.
The lay-time will cease during
. Thursdays/Fridays and Holidays.
The lay-time usually ceases from Wednesday 17:00 hrs. and resumes on Saturday at 08:00 hrs.
e- Fridays/Holidays Excluded (FHEX).
The lay-time usually ceases from Thursday at 12:00 hrs. and resumes on Saturday at 08:00 hrs. This term is equivalent to the term of SHEX.
f- Thursdays/ Fridays/ Holidays Included (TFHINC).
No exception periods are in effect and it is equivalent to the term of SSHINC.
g-Fridays/Saturdays/Holidays excluded (FSHEX) or Saturdays/Holidays excluded or Fridays/Saturdays/Holidays included.
A term that is in effect when calling Israel and some Arab countries and it is equivalent to SSHEX/ SHEX/SSHINC respectively.
For FSHEX, the lay-time usually ceases from Thursday 12:00 hrs. and resumes on Sunday at 08:00 hrs.

The expansion of the exception periods affects the 'freight rate', as the Owners should take into account and calculate in their Voyage Estimation the potential loss of time for completion of loading or discharging plus the possible loss of time until the N.O.R. is valid.
The type of 'holidays' (Charter Party holidays or Bimco holidays or official/local or legal/general holidays which are in effect at the ports of contact) which will be excluded as working days from the Charter Party should be clearly specified during the negotiations,
. as only the expression of 'holidays' in general should have no limitations to it.

If the ship is operating during the exception periods, the lay-time will count for the actual or half actual time used, otherwise it will remain ceased, in accordance with the provisions of the 'unless used' or 'even if used' charter party terms respectively.
The Port Authorities could order the ship to operate during the exception periods for its earlier departure, for various reasons, or due to Charterers'/Shippers' or Receivers' arrangements for overtime, for the purpose of quick despatch. In all cases, all the costs of any work in progress during the exception periods and

on overtime should be clearly stated in the Charter Party as being for the party ordering this operation and for the Charterers' account, even if ordered by the Port Authorities.

If the ship is in operation during the exception periods, the same could be covered and expressed in the Charter Party as follows:
a- Even If Used (abbreviated as 'EIU').
Even if, during the exception periods, the ship is
. in operation, the lay-time will remain ceased
and no time will be counted as lay-time.
In a few words, if Shippers or Receivers decide to work during weekends and on overtime for their own reasons, then the Charterers are the only party that would benefit from such an order. It is possible for the ship to be completed earlier and sail from the port by paying despatch money to the Charterers.
Owners should also take into consideration the fact that before a weekend or before the beginning of a long holiday or a strike in the port, the time could happen to have ceased only few minutes before the ship enters on demurrage. This could cost thousands of dollars, making it necessary for the Owners to exercise due diligence, when deciding on what to accept when working on this provision.
b- Unless Used (abbreviated as 'UU')
or Unless Used If Used Actual Time Used To Count (abbreviated as 'UUIUATUTC').
The full or actual time used during the exception periods will count as lay-time. The work during the exception periods will result in the entry of the ship on demurrage earlier or will save despatch, as the case may be.
- Unless Used If Used Half Actual Time Used To Count (abbreviated as 'UUIUHATUTC').
The actual time used during the exception periods should be equally shared between the Charterers and Owners. If, for example, the ship has operated for 12 hours, then only six hours will be counted as lay-time.

Obviously, these terms are not applicable when SSHINC/TFHINC terms are in effect.

The effect of the commencement of the lay-time/exception periods (non-working days) on Voyage Charter, could be illustrated as follows:

Charter Party terms:
- cargo for loading: 48,000 mtns
- daily loading rate: 8,000 mtns
- N.O.R. will be tendered within working hours from 08:00 to 17:00 from Mon-

day to Friday
and
- the Lay-time will commence counting from 14:00 hrs. of the same day, if the N.O.R. is tendered before noon, or at 08:00 hrs. of the next working day, if tendered in the afternoon.
As per the signed Statement of Facts (SOF):
- the ship has arrived/berthed and tendered the N.O.R. at the load port on Tuesday 7th May at 09:00 hrs.
- the ship has commenced the loading operation on 7th May at 10:00 hrs.
- the ship has completed the loading operation on Tuesday 14th May at 14:00 hrs.
- the ship has worked on Saturday 11th May from 08:00 hrs. until 18:00 hrs.
- the ship has worked on Sunday 12th May from 08:00 hrs. until 14:00h hrs.
- no rain periods made during the loading or other interruptions that could affect the counting of the lay-time
Taking into consideration the following parameters:
- the lay-time allowed for loading is 6 days (48,000 mtns: 8,000 mtns = 6 days).
- the N.O.R. tendered at 09:00 hrs. on 7th May and the lay-time will commence counting at 14:00 hrs. of the same day.
1ST case.
The Charter Party states:
- the actual time used before the commencement of lay-time will count as lay-time.
- time from Friday 17:00 hrs. until Monday 08:00 hrs. will not count as lay-time Unless Used –If Used the Actual Time Used to Count (UU-IUATUTC).
The lay-time used until completion of the loading operation will be calculated as follows:
-Tuesday 7th May from 10:00 hrs. to 24:00 hrs. (14 hrs.) / the lay-time should have commenced counting at 14:00 hrs., however, as per Charter Party the actual time used before the commencement of lay-time to count as lay-time.
-Wednesday 8th May from 00:01 hrs. to 24:00 hrs. (1 day).
-Thursday 9th May from 00:01 hrs. to 24:00 hrs. (1 day).
-Friday 10th May from 00:01 hrs. to 17:00 hrs. (17 hrs.).
-Saturday 11th May from 08:00 hrs. to 18:00 hrs. (10 hrs.)/ the Charter Party states UU-IUATUTC.
-Sunday 12th May from 08:00 hrs. to 14:00 hrs. (6 hrs.) / the Charter Party states UU-IUATUTC.
-Monday 13th May from 08:00 hrs. to 24:00 hrs. (16 hrs.).
-Tuesday 14th May from 00:01 hrs. to 14:00 hrs. (14 hrs.).
Total time used 5 days and 5 hours.
2ND case.

The Charter Party states:
- time used before the commencement of lay-time will not count as lay-time.
- time from Friday 17:00 hrs. until Monday 08:00 hrs. will not count as lay-time Even If Used (EIU).

The lay-time used until completion of the loading operation will be calculated as follows:
-Tuesday 7th May from 14:00 hrs. to 24:00 hrs. (10 hrs.) / the lay-time will commence counting at 14:00 hrs., even if the ship commenced the loading earlier.
-Wednesday 8th May from 00:01 hrs. to 24:00 hrs. (1 day).
-Thursday 9th May from 00:01 hrs. to 24:00 hrs. (1 day).
-Friday 10th May from 00:01 hrs. to 17:00 hrs. (17 hrs.).
-Saturday 11th May from 00:01 hrs. to 24:00:00 hrs. (0 hrs.)/ the Charter Party states EIU.
-Sunday 12th May from 00:01 hrs. to 24:00 hrs. (0 hrs.)/ the Charter Party states EIU.
-Monday 13th May from 08:00 hrs. to 24:00 hrs. (16 hrs.).
-Tuesday 14th May from 00:01 hrs. to 14:00 hrs. (14 hrs.).

Total time used 3 days and 23 hours.

The Owners of the 1ST case:
-will pay lower despatch money to the Charterers
-or in case of delay, the ship will enter on demurrage much earlier than the ship of the 2ND case.

10. Freight.

On Voyage Charter the 'freight' is the amount paid to Owners as compensation for the carriage of goods by sea from one place to another.

This amount of 'freight', as well as the possible demurrage incurred would cover the entire time required to execute the Voyage Charter,
including crew's wages and all costs directly associated with them (transportations, air-tickets, doctors etc.), also provisions/stores/ costs required for ship's maintenance/ insurances/fuel at sea and in ports/lubricants/port's disbursements and costs for passing the canals/straits/loans/office expenses etc. and all other expenses either necessary to make the ship seaworthy or to be incurred during the voyage.

The 'freight rate' is the most important negotiable point and the base for the conclusion of a fixture.

The required 'freight rate' it is subject to:
- the prevailing market at the time of fixture.
- the existing competition with other ships in the loading area.
- the distance between the loading and discharging port/s.

- the loading and discharging speed.
- the exception periods.
- the grade/type of cargo and its stowage factor.
- the prevailing bunkers' prices at the time of fixture.
- the port restrictions and ordinary port expenses.
- if any taxes/dues should be imposed on the ship/freight.
- if any extra or war insurance should be imposed on the ship.
- the existing commission of the brokers involved in the fixture.
- the ship's running expenses.

The Owners will decide if the achievable 'freight rate' is attractive for the ship, taking into consideration:

- the resulting net or gross daily income.
- the total duration of the voyage.
- if the income is in line with the existing market.
- the ship's repositioning at the discharging area.

For a biz to be considered attractive, the achieved 'freight' should at least cover the ship's 'running expenses', which are the standard daily expenses, resulting from the:

- class and flag commitments (special survey/dry-dock etc.).
- provisions, repairs, spares supplied to the ship.
- crew's wages, transportations, repatriations etc.
- insurance of the ship (H&M, War Risk, P&I Club costs etc.).
- lubricants and fresh water supplied to the ship.
- office expenses (staff wages/insurance/ rental costs etc.).
- installments on loans obtained from the banks for the purchase of the ship and

the costs calculated/levied for the Owners' account for each particular voyage, such as the:

- quantity of fuel, which will be consumed for the execution of the voyage.
- port disbursements including taxes/dues etc.
- passing through the canals/straits etc.
- extra and/or war insurances, as will be imposed by the insurance companies.
- existing brokerage commissions.

and all other costs, which must necessarily be covered by the Owners for the successful execution of the voyage.

The amount due as 'freight', resulting from the following calculation methods:
a- freight payable 'per metric ton' or 'per cubic metre' of the cargo loaded.
The ship will load the maximum possible contractual quantity of cargo and the resulting amount of due 'freight' will be set, when the agreed 'freight rate' be . multiplied

with the intaken Bill of Lading cargo quantity measured in metric tons. It is the most common method of payment.
For example:
Charter Party 'freight rate' USD 10.00 per metric ton of cargo loaded.
As per Bill of Lading cargo loaded 10,000 metric tons.
Gross (i.e., without any deductions) 'Freight' due to Owners
10,000 mtns x USD 10, 00 = USD 100,000
The actual quantity of cargo loaded should be determined by the draft survey carried out or by the findings of shore means, as will be presented by Shippers, but verified by the Master of the ship.
For cargo that is not in bulk form, but on pallets or is vehicles/projects/logs/ sawn timber etc., the resulting amount of freight due will be set when the agreed 'freight rate' is multiplied with the intaken Bill of Lading quantity measured in cubic meters.
For example:
Charter Party freight rate USD 50.00 per cubic metre of cargo loaded.
As per Bill of Lading cargo loaded 3,000 cbm.
Gross (i.e., without any deductions) 'Freight' due to Owners
3,000 cbm x USD 50, 00 = USD 150,000.
b-Freight payable 'on lumpsum' basis:
Due to the uncertain availability of enough quantity of cargo or due to an unconfirmed stowage factor, the Charterers and Owners could reach a Charter Party agreement in which the ship will be compensated with a 'lumpsum' amount, regardless of the quantity of cargo loaded.
The whole ship's cargo spaces are to the Charterers' entire disposal, who will load the ship in accordance with the available cargo quantity or until all holds 'full up', but always subject to port restrictions and ship's stresses.

The 'freight' which is due basis one load port/one discharge port, is abbreviated as
- 'freight USD... 1/1'.
If it covers one or two load ports and/or one or two discharge ports, then it will be abbreviated as
- 'freight USD.... 2/1' or '1/2' or '2/2' etc. as the case may be.
If, for the completion of loading or discharging, two or more ports will be used, then the due additional 'freight' should be agreed to be resulted
. on the 'entire' loaded quantity of cargo,
and not for the quantity intended for/from each port.
This additional amount of 'freight' will cover the added duration including:
- the additional fuel which will be consumed,
- the ship's running expenses,

- the additional port expenses,
- the daily ship's profit,

and

- the increased amount for the brokerage commissions.

A provision should be incorporated in the Charter Party, clearly stating that:
- the amount due as 'freight' is payable by the Charterers to Owners net of tax and bank charges.

-Forms of Payment:

The payment will be made within the days allowed as agreed and incorporated in the Charter Party

. after the 'sailing' of the ship from the port,

or

. after completion of loading', regardless of whether the ship has sailed or not and provided that the original Bills of Lading have been signed and released to the Shippers.

The effect of an 'after sailing' payment occurs, when the ship is ready to sail with all shipping documents signed and released, but the sailing is delayed for various and unexpected reasons (tidal port/weather stress/Port Authorities ban etc.), and therefore the payment of 'freight' will be delayed proportionally.

The provision of 'after completion of loading' implies obligations on the Charterers to arrange for remittance of 'freight' within the number of days allowed by the Charter Party, even if the sailing of the ship is delayed from the load port.

If the Bill/s of Lading have been marked as

a- 'freight payable as per Charter Party',

the Charterers will pay the 'freight' within the number of days allowed by the Charter Party, after the ship's sailing from the port or after the completion of loading, as will be agreed upon. The number of days allowed for payment is subject to the distance between the load and discharge port. For long distances between ports, the 'freight' payment will usually be made within 3-5 banking days, but when the distance is short, the payment should be arranged to be paid as soon as possible but, in any case,

. before breaking bulk (abbreviated as 'bbb').

If the Charter Party stipulates that 'the freight is payable bbb', it means that the Owners will receive the 'freight' before the hatches are open and the discharging operation is in progress.

However, when it stipulates 'freight payable within .... banking days but in case bbb', the Charterers must pay the 'freight' within the agreed banking days, but

if the Charterers' bank is unable to remit the 'freight' within these days (due to holidays or other reasons), the payment should be made in any case at the latest 'bbb'.
In the case of violation of the 'freight payment term'
. under the English Law,
the Owners have the right to lien the cargo.
The Owners should take advice from their legal department, on whether they should lien the cargo before discharge and keep it in the ship's holds or if they should discharge and store it in the port, awaiting the court's decision.
If marked as
b- 'freight prepaid',
under the English Law, the Owners have no right to lien the cargo for the non-payment of 'freight' and demurrage and are obliged to deliver the cargo and complete the voyage.
In this case, it should be agreed clearly in the Charter Party, that:
- the 'freight' is payable 100 percent without deductions of possible despatch at load port,
- the possible demurrage incurred at load port will be paid together with the payment of the 'freight',
- the Bills of Lading will not be released to any party without the Owners' prior written consent and in accordance with the terms of the Charter Party, the same should be kept in the custody of the attending Agents or Owners' representatives as agreed in the Charter Party, until the immediate payment has been irrevocably effected or the amount due has been credited to the Owners' bank account.
Another method of payment could also be agreed to be used as follows:
c- 'on delivery',
the 'freight' will be paid during the delivery of the cargo to Receivers or as otherwise agreed upon in the Charter Party.
d- 'upon arrival and before the delivery of the cargo,
the 'freight' is payable on tendering of the N.O.R. and before the commencement of the discharging operation.
e- '7 days before the arrival at the port of discharge',
the 'freight' is payable 7 days before the ship's arrival at the discharge port as well as many other less frequently used forms of payment.

The percentage of the 'freight' payment is directly linked with:
- the Bills of Lading and how they have been marked.
- the agreed demurrage clause and if 'free despatch' is in effect or not.
For Bills of Lading marked as 'freight payable as per Charter Party' and when the 'half despatch' term is in effect, the Charterers will seek to secure the payment

of possible despatch, which might be incurred at load and discharge port/s, by keeping a certain percentage of the due gross freight, which usually lies between 5 to 10 percent, i.e. the freight is payable 95 pct (or 90 pct, as the case may be) of the due gross freight etc.
This withheld 'freight' plus possible demurrage or possible less despatch incurred, will be paid to the Owners within an agreed period, after the completion of the voyage and after the Owners' presentation of all the required port documents/Time Sheets/S.O.F/final freight statement etc.
However, when marked as 'freight prepaid' and irrespective of the agreed demurrage term, or marked 'freight payable as per Charter Party' but with the 'free despatch' term, then there is no reason for the Charterers to withhold any amount and the contract term of 'freight is payable 100 pct' must be in effect in the Charter Party. However, the Charterers should reserve the right from the 100 percent gross payment, in order to deduct the agreed 'address commission' and/or the 'extra insurance' or other expenses which are for the Owners' account and will be covered by them.

- 'Freight' payment in loss of ship:

The 'freight' should be paid even if the ship is lost during loading or after signing the Bills of Lading, as will be agreed upon.
The following forms of agreement could be in effect:
a-'full freight deemed earned upon shipment discountless and non-returnable ship and or cargo lost or not lost' (abbreviated as 'ffdeusdanrsaoclonl').
The gross freight is due upon commencement of loading, whether the ship and/or cargo is lost or not.
b-'full freight deemed earned on shipment discountless and non-returnable ship and or cargo lost or not lost' (abbreviated as 'ffdeosdanrsaoclonl').
The gross freight is due on completion of loading, whether the ship and/or cargo is lost or not.
c- 'full freight deemed earned on signing Bills of Lading discountless and non-returnable ship and or cargo lost or not lost' (abbreviated as 'ffdeosboldanrsaoclonl').
The gross freight is due upon signing the Bill of Lading, whether the ship and/or cargo is lost or not.

-Deadfreight:

Under the terms of the Charter Party,
. the Charterers are the party liable for the payment of any deadfreight due to Owners.

If, the loading operation was completed, and the freight is payable per metric ton of cargo loaded (not on a lumpsum basis), but without the ship having loaded the declared by the Master contractual cargo quantity allowed under the terms of the Charter Party (the ship has short-laden),

. because Charterers or Shippers were unable, for various reason, to present the fully declared amount of cargo for loading,

or

. due to the Charterers' incorrect or inaccurate declaration of the cargo stowage factor (higher cargo stowage factor), which resulted to 'full up' the ship's cargo holds, but without obtaining the cargo weight expected for loading, the Owners then have the right to claim for dead-freight payment.

The payment of the deadfreight depends on

. whether the ship could actually load the declared

but not presented cargo quantity and was not prevented by the port restrictions/ship's stability etc.

When deadfreight payment is in effect, the lay-time allowed at all ports should be calculated

. on the paid quantity of cargo and not on the actual loaded quantity,

i.e., if the cargo loaded onboard is 5,000 mtns but the Charterers paid for 5,200 mtns, then the lay-time will be calculated based on a cargo quantity of 5,200 mtns.

The Charter Party should specify, if the Owners have the right to clause or delay the release of the Bills of Lading, until the Charterers accept and pay the alleged 'deadfreight'.

The Charterers should incorporate in the Charter Party, that the ship will not sail from the port, unless the Charterers or their Agents are notified at the time of occurrence. It should, however, be agreed that the time lost waiting for the Charterers' reply will count as lay-time.

11. Demurrage/Despatch.

The time used, in excess of the allowed lay-time under the terms of the Charter Party, that is required for completion of the specific operation in the contacted port, is referred to as

. 'demurrage'.

The Owners will be compensated for all the delays incurred, with a daily demurrage rate or pro rata for any part of the day, as agreed and incorporated in the Charter Party.

Time lost due to bad weather or strikes in the ports or exception periods in accordance with the Charter Party terms or failure of the ship's machinery/equipment or breakdown of the ship's cranes etc. will not count as lay-time and will

delay the completion of the operation, as well as the entry of the ship on demurrage. However, when the ship is actually entered on demurrage (demurrage commences), no exception periods are in effect and time lost due to bad weather/force majeure etc. will count as demurrage time.
The same is known as
. *'once on demurrage, always on demurrage'.*
The continued running of the time on demurrage, could be prevented:
- if the time lost is directly attributable to the Owners'/Master's fault, which resulted in the ship not being in full use.
- if it is clearly worded in the Charter Party that some specific periods will not count as lay-time, even if the ship is already on demurrage, i.e., shifting time from the anchorage to the berth, also time required the surveyors to carry out the draft survey etc.
The payment of the demurrage lies on and will
. only to be paid by the Charterers,
as will clearly be stated in the Charter Party, however, the Owners should keep the right to lien the cargo (subject to Bill of Lading and how it is marked), in the event of a Charterers' default.
The demurrage rate will be proposed by the Owners during the presentation of their initial offer and the Charterers will accept it or negotiate the same at a different rate.

The amount paid as demurrage, covers the ship's
- daily running expenses,
- the Owners' daily profit,
- the increased port expenses, due to the ship's prolonged stay in the port,
and
- the extra consumed quantity of fuels, due to ship's prolonged stay in the port.

The amount of demurrage is subject to brokerage commissions, which is payable by the Owners to the brokers involved in the particular fixture.
If the time actually needed and used for completion of the specific operation in the contacted port is shorter than the allowed lay-time, then, in accordance with the terms of the Charter Party, all time saved or working time saved, is referred to as
. 'despatch'.
The ship will be at the Owners' disposal for the next employment at a date earlier than had been calculated/planned, but the Charterers should be entitled to claim for despatch monies, which are usually calculated at half the demurrage rate and are certainly not subject to any brokerage commission.

The claims for demurrage/despatch will be settled after the lay-time calculation is made, which will be based on the signed Statement of Facts.
In the Charter Party, the demurrage/despatch term could be agreed as follows:
a- Demurrage USD .... free despatch.
b- Demurrage USD...... with half despatch for all working time saved both ends.
c- Demurrage USD .... with half despatch for all time saved both ends.
d-Demurrage USD.... /Despatch USD....... for all working time saved (despatch rate which does not reflect the half demurrage rate).
There are also agreements for small ships based on:
e-Demurrage USD .... per cargo ton loaded.

a= Demurrage USD.... free dispatch.
It is mainly in effect for fixtures made on small/handy sized ships. However, and when the market is good, it could be achieved for all sizes.
In general, and having agreed this term, the Owners are the only party, which will benefit from the ship's earlier date of completion.
b= Demurrage USD .... with half despatch for all working time saved.
It is mainly in effect for fixtures made on handy-max sized and larger ships. The working time saved at the load and/or discharge port will be multiplied with half the amount agreed as demurrage rate and the resulting amount will be paid as despatch monies to Charterers by the Owners.
c= Demurrage USD...... with half despatch for all time saved both ends.
A term, which could cost a lot to the Owners. The exception periods will be counted as time saved and the Owners can lose a huge amount if the demurrage rate is high.
Therefore, the Owners should squeeze the amount of the demurrage rate as much as possible, when they know that the port can maintain the contractual loading or discharging Charter Party speed and no delays are expected.
d=Demurrage USD.... /Despatch USD....... for all working time saved. (despatch rate
which does not reflect the half demurrage rate).
A fixed despatch rate, which does not necessarily, reflects the half demurrage rate. The Charter Party might have an agreed amount as demurrage rate, but as despatch the one/third of the demurrage rate or any other amount.

The effect of the above terms could be illustrated by the following simple lay-time calculation resulting to demurrage:

Charter Party terms:
- cargo quantity for loading: 32,000 mtns.
- daily loading rate: 8,000 mtns.

- demurrage rate: USD 8,000 pdpr / Despatch ....
- N.O.R. will be given within working hours from 08:00 to 17:00 from Monday to Friday.
- the lay-time will commence counting from 14:00 hrs. of the same day if the N.O.R. is tendered before noon and at 08:00 hrs. of the next working day if tendered in the afternoon.
- the time from Friday 17:00 hrs. until Monday 08:00 hrs. not to count even if used (SSHEX EIU).
- the time used before the commencement of lay-time not to count as lay-time.

As per the signed Statement of Facts (SOF):
- the ship arrived/berthed and tendered the N.O.R. at load port on Tuesday 7th May at 09:00 hrs.
- the ship commenced the loading operation on Tuesday 7th May at 14:00 hrs.
- the ship completed the loading operation on Friday 10th May at 14:00 hrs.
- no rain periods or other interruptions that could affect the counting of the lay-time have been inserted in the SOF.

Considering the following parameters:
- the lay-time allowed for loading is 4 days (32,000 mtns: 8,000 mtns = 4 days).
- the N.O.R. given at 09:00 hrs. on 7th May and the lay-time will commence counting at 14:00 hrs. of the same day, even if the loading operation commenced earlier, then the allowed lay-time for completion of the loading operation will be calculated as follows:
-Tuesday 7th May 10 hours (from 14:00 hrs. to 24:00 hrs.).
-Wednesday 8th May 24 hrs. (from 00:01 hrs. to 24:00 hrs.).
-Thursday 9th May 24 hrs. (from 00:01 hrs. to 24:00 hrs.).
-Friday 10th May 17 hours (from 00:01 hrs. to 17:00 hrs.) (the ship completed the loading at 14:00 hrs.).
-Saturday 11th May no work /exception period.
-Sunday 12th May no work/ exception period.
- Monday 13th May 16 hrs. (from 08:00 hrs. to 24:00 hrs.).
- Tuesday 14th May 05 hrs. (from 00:01 hrs. to 05:00 hrs.).

Total 96 hours (4 days). The lay-time expires on Tuesday at 05:00 hrs.

Time that will be used beyond Tuesday at 05:00 hrs. will count as demurrage, instead, time that is saved from completion of loading until Tuesday at 05:00 will count as despatch.

A per SOF the ship completed the loading on Friday 10th May at 14:00 hrs. Obviously despatch money is in effect.

The despatch amount due to Charterers:

a- Demurrage USD 8,000/free dispatch.

The ship completed earlier, however, the term 'free despatch' prevails. The Owners *will pay nothing* as despatch money and will benefit from the ship's

earlier departure.
b- Demurrage USD 8,000/half despatch for working time saved both ends.
From Friday at 14:00 hrs. (completion of loading) until Tuesday at 05:00 hrs. (allowed lay-time) the working lay-time saved will be calculated as follows:
-Friday 10th May from 14:00 hrs. until 17:00 hrs. = 3 hrs.
Friday 10th May from 17:00 hrs. until 24:00 hrs. = 0 hrs. -exception period.
-Saturday 11th May from 00:01 hrs. until 24:00 hrs. = 0 hrs. -exception period.
-Sunday 12th May from 00:01 hrs. until 24:00 hrs. = 0 hrs. -exception period.
-Monday 13th May from 00:01 hrs. until 08:00 hrs. = 0 hrs. -exception period.
Monday 13th May from 08:00 hrs. until 24:00 hrs. = 16 hrs.
-Tuesday 14th May from 00:01 hrs. until 05:00 hrs. = 5 hrs.
3 hrs+16 hrs.+ 5 hrs. = total 24 hrs./ 1day x USD 4,000 (half demurrage rate) = USD 4,000 are due to Charterers.
c- Demurrage USD 8,000/half despatch for all time saved.
From Friday at 14:00 hrs. (completion of loading) until Tuesday at 05:00 hrs. (allowed lay-time) the all-time saved will be calculated as follows:
-Friday 10th May from 14:00 hrs. until 24:00 hrs. = 10 hrs.
-Saturday 11th May from 00:01 hrs. until 24:00 hrs. = 24 hours.
-Sunday 12th May from 00:01 hrs. until 24:00 hrs. = 24 hours.
-Monday 13th May from 00:01 hrs. until 24:00 hrs. = 24 hrs.
-Tuesday 14th May from 00:01 hrs. until 05:00 hrs. = 5 hrs.
10 hrs.+ 24 hrs.+ 24 hrs. +24 hrs. + 5 hrs. = total 87 hrs./ 3, 6256 days x USD 4,000 = USD 14,502.40 are due to Charterers.
d- Demurrage USD 8,000/Despatch USD 2,500 for working time saved.
From Friday at 14:00 hrs. (completion of loading) until Tuesday at 05:00 hrs. (allowed lay-time) the working lay-time saved will be calculated as follows:
-Friday 10th May from 14:00 hrs. until 17:00 hrs. = 3 hrs.
Friday 10th May from 17:00 hrs. until 24:00 hrs. = 0 hrs. exception period.
-Saturday 11th May from 00:01 hrs. until 24:00 hrs. = 0 hrs. exception period.
-Sunday 12th May from 00:01 hrs. until 24:00 hrs. = 0 hrs. exception period.
-Monday 13th May from 00:01 hrs. until 08:00 hrs. = 0 hrs. exception period.
Monday 13th May from 08:00 hrs. until 24:00 hrs. = 16 hrs.
-Tuesday 14th May from 00:01 hrs. until 05:00 hrs. = 5 hrs.
3 hrs+16 hrs.+ 5 hrs. = total 24 hrs./ 1day x USD 2,500 = USD 2,500 are due to Charterers.

12. Original Bill of Lading.

The Bills of Lading will be mainly signed as
. 'freight payable as per Charter Party',
or as

. 'freight prepaid' or 'freight paid'

The Bills of Lading have the following key characteristics, which exist for both on Voyage and Time Charter parties:

A.

The Bill of Lading is a legal document of title that acts as proof of the transportation contract between the Carriers and the Shippers.

It is a document which includes the four following key elements of information:

- *the description of the type/grade of the cargo loaded.*
- *the date of shipment (i.e., the date of completion of the loading operation).*
- *by whom the cargo was loaded and to whom the cargo will be delivered.*
- *the conditions under which the transportation will be executed.*

By the Bill of Lading, a contract of carriage is created, and the parties are identified, i.e., Shippers, Receivers, Carriers etc. and the carriage is identified, i.e., the load and discharge port.

When signed, it should be the link between the Owners and the Buyers/Consignees of the cargo, who are to receive their cargo upon the ship's arrival at the discharge port and the presentation of one originally issued Bill of Lading.

B.

There are two basic types:

1. Straight Bill of Lading:

It is a non-negotiable document, which clearly states that the cargo is consigned to a named Consignee/ assigned to a specific party (as stated in the Letter of Credit. The Bills of Lading will only be endorsed by this named Consignee and the Shippers do not have the right to alter the Bill of Lading.

2. 'To Order' Bill of Lading:

It is a negotiable document, where the title to the cargo is conferred to the 'order' of the Shipper or to the order of a bank (it is usually the party named in the Letter Of Credit) and can be transferred by one party to another by endorsement, in order to assign a title to the next party, i.e. the goods are consigned to a non-designated party.

C.

It will be issued by the Shippers or by the ship's Agents on behalf of the Carriers only after the completion of the loading of the cargo and completion of the customs formalities. After signing, the Carrier, who will appear on the Bill of Lading, acknowledges receipt of the cargo, which will remain under his custody until the time of delivery to the named Consignees and the ship is allowed to sail from the port towards its final destination. However, the Carrier does not have to be the head Owner of the ship, but even the Charterers or the Freight Forwarders or any other party that actually assumes the responsibility for the carriage, even for matters that they have no direct control over (e.g., navigation,

safety etc.). Obviously, there should be no objection from the P&I Clubs about the name of the carrier, but Carriers unknown to the Owners, who are not covered by a reliable P&I Club coverage, or who have a bad reputation in the market,
. can spell big troubles for the ship.
D.
Regardless of who the Carriers are, the Master is the only person
. who will sign the Bill or authorize another party to sign, but always on his behalf.
E.
The Bill of Lading is frequently issued in a set of
. three originals in standardized forms.
In the port of discharge, the party who firstly presents
. an endorsed original Bill of Lading is entitled to receive the cargo from the ship. The other two issued originals will become invalid for the receipt of the cargo.
F.
The Bill of Lading will be issued in
. strict conformity with the Mate's Receipts,
with regard to the quantity, description and condition of the cargo, as well as the date of shipment.
The Master has no obligation to agree to sign Bills containing an inaccurate or incorrect description of goods.
He should also strictly refuse to sign any
. ante-/post-dated bill of lading.
After the signing, all three original Bills of Lading (a full set) will be returned to the Shippers/Agents. The Master will keep a non-negotiable copy on board.
G.
If the ship has sailed before the Bill is issued and the Charter Party so expressly states it, the Master will allow another party involved (usually the Charterers' or ship's Agents or the ship's P&I Club's representatives or as should be agreed during the negotiations) to sign the Bill, but always on his behalf. He will hand over to the authorized party
. a Letter of Authorization,
which will contain the full and proper instructions (as per Master's wording) and the conditions under which they will sign the Bill of Lading. The Letter of Authorization must be fully accepted and signed by the authorized party.
Many Charter Parties include the term that the Bill of Lading, before being signed and released by the authorized party, will be faxed/e-mailed directly to the Owners' office for approval.
H.
The most usable type of Bill of Lading contains the following standard 'boxes':

-the Shippers' name.
-the Consignees' name.
-the notify party.
-the ship's name.
-the load port.
-the discharge port.
-the apparent condition of the goods.
-the cargo description and the loaded weight.
-the place of issuance and the date of shipment.
-the number of original Bills of Lading issued.
-the Condition of Carriage.

I.

Normal practice is for the ship to discharge the cargo and deliver it to the person entitled, who before the commencement of the discharging operation, will first hand over to the Master an endorsed original Bill of Lading, as issued at load port.

Any deviation from this practice

. may be subject to the P&I Club's rules,

mainly due to the imminent risk of the cargo being delivered to the wrong people (wrongful delivery of goods) or in general that they are conducting an inappropriate delivery of goods.

If the Consignees are unable to hand over the original Bill of Lading to the Master, thus enabling them to take their cargo, they will present to the Master of the ship

. a Letter of Indemnity (a guarantee letter abbreviated as 'l.o.i.'),

which covers the Owners from the consequences of such a delivery of goods. Nevertheless, if there is no such provision/option in the Charter Party and it is not customary in the port of contact for the ship to discharge with a l.o.i., it is at the commercial decision of the Owners/Master, whether to accept or to reject such a way of delivery of goods.

Potential complications and delays at the discharge port, due to the Owners' refusal to discharge without presentation of the original Bill of Lading, will lead the Charterers to agree to have the option to discharge without the presentation of the original Bill of Lading, from the time of commencement of the negotiations. They will ask for themselves or the Consignees to have the option to provide, before the discharging commencement, a letter of indemnity duly signed by the providing party. Many times, the fixture

. is subject to acceptance of the same by the Owners.

The l.o.i. will usually be issued in a

. standardized form,

as drafted by the International Group of P&I Clubs for the Owners'/ operators' facilitation, which gives to Owners the right to appeal against the party issuing the same. Even if those standard forms are drafted by the P&I Clubs, the ship will remain subject to P&I Club's insurance, in case of wrong delivery of the cargo and in case the l.o.i. is found to be inadequate, the risk and loss will rest with the Owners. It is important that the issued l.o.i. should also be counter-signed by a first-class bank, which, however, almost never happens and in most of the times it is rejected by the Charterers. When a bank joins the l.o.i., it will insist on a monetary limit, which in any case must not be less than 200 percent of the market value of the cargo, at the time of cargo delivery. The bank will pay any amount up to the limit of its liability, provided the party that signed the l.o.i. fails to provide bail or other security.

A l.o.i. may also be issued for delivery of cargo in a port other than that stated on the Bill of Lading or for signing a 'clean on board' when the cargo is not in apparent good order and condition.

In general, Owners, before accepting an l.o.i., must obtain advice from the ship's P&I Club, as in fact for ANY acceptance of a l.o.i. might make the ship fall outside the P&I Club's insurance. The Charterers' reliability/credibility for accepting an l.o.i. must be to the Owners' full satisfaction.

J.

Sometimes the Shippers/Charterers will ask the Master to keep on board one original Bill of Lading from the total issued.

In this case, it is a rule that all originally issued Bills of Lading

. must be properly marked,

that an original Bill of Lading is kept on board and that the cargo delivery can be properly carried out with instructions from the Shippers/Charterers. However, this method of issuance is not recommended by the P&I Clubs and the ship may be subject to rules. The Master will deliver the original Bill of Lading relating to the intended Consignees for their endorsement. The Owners, in the case that the Bill is issued 'to order', have a risk of unauthorized cargo delivery, as other Consignees could claim for delivery of cargo against presentation of one original Bill of Lading (from the other two originally issued Bills of Lading that have circulated) that has gone through the regular bank channels.

K.

Some types of Bills of Lading omit some important terms, such as the Paramount clause, the Arbitration clause etc. Care should be taken when negotiating the terms of the Charter Party and the Owners' protective clauses should be clearly agreed upon. The Owners must ensure that they are not exposed to risks that exceed the terms of the Charter Party and there should be a provision in the Bills explicitly obliging the Charterers/Shippers

. to incorporate all the terms, conditions, liberties, warranties and exceptions of the Voyage Charter Party, including the lien and arbitration clause in the Bills.

That will allow the Owners to defend against Bill of Lading claims based on the Charter Party clauses and to be able to seek an indemnity under the Charter Party terms. It will also give notice to any innocent third-party holder of the Bills that the Bills

. are subject to the Charter Party terms and conditions.

The lien of the cargo against the Receivers for unpaid due freight/demurrage/ hires from the Charterers could be subject to the Owners proving that the Bill of Lading incorporates a lien clause that has the same effect as under the lien clause of the governing Charter Party, therefore it making necessary the incorporation of the Charter Party terms and conditions into the issued Bill of Lading.

When Congen form Bills are to be used, if they are of the 1994 form, they will include on the reverse side a provision expressly incorporating the law and jurisdiction clause of the Charter Party, of which the date appears on the front of the Bills. As a matter of English Law, this is normally sufficient for its terms and conditions to be in effect to the Bills. The provisions of the Hague/ Hague-Visby rules and any amendments thereto (the Paramount clause general), the New Both-to-Blame Collision clause, the New Jason clause, the General average clause and the War risk clause are to be incorporated into all issued Bills of Lading.

L.

Under a single Bill of Lading, the delivery of the cargo can only be made to one consignee. However, in bulk trades, the cargo might be requested to be split between several Receivers. If such a request is made by the Charterers, then the Owners should take care to ensure that they have followed the correct procedures.

The 1st full set (all 3/3) of the original Bills of Lading must be delivered to the Owners' hands before the Charterers issue the new Bills of Lading. The Charterers must not have the right to change the total quantity of cargo/cargo description and the date of shipment. The New Bills will conform to all terms and conditions and exceptions of the originally issued in load port Bill/s of Lading and will not prejudice the ship-Owners' rights. Once the replacement Bills of Lading have been issued, then the cargo that is stated in that particular Bill of Lading should only be delivered to the Receivers named in the new Bill of Lading.

There should be also a risk of one or more of the Receivers receiving more or less cargo than they must receive. A provision in the Charter Party should state that in the case of more than one Bill of Lading have been issued to various Consignees, the Owners/Master and ship will not be responsible for delivery of the cargo to each Bill of Lading holder, however and without prejudice to the Charter Party Interclub Agreement clause, the Owners remain responsible for the total intaken cargo quantity loaded and discharged.

The Charterers may request 'Delivery Orders' to be issued instead of splitting the Bills of Lading.

The cargo interests will trade the cargo during the carriage and will deliver it without requiring Bills of Lading. The same precautions should also be in effect for delivery of the cargo with 'Delivery Orders'.
M.
If the Bill of Lading is issued to the 'order of a bank', the bank must counter sign the l.o.i. for discharging without presentation of the original Bills of Lading or to confirm directly to the Owners their acceptance/permission to discharge the cargo, i.e., to confirm delivery instructions, even though the Charterers have performed this exchange with a l.o.i. As per the issued Bill of Lading, the bank is the holder of the Bills of Lading in question. The bank that made the payment in respect to the goods holds the Bills of Lading as security, pending payment in turn upon acceptance of the Bills and shipping documents under a Letter of Credit issued by the Receivers' bank. If the Receivers' bank rejects the documents, they will not make the payment and if the cargo has been discharged/ delivered, then the bank might have a claim against the ship/Owners, as the holder of the Bills of Lading.
If issued 'to order of shippers', then the Shippers endorse the Bill of Lading and order as to whom the cargo should be delivered.
N.
The issued original Bills of Lading must contain the correct cargo quantity, which has been loaded on board the ship.
At the load port, the Shippers
. are obliged to supply the loading figures and to warrant those figures.
The same should be presented either by a carried out
. draft survey or by the shore scale measurements.
Problems may arise at the load port, in the case that the shippers' figures exceed the ship's figures or the Receivers' figures are less than the ship's figures (Bill of Lading figures) at the discharge port.
In the port of shipment and if there is a difference on the loaded cargo quantity beyond what is considered normal and/or customary (the trade allowance is 0,5 percent on the loaded cargo quantity) and the Shippers insist that the Bill of Lading will be issued in accordance with their findings (basically that happens when Shippers supply shore scale measurements), but the Master or his representative
. has been able to determine the cargo quantity on board and there are reasonable grounds for disputing the accuracy of the shore figures, i.e., the draft survey was carried out in a calm sea, the ship was not rolling and pitching, there were no strong currents etc. and in general the draft survey was carried out in perfect conditions, under the Hague and Hague-Visby Rules Article III Rule 3
. he is not bound to accept the supplied figures to be inserted as intaken cargo quantity.

The ship should strictly follow the P&I Club's recommendations and at least the original Bill of Lading should be claused (marked) with
. the ship's figures.
As an alternative, some reservations could be made on the Bill of Lading such as 'weight unknown' or 'quantity and quality unknown', said to be' etc. or to accept a Letter of Indemnity by the Shippers or Charterers, but in fact the Owners should be aware that such reservations might not have effect in many jurisdictions and/ or the ship could be subject to the P&I Club's rules, as it has accepted incorrect figures.

In general, the Owners and Master take considerable risks, which might lead to the ship falling outside the P&I Club's Insurance, if accepting:
1.
a l.o.i. due to non-presentation of original Bills of Lading. The Owners are of the risk of wrongful delivery of cargo or, in general, that they are conducting an inappropriate delivery of goods.
2.
a l.o.i. for change of destination, without collection of the full set as issued in the load port. The Owners are of the risk that an original Bill of Lading is presented for the cargo in the port as stated in the original Bill of Lading.
3.
a l.o.i. for 'clean on board' for cargo which appears not to be in sound condition. The Master of the ship must only sign Bills of Lading which accurately describe the condition of the cargo, or, in any other case, the Owners are exposed to receive huge claims.
4.
to sign ante-/post-dated Bills of Lading.
. *Never do it.*
The date of shipment shown in the Bill of Lading will determine the value of the cargo in the sales contract. Ante-/post-dated Bills of Lading will affect the cargo value and the Owners could be held fully liable, with huge consequences.
The correct date of shipment is the date of completion of loading or otherwise, the date of completion of each parcel, if so, required by the Charterers. The Bill of Lading can be issued before the full completion of loading, provided that the Bill of Lading reflects the cargo loaded when issued.
5.
to sign a Bill of Lading containing an inaccurate or incorrect description of the cargo and cargo loaded figures. Reservations, which could be made on the Bill of Lading such as 'weight unknown' or 'quantity and quality unknown' or 'said to be' etc. or accepting a Letter of Indemnity by the Shippers or Charterers, but such reservations might not have effect in many jurisdictions and/or the ship

could be subject to the P&I Club's rules, as having accepted an inaccurate or incorrect cargo description and the cargo loaded figures.

6.

to sign a Bill of Lading incorporating the Cargo Value. The P&I Clubs reject ad valorem Bills of Lading.

7.

to sign a Bill of Lading not adhering to the originally issued Mate's Receipts with regard to the quantity, description, condition of the cargo, as well as the date of shipment.

The Charter Party should explicitly state that the original Bills of Lading will be issued in strict compliance with the Mate's receipts or Tally clerks.

8.

not to sign identical Bills of Lading which exist for the shipment, i.e., if all three (3/3) original Bills of Lading (the full set), as issued for the cargo on board the ship, have not been issued identically and have not been marked as 'original'.

9.

to sign outside of the designated box or place for signature in the Bill of Lading, which could be considered as a specific acceptance of the incorporation in the place of signature.

10.

to carry in the ship's bag one original Bill of Lading from the total issued. The Owners, in the case that the Bill is issued 'to order', have a risk of unauthorized cargo delivery, as other consignees could claim for delivery of the cargo against the presentation of one original Bill of Lading (from the other two issued and circulated original Bills of Lading) that has gone through the regular bank channels.

11.

to sign a Bill of Lading for deck cargo. The Owners, when loading deck cargo, except on certain specific circumstances (containers/logs etc.), might be in breach of the contract of carriage and the ship may fall outside of the P&I Club's rules.

The Master should avoid signing Bills of Lading (otherwise they must keep reservations until they are satisfied), which incorporate:

1.

A Charter Party date other than the Charter Party made by the head Owners.

If the Owners fix with the Operators who appeared in the market as Charterers, then, after completion of loading, the Agents might present a Bill of Lading incorporating the date of the head Charter Party, which was made between the Charterers (Operators) and the head Charterers.

However, the Owners, either on Voyage or on Time Charter, are not party to the Sub-Charter Party and ignorant of its terms.

The Owners must ensure that they are not exposed to risks that exceed the

terms of their Charter Party and there should be a provision explicitly obliging the Charterers/Shippers to incorporate all the terms, conditions, liberties, warranties, and exceptions of the Charter Party in the Bills made between Owners and Operators, including the lien and arbitration clause.
In the case that the Bills are issued without the incorporation of the Charter Party made by the head Owners, then that would be a breach of the Charter Party terms and the Owners can hold the Charterers fully liable for any consequences and losses/damages flowing from that and any Bills will be issued under protest. The Owners are, in the circumstances at least, entitled to see a full copy of the Voyage Charter Party made between the Charterers and head Charterers, as this is to be incorporated into the Bill and will form the terms of the contract of carriage. The Owners must ensure that the carriage is covered and that the contract is governed and construed in accordance with the English Law unless otherwise stated in the Charter Party and the disputes will be referred to arbitration in London unless otherwise stated in the Charter Party.
2.
Letter of Credit (L/C) numbers or references to Letter of Credit on the face of the Bill of Lading / ad valorem Bill of Lading.
The question remains whether a reference of Letter of Credit number in a Bill of Lading, would have the effect of making the BL an ad valorem bill, and could prejudice Owners' interests and P&I Club's cover.
This is by reason of the fact that inserting the L/C number on the face of the B/L, could be interpreted in a number of jurisdictions as incorporating the terms of the Letter of Credit, including the agreed values of the cargo therein and thus not being able to limit Owners' liability under the Hague -Visby Rules, should the occasion arise.
Some P+I Clubs recommend that the mere reference to a Letter of Credit on the face of the B/L does not make it an ad valorem B/L, and a problem should arise, only in cases where the value of the cargo has been specifically declared and inserted in the Bill of Lading.
In general, it would be always preferable to avoid the L/C reference numbers in the description of the goods.
In all cases, if the Owners have commercially decided to accept it and accommodate the Charterers, then a reservation should be taken in order to minimize the consequences. This reservation, along with the L/C number or the references, should preferably go in the top right of the Bill (just under 'reference no' of the 'Congenbill') and not in the box referring to the cargo description.
The reservation could be worded as follows (both are a precise copy from executed Bills of Lading):
'For information only and inserted at the request of the shippers/consignees/endorsees/holders/receivers of/under this bill of lading for their convenience and

administrative purposes to identify this shipment only and is not a declaration of value. The value of the cargo and the terms of any L/C are unknown to the carrier/ship/master /owners of the ship. This is not an ad valorem bill of lading.'
or:
'The letter of Credit details noted above are for reference purposes only. The letter of Credit itself and its terms and conditions are not expressly or impliedly incorporated or deemed to be incorporated into or from part of this Bill of Lading'.
When the value of the cargo is inserted, the reservation could be worded as follows (precise copy from a Bill of Lading):
'For shipper's information and reference purposes only. No representation as to the value of the cargo carried under this Bill of Lading is made by the carrier and this is not an ad valorem Bill of Lading'.
3.
As 'carrier' other than the actual Owners.
Carriers unknown to the Owners, who are not covered by a reliable P&I Club coverage, or have a bad reputation in the market, can give rise to huge consequences against the ship and the ship may be arrested at the port by third parties, who have claimed against the named carriers appearing in the original Bills of Lading for various reasons.

13. Dates of shipment/Lay-days/Cancelling.

The agreement on the Main terms and Charter Party details is subject to readiness of the cargo within the agreed lay-days. The date of shipment should be in compliance with the Shippers' readiness to load and the carriage of cargo for export near to the port.
The unreadiness of the cargo, the congestion at the port or the Charterers' difficulty to fix a ship in the right position could lead to the shipment's postponement.

The Lay-days/Cancelling (abbreviated as' lay-can' or 'l/c')
- is the range of dates, when the cargo is expected to be ready for export and within which the ship must be presented for loading.
'Lay-days' is the first day when the ship can tender the N.O.R. and the 'Cancelling' date is the last date when the N.O.R could be tendered until 24:00 hrs., unless otherwise stated in the Charter Party.

A N.O.R. tendered earlier than the first lay-day specified in the Charter Party or later than the last day, is invalid and the ship cannot be considered as an arrived ship and proceed for loading.
In a few words,

- the Charterers have no contractual obligation to accept the ship for an earlier loading and will additionally reserve the right, under the terms of the Charter Party, to cancel the fixture/ship in the case of a late arrival.
However, a provision in the Charter Party should state that an earlier arrival is to be accepted only after the Charterers' consent, and, if accepted, the conditions under which it should be made.
In addition, the cancelling clause will clearly state, when the ship will be considered as cancelled, in the event of a late arrival.
If the ship is delayed due to the fault of the Owners/ship (obviously, delays due to adverse weather conditions or delays in completion of discharge of previous discharge port beyond Owners' fault etc. are always excluded), the Owners are exposed to receiving claim from the Charterers for loses/ damages/extra expense (such as additional rental cost for the storage of cargo, terminal charges, sales contract penalties etc.) that they might suffer, due to the delay of the ship's arrival.

14. Determination of the loaded and discharged quantities of cargo/Draft's Survey.

At the load port the Shippers
. are obliged to supply the loading figures and to warrant those figures.
The same should be presented either by a carried out
. draft survey or by the shore scale measurements.
At the discharge port the Receivers are to determine the discharged cargo quantity
. by a carried out draft survey or by shore measurements or even by weighting the trucks at the customs house.
Problems could arise at the load port, in the case that the shippers' figures exceed the ship's figures or the Receivers' figures are less than the ship's figures at the discharge port and both are beyond a normal and/or customary difference (exceeding the trade allowance) on the loaded and/or discharged cargo quantity.
At the load port, some reservations which could be made on the Bill of Lading such as 'weight unknown' or 'quantity and quality unknown', 'said to be' etc. or accepting a Letter of Indemnity by Shippers or Charterers, might not be enough to cover the Owners and the ship could be subject to the P&I Club's rules.
However, if the cargo quantity is going to be determined only by a draft survey in both ends, it is common and for the avoidance of any arguments, for the Charterers'/Shippers' draft surveyors to conduct the draft survey jointly with the Master/Chief Officer of the ship and/or with the Owners' appointed draft surveyors.
At the discharge port, the Owners are not obliged to accept the shore measurements or to be involved in the weighting of the trucks. In all cases, the Master should carry out his own draft survey and the ship's P&I Club's assistance and/

or recommendations should be requested, especially in places and Charter Parties, where the draft survey is not to be taken into consideration for the determination of the cargo quantity.
The ship must be in possession of all ship's plans, enabling the surveyors to carry out a correct draft survey. Owners/Master should take care, before the ship approaches the ports, for the plimsoll marks and draft marks on the port and starboard side, bow, amid ship and stern are to be clearly cut and marked on shell platting and to be always readable.
For an accurate draft survey, the ship's tanks must be empty or fully pressed, otherwise, the Master should inform the surveyors of the remaining ballast on them. During the time that the draft survey is being carried out, the bunkering operation or the ballasting/de-ballasting of the ship's tanks or generally the shifting of liquids from one tank to the other is not allowed.
The
. wrong/improper measurement
of the cargo quantity onboard the ship in load port should result in the Owners sustaining heavy delays/expenses, due to the short or surplus cargo delivery at the port of discharge.
In 'difficult' areas/ports, it is recommended that the ship's P&I Club's appointed surveyor should represent the Owners.

The Owners/ship is the only contracting party that is mainly liable for cargo claims to third parties on Voyage Charter.

The ship must be covered by a first-class P&I Club, which should be ready to issue a letter of undertaking upon request, in order to release the ship of the detention of an alleged cargo claim (in most of the times for cargo shortage), as might be presented by the relevant parties in the discharge port.
The time that is required for the surveyors to carry out the operation, or the adverse weather conditions/rough seas or even the late appearance of the surveyors, could cause significant delays in the sailing of the ship and Owners could suffer not only from the lost time, but also from the increased port expenses (tugs/ pilots working overtime) or even due to nighttime not allowing the ship to sail until daylight.

15. Age of the ship-Extra insurance (abbreviated as 'exins') or additional premiums (abbreviated as 'ap').

Age of the ship:

When the ship exceeds 15 years of age it is characterized as an 'over-aged' ship

and could be excluded from certain specific biz due:
- to the Letter of Credit terms,
- to heavy extra insurance which is imposed on the ship by the cargo insurance companies.
When the market drops, many L/C require non over-aged ships, as plenty of them are available in the market. In such a market, the Charterers, even if there is no age limit imposed by the L/C, search and fix the youngest possible ship, close to their 'freight ideas'. However, in a promising market where there are businesses available for all ships of any age, the L/C and Charterers conform accordingly.

Extra insurances (abbreviated as 'exins'):

For overaged ships, a cost for 'extra insurances' could be in effect, beyond those normal insurances in effect even for non-overaged ships. For older ships (more than 20 or 25 years old), the extra cargo insurance imposed could be heavier. To enable a fixture to be solved, the Charterers could request the Owners to cover the same in full against vouchers, or to contribute with a lumpsum amount on them. This affects the ship's daily income and if the imposed exins are not reasonable, it could lead the Owners to dropping the biz.
Extra insurance could also be imposed due to the ship's flag/ownership etc.

The extra insurance term could be expressed as follows:
a-free of extra insurance (abbreviated as 'free exins').
b-full extra insurances will be for the ship's account.
c-extra insurance for the Charterers account, but the Owners to contribute with USD..... (or Owners will pay maximum USD ...as extra insurance).

a-free of extra insurance (abbreviated as 'free exins').
The Owners will pay nothing even if the ship is overaged. This term is essential for the Owners of almost all coasters/small handy sized ships, which would otherwise be difficult to be fixed.
b-full extra insurances will be for the ship's account.
The Owners will fully cover the cost, to such amount as will be imposed on the ship. However, original vouchers should be asked to be presented to the Owners by the Charterers. When such a term is agreed upon, Charterers will obviously seek to deduct the relevant amount from the initial 'freight' payment.
c-extra insurance for the Charterers account, but the Owners to contribute with USD.... (or Owners will pay maximum USD ...as extra insurance).
The Owners will pay part of the imposed extra insurance and the balance will be covered by the Charterers. Again, original vouchers should be asked to be pre-

sented to the Owners. Charterers usually deduct the relevant agreed amount from the initial 'freight' payment.

16. Stevedore damage.

For the majority of the Voyage Charters and especially for small sized ships, the Charter Party incorporates the term that the stevedore damage caused to the ship should be settled directly between the Owners and the respect stevedoring company.
However, and in the case that the ship intends to load heavy and/or long pieces of cargo or wood or steel cargo in general, including pig iron/scrap or to discharge with a very large grabs, but its hatches that are not as wide, then it is possible for the ship to sustain heavy stevedore damage, resulting in the ship being considered inoperative and un-seaworthy.
Generally, the damage mainly occurs during the loading or discharging with the use of shore grabs or when using payloaders/bulldozers in the holds of the ship for the purpose of stowing or removing the cargo from the sides of the ship and carrying it to places easily accessible to the grabs.
Damage could be caused to the ship's holds' ladders, frames, hoppers, tank top, hatchcovers etc. and to the cargo.
Since such damage to the ship is not covered under the ship's Hull and Machinery insurance, the stevedoring clause becomes an important clause, and the Owners must take extra care and agree to a full and comprehensive clause.
The stevedore damage should be categorized as follows:
- slight damage, which can be repaired on the spot by stevedores without causing any delays to the ship's sailing
- damage not affecting the seaworthiness/cargoworthiness of the ship and which can be repaired by the crew at a later stage or together with the ship's next dry-docking. The damage will be estimated and be paid by the stevedores prior to sailing or the Charterers will be liable for the payment when they are repaired.
- damage affecting the ship's class and/or seaworthiness and/or cargo worthiness and which must be repaired prior to sailing from the port and in any case prior to loading the next cargo.
Care should be taken in the ship's holds, where all vulnerable parts must be properly protected by the Owners for possible damage from the Shippers'/Receivers' means of operation.

The Charter Party clause could be worded as follows:
The Stevedoring company will be appointed and paid by the Charterers, but will work under the supervision of the Master, if it is possible.

Charterers will be liable and responsible for any damage caused by the stevedores to the cargo or ship or its fittings at the ports of contact during this Voyage Charter, including hidden damage, which should be reported as soon as detected, but in any case, before the ship's sailing from the last discharge port.
However, the Owners guarantee that all vulnerable parts in the ship's holds will be properly protected under their time and expenses.
If the ship suffers stevedore damage, the Master of the ship:
1. will report to the Charterers or their Agents in writing within 24 hours of occurrence or from the time of discovery, but in any case, before the ship sails from the port.
2. will hold the stevedores or the respective parties responsible and liable for any damage, in writing, with copies sent to the Charterers or their Agents. He will work with the Charterers or their Agents and will make every effort to reach a written acknowledgement from the parties responsible for causing the damage, unless the damage has in the meantime been repaired. If the Master is unable to obtain acceptance of liability and responsibility from the party who caused the damage, he is not allowed to sail from the port before coming into contact with the Charterers, allowing them to sort the matters out and the time lost will continue to count as lay-time or time on demurrage.
3. will try to get stevedores to repair the damage and try to settle the matter directly with them in the first stage and the time lost will continue to count as lay-time or time on demurrage.
For stevedore damage which:
a). affects the effectiveness of the ship and the safe operation and/or safety of the crew and/or affect the trading capabilities and/or seaworthiness and/or cargo worthiness and/or is subject to classification requirements, then such damage as will be ascertained by the ship's Class surveyor, who will be appointed by the Owners at the Charterers' cost, is to be repaired by the Charterers at their own expenses/risk, to the Master's and/or Class surveyor's entire satisfaction and the time lost will continue to count as lay-time or time on demurrage until completion of all repairs and the 'port clearance' is issued and delivered to the ship.
Charterers are to reimburse the Owners for the Class Surveyor's fees, once the Owners have submitted their final freight invoice, supported with the original vouchers.
The damage will be repaired at the place of occurrence, if it is possible and there are adequate facilities for permanent repairs, or otherwise, temporary repairs are to be made on the spot or as when detected, at the Charterers' risk and expenses and to the Master's and/or Class surveyor's satisfaction and the time lost will continue to count as lay-time or time on demurrage. The ship will sail or shift to the nearest shipyard, as chosen by the Charterers, with fuel consumed at sea

or during shifting to be for the Charterers' account, which will be settled together with the Owners' final freight statement, in order to complete the repairs.
The time lost during the sailing or shifting to the shipyard and until completion of the works with the damage being repaired to the Master's and/or Class surveyor's entire satisfaction and the 'port clearance' is issued and delivered to the ship or, at the Owners' option, at equidistance from the discharge port, will count as lay-time or time on demurrage.
Shipyard's expenses and/or port expenses, including in and out tugs and/or pilots and/or agency fees etc., will be entirely for the Charterers' account.
b). does not affect the effectiveness of the ship and the safe operation and/or the safety of the crew and/or the trading capabilities and/or the seaworthiness or cargo worthiness and/or is subject to classification requirements and cannot be repaired by the stevedores, the Owners then undertake to accept the ship's sailing, in accordance with the terms of the Charter Party.
However, the said damage is to be assessed before the ship's departure from the discharge port and the costs to be agreed upon between the Master, Owners, Charterers and the respective survey company and to be paid together with the Owners' final freight statement. The survey company will be mutually appointed by the Owners and Charterers, if considered necessary, at the Charterers', Owners' equal expenses and will promptly inform the Charterers, Owners with the damage report, which will be issued before the ship sails, with a copy of all correspondence exchanged in this connection. The time lost will continue to count as lay-time or time on demurrage, until the 'port clearance' is issued and delivered to the ship.
Owners have the option of having the damage repaired at their next convenient time or at the ship's next Dry-Docking.
In the case that the Master, Owners, Charterers cannot reach an agreement on the cost with the respective survey company, the Charterers will then undertake to reimburse for the actual cost of the repairs, but not for the time lost, upon presentation of the original bills from the repairers or dockyard, that are consistent with the surveyors' report on the extent of such damage.
c). is small and has been repaired by the crew, the Charterers will reimburse the Owners for the cost on request, but not for the time lost and upon presentation of the Owners' repair statement, which is acceptable even by fax or e-mail.
However, and in any case:
- the Charterers are responsible for any directly related expenses resulting therefrom and incurred thereby and must pay compensation for any damage caused to the ship and/or restore the ship to its initial condition, under their own time and expenses, whether or not any payment has been made by the respective parties and/or stevedores to them.
- the Charterers will not to be liable for damage when the Master has accepted a

statement in writing from the stevedores and/or respective parties that the operations were carried out without any damage, or that the repairs were made to his satisfaction.

17. Taxes/Dues.

Some countries impose certain taxes/dues, which are calculated/levied on the export cargo which is loaded onboard and/or on import cargo, as well as, on the ship and/or on the freight beneficiary calculated on the Charter Party freight (freight tax). The taxes/dues are payable to income tax authorities, before the ship's sailing from the port, through the attending Agents and are actually calculated separately from the ordinary charged ship's port expenses. The Owners, before fixing the ship, should request the Agents of the port to inform them of any taxes that are in effect on the ship/freight/cargo/ownership etc. enabling them to adjust their offer to the Charterers accordingly.

18. Fumigation.

Fumigation work could be ordered because of
-the illness of the ship's crew,
-the ship's previous trading,
-the cargo holds/hatches rejection, due to live insects in/on them,
-the Charterers' cargo loaded or intended to be loaded.
All fumigations should be performed as per IMO regulations/recommendations and the Owners/ Master should take all the required precautions for the safety of the ship's crew.

The Port Authorities could request for the ship's crew to be away from the ship during the fumigation, in which case if the fumigation has been ordered by the Charterers/Shippers for the cargo loaded or intended to be loaded, then all lodging expenses should be agreed to be for the Charterers' account and the time to continue to count as lay-time or time on demurrage.
The fumigation Authorities/fumigation Company should provide in writing the fumigation plan which will take place on ship. After completion it should also provide in writing full and detailed instructions for the sea voyage to the Master for the crew's safety and for the period of time that the ventilation of holds is not allowed, as well as, when the holds must be opened, enabling the discharging operation to commence for the crew's/stevedores' safety.
In general, and as per the P&I Club's recommendations the fumigators should provide written documentation in respect to the following:

- Pre-fumigation inspection certificate
- Standard safety recommendations for ships with fumigated grain cargoes.
- Manufacturers' information or safety data sheet.
- First aid and medical treatment instructions.
- Fumigation certificate.
- Fumigation plan.
- Statement of vessel suitability for fumigation and fumigant application compliance.
- Detailed Precautions and procedures during voyage and for the period of time that the ventilation of holds is not allowed, as well as, when the holds must be opened enabling the discharging operation to commence for the crews'/stevedores' safety and all information required for the crew's safety during the voyage.
- Precautions and procedures during discharge.

The Owners should officially seek the ship's P&I Club's recommendations about the fumigation procedures or to how to clause the Bill of Lading, in case the cargo appears to not be in sound condition after the fumigation has been performed.

The Charter Party clause could be worded as follows:
Owners are responsible for all costs, extra expenses, lodging expenses, delays, detention that may arise, if fumigation is ordered because of:
-live or dead insects found in the ship's holds or compartments during the holds' inspection
-Officers' and/or Crew's illness
-ship's previous trade
The lay-time or time on demurrage will cease from the time of which the fumigation is ordered, until the fumigation certificate is granted, fully certifying that the ship is in full operational condition and free to load its cargo.
However, for all other ordered fumigations by the Charterers and/or Shippers and/ or by any other party and for any reason, which are related to the cargo intended to be loaded or already loaded, the Charterers remain responsible for all costs and extra expense incurred, including port expenses, delays, risks, detention that may arise and the time lost for fumigation will continue to count as lay-time or time on demurrage.
If the Port Authorities or the Master deem it necessary for the Officers and/or crew to stay ashore, all lodging expenses then, including expenses for transportation and meals will be entirely for the Charterers' account, which are payable directly to Agents and the time lost will continue to count as lay-time or time on demurrage.

All fumigations will be performed as per IMO regulations and recommendations and under the Master's supervision with respect to crew safety.
If the cargo is fumigated after the completion of loading, then the fumigation Authorities and/or fumigation company will provide in writing:
- full and detailed instructions for the sea voyage, to the Master's satisfaction,
- it will state the period of time that the ventilation of holds is not allowed,
-when the holds must be opened, enabling the discharging operation to commence, for the Crew's and/or Stevedores' safety
and
- all other information required for the Crew's safety during the sea voyage and during the discharging operation.
The Master will record in the ship's log all the operations and in the case that the cargo that was loaded appears not to be in sound condition and has been damaged or contaminated after the performed fumigation, then the original Bills of Lading will be issued strictly in accordance with the ship's P&I Club's recommendations.
The Charterers will be liable and responsible for any cargo damage/impact on the cargo quality whatsoever, as a result of fumigation of the cargo and/or cargo holds and will hold the Owners harmless and keep them indemnified against all claims whatsoever, which may arise and be made by any party against them.
If the ship is detained by any party and generally its sailing is not allowed from the port, then the Charterers undertake the responsibility to release the ship, by taking all the necessary and appropriate measures (issuance of security etc.), as the case may be, or arise and any expenses incurred, including but not limited to additional port expenses/Agency fees etc. will be for the Charterers' account payable on request and upon presentation by the Owners of the original vouchers, which are acceptable even by fax or e-mail. The time lost will continue to count as lay-time or time on demurrage until the ship is free to sail from the port and the 'port clearance' is issued and delivered to the ship.

19. Agents.

When a ship arrives at a port for cargo operation or for any other reason (repairs, to take bunkers or spares, to change the crew etc.) the appointment of Agents is necessary and mandatory.
They will take care of all issues, along with facilitating the smooth process of the ship in the port and its sailing by arranging tugs/pilots/port clearance etc. They will provide useful information on the prevailing restrictions at the port, on facilities, on the cargo etc. and keep the interested parties fully informed of the ship's movements and the progress of the loading and discharging operation.
They will arrange for the documentation of the cargo, for crew changes (trans-

portations, visas etc.) and hospitalization, for providing cash to the Master, as well as many other tasks.
One of their main duties is to issue and sign along with the Master and sometimes with the Shippers or Receivers, the Statement of Facts (S.O.F.), which records all the events from the time of arrival at the port until the last pilot disembarks from the ship, after completion of the operation.
Most of the times the Agents will issue the original Bill of Lading on behalf of the Owners and will be the party who will sign the Bill of Lading on the Master's behalf (subject to Charter Party agreement).

In the Charter Party the appointment of the Agency will be agreed as follows:
a- Owners' Agents at Both Ends (abbreviated as 'OABE')
b- Charterers' Agents at Both Ends (abbreviated as 'CHABE')
c- Owners' Agents at Load port /Charterers' agents at Discharge port (or the opposite)

a- Owners Agents at Both Ends:
The agents, who will be attending, will have been selected and appointed by the Owners. This term is mainly achievable for ships of small sizes or when liner terms are in effect. The Owners have the flexibility to negotiate with various agency offices in every intended port and should feel more comfortable that their interests will be well protected. Additionally, they have the advantage of minimizing certain costs or even reducing the tariff of the standard charged agency fees.
b- Charterers' Agents at Both Ends:
The Agents, who will be attending, will have been selected by the Charterers and appointed by the Owners. In fact, the Shippers or Receivers are the party who will select the Agents, acting on their confidence that everything will run without problem at the port, with the cargo being loaded or delivered to them smoothly. Especially at the loadport, the Agents are the party who will issue and will most of the times sign the original Bills of Lading on the Master's behalf, and therefore the Agents should be well known to the Shippers. In this case there are not many chances for the Owners to negotiate the costs and they will be required to pay the same according to the port tariffs.
Protective agents:
If the Agents selected by the Charterers are not known to the Owners and they consider that their interests are not being protected, Owners have the option of selecting and appointing other Agents of their own choice, who will act as the Owners'
. 'protective Agents',
unless the same is not allowed under the terms of the Charter Party.

The 'protective Agents' will take care of the port charges and all husbandry issues will be completed through them. The Agency fees should be limited, however, the agency fees for the officially appointed Agents will be paid in accordance with the port tariffs.
c- Owners' Agents at Load port/Charterers' Agents at Discharge port (or the opposite):
Owners will select and appoint the attending Agents at the load port and the Charterers at the discharge port or the opposite.

20. Notices.

Charterers and Shippers or Receivers should be informed before the ship's arrival at the port of operation, of its 'estimated time of arrival' (abbreviated as 'e.t.a.').
The notices at the load port should assist Shippers in the correct time of presenting the cargo at the port for loading and/or for arranging for the stand-by of stevedores/trucks etc. for the immediate commencement of the respect operation. At the discharge port, for the Receivers' readiness to receive the cargo and present in a timely manner the means of cargo transportation from the port to their mines (by lorries/wagons etc.).
It will be initially tendered as an
. 'approximate' and then finally as a 'definite' notice.
With the 'on fixing' or the 'approximate' notices, the ship's position is still uncertain and may change at any time. Therefore, reservations should be kept by the Master/Owners even when tendering the 'approximate' notices that given in good faith basis and according to the current available information and subject to unforeseen circumstances/weather condition and subject to completion of cargo discharging operation at its current port.
On the other hand, the 'definite' notices, that are usually less than three days before the expected date of the ship's arrival, declare that the ship's position is clear and if a delay occurs then it will have been due to unforeseen circumstances (continued adverse weather conditions, accidents etc.). The 'definite' notices increase the responsibilities of the party tendering the same.
If the notices have not been given correctly or omitted by the party responsible, the other party may be entitled to submit claim for losses suffered, due to breach of this clause.

The Charter party for any purpose and for the interested parties to be properly informed and protected could incorporate that:

Written notices of the ship's expected arrival at the load port will be tendered

to the Charterers and Agents on fixing, then .../ .../.../... days approximate and .../.../... hours definite notices.
Owners'/Master's tendered 'approximate' notice, with the ship's estimated time of arrival at load port has changed by more than 1 day, then the Master/Owners are to keep the Charterers and attending Agents closely informed on the ship's exact position/full itinerary and cause of delay.
When the Master tenders the definite notices, he will state the expected time of arrival/draft on arrival, the anticipated draft on sailing and the expected loadable cargo quantity.
Owners/Master will remain responsible for the accuracy of the tendered 'definite' notices, which can be changed only by adverse weather conditions and/or unforeseen circumstances and/or delays in completion of cargo discharging operations at the current port, always beyond Owners'/Master's/ ship's fault. If has changed by more than 12 hours, then the Master/Owners are to keep the Charterers and attending Agents closely informed on the ship's exact position, full itinerary, and cause of delay.
If any willful misrepresentation of the definite notice is tendered by the Owners/ Master of the ship to the Charterers and/or Agents for the position and readiness of the ship, they will remain responsible and liable for any and all consequences/ time lost/loss of turn for berthing until the berth becomes available again/additional costs including, but not limited to, the cost for cargo storage in the port, for the cancelled stevedoring shift etc.
On sailing from the load port, the Owners/Master will tender notices to the Charterers and Agents at the discharge port with the estimated time of arrival, the quantity of cargo loaded and the anticipated ship's draft on arrival.
.../.../... days approximate and .../.../... hours definite notices will be tendered thereafter.
The Charterers and Agents are to be kept by the Master/Owners continuously and fully advised of any change in the ship's position and its estimated time of arrival at the discharge port. When the Master tenders the definite notices, he will state the expected time of arrival and Owners/Master will remain responsible for the accuracy of the tendered 'definite' notices, which can be changed only by adverse weather conditions and/or unforeseen circumstances, always beyond Owners'/ Master's/ ship's fault.

21. The type of Charter Party/ brokerage commissions.

The type of Charter Party should be to the Owners' knowledge, as soon as possible, as well as, if a private or a standard form will be in use.
There are various types of Charter Parties that have been adopted or issued or recommended by various maritime organizations, covering the Voyage Charter

and the type/grade of cargo for carriage.
These offerings have the advantage that their printed terms are known to the parties interested and can assist in the easier conclusion of the fixture.
The Gencon Charter Party is mainly adopted for general cargoes, such as bagged cement etc., and provides favourable terms for the Owners. For grain cargo the Norgrain or the Synacomex will be in use, the Sugar Charter Party for sugar cargo, the Phosphate Charter Party (Africanphos) for the phosphate cargo, the Ferticon/Fertivoy for fertilizer cargo etc.
There are also private Charter Party forms, as issued by various large Charterers' houses.
The Owners are the only party paying the nominal 1, 25 percent to each broker involved in the fixture, from the gross freight/dead-freight/demurrages paid by Charterers and the address commission (usually 2,5 percent for Voyage Charter) as will be charged by the Charterers.
The involvement of many brokers is not preferred by the Owners, as the total commission will not be kept to a minimum and will affect the freight to a great extent (heavy deductions) and will therefore decrease the daily income and profit of the ship.
A provision could be incorporated on the Voyage Charter Party, stating that if the ship does not execute the voyage and depending on the party at fault, at least one third of the brokerage on the estimated amount of freight must be paid by the Owners or Charterers (cancellations due to force majeure reasons will be excluded), as compensation for the expenses and provided work.

VOYAGE ESTIMATION-DAILY REVENUE PERFORMANCE

The calculation of the Voyage Estimation, which will result in the expected daily revenue performance of the ship, will be made by the Owners and it is the essential step before the Owners make an offer to the Charterers.
The older manual methods of calculation have been replaced by automated systems that are faster, eliminate mistakes and provide additional information, maps etc. The Owners will easily check the initial Voyage Estimation against the actual voyage outcome, after completion of the discharge/voyage.

For the Owners to be able to make the Voyage Estimation, which is based on the Charterers' 'freight ideas' as given by them or on the basis of the rate required to execute the voyage, they should know at least:
- the ship's daily quantity of fuel consumption at sea/port.
- the existing fuel prices at the area of the shipment.
- the ship's cargo capacity for the intended cargo.
- the cargo details (trimming, stowage factor etc.).

- the port of loading and discharging and the rotation if more than one load or discharge port applies.
- the distance/s between the ports.
- the port expenses and the possible taxes/dues that are in effect.
- the loading and discharging speed and the exception periods.
- the existing brokerage commissions.

Owners should take into consideration the 'running expenses' of the ship, which are the standard daily expenses, such as:
- crew's wages, provisions, repairs, spares, repatriations etc.
- insurances of the ship (H&M, War Risk, P&I Club etc.).
- lubricants and fresh water.
- office's expenses (staff wages/insurance/rental etc.).
- installments on loans obtained from the banks for the purchase of the ship.

The 'cargo order' entered the market and reached the Owners as follows:
-Account ....
Cargo: 50,000/10 pct moloo bulk nickel ore
1sp/sb Thessaloniki Greece 12,5 mtrs sw draft / 1sp/sb Shanghai China no port restrictions
12,000 mtns pwwd sshinc/ 20,000 mtns pwwd sshinc
12 hours tt / 12 hours tt
Lay/can ......
Gearless/ oa ok
3,75 ttl commission
Ideas USD 35 pmtn

The basic terms are known, allowing the Owners to draft the initial Voyage Estimation and make an offer to Charterers.
However, they should have their reservations for checking the actual port expenses and whether any further port restrictions except the draft at the load port are in effect.
The interested ship provides the following characteristics:
-About 58,500 dwt on about 12,20 mtrs sswd.
-Speed/consumption: ballast 13,50 knots on 28 mtns VLSFO (Very Low Sulphur Fuel Oil).
laden 13,50 knots on 31 mtns VLSFO.
at sea/port/idle 2,5 mtns LSMGO (Low Sulphur Marine Gasoil).
Ship's present position: Alexandria Egypt under discharge.
The fastest route to execute the voyage is: Alexandria-Thessaloniki-via Suez Canal-Shanghai port.
The ship can load all the available quantity of cargo, i.e. 55,000 mtns.
Charterers' 'freight ideas' is USD 35 pmtn of cargo loaded.

The gross amount of freight is 55,000 mtns x USD 35 = USD 1,925.000
The calculation will be made as follows:
a). Distance in nautical miles
Distance:
Alexandria-Thessaloniki : 677 nmiles.
Thessaloniki-Suez-Shanghai: 8,063 nmiles.
Total 8,740 nmiles.
The ship's actual service speed in fair weather conditions is 13,50 knots. The weather factor and the adverse currents should be taken into consideration. For such a long trip an average service speed of 13 knots should be calculated.
Daily sailing speed 13 knots x 24 hrs. = 312 nmiles therefore 8740 nmiles : 312 nmiles = 28 days.
Or faster and easier: 8,740 miles: 13 knots: 24 hrs. = 28 days.
The sailing time from Alexandria to Thessaloniki to Shanghai through the Suez Canal is 28 days.
b). Loading/Discharging speed:
-Load.
The ship will load all the available cargo quantity of 55,000 mtns.
The loading rate is 12,000 mtns sshinc. No exception periods should be calculated.
55,000 mtns: 12,000 mtns daily = 4,58 days loading.
-Discharge.
The discharging speed is 20,000 mtns sshinc. No exception periods should be calculated.
55,000 mtns: 20,000 mtns daily = 2,75 days.
Total days required for loading/ discharging = 4,58 + 2,75 = 7,33 days.
c). Delays.
The following delays should be taken into consideration and be calculated accordingly:
In both ends the lay-time will commence 12 hours after tendering the Notice of Readiness (turn time). 12 hrs.+ 12 hrs. = 24 hrs. = 1 day.
Delay of crossing through the Suez Canal.
Suez Canal delay= 1 day.
Fuel should be required to be supplied to the ship, in order to execute the voyage. En route to Shanghai the most convenient bunkering stations with insignificant deviation are the ports of Suez and Singapore.
Delay for the supply of bunkers: 12 hours.
Total delays 1 day + 1 day + 12 hours = 2,5 days.
No rain or other delays at sea/port have been calculated at both ends. However, every Owner calculates a few days (usually 2/3 days) in addition to the actual days, in order to cover any unforeseen circumstances.

d). Fuel Cost:
-VLSFO calculation:
Due to the different consumption when the ship is trading in an empty or laden condition, a separate calculation should be made.
Steaming time from Alexandria to Thessaloniki in empty condition:
Distance 677 miles: 13 knots: 24 hrs. = 2,17 days.
VLSFO bunker consumption (empty condition) 28 mtns daily.
Price for VLSFO say USD 460 pmtn then the total VLSFO cost is:
2,17 days x 28 mtns daily x USD 460 = USD 27,949.
Steaming time from Thessaloniki to Shanghai in laden condition:
Distance 8,063 nm: 13 knots: 24 hrs. = 25,84 days.
IFO bunker consumption (laden condition) 31 mtns.
25,84 days x 31 tons x USD 460 ptn = USD 368,478.
Total VLSFO consumption cost for the voyage is: USD 396,427.
-LSMGO calculation:
The ship burns LSMGO 2,5 mtns at sea/port/idle. The total voyage duration should be taken into consideration as follows:
Sailing time 28 days + l/d 7,33 days + delays 2,5 days = grand total duration 37,83 days, i.e., 38 days
Therefore 38 days x 2,5 mtns LSMGO = 95 mtns LSMGO.
For port approaches/manoeuvres (entering/sailing ports) and for crossing the Suez Canal, the ship will burn additional LSMGO bunkers. i.e., load port 3 tons+ discharge port 3 tons + bunkering port 3 tons + Suez Canal 8 tons = total additional 17 mtns LSMGO
Total LSMGO consumption 95 mtns + 17 mtns = 112 mtons.
Price for LSMGO say USD 760 pmtn then the total LSMGO cost is:
112 mtons x USD 760 per mton = USD 85,120
Grand total VLSFO + LSMGO cost = USD 396,427 + USD 85,120 = USD 481,547
e). Port expenses/Canal Expenses:
Port expenses plus the cost of crossing the Suez Canal are approximately the following:
Load port: USD 35,000
Discharge port: USD 75,000
Bunkering station: USD 800 (agency fees/ launch etc.)
Suez Canal: USD 175,000
Total expenses USD 285,800.
f). Commissions:
The ship should load 55,000mtns. Charterers' ideas USD 35 per metric ton. The total brokerage commission is 3,75 pct on gross freight.
USD 35 pmtn x 55,000 mtns = USD 1,925.000 x 3,75 pct comm. =total commission USD 72,187.

To summarize:
'Freight' USD 35 pmtn x 55,000 mtns cargo = Gross 'Freight' USD 1,925.00
Total duration: 38 days
Expenses
a.: USD 396,427
d.: USD 85,120
e.: USD 285,800
f.: USD 72,187
Grand total expenses USD 839,534
Gross 'Freight' USD 1,925.000 – Total Expenses 839,534 = Net 'Freight' USD 1,085.466
The Net freight will be divided with the total duration of the voyage, i.e.
USD 1,085.466: 38 days = USD 28,564 daily.
The ship earns USD 28,564 daily gross (Time Charter Equivalent per day)
The income of USD 28,564 daily is Gross and the ship's 'running expenses' should be deducted accordingly.
Every Owner makes his own Voyage Estimation with the above or different way of calculation.

The 'cargo orders', which are based on Voyage Charter could enter the market as follows:

1.
Account (abbreviated as 'a/c') .... (Charterers name)
28/38,000 mtns rock phos in bulk (phos=phosphate cargo)
Casablanca/Pasadena 35 fw
16/22 Dec 2...
6/7 ttl days depending size
Fri 5pm-Monday 8am satshex eiu bends
Max loa 199 mtrs/beam 29,25 mtrs/ wltohc 38 feet
5 ttl incl 3,75 add comm

At the top of the cargo order the Charterers' name should be stated as a guide to the Owners. Many times, for commercial reasons, the name of the Charterers might not be disclosed and be replaced with the expression:
- 'fcc'
which means that the order is for the account of 'first class Charterers'.
The brokers involved in this biz, inform the market that, although the name of the Charterers is not or cannot be disclosed in the open market for commercial reasons, it is in fact dealing with a 'first class' name and the market should not worry about anything serious. However, this could include a wide range of re-

sponsibilities for the brokers
. such as assuring the market that it is dealing with A1 people.
In any case, the Charterers' name must be revealed upon the ship's commitment with them or upon their first counter offer.
Charterers with little activity, who have not been well established in the market, should be avoided being qualified as 'fcc'.
The expression as:
- 'direct',
or even
- 'direct/close',
could be enough for brokers quoting the biz and it lies on the Owners to check the Charterers' credibility and financial standing in the market.
As per the order, a handy/handymax-sized ship is required which should be capable of loading a cargo quantity of 28,000 to 38,000 metric tons, at the declared port restrictions (draft-loa-beam-wltohc).

A. Which ships are workable?
The order states:
28/38,000 mtns rock phos in bulk
Casablanca/Pasadena 35 fw
Sid/BC Max loa 199 mtrs/beam 29,25 mtrs/ wltohc 38 feet
16/22 Dec 2....
All singledeckers/bulk carriers that can accommodate a cargo quantity of 28,000 to 38,000 metric tons at the prevailing available draft at a discharge port of 35' of fresh water, could be considered workable by the Charterers.
No age restriction has been disclosed and geared or gearless ships are all workable.
The proposed ships should also comply with the rest of the port restrictions, expressly with the maximum length over all (LOA) of 199 meters, a beam maximum of 29,25 meters and the distance (the height) from the water line to the top of the hatch coaming (abbreviated as 'wltohc') which should not exceed 38 feet (about 11,60 meters).
Even if the Charterers have disclosed the port restrictions, the Owners remain responsible for checking themselves through the Agents at both ends for actual or further port restrictions which could prevent the ship's approach, unless the same is guaranteed by the Charterers through the progress of the negotiations.
B. Which ships have priority in working the biz?
The ship which:
a. is younger,
b. could meet the lay-days required for the shipment,
c. could accommodate the maximum quantity of cargo,

has priority in working the biz. The negotiated 'freight rate' will determine whether the Charterers should also look for alternatives which could offer cheaper rates.

C. Loading and discharging locations/type of cargo to be loaded.

The order states:

18/38,000 mtns rock phos

Casablanca/Pasadena 35 fw

The 'first port' as will be stated in the order is always

. the port of loading

and the port after the slash

. is the port of discharge.

The ship will load from the port of Casablanca located in Morocco.

The port of discharge is in Pasadena located in the Mississippi river in the United States of America.

The cargo is rock phosphate in bulk, a non-dangerous mineral fertilizer cargo, which in any case should not be exposed to rain. The cargo is heavily stowed into the ship's holds and the Master will load the ship according to the ship's maximum permissible weight on the availability of the draft prevailing in both ports.

D. Loading and discharging rate/lay-time calculation.

The order states:

6/7 ttl days depending size

Fri 5pm-Monday 8am satshex eiu bends

The loading and discharging speed is subject to the loadable cargo quantity.

For smaller sizes 6 total days should stand and for larger ship sizes, 7 total days. Saturdays/Sundays/Holidays are exception periods and the lay-time ceases to count from Friday at 17:00 hrs. until next Monday at 08:00 hrs. even if be used.

E. Type of Charter Party/total commission.

The order states:

5 ttl incl 3,75 add comm.

The type of Charter Party has not been disclosed and the same should be asked by the Owners. The commission is 5 percent to the broker quoting the biz which includes 3,75 percent as an address commission. The commission should be increased by 1,25 pct for any other broker involved in the fixture.

2.

Account......

Rouen/Rotterdam

3,000 mtns upto 6,000 mtns bulk wheat about 44'

Next week/onwards

Load/discharge dep size

Synacomex 90 with 4 pct

SID-Max 25 years – gearless ok

A. Which ships are workable?

The order states:

3,000 mtns 6,000 mtns bulk wheat about 44'

Next week/onwards

SID-Max 25 years – gearless ok

All singledecker ships (subject to port restrictions), which are of a maximum of 25 years, geared or gearless and could provide a cargo capacity between 3,000 mtns and 6,000 mtns, are potentially workable. Obviously, the ship, in order to load the minimum cargo quantity, must provide a minimum grain capacity of 132,000 cbft in cargo holds, i.e., 3,000 mtns x 44' = about 132,000 cbft grains. Charterers can load the ship during the next forthcoming week by the time the order entered the market or even a few days later.

B. Which ships have priority in working the biz?

The ship which

a. is younger,

b. could meet the lay-days required for the shipment,

c. could accommodate the maximum possible cargo quantity,

has the priority in working the biz.

C. Loading and discharging locations/type of cargo to be loaded.

The order states:

Rouen /Rotterdam

bulk wheat about 44'

Charterers will load the ship from the port of Rouen situated on the river Seine in northern France. The river ports are situated about 70/80 nautical miles from the estuary and the range is all tidal.

The Owners, before fixing, should be aware of the inbound and outbound allowable drafts according to the monthly tide tables, as issued by the pilots and Port Authorities.

The port of discharge is the port of Rotterdam, one of the biggest and busiest ports in the world, with a huge volume of cargoes loading and discharging.

The cargo is bulk wheat, one of the most common goods transported. Charterers have informed that the stowage factor of the cargo is 44', therefore, the ship must provide a minimum grain capacity of 132,000 cubic feet in the cargo holds, in order to load the minimum cargo quantity (3,000 mtns).

D. Loading and discharging rate/lay-time calculation.

The order states:

Load/discharge dep size

Charterers have not disclosed the loading and the discharging speed. Owners cannot offer for the biz until they show their interest and present the characteristics of their ship to the Charterers. Depending on the size, the Charterers will inform the respective loading/discharging rates or the total days, which will stand for this size of ship.

E. Type of Charter Party/total commission.

The order states:

Synacomex 90 with 4 pct

The Synacomex 90 will be used, which is an amended Charter Party of the older Synacomex. It is the most common Charter Party in use, when loading bulk grain/ agricultural products etc. on Voyage Charter.

The commission is 4 percent in total to the brokers who quote the business. The commission should be increased by 1,25 percent for any other broker involved in the fixture. The address commission is not declared, and the Owners have the right to request to receive a breakdown of the total commission, in order to prevent any overcharged commission, i.e., more than 1,25 percent for each broker involved.

3.

Account....

Cargo quantity 125,000 mtn 10pct moloo iron ore in bulk

Loading port 1sb 1sp out of Tubarao/Ubu/Sepetiba Bay at chopt

Discharging port 1sb 1sp Kaohsiung

Laydays Dec 25 2.... - Jan 5 2.... (try earlier)

Loading terms: 40,000 mtns per wwd shinc

12 hrs T/T unless earlier commence in which case ATUTC - Discharging terms: 38,000 mtns per wwd shinc

12 hrs T/T unless sooner commence in which case HATUTC

N.O.R. to be tendered atdnshinc at both ends

C/P details based on Charterers proforma

Chabe

All subs to be reconfirmed within 2 working days after main terms fixed

5 ttl com

A. Which ships are workable?

The order states:

Cargo quantity 125,000 mtn 10pct moloo iron ore in bulk

Laydays Dec 25 2... - Jan 5 2.... (try earlier)

On the basis of 10 percent of an upward or downward adjustment on the basic cargo quantity, the available amount of cargo is minimum 112,500 mtns to the maximum 137,500 mtns.

All Capesize ships which are in position and are interested in the biz with a dwat

from about 116,000 metric tons to about 140,000 metric tons or more, are potentially workable.
Charterers have not expressed any age limitation.
B. Which ships have priority in working the biz?
There should be a priority to the younger ships, which can meet the lay-days required for the shipment and can accommodate the maximum available quantity.
C. Loading and discharging locations/type of cargo to be loaded.
The order states:
Loading port 1sb 1sp out of Tubarao/Ubu/Sepetiba Bay at chopt
Discharging port 1sb 1sp Kaohsiung
iron ore in bulk
The load port will either be the port of Tubarao or Ubu (Ponta Ubu) or Sepetiba Bay, all located in Brazil. It is to Charterers option which port will be nominated as load port ('Charterers option' abbreviated as 'chopt') and the time of nomination will be mutually agreed during the progress of the negotiations.
The discharge port will be the port of Kaohsiung, which is located in Taiwan.
One berth will be used at each port of contact.
The cargo is iron ore in bulk. It is a mineral cargo which may be liquefied during the voyage, resulting in cargo shift and possible capsizing of the ship. Care must be taken by the Master/Owners before the commencement of the loading and especially in the areas where the cargo is exposed to rain without proper protection. The moisture content of the cargo must be measured/determined by the recognised international methods, which in any case must not be in excess of their Transportable Moisture Limit (TML).
In Brazil, due to very fast loading of such types of cargo and generally cargoes with a stowage factor above 01 ton/01 cbm, for ships aged equal or more than 18 years old, the condition survey is compulsory before the commencement of the loading (DPC Survey). This survey is valid for one year from the time of inspection.
D. Loading and discharging rate/lay-time calculation.
The order states:
Loading terms: 40,000 mtns per wwd shinc
12 hrs T/T unless earlier commence in which case ATUTC
Discharging terms: 38,000 mtns per wwd shinc
12 hrs T/T unless sooner commence in which case HATUTC
N.O.R. to be tendered atdnshinc at both ends
Load port:
The loading rate is 40,000 metric tons per weather working day Sundays/Holidays included. Therefore, all the Weekends/Holidays are to be considered as normal working days.

Due to very fast loading in Brazilian ports, the ship must load strictly as per provisions of SOLAS Chapter VI, regulation 7.3 (loading/unloading and stowage of solid bulk cargoes) and the Code of Practice for the Safe Loading and Unloading of Bulk Carriers (BLU Code). The Master and the terminal's representative should agree on a plan, which shall ensure that the permissible forces and moments on the ship are not exceeded during the whole loading.

The lay-time at the load port will commence counting at the end of the Turn Time of 12 hrs., unless the ship commences earlier. The Actual time used before the commencement of the lay-time will count as lay-time (abbreviated as 'ATUTC').

The Notice of Readiness (N.O.R.) can be tendered at all times, day, night, Sundays, Holidays included (abbreviated as 'atdnshinc').

Discharge port:

The discharging rate is 38,000 metric tons per weather working day, Sundays, Holidays included.

All the Weekends/Holidays are to be considered as normal working days.

The lay-time will commence counting at the end of the Turn Time of 12 hours, unless the ship commences earlier, in which case the Half Actual time used to count (abbreviated as 'HATUTC') as lay-time.

The Notice of Readiness (N.O.R.) can be tendered at all times, day, night, Sundays, Holidays included (atdnshinc).

<u>E. Type of Charter Party/total commission.</u>

The order states:

C/P details based on Charterers proforma.

Chabe

The type of the Charter Party is not disclosed. The Charter Party details will be worked on the Charterers' Charter Party proforma. The Owners should request a previously executed one to be presented.

The Charterers will select and appoint the attending Agents as 'Charterers Agents Both Ends' (abbreviated as 'chabe') which will be in effect.

PART II

CHAPTER II

HOW TO MAKE AN OFFER FOR A VOYAGE CHARTER

The forms of bids on Main Terms for Voyage Charter are much more complicated compared to those that are based on Time Charter. For Owners who agree to Charter their ships with this method of fixing (on Voyage Charter), the negotiations on the Main Terms constitute the largest and most important part of their efforts to conclude the fixture.
All the
. important terms should be discussed during the Main Terms
and nothing should be left to the Charter Party details (Charter Party proforma), which can later badly affect the results of the Voyage Estimation. This way of negotiations allow the Owners to make much more accurate calculations and take decisions before finalizing the Main Terms, whether they take the biz and further proceed to conclude the fixture or they leave it and continue searching for a better biz.
The Owners' calculation on the Voyage Estimation should be based on the most accurate data and in the absence of unforeseen circumstances, the expected daily revenue performance of the ship should be as close as possible to the results obtained from the final voyage analysis.
The Owners' initial firm offer and the Charterers' or Owners' following counter offers, should be negotiated with all the included important terms, as each term has its own validity and actually counts separately by adding or subtracting money.
The 'freight rate' is the most important negotiable point and the base for the conclusion of a fixture, however, it is not the only term to designate a biz as 'good'. It is just a good 'starter'. A successful fixture is made when a good 'freight

rate' is achieved and it is in line with the rest of the negotiable Main Terms. For example, if a very good 'freight rate' is achieved, but the provision of the 'w/w/ w/ w' has been omitted or otherwise a 'berth terms' Charter Party is fixed, the expected good profit could turn into huge losses, if the port/s are congested and the tendered N.O.R. is invalid until berthing.
The presentation of the ship and terms from the Owners to the Charterers, for their circulated 'cargo order', could be expressed in one of the following three forms (all in writing):
1). as a 'firm bid' or 'firm offer'.
For a biz which is 'firm' (the Letter of Credit is in order) and it is to the Owners interest, the ship's workability could be checked from the data given by the Charterers. If it is workable and 'free' to work any biz (there is no commitment to any other biz), the firm offer will be passed on and the negotiations on the presented terms will be in progress. The 'Firm offer' and subsequent 'counter offers' keep the Owners and Charterers committed for a specific period of time. It is not fair for the Owners to be working two cargo orders for the same ship, at the same time or for the Charterers to be working two ships for the same cargo order.
2). as an 'indication'.
For a business which has not yet materialized or if the ship is not 'free' to work on a 'firm' basis, the Owners will only make an "indication". There is no contractual or ethical obligation, but the Charterers simply take a note of the Owners' intentions for the biz at this time. The 'freight rate' may change when the Owners make a bid at a later stage, as it is subject to the market's direction and the availability of other firm cargoes in the area.
3). as a 'firm indication'.
There are no commitments and it just shows that the Owners are ready to offer for the biz, in accordance with the presented terms and to alter the same into a 'bid', upon the Charterers' readiness to work.
The 'indication' and 'firm indication' should include:
-the ship's main applicable description for the Voyage Charter
-the description and quantity of cargo and the allowed tolerance.
-the load port/s.
-the discharge port/s.
-the lay-can.
-the loading/discharging rate and the exception periods.
-the 'freight rate'.
-the total brokerage commissions.

The firm negotiations and the subsequent fixture will be concluded in two parts:
a. the Main Terms

The Owners' bid with its negotiated terms represents the Main Terms. The conclusion of the Main Terms is the agreement that is reached on some basic terms. Thus, the ship is theoretically near the conclusion of its fixture. There are no standard forms which could include standard recognised/ accepted terms by all the interested parties. Each voyage has its peculiarities by adding terms for the needs of the shipment and some of them are often modified and updated in accordance with the Arbitration awards that provide additional protection for both sides.

b. the Charter Party Details.

Following the agreement on the Main Terms, the parties that are involved, depending on the law which is in effect during the negotiations or according to shipping ethics, will be obliged or will be required to further proceed to the negotiation of the Charter Party proforma and its additional terms and conditions which are referred to as 'details'.

The Charterers will produce their Charter Party proforma (usually an already executed one that is based on a standard type of Charter Party), which is in conformity with the intended Voyage and cargo for transportation, for the likely comments of the Owners.

The intended type of Charter Party should be known to the Owners from the beginning of the negotiations, as the already known printed terms could make getting the fixture easier.

The most common Charter Party in use is the Gencon Charter Party, which is adopted for the transportation of general cargoes, minerals, cement etc. For grain cargoes the Norgrain or the Synacomex will be in use, the Sugar Charter Party for sugar cargo, the Phosphate Charter Party (Africanphos) for phosphate cargo, the Ferticon/Fertivoy for fertilizer cargo etc.

There are also private forms, but if the Owners have agreed on all the important terms during the negotiations of the Main Terms, they shouldn't worry about anything serious when they start working on the 'details'.

The proforma that is based on a standard type of Charter Party consists of two parts:

a). the Main Body

The Main Body of a standard type of Charter Party is a standard wording as adopted or issued or recommended by various shipping organizations and amended from time to time.

These terms set the limit of obligations and responsibilities of each contracting party. All printed terms are basically negotiable, and the Charterers/Owners can make their deletions/amendments/ additions in accordance with the prevailing trading conditions and circumstances.

b). the Rider or Additional Clauses
The Rider or Additional clauses cover any omissions of the main body or add new regulations as well as, other specific/special terms that are in effect on the voyage and/or to protect the Owners' and/or the Charterers' interests. Many terms allocate liabilities and responsibilities for specific risks to each contracting party and try to make each case "crystal clear', without doubts or room for any disagreement, which might otherwise lead to extra costs and lengthy litigations.

For the terms of the Charter Party nothing is taken for granted, fix it with crystal clear terms so as to prevent the contracting parties from referring to arbitration.

Until conclusion of the negotiations and subject to prevailing law to negotiations, each party, in case of disagreement or for other serious reasons, has the right to withdraw its interest from the biz/ship.
Particularly, under the terms of the English law and until the further completion of successful negotiations on the Charter Party 'details', there is no contract, and each party could 'escape' from the negotiations. However, this will be contrary to shipping ethics and can cost the Owners or Charterers, their reputation. Under the terms of the American law, the consent of 'subject to the details' on the Main Terms binds both parties to work on the same and to make an attempt to conclude the negotiations.
The written abbreviations that are used during the negotiations, including those on rider/ additional clauses, do not exist when the Charter Party is due to be issued. Whole words must be entered, which would in this case minimize disputes on any misunderstandings due to abbreviations.

Before the interested parties come to a conclusion of the agreement (final fixture), the following procedures should be followed:
a). the Charterers should request from their brokers to search for a suitable sizewise and datewise ship, which can load the intended type/grade/form of cargo and the minimum available cargo quantity, at the prevailing port restrictions. This is the first move before any further moves take place.
b). the ship should be in the right position and should agree to execute this particular biz on Voyage Charter with these Charterers.
c). pursuant to the customary procedure, a full offer should be made by the Owners to the Charterers, beginning the progress of the negotiations on the Main Terms which will hopefully be concluded.

The 'offer' will be given for the Charterers' consideration and for their subsequent acceptance or response within a specific period, after which
. it cannot be revoked or asked to be modified by the Owners.

The Charterers' clean acceptance, depending on the law applicable to the negotiations, could bind the Owners and Charterers to further develop on the Charter Party details.

The reply to the Owners' offer (as well as all subsequent replies from both parties), always within a specified period, is expressed
. as 'counter offer'
which should be given on an Accept/Except basis, i.e., the negotiating parties respond just with the terms they want to amend/ alter.
All other terms are to be considered as 'accepted' and any reference back to them for re-negotiation should be considered
. as 'back-trade',
which might result in the termination of the negotiations.

The negotiations on the Main Terms and the Owners'/Charterers' commitment could be terminated:
-if the Charterers did not reply at all to the Owners' offer for various reasons.
-if the Charterers decline the Owners' offer and send a binding offer for the Owners' acceptance/ counter-offer that is declined by the Owners.
-if any party replies out of the specified period, throughout the negotiations.
-if any party fails to clean accept the offer or the counter offer as given from the other party throughout the negotiations.

It would be unfair/contrary to the ethics of shipping and would cost the Owners' and Charterers' reputation, if the ship has been dually committed on a 'firm' basis or the Charterers have been working two ships simultaneously or they drop the negotiations for another cheaper ship or the Owners drop it for a better biz, without giving to the other party the
. 'first refusal'
i.e., a last chance to fix together on their terms before drop the negotiations.

When the negotiations are concluded (with the Charter Party details in order), the Charterers must confirm the final fixture within the deadline for the lifting of 'subjects', as will be agreed during the negotiations of the Main Terms.
-Subjects' or 'subs' for short,
are the Charterers' duly kept reservations to conclude the fixture.
They will present the ship, as the case may be, to the Shippers/Sellers and/or to Consignees/ Buyers, for approval of the ship's characteristics and will make the cargo ready (called as 'stem'= subject to enough merchandise).

If the Owners are to bid, they must be aware of:

a- the Charterers' name.
At the top of the cargo order, the Charterers' name should be disclosed as a guide to the Owners.
However, many times for commercial reasons, the name of the Charterers might not be disclosed and be replaced with the expression:
. 'fcc'
which means that the order is for the account of "first class' Charterers".
The brokers involved inform the market that, although the name of the Charterers is not or cannot be disclosed in the open market for commercial reasons, it is in fact dealing with a first-class name and the market should not worry about anything serious. However, this could include a wide range of responsibilities for the brokers
. such as assuring the market that it is dealing with A1 people,
who, if they eventually step down from their obligations, could cause them problems.
In any case, the Charterers' name must be revealed upon the ship's commitment with them or upon their first counter offer.
Charterers with little activity, which have not been well established in the market, should be avoided being qualified as 'fcc'.
The expression as:
. 'direct' or even 'direct/close'
could be enough for brokers quoting the biz and it lies on the Owners to check the Charterers' credibility and financial standing in the market.

b-The cargo form.
The form of the cargo for transport will determine the type of ship, which is the most workable for the particular biz. For example, for logs as cargo, the' logger' type is the ideal ship but if the cargo is in bags/pallets, then a multipurpose ship is preferred. For cargoes in bulk form, the singledeckers/ bulkcarriers (sid/bc) are the most workable ships.

c- the type and grade of the cargo/cargo quantity and its tolerance/stowage factor.
The basic cargo quantity and its actual stowage factor, in combination with the provided tolerance on quantity (the upwards or downwards adjustment) will determine the actual required ship size. When the cargoes are dangerous, belonging to the 'Group B' of the IMSBC Code, the ship should be approved and certified by its class for transportation of such types of cargoes.

d- the ports of loading/discharging and the existing port restrictions/port disbursements (port expenses).

The load and discharge ports or the fixed ranges should be known to the Owners. The distance between the ports is the first calculation point when the Owners make the initial Voyage Estimation. Obviously, the longer distance means extra fuel cost, additional running expenses and generally, a longer total duration rapidly increases the rate of the freight. Owners, before fixing, must be aware of the prevailing restrictions in the ports and of all information that Charterers have updated the market regarding the port restrictions and/or on taxes, etc., as well as, for the existing port expenses in all ports.

e- If there is any age limit to the ship.
If it is not required by the Letter of Credit terms, the fixture could be concluded, provided that the Owners accept to cover part or all of the extra insurance, as will be charged by the cargo insurance companies.

f- when the cargo is expected to be available and ready for shipment.
The readiness of the cargo, along with the available berth at this time, will determine the dates of the shipment. If the Owners request for an earlier or later shipment, then the Charterers will check if the ship's proposed dates for loading could be accepted by the Shippers and is allowed by the Letter of Credit terms.

g- the loading/discharging rate and the exception periods.
The distance between the ports, along with the speed of loading/discharging and the exception periods, will determine the duration of the voyage and will greatly affect the required freight rate.

h- the type of Charter Party and the existing total brokerage commissions.
The type of the Charter Party should be to the Owners' knowledge as soon as possible, as well as, whether a private or a standard form will be in use. The involvement of many brokers is not preferred by the Owners, as the total commission will not be kept to a minimum and will greatly affect the freight (heavy deductions) and therefore the daily income and profit of the ship.

In short, before the Owners send their offer, they should be aware of:
-the Charterers' name, otherwise relevant reservations should be kept,
-the cargo form,
-the type of cargo/cargo quantity and its tolerance/stowage factor,
-when the cargo is expected to be ready for shipment (lay-can),
-the loading/discharging ports or ranges and the port restrictions,
-the estimated port disbursements,
-the rate (speed) of loading/discharging and the exception periods,
-the type of the Charter Party and the existing total brokerage commissions,

and
-if there is any age limit which affects the workability of the ship.

During the negotiations, the Owners will decide if the particular biz is attractive and will endeavour to fix it, always taking into consideration:

- *the net daily income along with the duration of the voyage*
- *if the income is in line with the existing market*
- *expectations of the ship's repositioning in the area of discharge*

PREPARING TO OFFER.

The 'cargo order' has been circulated as follows:

Account ....
40,000 mtns bulk wheat about 44 with 10pct moloo
1 / 2 sb Cza/1 /2 sb Tarragona and/or 1 /2 sb Valencia
Lay-can 20/25 January 2....
4,000 sshex e.i.u./10,000 shinc
Reversible lay-time
Syna 90 chartrs proforma with 3,75pct ttl incl 2,50 pct add comm
Sid/BC -grd/grls max 25 years
chabe
ideas USD 19.00 pmtn 1/1

This particular 'cargo order' contains all data, allowing the Owners to check the ship's workability and to 'get the ball rolling' by sending a firm offer to the Charterers, through the preferred brokerage channel by those who spread the biz in the market, or, otherwise, through the channel that first called them and committed the ship for this particular biz.

THE OFFER.

The Owners' offer should basically include:
-the ship's basic description, as will apply for the particular voyage.
-the name of the Charterers or will otherwise keep the relevant reservations.
-when the 'subjects' will be lifted.
-the cargo description/quantity and tolerance.
-the dates for accepting the ship for loading.
-the load port and the maximum number of berths allowed for use.
-the discharge port and the maximum number of berths allowed for use.

-the 'freight rate'/when the payment will be made/how the Bills of Lading will be marked.
-if and when the freight is due in case the ship and/or cargo is lost.
-the loading rate and the exception periods.
-the discharging rate and the exception periods.
-when the Notice of Readiness is valid and when the lay-time will start to count.
-the provision of the w/w/w/w.
-the demurrage/despatch rate and if the lay-time between the ports is reversible or not.
-which party will cover the possible imposed taxes/ dues etc., on cargo/ship/ freight.
-which party will cover the imposed extra insurances on the age/flag/ownership etc.
-who will select the attending Agents in the ports of contact.
-who will bear the cost and loss of time during several operations in the ports.
-the number of notices which will be sent before the ship's arrival at the ports.
-if it will discharge on presentation of the original Bills of Lading.
-the stevedoring damage clause.
-the determination of the loaded and discharged cargo quantity.
-which law and arbitration forum will apply to negotiations and to the eventual fixture.
-the type of the Charter Party in use and the existing total brokerage commissions.

The Owners' offer could start as follows:

The name of the Charterers is Messrs. 'Eagle Trading and Shipping Inc'.
The name of the ship is 'MV TAXIARCHES'.
At the top of the message, the Owners should state which 'cargo order' (biz) they are referring to.

Ref Account Eagle Trading and Shipping Inc
40,000 mtns bulk wheat
Constanza/ Spain
MV TAXIARCHES

Reference is made to the above mentioned 'cargo order'.
The Owners are in the position to firm bid for reply here within 1 hour as follows:
or
The Owners offer you firm for 1 hour reply as follows:
or in any other form which clearly states that the ship is being offered and com-

mitted with them for a specific period of time.

The Owners continue:
"
MV TAXIARCHES
....
....
"

Comments:
The ship's main description was given, as will apply for this particular voyage. The ship's description, as provided by the Owners, must be an accurate representation of the ship's characteristics, which can only be modified with the usual and acceptable shipping interpretation of the word 'about'. The Owners should be kept responsible and liable for any consequences/delays/ loss/expenses that may arise, when providing any misleading information on the ship's characteristics.
When an agreement is to be made on free port disbursements ('free d/a') to the ship (the Charterers will cover the port costs) or when ranges instead of ports have been specified, then certain additional characteristics should be asked to be presented to the Charterers.
When general cargoes/logs/steel/pallets, etc. are to be loaded, the holds' dimensions/hatch dimensions /tank top strength, etc. are also to be presented by the Owners.
The ship's basic description should include at least:
- the ship's name and ex name/s.
- its type.
- when it was built.
- the flag.
- the summer dwat and at which summer draft.
- the loa (length over all)/the beam.
- the Class.
- the Call sign.
- the imo number.
- the P&I Club.
- the total Grain capacity/the total Bale Capacity.
- the number of holds and number of hatches.
- the cargo gear description (for geared ships).

The Owners continue:
<u>*1. Account (Charterers' name)*</u>

The order states:
Account Eagle Trading and Shipping Inc.

The offer could be given as follows:
For:
1.
Account Messrs. Eagle Trading and Shipping Inc. Sub Charterers approval by the Owners. Please advise of the Charterers' full background, where they are based and recent fixtures, for the Owners' approval.

Comments:
Only the name of the Charterers does not commercially cover the Owners. Charterers have been asked to provide:
- their full background which should include, among others, the history of the company/when it was established and its activities,
- the place where the company is based,
- a list of recent fixtures with other Owners.

Charterers must send the name of the ships/the owning or managing companies/through which brokers they have fixed, as well as their contact numbers. The Owners should check through the Owners or brokers and the Charterers will be approved only when the Owners have been fully satisfied on their reliability and performance.
Until the Charterers are approved, they will remain subject to the Owners' approval during the negotiations.
In case of dissatisfaction, the Owners have the right
. to drop the negotiations at any time,
as no one should take the risk to fix with someone, who would step down from full or partial fulfilment of their contractual obligations.

The Owners continue:
2. Subjects
The offer could be given as follows:
2.
All subs will be lifted within 2 hours after fixing the main terms.

Comments:
The Owners have offered that all subjects (abbreviated as 'subs') which should apply for this fixture will be lifted
. within two hours after the agreement of the Main Terms.
If it is accepted, the Charterers should declare the same, even if the allowed time of 2 hours to lift the subs runs outside of the normal working office hours

or even if the Charter Party details have not yet been discussed and fully agreed. Charterers may counter reply based on a different time, which should comply with the actual time that is required to present the ship to parties involved and make the cargo ready, with all relevant approvals.
Usually and logically, the presentation of the ship takes place only
. after the conclusion of the Main Terms
and when the Charter Party details are in order (all negotiations have been concluded) and not earlier. The Charterers might otherwise be in a difficult position on the rest of the terms, which remain for negotiation or towards the Shippers/ Receivers, when the subs are lifted but the ship remains unfixed or it eventually fails for various reasons.

The Owners continue:
3. Cargo description- stowage factor-flexibility on cargo quantity.
The order states:
40,000 mtns bulk wheat about 44' with 10pct moloo

The workable ship should be able to load a minimum of 36,000 mtns and a maximum of 44,000 mtns.
The provided tolerance (upward or downward adjustment) gives the Owners the privilege to:
- load any quantity of cargo within the allowable edges.
- bring the ship to a seaworthy condition.
- protect themselves from any short laden cargo quantity and its consequences.

The Owners' offer should basically include:
- the basic quantity of cargo.
- the upward or downward adjustment.
- the type/grade of cargo and in what form (bulk or in bags etc.) it will be transported.
- the stowage factor.

The exact loading quantity of cargo is subject to:
- the actual quantity of fuels on board the ship.
- the trimming and actual stowage factor of the cargo in the ship's holds.
- the stress of the ship.
- the actual depth and salinity of the waters in the port/berth.

The offer could be given as follows:
3.
cargo 40,000 metric tons bulk wheat, one grade. The Charterers guarantee that

cargo stows maximum 44 in the ship's holds. Cargo quantity with 10 pct more or less at the Owners' option.
Or for short:
cargo 40,000 mtns bulk wheat, one grade, with 10 pct moloo. Charterers gtee that the cargo stows max 44' in the ship's holds.

Comments:
The Owners have offered:
-and have accepted the basic cargo quantity and the tolerance, as circulated in the market by the Charterers.
However, they ask:
-the Charterers to guarantee that the cargo stows a maximum of 44' in the ship's holds. This should assist the Owners in calculating as precisely as possible the actual loadable cargo quantity and make an as accurate as possible calculation when working on the Voyage Estimation.
With a guaranteed stowage factor, the Owners should have the right to present claim for 'dead-freight',
if the ship fails to load the expected cargo weight, due to a higher (inaccurate) stowage factor.

The Owners continue:
4. Load ports-berths /Discharge ports-berths
The order states:
1 / 2 sb Cza/1 /2 sb Tarragona and/or 1 /2 sb Valencia

As per the 'cargo order' the Charterers:
-will load from one or two safe berths from the port of Constanza, which is located in Romania,
and
-will discharge in one or two safe berths in the port of Tarragona and/or Valencia, both located in Spain.
No information is given on the available drafts and/or any other possible prevailing port restrictions in all ports. Nevertheless, the ports are known that have the ability and the capacity to load/discharge much bigger ships, allowing the Owners to offer without delay and holding only a reservation to check the ports through the attending Agents.
The Owners' offer should basically include:
-the loading and discharging port/s and the rotation of the same.
-the number of berths allowed to be used, as well as who will bear the cost and lost time if two or more berths are used.
-if there are any port restrictions, something which can be confirmed and guar-

anteed by the Charterers.
-when the second discharge port will be nominated.
-a reservation to check the port expenses in effect on all ports.

The offer could be given as follows:
4.
1 safe port/1 safe berth Constanza Romania as loadport.
1safe port/1 safe berth Tarragona Spain and/or 1 safe port/1 safe berth Valencia Spain as discharge port/s.
If two discharge ports are used, the ship will discharge in rotation at the Owners' option and the cargo quantity for each discharge port should be in accordance with the ship's stress and the Master's approval.
The second discharge port, if used, will be declared before the completion of the loading and the issuance of the Bill of Lading.
Always afloat, always accessible at all ports of contact.
Charterers guarantee that there are no port restrictions for the ship, including, but not limited to, draft/l.o.a./beam/distance from water line to top of hatch-coaming etc. at all ports of contact.
Sub Owners check the prevailing port expenses at all ports of contact.

Or for short:
4.
1spsb Constanza Romania as loadport.
1spsb Tarragona and/ or 1spsb Valencia both in Spain as dischport/s.
If two discharge ports are used, the ship will discharge in rotation at the oo and the discharged cargo quantity at each port should be in accordance with the ship's stress and the Master's approval.
The second dischport, if used, will be declared before the completion of the loading and the issuance of the Bill of Lading.
AAAA at all ports of contact.
The Charterers guarantee that there are no port restrictions for the ship, including, but not limited to, draft/loa/beam/distance from water line to top of hatch-coaming etc. at all ports of contact.
Sub Owners check the prevailing port expenses at all ports of contact.

Comments:
The Owners have offered:
- one safe berth for the ship to operate in each safe port of contact. If it is accepted by the Charterers, then, if for any reason an additional berth is used, the shifting expenses (tugs/pilots/cost for mooring and unmooring etc.) and lost time will be for the Charterers' account.

Besides, the Owners can
- claim for the cost of the additional fuels consumed and possible crew's work in overtime.
- the designated ports and berths to remain
. 'always afloat', i.e., that the ship will not touch the bottom at any time during the loading or discharging operation
. and 'always accessible', i.e., that the ship will have direct access to the berth upon arrival in the port.
- the country of which each port is located.
- the second discharge port to be declared before the completion of the loading operation and before the relevant Bills of Lading have been issued.
In this case:
- the extra freight for the use of the second discharge port will be paid together with the initial freight payment and not at a later stage,
- the approximate duration of the voyage and the ship's next date of opening should be predictable much earlier,
- the Bills of Lading will be issued at the loading port directly for one or two discharge ports.
-that if two discharge ports are used, the ship will discharge such amount of cargo quantity at first port, as per the ship's stress and the Master's approval.
- the Charterers to guarantee that there are no port restrictions at all ports.
- by holding a reservation to check the port expenses at all ports of contact

The Owners continue:
5. Lay-can
The order states:
Lay-can 20/25 January 2....

Laydays/Cancelling (abbreviated as 'Lay-Can' or shorten 'L/C') is the span of dates, within which the ship must be presented for loading. 'Lay-days' is actually the first day, when the ship can tender the N.O.R. The 'Cancelling' date is the last date and the N.O.R could be tendered until 24:00 hrs, unless has been otherwise agreed on during negotiations.

The Owners' offer should basically include:
-the span of dates within which the ship must be presented, ready in all respects for loading.

The offer could be given as follows:
5.
Laydays-Cancelling 20/25 January 2....

or shorten:
L/C 20/25 Jan 2... 24:00 hrs.

Comments:
The Owners have offered:
-for the ship to be presented for loading within the dates starting from the 20th until the 25th of January 2.... and until 24:00 hrs., which are the same dates as have been circulated by the Charterers. There should obviously not be any problem, however, the Charterers could request from the Owners to declare the ship's full itinerary/its estimated time of arrival (e.t.a.) and the same should be incorporated in the Charter Party.

The Owners continue:
6. Freight.
The order states:
-ideas USD 19.00 pmtn 1/1.

The Charterers' ideas, as indicated in the market, is USD 19.00 per metric, on the basis of one load port/one discharge port ('1/1' for short). If a second discharge port will be used, an additional 'freight' on the entire cargo quantity will be paid.
Logically, when the negotiations are in progress, the Charterers will pay quite or much more (depending on the prevailing market) than their ideas, or otherwise, they should have stated from the beginning that the given rate is for fixing, i.e., 'fixing rate usd 19 pmtn 1/1'.

The Owners' offer should basically include:
-the 'freight rate' which will be multiplied by the loaded cargo quantity and will result in the due 'freight' basis 1/1.
-the additional 'freight rate' for the possible use of a second discharge port.
-who will pay for the stevedoring/cargo expenses ('free in/out' terms) and on whose account will the extra cargo trimming be if it is requested by the Master of the ship.
-when the due 'freight' will be made to the Owners' bank account and if the Owners accept deductions to be made on the gross 'freight'.
-the type of Bill of Lading and how it will be marked.
-when the demurrage will be paid and how the port documents/Owners' 'freight' and demurrage invoices will be forwarded to the Charterers.
-the time from which the 'freight' is due for payment, regardless of whether the ship and/or cargo is lost or not.

The offer could be given as follows:

6.

Freight USD 21,00 pmtn basis 1/1 free in/out trimmed (in short fiot) payable 100pct net of tax and bank charges into Owners' nominated bank account within 2 banking days after completion of loading and presentation/signing/ releasing of Congenbill edition 1994 Bill of Lading marked as 'freight payable as per Charter Party'.

If a second discharge port is used, the Charterers will pay additional freight of USD 2.50 pmtn for the entire quantity of the loaded cargo.

Possible Demurrage will be settled and be paid to the Owners by the Charterers within 3 banking days after completion of the discharging operation and after the Owners' presentation of all relevant documents, such as the N.O.R./S.O.F. and the Owners' Time Sheets from all ports, together with their final freight and demurrage invoices, which all are accepted even by fax. The original documents will be forwarded by mail/courier to the Charterers' offices as soon as possible.

If there is a dispute on the presented Time Sheets, the Charterers will pay the undisputed amount of demurrage without any delay and the disputed amount will be settled amicably or will be referred to Arbitration.

The lay-time computation for all ports will be based only on the issued S.O.F., which is duly signed by the Master and Agents in every port.

Full freight deemed earned, upon shipment discountless and non-returnable ship and or cargo lost or not lost.

Comments:

The Owners have offered:

-as 'freight rate' USD 21,00 per metric ton of loaded cargo which is based on one load and one discharge port (1/1).

-the cargo to be loaded/discharged free of expenses to the Owners and the cargo trimming (and extra trimming if it is required) will be for the Charterers' account (fiot).

-that the payment will be made 100 percent net of tax and bank charges within 2 banking days after completion of the loading operation and after the signing and releasing of the Bills of Lading. The Owners are trying to secure the payment as early as possible and are asking to be paid after completion of the loading and not after the ship's sailing from the port.

-that the issued Bill of Lading should be the Congenbill edition 1994 with its standard conditions of carriage. It will be marked as 'freight payable as per Charter party' in which case, and, if the Charterers fail on their financial obligations, the Owners have the right to lien the cargo.

-that for the possible second discharge port, an additional 'freight rate' of USD 2,50 pmtn will be paid on the entire loaded cargo quantity.

-that the Charterers, within 3 banking days after completion of discharge and upon the Owners' presentation of the required documents, must pay for the possible demurrage incurred and close the file. The lay-time will be calculated based only on the issued S.O.F.
and
-that the 'freight' will become due upon shipment, i.e., upon commencement of loading and it is discountless and non-returnable, whether the ship and/or cargo is lost or not.

The Owners continue:
7. Loading and discharging speed-N.O.R.-Commencement of lay-time
The order states:
4,000 sshex e.i.u/10,000 shinc.

The term 'sshex even if used' at load port usually declares that the time will cease to count from Friday afternoon (17:00 hrs.) until Monday morning (08:00 hrs.). The lay-time will not count even if the port/ship has been working during this period.
At the discharge port the term 'shinc' declares that the lay-time will continue to count even during weekends and/or holidays, regardless of whether the ship/ port actually worked or not.
Tendering the Notice of Readiness is one of the main tasks of the Master, as it well-defines/ declares the arrival of the ship at the port and its readiness for the purpose of call.
The effective N.O.R. should be tendered and re-tendered if necessary:
- within the Charter Party laydays-cancelling dates at load port,
- when arriving at the commercial limits of the port, or at the place as will be instructed by the Port Authorities to wait, or at a berth as specified in the Charter Party,
- the ship is ready in all respects to operate for the purpose of call,
- within the working days and working hours, as stipulated in the Charter Party.

The Owners' offer should basically include:
-the loading and discharging rate,
-the excepted periods,
-if the lay-time will count while the ship is operating during the excepted periods,
-when the Master has the right to tender the Notice of Readiness, by what means and at what time,
-when the lay-time will commence counting and when it will cease,
-if the time used before the commencement of lay-time will count as lay-time.

The offer could be given as follows:

7.

Loading rate/lay-time:

the cargo will be loaded free of risks and expenses to the ship at the rate of 5,000 metric tons daily or pro rata, per weather working day of 24 consecutive hours, Sundays and Holidays excluded, unless used, in which case the actual time used to count as lay-time.

Time from Saturday noon until Monday 08:00 hrs. or from 17:00 hrs. on a day preceding a Holiday until 08:00 hrs. of the next working day, not to count as lay-time, unless used, in which case the actual time used to count as lay-time.

The Notice Of Readiness will be tendered in writing at berth or in case of congestion or for any other reason beyond the Owners'/ship's fault the ship's berthing is not allowed upon arrival, from the anchorage or from the place where it will be instructed to wait within or out of the port's commercial limits, by any means of telecommunication (cable/ telex/e-mail/fax) to the Agents, during the official working hours from 08:00 hrs.- up to and including 17:00 hrs. from Monday to Friday and from 08:00 hrs.- up to and including 12:00 hrs. on Saturday. The lay-time will commence counting at 14:00 hrs. if the Notice of Readiness is tendered before noon or at 08:00 hrs. the next working day, if the Notice of Readiness is tendered in the afternoon, whether in berth or not, whether in free pratique or not, whether customs are cleared or not, whether in port or not.

The 'NOR is deemed to be tendered and accepted from the time of tendering and the lay-time will commence counting in accordance with the terms of the Charter Party.

Time used before the commencement of lay-time will count as lay-time.

-Discharging rate/lay-time:

the cargo will be discharged in one or two ports, free of risks and expenses to the ship, at the rate of 12,000 metric tons daily or pro rata, per weather working day of 24 consecutive hours, Saturdays, Sundays, and Holidays included. The Notice Of Readiness at the first or sole discharge port will be tendered in writing at berth or in case of congestion or for any other reason beyond the Owners'/ ship's fault the ship's berthing is not allowed upon arrival, from the anchorage or from the place where it will be instructed to wait within or out of the port's commercial limits, by any means of telecommunication (cable/ telex/e-mail/ fax) to Agents during the official working hours from 0800 hrs.-up to and including 1700 hrs. from Monday to Sunday. The lay-time will commence counting at 14:00 hrs. if the Notice of Readiness is tendered before noon or at 08:00 hrs. of the next working day, if the Notice of Readiness is tendered in the afternoon, whether in berth or not, whether in free pratique or not, whether customs are cleared or not, whether in port or not.

The 'NOR is deemed to be tendered and accepted from the time of tendering

and the lay-time will commence counting in accordance with the terms of the Charter Party.
Time used before the commencement of lay-time will count as lay-time.
If a second discharge port is used, the NO.R. will be tendered in writing at berth or in case of congestion or for any other reason beyond the Owners'/ship's fault the ship's berthing is not allowed upon arrival, from the anchorage or from the place where it will be instructed to wait within or out of the port's commercial limits, by any means of telecommunication (cable/ telex/e-mail/fax) to the Agents and the lay-time will commence counting upon tendering at any time day or night from Monday to Sunday, whether in berth or not, whether in free pratique or not, whether customs are cleared or not, whether in port or not.

Comments:
The Owners have offered:
-with terms which do not fully comply with the Charterers' cargo order.
The Owners will:
-try to achieve higher loading or discharging rates and especially when it is known that the port/s can provide a faster loading/discharging operation
-try to obtain certain favourable terms relative to the exception periods.

The Owners continue:
8. Demurrages-Despatches.

The time used at the port, in excess of the allowed lay-time under the terms of the Charter Party, that is required for completion of the specific operation at the contacted port, is called
- 'demurrage'.
The Owners shall be compensated for all the delays incurred, with a daily demurrage rate or pro rata for any part of the day, as agreed and incorporated in the Charter Party.
If the time actually needed and used for completion of the specific operation at the contacted port is shorter than the allowed lay-time in accordance with the terms of the Charter Party, then all-time saved or working time saved, is called
- 'despatch'.

The lay-time shall cease and the entry of the ship on demurrage could delay, because of lost time due to bad weather conditions, strikes in ports, exception periods, breakdowns of the ship's machinery/ equipment or breakdowns of the ship's cranes, etc. and generally, lost time due to any fault of the ship shall not count as lay-time.
However, when the ship enters on demurrage, no exception periods are in ef-

fect and lost time due to bad weather/force majeure will count as lay-time. The same is known as:

. *'once on demurrage, always on demurrage'.*

The agreed demurrage rate will cover:
-the ship's daily running expenses.
-the Owners' daily profit.
-the increased port expenses and the extra fuel consumed due to the ship's prolonged stay in the port.

The demurrage term could be agreed as follows:
-Demurrage USD .... free dispatch.
-Demurrage USD.... with half despatch for all working time saved in both ends.
-Demurrage USD .... with half despatch for all time saved in both ends.
-Demurrage USD.... /Despatch USD....... for all working time saved (despatch rate which does not reflect half the demurrage rate).
Sometimes, when the Charterers know that the ship will load and discharge in full despatch, they could try to achieve a very high demurrage rate under the term of 'half despatch', which reflects a decent amount to be returned to them as despatch money.
The Owners, especially when working smaller sized ships, should always commence the negotiations by offering the term 'free despatch' to be in effect.

The offer could be given as follows:
8.
Demurrage USD 15,000 per day, pro rata free despatch (or for short: Dem USD 15,000 pdpr FD)
The lay-time between the load and discharge port and between the discharge ports will be non-reversible.
The demurrage claim will be settled between the Owners and Charterers and the Charterers will be solely responsible for paying the demurrage to the Owners.
The Laytime computation will be based only on the signed Statement Of Facts.

Comments:
The Owners have offered:
-basis free despatch. No despatch money will be paid to the Charterers due to the earlier completion of the ship's operation.
-the lay-time to be non-reversible between the load and discharge ports and between the discharge ports, i.e., for each port of contact, the used time for the operation's completion will be calculated separately and a separate Time Sheet

will be issued.
-the Charterers are the only party responsible for paying the possible demurrage incurred, at both load and discharge ports.

The Owners continue:
9. Determination of the loaded and discharged quantities of cargo/Draft's Survey.

At the load port, the Shippers are obliged to supply the loading figures and to warrant those figures.
The same should be presented, either by a carried out draft survey on the ship or by the shore scale measurements.
Problems may arise at the load port, in the case that the shippers' figures exceed the ship's figures, or the Receivers figures are less than the ship's figures at the discharge port and both are beyond a normal and/or customary difference on the loaded and/or discharged cargo quantity.
In view that, on Voyage Charter, the Owners/ship
- are the only contracting party liable for an alleged cargo shortage claim, as could be presented at the discharge port by the Consignees, which would result in lost time and additional costs, the ship must be covered
- by a first-class P&I Club and special care should be taken when discussing the determination of the loaded and discharged quantities of cargo.
The Owners' offer should basically include the method of determination of the cargo quantity in the ports of contact.

The offer could be given as follows:
9.
Loaded and discharged cargo quantities will be ascertained by a joint draft survey which will be held between the Owners and Charterers/Shippers at the load port and between the Owners and Charterers/Receivers at the discharge port. Each party will appoint and pay for their surveyors. It is the Owners' option to appoint the ship's Master or Chief Officer as their surveyor. The appointed surveyors are to be allowed to have full access to the ship for the purpose of the draft survey.
One draft survey report will be issued which will be signed by all the surveyors involved.
The draft survey will be carried out in berth at all ports, upon completion of the loading and discharging operation. Expenses incurred to hire a launch or tug in order for the draft survey to be carried out in berth or in the anchorage will be for the Charterers' account and time lost will count as lay-time or time on demurrage. At the load and discharge port, the shore scale figures are not to be tak-

en into consideration and the ship will load and discharge according to the findings of the carried out draft survey. The Owners/ Master/Ship will not be held responsible for any discrepancies which may arise between the draft survey and the shore figures and it is explicitly agreed that the Bill of Lading will be issued according to the draft survey findings. Delays incurred due to discrepancies between the draft survey and shore figures will count as lay-time or time on demurrage.

Comments:
The Owners have offered:
-that the loaded and discharged quantity of cargo will be ascertained by a joint draft survey which will be held between the Owners/Charterers/Shippers and the Receivers at the load and discharge port, respectively.
-shore figures shall not be taken into consideration and incurred delays due to discrepancies between the draft survey and shore figures will count as lay-time.

The Owners continue:
10. Time counting during several operations in the port.

Various operations and stoppages could be taking place in the ports and the Owners/Charterers should agree and clarify who will bear the costs and lost time during these operations.
Every lost minute could greatly affect the time in which the ship will enter demurrage and it is necessary for the best known/predictable/customary operations in the port to be specified and to be entirely clear on whether the time will count or not during such stoppages.

The offer could be given as follows:
10.
During the contact of the ship with any port for operation under the terms of this Charter Party the:
a- first shifting time from the anchorage or standby berth or pilot station to the load or discharge berth will not count as lay-time unless the ship is already on demurrage.
b- time required for the initial/intermediate/final draft survey in all ports will count as lay-time.
c- time required for the opening and closing of hatches will count as lay-time.
d- national or state holidays to apply only as holidays. Local or part holidays will count as lay-time.
e- time used for trimming or extra trimming will count as lay-time and will be arranged and be paid by the Charterers.

f- time used for the shifting of shore cranes between the cargo holds will count as lay-time.
g- time used for the ship's shifting alongside quay by mooring lines in order to make all hatches available for the operation will count as lay-time but will not be considered as shifting unless tugs or pilots are used, in which case it will count as shifting.
h- time used for placing/removing/manoeuvring of the bulldozers in the ship's cargo holds will count as lay-time.
i- time used for holds' or cargo fumigation before or after the completion of loading will count as lay-time.
j- time lost due to the Charterers, or Charterers' Agents or the terminal operator or the Shippers or Receivers or any Authority prohibiting, impeding or restricting the loading or discharging operation beyond the Owners'/ship's fault, will count as lay-time.
k- cargo documents including the Bill/s of Lading will be presented to the Master within 2 hours after the completion of the operation in each port, but provided that the draft survey has been carried out. Upon expiration of those 2 hours and if all cargo documents including the Bill/s of Lading have not been presented, the lay-time will commence counting again, until all documents including the Bill/s of Lading have been presented accordingly. If the ship will be instructed to vacate the berth and shift to another waiting berth, then the shifting expenses will be for the Charterers' account and the time lost will count as lay-time or time on demurrage.
However, if the ship is already on demurrage, then all-time waiting for documents including Bill/s of Lading will count as lay-time.
l- at the loading and discharging port/s, any shifting beyond the number of shifting allowed by the Charter Party and if any overtime is ordered, the time and the expenses will be for the account of the party ordering the same, but if it is ordered by the Port Authorities, the expenses will be for the Charterers' account and the time lost will continue to count as lay-time or time on demurrage
m- the ship will be left in a seaworthy and safe stress/trim condition to the Master's satisfaction during any shifting which will take place between the loading berths, as well as, between the discharging berths and between the discharging ports and the time lost will continue to count as lay-time or time on demurrage.

Comments:
The Owners have offered:
-terms which include certain specific stoppages during the operation in the ports and clarify if the time lost will count as lay-time and who will bear the cost of same.

The Owners continue:
11. Extra insurance (abbreviated as 'exins') or additional premiums (abbreviated as 'ap').

Extra cargo insurance could be in effect due:
-to the age of the fixed ship (more than 15 years old),
-to the flag/ownership of the fixed ship.
For older ships (more than 20 or 25 years old), the imposed extra cargo insurance could be heavier. Enabling a fixture to be solved, the Charterers could request from the Owners to cover the same in full against vouchers or to contribute with a lumpsum amount on them. This affects the ship's daily income and if the imposed exins is not reasonable, it could lead to the biz being dropped by the Owners.
This term could be agreed in one of the following forms:
-free extra insurance (free exins).
-full extra insurance for the ship's account (the full cost as will be imposed by the insurers).
-extra insurance for the Charterers' account but the Owners will contribute with USD..... or the Owners will pay maximum USD ...as extra insurance.

The offer could be given as follows:
11.
Extra insurances (or additional premiums) due to the ship's age/flag/ownership etc. will be for the Charterers' account.
or in short:
Exins (or Ap) due to ship's age/flag/ownership etc. will be for the Charterers' account
or
Free of exins (or ap) due to the ship's age/flag/ownership etc.

Comments:
The Owners have offered:
-that all extra insurance (or additional premiums) which will be imposed by the cargo insurances will be for the Charterers' account.

The Owners continue:
12. Taxes/Dues.

Some countries impose taxes/dues which are calculated/levied on cargo loaded, as well as, on the ship and on Charter Party freight. The Indian and Turkish taxes on the Charter Party 'freight' and the Brazilian Inframar taxes/T3 tariff in Spain which are charged on the cargo transported are some of the most known

taxes in the market.
The Owners, before fixing the ship, should request the Agents of the port to inform them,
- if any taxes are applicable on the ship/freight/cargo etc. that could be charged for their account.
The known taxes in the countries which the ship will contact should be specifically stated in the Owners' offer and incorporated in the Charter Party terms.

The offer could be given as follows:
12.
Any taxes/dues/wharfage dues/charges/duties calculated/levied on cargo, including, but not limited to the T3 tariff in Spain, a/o on cargo quantity loaded a/o on Charter Party freight will be for the Charterers' account at both ends. The Owners will pay only for the ordinary port expenses at all ports of contact.

Comments:
The Owners have offered:
-and have covered themselves of all possible applicable taxes/dues etc. in the ports of contact,
-also ask the Owners to pay only for the ordinary port expenses at all ports of contact.

The Owners continue:
13-Agents.

The order states:
Chabe, i.e., Charterers' Agents at both ends.

When a ship arrives at a port for cargo operation, the appointment of Agents is necessary and mandatory. They will take care of all issues, along with handling the smooth processing of the ship in port and its sailing by arranging tugs/pilots/port clearance etc. and by issuing all relevant cargo documents. They will provide useful information about the prevailing restrictions at the port, on facilities, on the cargo, etc. and will keep the interested parties fully informed on the ship's movements and the progress of the loading and discharging operation.
When on Voyage Charter, there is always an argument between the negotiating parties, as to who will select the attending Agents in all ports. If the Owners' Agents apply, the Owners have the flexibility of negotiating with various agency offices in every intended port and should feel more comfortable that their interests will be well protected.
Additionally, they take the opportunity to minimize certain costs or to even re-

duce the tariff of the standard imposed agency fees or to negotiate for the cost of the hired tugs.

The offer could be given as follows:
13.
The attending Agents will be selected and be appointed by the Owners at both ends.
Or in short:
Owners' Agents at both ends.

Comments:
The Owners have offered:
-that the Agents will be selected and appointed by them.

The Owners continue:
14.Arbitration.

The place of arbitration and the prevailing law should be discussed and agreed since the negotiations of Main Terms.
The main arbitration venues are
- in London with English Law to apply
and
- in New York with U.S. Law to apply
and recently Bimco has added suitable clause also for Singapore and Hong Kong.

The most acceptable and usable clause for arbitration purposes is the Bimco's published arbitration clause called the 'dispute resolution clause', either for the London venue or for New York or even outside of these two venues. They also provide a mediation clause.

The offer could be given as follows:
14.
Arbitration in London. English law will apply to the negotiations and to the eventual agreement.

Comments:
Another jurisdiction may be proposed and agreed between the Owners and Charterers, but the above is the most acceptable to all parties. If the negotiations are made under English Law, the fixture is concluded and responsibilities should be portioned only after full agreement has been reached, i.e., on main terms & Charter Party's details and when 'hands are given'.

The general average term and under which rules it will be adjusted, can be discussed during the negotiations on the Charter Party details.

The Owners continue:
15. Discharging against Original Bill of Lading.

The absolute majority of the Charterers will request through the negotiations to have the option to present a letter of indemnity (abbreviated as 'l.o.i.') in exchange for discharging the cargo to the Receivers without the presentation of the original Bill of Lading.
The normal practice is for the ship to discharge the cargo and deliver it to the entitled persons, who, before the commencement of discharge, hand over to the Master an original bill of lading as issued at the load port and has been duly endorsed by them. Any deviation from this practice could be subject to P&I Club's rules, mainly due to imminent risk of the cargo being delivered to the wrong people (wrongful delivery of goods).
A l.o.i. may also be issued by the Charterers, for delivery of cargo at a port other than that stated on the Bill of Lading or if the Owners are asked to sign a 'clean on board' while the cargo is not in apparent good order and condition.
In general, the Owners, before accepting any l.o.i., must be fully satisfied with the reliability/ credibility of the Charterers, as the ship, by such ways of work, may be subject to P&I Club's rules.

The offer could be given as follows:
15.
The ship will discharge only on presentation of the original bill of lading.

Comments:
The Owners have offered:
-the ship to discharge only on presentation of original Bills of Lading.
If the ship arrives in the port, but, due to missing original Bills of Lading, the commencement of the discharging operation is delayed, the tendered N.O.R. will be considered valid and the lay-time will commence counting in accordance with the terms of the Charter Party.

The Owners continue:
16. Notices of ship's arrival at load/discharge port.

Before the arrival at the load port and after the departure towards the discharge port, the Master must tender approximate and definite notices to the

Charterers/Agents and to the specified parties in accordance with the terms of the Charter Party.
The ship's position will be made known to all interested parties, enabling them to be ready for operation upon the ship's arrival.

The offer could be given as follows:
16.
Please advise on what notices will be tendered and to whom.

The Owners continue:
17. Stevedoring clause.

Damage in cargo holds/hatches is mainly incurred during loading or discharging by the use of shore grabs or while stevedores using payloaders/bulldozers in the ship's holds for the purpose of stowing or removing the cargo from the sides of the ship and carrying it to places more easily accessible by the grabs.
Damage could be caused to the holds' ladders, frames, hoppers, tank top, hatchcovers etc.
Care should be taken inside the ship's holds, where all vulnerable parts must be properly protected by the Owners for possible damage from the Shippers'/Receivers' means of operation.
The Owners' offer should basically include:
in the event that stevedoring damage is caused to the ship's holds/ hatches/fittings/cargo etc.,
-the measures that the Master should take,
-how/when the concerned parties must be notified,
-for whose account will the resulting delays in sailing be and who will pay for the damage occurred.
The ship may suffer severe damage in the ports of contact, resulting in making it inoperative and un-seaworthy.

The offer could be given as follows:
17.
The Stevedoring company will be appointed and paid by the Charterers, but will work under the supervision of the Master, if it is possible.
Charterers will be liable and responsible for any damage caused by the stevedores to the cargo or ship or its fittings at the ports of contact during this Voyage Charter, including hidden damage, which should be reported as soon as detected, but in any case, before the ship's sailing from the last discharge port.
If the ship suffers stevedore damage, the Master of the ship:
1. will report to the Charterers or their Agents in writing within 24 hours of oc-

currence or from the time of discovery, but in any case, before the ship sails from the port.
2. will hold the stevedores or the respective parties responsible and liable for any damage, in writing, with copies sent to the Charterers or their Agents. He will work with the Charterers or their Agents and will make every effort to reach a written acknowledgement from the parties responsible for causing the damage, unless the damage has in the meantime been repaired. If the Master is unable to obtain acceptance of liability and responsibility from the party who caused the damage, he is not allowed to sail from the port before coming into contact with the Charterers, allowing them to sort the matters out and the time lost will continue to count as lay-time or time on demurrage.
3. will try to get stevedores to repair the damage and try to settle the matter directly with them in the first stage and the time lost will continue to count as lay-time or time on demurrage.
For stevedore damage which:
a). affects the effectiveness of the ship and the safe operation and/or safety of the crew and/or affect the trading capabilities and/or seaworthiness and/or cargo worthiness and/or is subject to classification requirements, then such damage as will be ascertained by the ship's Class surveyor, who will be appointed by the Owners at the Charterers' cost, is to be repaired by the Charterers at their own expenses/risk, to the Master's and/or Class surveyor's entire satisfaction and the time lost will continue to count as lay-time or time on demurrage until completion of all repairs and the 'port clearance' is issued and delivered to the ship.
Charterers are to reimburse the Owners for the Class Surveyor's fees, once the Owners have submitted their final freight invoice, supported with the original vouchers.
The damage will be repaired at the place of occurrence, if it is possible and there are adequate facilities for permanent repairs, or otherwise, temporary repairs are to be made on the spot or as when detected, at the Charterers' risk and expenses and to the Master's and/or Class surveyor's satisfaction and the time lost will continue to count as lay-time or time on demurrage. The ship will sail or shift to the nearest shipyard, as chosen by the Charterers, with fuel consumed at sea or during shifting to be for the Charterers' account, which will be settled together with the Owners' final freight statement, in order to complete the repairs.
The time lost during the sailing or shifting to the shipyard and until completion of the works with the damage being repaired to the Master's and/or Class surveyor's entire satisfaction and the 'port clearance' is issued and delivered to the ship or, at the Owners' option, at equidistance from the discharge port, will count as lay-time or time on demurrage.
Shipyard's expenses and/or port expenses, including in and out tugs and/or pilots and/or agency fees etc., will be entirely for the Charterers' account.

b). does not affect the effectiveness of the ship and the safe operation and/or the safety of the crew and/or the trading capabilities and/or the seaworthiness or cargo worthiness and/or is subject to classification requirements and cannot be repaired by the stevedores, the Owners then undertake to accept the ship's sailing, in accordance with the terms of the Charter Party.
However, the said damage is to be assessed before the ship's departure from the discharge port and the costs to be agreed upon between the Master, Owners, Charterers and the respective survey company and to be paid together with the Owners' final freight statement. The survey company will be mutually appointed by the Owners and Charterers, if considered necessary, at the Charterers', Owners' equal expenses and will promptly inform the Charterers, Owners with the damage report, which will be issued before the ship sails, with a copy of all correspondence exchanged in this connection. The time lost will continue to count as lay-time or time on demurrage, until the 'port clearance' is issued and delivered to the ship.
Owners have the option of having the damage repaired at their next convenient time or at the ship's next Dry-Docking.
In the case that the Master, Owners, Charterers cannot reach an agreement on the cost with the respective survey company, the Charterers will then undertake to reimburse for the actual cost of the repairs, but not for the time lost, upon presentation of the original bills from the repairers or dockyard, that are consistent with the surveyors' report on the extent of such damage.
c). is small and has been repaired by the crew, the Charterers will reimburse the Owners for the cost on request, but not for the time lost and upon presentation of the Owners' repair statement, which is acceptable, even by fax or e-mail.
However, and in any case the Charterers are responsible for any directly related expenses resulting therefrom and incurred thereby and must pay compensation for any damage caused to the ship and/ or restore the ship to its initial condition under their own time and expenses, whether or not any payment has been made by the respective parties and/or stevedores to them.

Comments:
The Owners have offered:
-and trying to protect themselves from slight damage, from damage not affecting the seaworthiness/ cargoworthiness etc. and for damage affecting the ship's class/seaworthiness/cargoworthiness etc.

The Owners continue:
18. Type of Charter Party-Commission.

Normally, the type of Charter Party which will be used should be known to the market from the time when the order quoted in the market. The Owners should be aware of certain terms that exist in this type of Charter Party.

The order states:
Syna 90 chartrs proforma with 3,75pct ttl incl 2,50 pct add comm.

The offer could be given as follows:
18.
Sub further terms/details based on Synacomex 90 Charter Party, Charterers' executed proforma with 3,75pct ttl yr end including 2,5pct address commission.

Comments:
The Owners accept the Synacomex 90 Charter Party to apply, but the proforma will be based on an executed one.

<u>TO SUMMARIZE THE OWNERS' OFFER:</u>

The cargo entered the market as follows:
Account ....
40,000 mtns bulk wheat about 44 with 10pct moloo
1 / 2 sb Cza/1 /2 sb Tarragona and/or 1 /2 sb Valencia
Lay-can 20/25 January 2....
4,000 sshex e.i.u./10,000 shinc
Reversible lay-time
Syna 90 chartrs proforma with 3,75pct ttl incl 2,50 pct add comm
Grd/grls max 25 years
chabe ideas USD 19.00 pmtn 1/1.

And the offer could be given as follows:

Ref Account Eagle Trading and Shipping Inc
40,000 mtns bulk wheat
Constanza/Spain
<u>MV TAXIARCHES</u>

Reference is made to the above-mentioned cargo order.
The Owners are in the position to firm bid for reply here within 1 hour as follows:
MV TAXIARCHES
....
....
For:

1.

Account Messrs. Eagle Trading and Shipping Inc. Sub Charterers approval by the Owners. Please advise of the Charterers' full background, where they are based and recent fixtures, for the Owners' approval.

2.

All subs will be lifted within 2 hours after fixing the main terms.

3.

Cargo 40,000 mtns bulk wheat, one grade, with 10 pct moloo. Charterers gtee that the cargo stows max 44' in the ship's holds.

4.

1safe port/1 safe berth Constanza Romania as loadport. -1 safe port/ 1 safe berth Tarragona Spain and/ or 1safe port/ 1safe berth Valencia Spain as discharge port/s.

If two discharge ports are used, the ship will discharge in rotation at the Owners' option and the cargo quantity for each discharge port should be in accordance with the ship's stress and the Master's approval.

The second discharge port, if used, will be declared before the completion of the loading and the issuance of the Bill of Lading.

Always afloat, always accessible at all ports of contact.

Charterers guarantee that there are no port restrictions for the ship, including, but not limited to, draft/loa/beam/distance from water line to top of hatch-coaming etc. at all ports of contact.

Sub Owners check the prevailing port expenses at all ports of contact.

5.

Laydays-Cancelling 20/25 January 2.... 24:00 hrs.

6.

Freight USD 21,00 pmtn basis 1/1 free in/out trimmed (in short fiot) payable 100pct net of tax and bank charges into Owners' nominated bank account within 2 banking days after completion of loading and presentation/signing/ releasing of Congenbill edition 1994 Bill of Lading marked as 'freight payable as per Charter Party'.

If a second discharge port is used, the Charterers will pay additional freight of USD 2.50 pmtn for the entire quantity of the loaded cargo.

Possible Demurrage will be settled and be paid to the Owners by the Charterers within 3 banking days after completion of the discharging operation and after the Owners' presentation of all relevant documents, such as the N.O.R./S.O.F. and the Owners' Time Sheets from all ports, together with their final freight and demurrage invoices, which all are accepted even by fax. The original documents will be forwarded by mail/courier to the Charterers' offices as soon as possible.

If there is a dispute on the presented Time Sheets, the Charterers will pay the undisputed amount of demurrage without any delay and the disputed amount will

be settled amicably or will be referred to Arbitration.
The lay-time computation for all ports will be based only on the issued S.O.F., which is duly signed by the Master and Agents in every port.
Full freight deemed earned, upon shipment discountless and non-returnable ship and or cargo lost or not lost.
7.
Loading rate/lay-time:
the cargo will be loaded free of risks and expenses to the ship at the rate of 5,000 metric tons daily or pro rata, per weather working day of 24 consecutive hours, Sundays and Holidays excluded, unless used, in which case the actual time used to count as lay-time.
Time from Saturday noon until Monday 08:00 hrs. or from 17:00 hrs. on a day preceding a Holiday until 08:00 hrs. of the next working day, not to count as lay-time, unless used, in which case the actual time used to count as lay-time.
The Notice Of Readiness will be tendered in writing at berth or in case of congestion or for any other reason beyond the Owners'/ship's fault the ship's berthing is not allowed upon arrival, from the anchorage or from the place where it will be instructed to wait within or out of the port's commercial limits, by any means of telecommunication (cable/ telex/e-mail/fax) to the Agents, during the official working hours from 08:00 hrs.- up to and including 17:00 hrs. from Monday to Friday and from 08:00 hrs.- up to and including 12:00 hrs. on Saturday. The lay-time will commence counting at 14:00 hrs. if the Notice of Readiness is tendered before noon or at 08:00 hrs. the next working day, if the Notice of Readiness is tendered in the afternoon, whether in berth or not, whether in free pratique or not, whether customs are cleared or not, whether in port or not.
The 'NOR is deemed to be tendered and accepted from the time of tendering and the lay-time will commence counting in accordance with the terms of the Charter Party.
Time used before the commencement of lay-time will count as lay-time.
-Discharging rate/lay-time:
the cargo will be discharged in one or two ports, free of risks and expenses to the ship, at the rate of 12,000 metric tons daily or pro rata, per weather working day of 24 consecutive hours, Saturdays, Sundays, and Holidays included. The Notice Of Readiness at the first or sole discharge port will be tendered in writing at berth or in case of congestion or for any other reason beyond the Owners'/ship's fault the ship's berthing is not allowed upon arrival, from the anchorage or from the place where it will be instructed to wait within or out of the port's commercial limits, by any means of telecommunication (cable/ telex/e-mail/fax) to Agents during the official working hours from 0800 hrs. up to and including 1700 hrs. from Monday to Sunday. The lay-time will commence counting at 14:00 hrs. if the Notice of Readiness is tendered before noon or at 08:00 hrs. of the next

working day, if the Notice of Readiness is tendered in the afternoon, whether in berth or not, whether in free pratique or not, whether customs are cleared or not, whether in port or not.
The 'NOR is deemed to be tendered and accepted from the time of tendering and the lay-time will commence counting in accordance with the terms of the Charter Party.
Time used before the commencement of lay-time will count as lay-time.
If a second discharge port is used, the NO.R. will be tendered in writing at berth or in case of congestion or for any other reason beyond the Owners'/ship's fault the ship's berthing is not allowed upon arrival, from the anchorage or from the place where it will be instructed to wait within or out of the port's commercial limits, by any means of telecommunication (cable/ telex/e-mail/fax) to the Agents and the lay-time will commence counting upon tendering at any time day or night from Monday to Sunday, whether in berth or not, whether in free pratique or not, whether customs are cleared or not, whether in port or not.
8.
Demurrage USD 15,000 per day, pro rata free despatch (or for short: Dem USD 15,000 pdpr FD)
The lay-time between the load and discharge port and between the discharge ports will be non-reversible.
The demurrage claim will be settled between the Owners and Charterers and the Charterers will be solely responsible for paying the demurrage to the Owners.
The Laytime computation will be based only on the signed Statement Of Facts
9.
Loaded and discharged cargo quantities will be ascertained by a joint draft survey which will be held between the Owners and Charterers/Shippers at the load port and between the Owners and Charterers/Receivers at the discharge port. Each party will appoint and pay for their surveyors. It is the Owners' option to appoint the ship's Master or Chief Officer as their surveyor. The appointed surveyors are to be allowed to have full access to the ship for the purpose of the draft survey.
One draft survey report will be issued which will be signed by all the surveyors involved.
The draft survey will be carried out in berth in all ports, upon completion of the loading and discharging operation. Expenses incurred to hire a launch or tug in order for the draft survey to be carried out in berth or in the anchorage will be for the Charterers' account and time lost will count as lay-time or time on demurrage.
At the load and discharge port, the shore scale figures are not to be taken into consideration and the ship will load and discharge according to the findings of

the carried out draft survey. The Owners/ Master/Ship will not be held responsible for any discrepancies which may arise between the draft survey and the shore figures and it is explicitly agreed that the Bill of Lading will be issued according to the draft survey findings. Delays incurred due to discrepancies between the draft survey and shore figures will count as lay-time or time on demurrage.

10.

During the contact of the ship with any port for operation under the terms of this Charter Party the:

a- first shifting time from the anchorage or standby berth or pilot station to the load or discharge berth will not count as lay-time, unless the ship is already on demurrage.

b- time required for the initial/intermediate/final draft survey in all ports will count as lay-time.

c- time required for the opening and closing of hatches will count as lay-time

d- national or state holidays to apply only as holidays. Local or part holidays will count as lay-time.

e- time used for trimming or extra trimming will count as lay-time and will be arranged and be paid by the Charterers.

f- time used for the shifting of shore cranes between the cargo holds will count as lay-time.

g- time used for the ship's shifting alongside quay by mooring lines in order to make all hatches available for the operation will count as lay-time but will not be considered as shifting unless tugs or pilots are used, in which case it will count as shifting.

h- time used for placing/removing/manoeuvring of the bulldozers in the ship's cargo holds will count as lay-time.

i- time used for holds' or cargo fumigation before or after the completion of loading will count as lay-time.

j- time lost due to the Charterers, or Charterers' Agents or the terminal operator or the Shippers or Receivers or any Authority prohibiting, impeding, or restricting the loading or discharging operation beyond the Owners'/ship's fault, will count as lay-time.

k- cargo documents including the Bill/s of Lading will be presented to the Master within 2 hours after the completion of the operation in each port, but provided that the draft survey has been carried out. Upon expiration of those 2 hours and if all cargo documents including the Bill/s of Lading have not been presented, the lay-time will commence counting again, until all documents including the Bill/s of Lading have been presented accordingly. If the ship will be instructed to vacate the berth and shift to another waiting berth, then the shifting expenses will be for the Charterers' account and the time lost will count as lay-time or

time on demurrage.
However, if the ship is already on demurrage, then all time waiting for documents including Bill/s of Lading will count as lay-time.
l- at the loading and discharging port/s, any shifting beyond the number of shifting allowed by the Charter Party and if any overtime is ordered, the time and the expenses will be for the account of the party ordering the same, but if it is ordered by the Port Authorities, the expenses will be for the Charterers' account and the time lost will continue to count as lay-time or time on demurrage.
m- the ship will be left in a seaworthy and safe stress/trim condition to the Master's satisfaction during any shifting which will take place between the loading berths, as well as, between the discharging berths and between the discharging ports and the time lost will continue to count as lay-time or time on demurrage.
11.
Extra insurances (or additional premiums) due to the ship's age/flag/ownership etc. will be for the Charterers' account.
12.
Any taxes/dues/wharfage dues/charges/duties calculated/levied on cargo, including, but not limited to the T3 tariff in Spain, a/o on cargo quantity loaded a/o on Charter Party freight will be for the Charterers' account at both ends. The Owners will pay only for the ordinary port expenses in all ports of contact.
13.
The attending Agents will be selected and appointed by the Owners at both ends.
14.
Arbitration in London. English law will apply to the negotiations and to the eventual agreement.
15.
The ship will discharge only on presentation of the original bill of lading.
16.
Please advise on what notices will be tendered and to whom.
17.
The Stevedoring company will be appointed and paid by the Charterers, but will work under the supervision of the Master, if it is possible.
Charterers will be liable and responsible for any damage caused by the stevedores to the cargo or ship or its fittings at the ports of contact during this Voyage Charter, including hidden damage, which should be reported as soon as detected, but in any case, before the ship's sailing from the last discharge port.
If the ship suffers stevedore damage, the Master of the ship:
1. will report to the Charterers or their Agents in writing within 24 hours of occurrence or from the time of discovery, but in any case, before the ship sails from the port.
2. will hold the stevedores or the respective parties responsible and liable for any

damage, in writing, with copies sent to the Charterers or their Agents. He will work with the Charterers or their Agents and will make every effort to reach a written acknowledgement from the parties responsible for causing the damage, unless the damage has in the meantime been repaired. If the Master is unable to obtain acceptance of liability and responsibility from the party who caused the damage, he is not allowed to sail from the port before coming into contact with the Charterers, allowing them to sort the matters out and the time lost will continue to count as lay-time or time on demurrage.

3. will try to get stevedores to repair the damage and try to settle the matter directly with them in the first stage and the time lost will continue to count as lay-time or time on demurrage.

For stevedore damage which:

a). affects the effectiveness of the ship and the safe operation and/or safety of the crew and/or affect the trading capabilities and/or seaworthiness and/or cargo worthiness and/or is subject to classification requirements, then such damage as will be ascertained by the ship's Class surveyor, who will be appointed by the Owners at the Charterers' cost, is to be repaired by the Charterers at their own expenses/risk, to the Master's and/or Class surveyor's entire satisfaction and the time lost will continue to count as lay-time or time on demurrage until completion of all repairs and the 'port clearance' is issued and delivered to the ship.

Charterers are to reimburse the Owners for the Class Surveyor's fees, once the Owners have submitted their final freight invoice, supported with the original vouchers.

The damage will be repaired at the place of occurrence, if it is possible and there are adequate facilities for permanent repairs, or otherwise, temporary repairs are to be made on the spot or as when detected, at the Charterers' risk and expenses and to the Master's and/or Class surveyor's satisfaction and the time lost will continue to count as lay-time or time on demurrage. The ship will sail or shift to the nearest shipyard, as chosen by the Charterers, with fuel consumed at sea or during shifting to be for the Charterers' account, which will be settled together with the Owners' final freight statement, in order to complete the repairs. The time lost during the sailing or shifting to the shipyard and until completion of the works with the damage being repaired to the Master's and/or Class surveyor's entire satisfaction and the 'port clearance' is issued and delivered to the ship or, at the Owners' option, at equidistance from the discharge port, will count as lay-time or time on demurrage.

Shipyard's expenses and/or port expenses, including in and out tugs and/or pilots and/or agency fees etc., will be entirely for the Charterers' account.

b). does not affect the effectiveness of the ship and the safe operation and/or the safety of the crew and/or the trading capabilities and/or the seaworthiness

or cargo worthiness and/or is subject to classification requirements and cannot be repaired by the stevedores, the Owners then undertake to accept the ship's sailing, in accordance with the terms of the Charter Party.
However, the said damage is to be assessed before the ship's departure from the discharge port and the costs to be agreed upon between the Master, Owners, Charterers, and the respective survey company and to be paid together with the Owners' final freight statement. The survey company will be mutually appointed by the Owners and Charterers, if considered necessary, at the Charterers', Owners' equal expenses and will promptly inform the Charterers, Owners with the damage report, which will be issued before the ship sails, with a copy of all correspondence exchanged in this connection. The time lost will continue to count as lay-time or time on demurrage, until the 'port clearance' is issued and delivered to the ship.
Owners have the option of having the damage repaired at their next convenient time or at the ship's next Dry-Docking.
In the case that the Master, Owners, Charterers cannot reach an agreement on the cost with the respective survey company, the Charterers will then undertake to reimburse for the actual cost of the repairs, but not for the time lost, upon presentation of the original bills from the repairers or dockyard, that are consistent with the surveyors' report on the extent of such damage.
c). is small and has been repaired by the crew, the Charterers will reimburse the Owners for the cost on request, but not for the time lost and upon presentation of the Owners' repair statement, which is acceptable, even by fax or e-mail.
However, and in any case the Charterers are responsible for any directly related expenses resulting therefrom and incurred thereby and must pay compensation for any damage caused to the ship and/ or restore the ship to its initial condition under their own time and expenses, whether or not any payment has been made by the respective parties and/or stevedores to them.
18.
Sub further terms/details based on Synacomex 90 Charter Party, Charterers' executed proforma with 3,75pct ttl yr end including 2,5pct address commission.

CHARTERERS' COUNTER REPLY TO OWNERS' LAST:

The Charterers:
- will either agree to work on the Owners' offer and make a clean acceptance or counter offer on an accept/except basis
-or will decline the Owners' offer and forward a full re-offer for the Owners' acceptance or counter reply.

On the basis that the Charterers have accepted to work on the received offer, a counter offer on an Accept/Except basis could be given as follows:

(= terms not touched, have been fully accepted by Charterers=)

Ref MV TAXIARCHES
Account Eagle Trading and Shipping Inc
Constanza /2 Spain

Thanks for the Owners' offer on which Charterers counter offer on an A/E basis for reply latest at 14:00 hrs. London time today as follows:
or
Charterers a/e for 30 mnts reply as follows:
or
by any other form stating that the Charterers' counter offer is on an accept/except basis to the Owners' offer and that they are committed on the ship for a specific time.

Comments:
The Owners should respond:
-timely within the period that is allowed by the Charterers,
-or will ask for an extension to reply as they need additional time.

The Charterers continue:
-Please advise of full name/style/contact numbers of Head Owners/Managers and Disponent Owners.

Comments:
The Charterers asking for the full Charter Party chain to be revealed, i.e. all parties involved to be revealed along with the Owners' next counter offer.

The Charterers continue:
Owners guarantee that during the currency of this voyage:
a- the ship will not change ownership and/or class and will not be sold for scrap during or after this voyage without the Charterers' written consent.
b- the ship will proceed with the utmost despatch directly to the discharge port/s and will not deviate for any purpose whatsoever without the Charterers' consent, excluding bunkering en route and to the nearest bunkering station or for saving life.
c- the ship is fully covered by the P&I Club (as a member of an International Group of P&I Clubs) and has been covered for ship oil pollution risks and wreck removal risks and will remain so throughout the duration of this charter. In this

respect, the Owners will furnish the Charterers with the P&I Club's written confirmation, if and whenever it is requested by them.
d- the ship is fully classed as Lloyd's 100A1 or equivalent with a classification society being a full member of the International Association of Classification Societies (IACS) throughout the duration of this charter. In this respect, the Owners will furnish the Charterers with the Classification society's written confirmation, if and whenever it is requested by them.
e- the ship's H&M insurance will be fully maintained and will not be changed. In this respect, the Owners will furnish the Charterers with the H&M underwriters' written confirmation, if and whenever it is requested by them.
f- the ship has no outstanding recommendations, restrictions or any other requirements of any nature in respect to the flag state, port state control, or classification inspections that may affect this voyage and the Charter Party.
g- the ship is suitable for entering and berthing at the specified ports, as well as loading/ discharging the minimum contractual cargo quantity. The ship will comply with all applicable laws and regulations and is in all respects eligible for trading at the specified ports. The ship has all valid and up-to-date certificates and other documents on board required for this voyage.
h- the ship can open/close its hatches with its own means.
i- the ship has adequate mooring lines and will be tending them properly, whilst at the loading or discharging port.
j- the ship will give, free of expenses to the Charterers, the full use of its lighting on its deck and in its cargo holds, that will be adequate for all cargo operations.
k- the ship will comply with all applicable laws and regulations and is in all respects eligible for trading to the ports, places or countries specified in this Charter Party. The ship has all valid and up-to-date certificates and other documents on board that are required for this voyage.
l- the ship is fully fitted for the loading of grains according to the SOLAS 1974 regulations and any subsequent amendments thereto without requiring any shifting boards. The Owners guarantee that the ship is suitable for entering, berthing, and loading/discharging the contractual cargo quantity of wheat in bulk.
m- both, the ship and the company fully comply with ISPS/ISM requirements and will remain so throughout the duration of this charter.
n- the ship is sid/bc, steelfloored throughout, with hoppers port/starboard, with the engine/ bridge aft and its holds/hatches/tank tops are clear and unobstructed without a centreline bulkhead, walkways, centrebeams and free of all/any obstructions which might incur difficult access to the cargo during loading/discharging. Cargo will not be loaded in deep tanks, sidetanks, wing tanks, bunker spaces or any other confined spaces unreachable to the means of loading and discharging.
o- the ship has not carried mammalian meat, bone meal, fishmeal, asbestos cul-

let, or nuts for the past 12 months.
p- the ship has not been blacklisted/boycotted or arrested due to the ship's flag/ ownership/crew employment/age/past trading or for any other reason from local authorities and or unions in all countries designated under this Charter Party.
q- the ship is free of all encumbrances, arrests, and maritime liens.
r- the ship has no .... (insert countries which may affect the voyage) interests and is not blacklisted by Arab league countries.
s- the description of the ship as provided by the Owners is an accurate representation of the characteristics of the ship which can only be modified with the usual and acceptable shipping interpretation of the word 'about'. This clause shall override any inclusion of the words 'without guarantee' in the ship's description. The Owners are responsible and liable for any consequences/ delays/ losses/expenses that may arise when providing any misleading information on the ship's characteristics and on the terms agreed in this Charter Party.
t- the ship is ITF compliant or bona fide trade union's agreement or equivalent and will remain so throughout the duration of this charter and shall remain responsible for all consequences and/or delays if the ship is sailing under a flag of 'convenience'.
u- the ship, throughout the duration of the Charter Party will have on board and make available to the Charterers or their Agents or to the appointed surveyors, the capacity's plan, the deadweight and displacement scales, the hydrostatic curves and tables of displacement, the calibration scales for all ship's tanks and trimming correction tables as well as all sounding tables and all sounding tubes which will be properly maintained, in good condition and free from impediments. Plimsoll marks amidships and draft marks on port and starboard side, bow, amid ship and stern are always to be clearly cut and marked on shell platting and readable.

Comments:
The Owners should guarantee certain important points which are necessary for making the ship workable for the shipment. In terms which the ship does not comply with, the Owners should clearly indicate to the contrary, otherwise, in the case that the ship is fixed, claims could arise against them with huge consequences.

The Charterers continue:
1.
Account Eagle Trading and Shipping Inc of British Virgin Islands
(Charterers advise their background)
Recent fixtures:
MV .... Fixed through .... Ph. ...... Owners/brokers ph. ....

MV .... Fixed through .... Ph. .... Owners/brokers ph.............
Etc. etc.
(Charterers' recent fixtures are advised for the Owners' consideration/checking).

Comments:
The Charterers provide their full background/full name etc. Some recent fixtures have also been provided along with the brokers involved that will allow the Owners to make a quick search of the reliability/performance of the Charterers.

The Charterers continue:
2.
Sub Shippers'/Receivers' approval and stem which will be declarable at the latest 2 hours after the fixing of the main terms with the Charter Party details in order.

Comments:
The Charterers reject the Owners' proposal as to the time of lifting of subjects (Owners' offer was that all subs should be lifted within 2 hours after fixing the main terms).

The fixture is subject:
- to readiness of the cargo (stem= subject to enough merchandise)
- to the Shippers' and Receivers' approval to the ship's characteristics/Owners etc.

The subs will be lifted within 2 hours after the conclusion of the Main Terms& Charter Party details. The Charterers should reply if the 'subs' are lifted or not, even if the time is outside normal office hours.

The Charterers continue:
3.
The cargo stows about 44', but wog. No other cargo is allowed to be loaded.

Comments:
The Charterers, in comparison to the Owners' offer:
-accept the basic cargo quantity,
-accept the tolerance,
-refuse to guarantee the stowage factor of the cargo, and counter that it will not only be on an 'about' basis, but also 'without guarantee',
-the cargo will be transported as sole cargo (no part cargo).

The Owners' reaction on the cargo stowage should be one of the following:
-They will accept it and rely on this type of cargo normally stowing about 44, as well as, seeking cargo information from the attending Agents.
- They will try to delete the word 'wog'.
- They will insist to their offer.

The Charterers continue:
4.
1 safe port -1/2 safe berth/s Constanza Romania always afloat, as loadport.
1 safe port-1/2 safe berth/s Tarragona Spain always afloat and/ or at chopt
1safe port-1/ 2 safe berth/s Valencia Spain always afloat, as discharge port/s.
If two discharge ports are used, the ship will discharge in rotation at the Owners' option and will be declared before the completion of the loading and the issuance of the Bill of Lading. The cargo quantity for each discharge port should be in accordance with the ship's stress and the Master's approval.
If the second berth is used at all port/s, the shifting expenses, including officers'/ crew's overtime and consumed bunkers will be for the Owners' account, but time will continue to count as lay-time.
Owners are responsible for checking and satisfying themselves with all port/ berth restrictions and regulations at both ends.
Delete 'Sub Owners check the prevailing port expenses at all ports of contact' with the Owners' next counter offer.

The above counter offer could also be given as follows, with the Owners' untouched terms to have been accepted by the Charterers:
4.
The port of discharge and the use of the second discharge port will be at the Charterers' option.
1/2 sb in all ports with shifting expenses, including officers'/crew's overtime and consumed bunkers will be for the Owners' account, but time will continue to count as lay-time.
Only always afloat at all ports of contact.
Owners are responsible for checking and satisfying themselves with all port/ berth restrictions and regulations at both ends.
Delete 'Sub Owners check the prevailing port expenses at all ports of contact' with the Owners' next counter offer.

Comments:
The Charterers, in comparison to the Owners' offer:
-Request to have the right to use a second berth in all ports. The shifting expenses, including fuel consumed and crew overtime will be for the Owners' account,

but the lay-time will not cease during the shift.
-Request the number of discharge ports to be at their option.
-Accept the cargo for each discharge port to be in accordance with the ship's stress and to the Master's approval.
-Accept, if there are two discharge ports, for the ship to discharge in rotation according to the Owners' choice.
-Request only the term of 'always afloat' to be in effect in all ports. The Owners' addition 'always accessible' is rejected.
-Request the Owners' to retain responsibility for control of the ports'/berths' restrictions to all nominated ports.
-Request the Owners to lift their subject of checking the port expenses when counter offering.

The Charterers continue:
5.
Please advise of ship's full itinerary/e.t.a. at load port. The ship's itinerary/e.t.a. will be part of and incorporated in the Charter Party.

Comments:
The Charterers request for the ship's full itinerary and the estimated arrival time (abbreviated as 'e.t.a.') at the load port to be advised and the Owners' reply will form part of the terms and be incorporated in the Charter Party.
Obviously, the Owners' responsibilities will be increased when replying.

The Charterers continue:
6.
Freight USD 19.25 pmtn 1/1 free in/out spout/grabbed trimmed payable 90pct net of tax and bank charges into the Owners' nominated bank account within 5 banking days after the ship's sailing and signing/releasing of Congebill edition 1994 Bill of Lading marked as 'freight payable as per Charter Party', less the address commissions, less the estimated despatch at load port, less the Owners' contribution towards extra insurances and less any amount paid by the Charterers on the Owners' behalf at load port.
It is the Charterers' option the Bill of Lading to be marked as 'Freight Prepaid' in which case the original Bill of Lading will be kept in the Agents' custody and will be released to the Shippers upon the Charterers' confirmation of freight payment.
If a second discharge port is used, the Charterers will pay an additional freight of USD 1,50 pmtn on the entire cargo quantity.
After completion of the discharging operation, the Owners will courier/mail their original final freight statement/Time Sheets for all ports, supported by all the

original load/discharge documents (N.O.R./S.O.F. etc.). The balance freight plus demurrages in both ends or minus despatch at discharge port/s or any amount due to the Owners or Charterers will be paid within 10 banking days after receipt of all documents.
The Owners are not entitled to claim for dead-freight unless a proper notice has been given and acknowledged by the Charterers/Shippers before sailing. However, and if there is a dispute, the Owners and/or Master have not the right to clause or delay the release of the presented Bills of Lading. If the Charterers are liable for dead freight, then the time allowed for the loading and discharging will be calculated based on the paid cargo quantity and not on the actual loaded cargo quantity.
If there is a dispute on the presented Time Sheets, the Charterers will pay the undisputed amount within 10 banking days together with the balance freight and the disputed amount will be settled amicably or will be referred to Arbitration.
The lay-time computation in all ports will be based only on the S.O.F. as issued in every port and duly signed by the Master and Agents. The Master will endeavor to have it also signed by the Shippers and Receivers at the load and discharge port, respectively.
Full freight deemed earned on signing the Bill of Lading discountless and non-returnable ship and or cargo lost or not lost.

Comments:
The Charterers, in comparison to the Owners' offer:
-accept 'free in/out' but reject 'trimmed'. The cargo will be trimmed as the loading facilities (spout/ grab) can provide. The Master's request for additional trimming will be for the Owners' time/expenses.
- the 'freight' will be paid 90 percent (the Owners' offer was 100 percent) and the Charterers are entitled to deduct from the initial freight payment the address commission, the estimated possible despatch incurred at the load port, the extra insurances as will be agreed, and any amount paid on the Owners' behalf at the load port.
-request the freight to be paid within some days after the ship's sailing from the port (the Owners' offer was 'after completion of loading').
-accept Congebill edition 1994 type of Bill of Lading to be issued.
-request to have the option to mark the Bill of Lading 'freight prepaid' but will be released after the 'freight' payment.
-request to pay USD 1,50 pmtn (Owners' offer was at USD 2,5 pmtn) on entire cargo quantity for the use of a possible second discharge port.
-in case the ship is short-laden, should notice the Charterers/Shipper before sailing.
-request the balance freight & possible demurrage to be paid 10 days after the

Owners' presentation of all relevant documents.
-request the Master to endeavour to get the Shippers' and Receivers' signature on the issued original S.O.F.
-request the freight to be due after the signing of the bill of lading even if the cargo and/or the ship is lost or not lost (Owners' offer was 'upon shipment').

The Charterers continue:
7.
Loading rate/lay-time:
the cargo will be loaded free of risks and expenses to the ship at the rate of 4,000 metric tons daily or pro rata, per weather working day of 24 consecutive hours, Saturdays, Sundays, and Holidays excluded even if used.
Time from Friday 17:00 hrs. until Monday 08:00 hrs. or from 17:00 hrs. on a day preceding a Holiday until 08:00 hrs. of the next working day, not to count as lay-time, even if used.
Valid Notice Of Readiness will be tendered in writing at berth, or, in case of congestion or for any other reason beyond the Owners'/ship's fault, the ship's berthing is not allowed upon arrival, from the anchorage or from the place where it will be instructed to wait within or out of the port's commercial limits, by any means of telecommunication (cable/ telex/e-mail/fax) to the Agents, during the official working hours from 08:00 hrs.- up to and including 17:00 hrs. from Monday to Friday and the lay-time will commence counting at 08:00 hrs. the next working day, whether in berth or not, whether in free pratique or not, whether customs are cleared or not, whether in port or not.
If the ship berths upon arrival, a valid N.O.R. will be tendered only after all preloading surveys including holds' inspections are fully passed by the Charterers' and/or Shippers' surveyors and/or by a national cargo bureau and/or local grain surveyors, and provided the ship is at the berth and the free pratique is granted.
In the case of congestion or for any other reason beyond the Owners'/ship's fault, the ship's berthing is not allowed upon arrival, then after the ship's berthing, lost time for the preloading surveys including lost time for holds' inspections and to grant free pratique, will not count as lay-time or time on demurrage.
The 'NOR is deemed to be tendered and accepted from the time of tendering and the lay-time will commence counting in accordance with the terms of the Charter Party.
Time used before the commencement of lay-time will not count as lay-time.
Discharging rate/lay-time:
the cargo will be discharged in one or two ports, free of risks and expenses to the ship, at the rate of 10,000 metric tons daily or pro rata, per weather working day of 24 consecutive hours, Saturdays, Sundays, and Holidays included.

A valid Notice Of Readiness at the first or sole discharge port will be tendered in writing at berth or, in case of congestion or for any other reason beyond the Owners'/ship's fault, the ship's berthing is not allowed upon arrival, from the anchorage or from the place where it will be instructed to wait within or out of the port's commercial limits, by any means of telecommunication (cable/ telex-/e-mail/fax) to the Agents during the official working hours between 08:00- up to and including 1700 hrs. from Monday to Sunday and the lay-time will commence counting at 08:00 hrs. of the next working day, whether in berth or not, whether in free pratique or not, whether customs are cleared or not, whether in port or not.
If the ship berths upon arrival, a valid N.O.R. shall be tendered only after all pre-discharging surveys/ inspections are fully passed by the Charterers' and/or Receivers' surveyors and/or by a national cargo's bureau and/or local grain surveyors and provided the ship is at berth and the free pratique is granted.
In the case of congestion or for any other reason beyond the Owners'/ship's fault, the ship's berthing is not allowed upon arrival, then after the ship's berthing, the lost time for the pre-discharging surveys, including the time to grant free pratique, will not count as lay-time or time on demurrage.
The 'NOR is deemed to be tendered and accepted from the time of tendering and the lay-time will commence counting in accordance with the terms of the Charter Party.
Time used before the commencement of lay-time will not count as lay-time.
If a second discharge port is used, the NO.R. will be tendered in writing upon arrival at berth or, in case of congestion or for any other reason beyond the Owners'/ ship's fault, the ship's berthing is not allowed upon arrival, from the anchorage or from the place where it will be instructed to wait within or out of the port's commercial limits as will be instructed to wait, by any means of telecommunication (cable/ telex/e-mail/fax) to the Agents, at any time day or night from Monday to Sunday. The lay-time will commence counting upon the first official resumption of work at the port, whether in berth or not, whether in free pratique or not, whether customs are cleared or not, whether in port or not.
All time used for pre-discharging surveys, including free pratique, will count as lay-time unless the pre-discharging surveys, including free pratique, have been failed, then the time from failure until fully passing is not to count as lay-time or time on demurrage.

Comments:
The Charterers, in comparison to the Owners' offer:
At the load port, request:
-4,000 mtns as loading rate (Owners' offer was at 5,000 mtns)
-Saturday to be an exception day and the lay-time will cease from 17:00 hrs. on

Friday (the Owners' offer was that Saturday is a working day until noon time).
-the lay-time will not count during the exception periods, even if used (Owners' offer was 'unless used').
-the N.OR. to be tendered from Monday to Friday from 08:00 to17:00 hrs. (Owners' offer had included Saturday from 08:00 to 12:00 hrs.).
-the lay-time to commence counting the next working day at 08:00 hrs. if the N.O.R. is given within working hours of the previous working day (Owners' offer was 08:00/14:00 hrs. term).
-the N.O.R. at berth to be tendered only after the pre-loading surveys are fully passed -the time used before the commencement of lay-time not to count (Owners' offer was to count).
At discharge port, request:
-10,000 mtns to apply as discharging rate as per their initial cargo order (Owners' offer was at 12.000 mtns)
-the lay-time to commence counting the next working day at 08:00 hrs if the N.O.R. is given within working hours of the previous working day (Owners' offer was 08:00/14:00 hrs. term).
-the N.O.R. at berth to be tendered only after the pre-discharging surveys are fully passed.
-the time used before the commencement of lay-time not to count (Owners' offer was to count).
-if a second discharge port is used, the lay-time to commence upon the first official resumption of work (Owners' offer was to count upon arrival).

The Charterers continue:
8.
Demurrage USD 13,000 per day, pro rata half despatch for all working time saved at both ends.
Lay-time to be non-reversible between the load/discharge port but to be reversible between the discharging ports.

Comments:
The Charterers, in comparison to the Owners' offer:
-request as dem rate USD 13,000 (the Owners' offer was 15,000).
-request half despatch (the Owners' offer was 'free despatch').
-the Charterers accept the lay-time to be non-reversible between the load and discharge port but if a second discharge port is used, then the lay-time will be reversible between the two discharge ports, i.e., a Time Sheet will be issued for the load port and a unique Time Sheet for both discharge ports.
-accept that they are the sole party responsible for the payment of the demurrage to the Owners.

- the Charterers accepted the lay-time to be only computed on the signed S.O.F.

The Charterers continue:

9.

At no time during the draft survey (initial/intermediate or final) in all ports, should the liquids be shifted, taken in, or pumped out without prior advice to the Charterers and all draft surveyors. If it is required, the Master is to forward to the Charterers/Agents, upon arrival at the loading and discharging port and before the commencement of the loading and discharging operation, a certificate indicating all the remaining ballast in the ship's tanks.

Comments:

The Charterers, in comparison to the Owners' offer:

-accept the Owners' clause, but the Charterers add that the Master, during the draft survey, is not allowed to shift any liquids, take in or pump out ballast water, which can eventually affect the correct measurement of loaded or discharged cargo.

The Charterers continue:

10.

a- first shifting time from the anchorage or standby berth or pilot station to the load or discharge berth will not count as lay-time, even if the ship is on demurrage.

b- time required for the initial/intermediate/final draft survey in all ports not to count as lay-time, even if the ship is on demurrage.

c- required time for the opening and closing of hatches will count as lay-time except for the first opening and last closing, where time will not count as lay-time, even if the ship is on demurrage.

d- national or state/local holidays as specified in the current bimco's calendar to apply only as holidays.

e- the ship will load with trim as the spout/grab can provide under the Charterers' time/expenses. Extra trimming required by the Master of the ship will be arranged and paid for by the Owners at their time/risk.

g- time used for the ship's shifting alongside quay by mooring lines in order to make all hatches available for the operation will count as lay-time. The shifting alongside will not be considered as berth shifting unless tugs are used in which case it will count as shifting. In all cases, the Officers', and Crews' expenses/overtime will be for the Owners' account.

l- the ship's Officers and Crew's overtime will be for the Owners' account.

n- if the loading and/or discharging operation is interrupted or extended due to the insufficiency of the ship's ballast/de-ballast pump capacity in relation to the

loading/discharging capacity, then such time lost will not count as lay-time and extra expenses incurred will be for the Owners' account.
o- the Master of the ship will liaise with the stevedores to load/distribute the cargo in the cargo holds and discharge the ship within the safe operational limits and as per the SOLAS requirements. Time lost and expenses incurred due to interruption of the loading or discharging operation for stability reasons or for not adhering to the SOLAS provisions will be for the Owners' account.
p- the Owners are responsible and liable for all consequences, including, but not limited to, time lost and extra expenses incurred for unloading any overloaded cargo.
q- the ship will supply lights on board, when it is required, to lighten the deck's complete surface during the day and night, free of expenses to the Charterers.
r- all time lost and extra expenses incurred, due to the Owners' failure to provide the Agents with sufficient funds for covering the port expenses and other Owners' expenses incurred, before the ship's sailing from the port, will be for the Owners' account.
s- all time lost due to labourers boycotting the ship, because of its flag/ownership/registry/crew will not count as lay-time or time on demurrage and extra expense incurred will be for the Owners' account.
t- the holds' ladders must be in a good state of repair and in compliance with the safety at work and health act, otherwise, time lost will not count as lay-time on time on demurrages.
u- if the ship fails to pass the Port State Control inspection, time lost and expenses incurred in order to comply with recommendations or repairs made by the Authorities will not count as lay-time or time on demurrage.

Comments:
The Charterers, in comparison to the Owners' offer:
-accept some points (all points not touched have been accepted).
-amend some others according to their needs.
-add new points.

The Charterers continue:
11.
Extra insurance due to the ship's flag/ownership etc. will be in full for the Owners' account. Extra insurance due to the ship's age will be for the Charterers' account, but the Owners will contribute with a lump-sum amount of USD 10,000.

Comments:
The Charterers, in comparison to the Owners' offer:
-the imposed extra insurance due to the ship's age is to be for the Charterers'

account, however, the Owners will contribute with a lump-sum amount of USD 10,000. Obviously, the extra insurance should be more than USD 10,000.
-all other extra insurance, if applicable, is to be for the Owners' account in full (the Owners' offer was that all extra insurance applicable to be for the Charterers' account, even if it is due to the ship's age/ownership/flag etc.).

The Charterers continue:
12.
Any taxes/dues/wharfage dues/charges/duties calculated/levied on cargo including but not limited to the T3 tariff in Spain a/o on loaded cargo quantity a/o on Charter Party freight will be for the Charterers' account at both ends.
Any taxes/dues/wharfage dues/charges calculated/levied on ship/flag/ownership will be for the Owners' account at both ends.
The Owners will pay port expenses at both ends as per the port tariffs.

The above counter offer could also be given as follows, with the Owners' untouched terms to have been accepted by the Charterers:
12.
Any taxes/dues/wharfage dues/charges calculated/levied on ship/flag/ownership will be for the Owners' account at both ends.
The Owners will pay for the charged port expenses at both ends as per the port tariffs.

Comments:
The Charterers, in comparison to the Owners' offer:
-request the taxes/dues/wharfages/charges calculated/levied on ship/flag/ownership to be for the Owners' account.
-accept when calculated/levied on cargo/freight to be for their account
-request the Owners to pay the port expenses as per the port tariffs.

The Charterers continue:
13.
The attending Agents will be selected by the Charterers at both ends and the Owners will appoint the same.

Comments:
The Charterers, in comparison to the Owners' offer:
-request the Agents to be selected by them and to be appointed by the Owners (Owners' offer was to be selected and appointed by them).

The Charterers continue:

15.
In the case that the original Bills of Lading are not available at the discharge port, the Master/ Owners can allow the discharge/release of the entire cargo quantity without presentation of the original Bill of Lading in exchange for the Charterers' single L.O.I., as per the Owners' P&I Club's wording, signed only by the Charterers.

Comments:
The Charterers, in comparison to the Owners' offer:
-request to have the right for the cargo to be discharged and released without the presentation of the original Bill of Lading (the Owners' offer was to discharge only on presentation of the original Bill of Lading).

The Charterers continue:
16.
Written notices of the ship's expected arrival at the load port will be tendered to the Charterers and Agents on fixing, then .../ .../.../... days approximate and .../.../... hours definite notices.
Owners'/Master's tendered 'approximate' notice, with the ship's estimated time of arrival at load port has changed by more than 1 day, then the Master/Owners are to keep the Charterers and attending Agents closely informed on the ship's exact position/full itinerary and cause of delay.
When the Master tenders the definite notices, he will state the expected time of arrival/draft on arrival, the anticipated draft on sailing and the expected loadable cargo quantity.
Owners/Master will remain responsible for the accuracy of the tendered 'definite' notices, which can be changed only by adverse weather conditions and/or unforeseen circumstances and/or delays in completion of cargo discharging operations at the current port, always beyond Owners'/Master's/ ship's fault. If has changed by more than 12 hours, then the Master/Owners are to keep the Charterers and attending Agents closely informed on the ship's exact position, full itinerary, and cause of delay.
If any willful misrepresentation of the definite notice is tendered by the Owners/ Master of the ship to the Charterers and/or Agents for the position and readiness of the ship, they will remain responsible and liable for any and all consequences/ time lost/loss of turn for berthing until the berth becomes available again/additional costs including, but not limited to, the cost for cargo storage in the port, for the cancelled stevedoring shift etc.
On sailing from the load port, the Owners/Master will tender notices to the Charterers and Agents at the discharge port with the estimated time of arrival, the quantity of cargo loaded and the anticipated ship's draft on arrival.

.../.../... days approximate and .../.../... hours definite notices will be tendered thereafter.
The Charterers and Agents are to be kept by the Master/Owners continuously and fully advised of any change in the ship's position and its estimated time of arrival at the discharge port. When the Master tenders the definite notices, he will state the expected time of arrival and Owners/Master will remain responsible for the accuracy of the tendered 'definite' notices, which can be changed only by adverse weather conditions and/or unforeseen circumstances, always beyond Owners'/ Master's/ ship's fault.

Comments:
The Charterers state the required notices and to whom they will be given. However, they hold the Owners responsible for any willful misrepresentation of the definite notice which could be tendered by the Owners/Master of the ship.

The Charterers continue:
17.
The stevedores, although appointed by the Charterers/Shippers or their Agents at the loading port and by the Charterers/Receivers or their Agents at the discharging port respectively, will be deemed servants of the Owners and will be under the direction and control of the Master.
Claims for stevedore damage is to be settled directly between the Owners and stevedores, but the Charterers are to render all possible assistance and in case the Owners cannot settle with stevedores the Charterers to remain ultimately responsible.

Comments:
The Charterers, in comparison to the Owners' offer:
-reject in full the Owners' proposed stevedoring clause.
-request the stevedoring claims to be settled directly between the Owners and stevedores.
-accept to be ultimately responsible in case the Owners cannot settle with the stevedores.
For such a type of cargo and for loading/discharging with normal means of operation, the Charterers' proposal to remain ultimately responsible could be accepted by the Owners. However, when the cargo is pig iron, scrap, etc. or it is loaded/discharged with big grabs, etc., a full stevedoring term must be agreed.

The Charterers continue:
19.
Holds' inspection:

The ship, upon arrival at the first or sole load port, must have all cargo holds and hatches free of loose rust flakes/rust scales and loose scales, free from salt and free from residues of previous cargo. It will also be clean/clear/fresh water washed and dried and free from insects and odours and in all respects ready and suitable for loading the Charterers' cargo, according to the rules and regulations prevailing at the port of loading.
If the ship fails the preloading surveys, including the holds' inspections, the Charterers and/or Shippers and/or Port Authorities have the right to order the ship to shift from the berth to anchorage or to a waiting place as deemed necessary for commercial and/or safety reasons, due to the ship's inability to pass the preloading surveys and inspections. All shifting expenses incurred, including tugs/pilots etc. to anchorage/waiting place and then back to berth will be for the Owners' account and the time lost will not count as lay-time or time on demurrage.
The Owners will take immediate corrective steps to expedite the readiness of the ship-cleaning, as fast as possible, either with the Crew of the ship, if it is permitted by the regulations/port labour unions, or with the shore labourers, in order to meet the surveyors' requirements at their cost/risk.
If the ship fails to pass re-inspection within 72 hours from the first inspection, the Charterers have the option to immediately declare the cancellation of this Charter Party, which will be considered null and void and no claim will be pursued, either by the Owners or Charterers or any third party. All in/out of the port expenses, including Agency fees, will be for the Owners' account.
Hatch covers' inspection:
The ship's hatch covers will be watertight and will be maintained in top condition. The Charterers/ Shippers may order, through their surveyors in the berth or in anchorage, for an ultrasonic test and/or a hose test or Chalk test on all hatch covers, but the Charterers' requirements will not exceed the ship's class rule. The costs of the survey will be for the Charterers' account, but the Master must give every facility to the Charterers and their surveyor, in order to conduct such tests and the time lost will count as lay-time or time on demurrage.
In the event that the ship has failed the hatch inspection, the Owners will immediately arrange for shore labourers, or, at their discretion, for the ship's crew, provided it is permitted by regulations/ port labour unions, to make repairs in the anchorage or in berth if allowed by the Shippers/Port Authorities. Time lost/ costs for this operation and extra expenses incurred, including the shift of the ship to anchorage and back to the berth will be for the Owners' account.
If the ship fails to pass re-inspection within 72 hours from the first inspection, the Charterers have the option to immediately declare the cancellation of this Charter Party which will be considered null and void and no claim will be pursued either by the Owners or Charterers or any third party. All in/out of the port expenses including Agency fees will be for the Owners' account.

Comments:
Charterers have added a new term and request the cargo holds/hatches upon arrival at the load port to be clean, washed down and dried, free of rust, free of loose rust flakes/scales and residues of previous cargo. If the ship fails the preloading surveys, including the holds' inspection, the Charterers or the Port Authorities have the right to order the ship to shift in the anchorage and lost time & expenses incurred will be for the Owners' account.
Charterers reserve the right, before the commencement of loading, to perform an ultrasonic test and/or a hose tests or Chalk tests on all hatch covers. If a hatch cover is found defective, the Owners will have to repair it immediately under their time/expenses. In the case of a second rejection, the Charterers have the option to immediately declare the cancellation of the Charter Party.

TO SUMMARIZE THE CHARTERERS' COUNTER OFFER:

Ref MV TAXIARCHES
Account Eagle Trading and Shipping Inc
Constanza /2 Spain

Charterers a/e for 30 mnts reply as follows:
-Please advise of full name/style/contact numbers of Head Owners/Managers and Disponent Owners.

Owners guarantee that during the currency of this voyage:
a- the ship will not change ownership and/or class and will not be sold for scrap during or after this voyage without the Charterers' written consent.
b- the ship will proceed with the utmost despatch directly to the discharge port/s and will not deviate for any purpose whatsoever without the Charterers' consent, excluding bunkering en route and to the nearest bunkering station or for saving life.
c- the ship is fully covered by the P&I Club (as a member of an International Group of P&I Clubs) and has been covered for ship oil pollution risks and wreck removal risks and will remain so throughout the duration of this charter. In this respect, the Owners will furnish the Charterers with the P&I Club's written confirmation, if and whenever it is requested by them.
d- the ship is fully classed as Lloyd's 100A1 or equivalent with a classification society being a full member of the International Association of Classification Societies (IACS) throughout the duration of this charter. In this respect, the Owners will furnish the Charterers with the Classification society's written confirmation, if and whenever it is requested by them.

e- the ship's H&M insurance will be fully maintained and will not be changed. In this respect, the Owners will furnish the Charterers with the H&M underwriters' written confirmation, if and whenever it is requested by them.
f- the ship has no outstanding recommendations, restrictions or any other requirements of any nature in respect to the flag state, port state control, or classification inspections that may affect this voyage and the Charter Party.
g- the ship is suitable for entering and berthing at the specified ports, as well as loading/ discharging the minimum contractual cargo quantity. The ship will comply with all applicable laws and regulations and is in all respects eligible for trading at the specified ports. The ship has all valid and up-to-date certificates and other documents on board required for this voyage.
h- the ship can open/close its hatches with its own means.
i- the ship has adequate mooring lines and will be tending them properly, whilst at the loading or discharging port.
j- the ship will give, free of expenses to the Charterers, the full use of its lighting on its deck and in its cargo holds, that will be adequate for all cargo operations.
k- the ship will comply with all applicable laws and regulations and is in all respects eligible for trading to the ports, places or countries specified in this Charter Party. The ship has all valid and up-to-date certificates and other documents on board that are required for this voyage.
l- the ship is fully fitted for the loading of grains according to the SOLAS 1974 regulations and any subsequent amendments thereto without requiring any shifting boards. The Owners guarantee that the ship is suitable for entering, berthing, and loading/discharging the contractual cargo quantity of wheat in bulk.
m- both, the ship and the company fully comply with ISPS/ISM requirements and will remain so throughout the duration of this charter.
n- the ship is sid/bc, steelfloored throughout, with hoppers port/starboard, with the engine/ bridge aft and its holds/hatches/tank tops are clear and unobstructed without a centreline bulkhead, walkways, centrebeams and free of all/any obstructions which might incur difficult access to the cargo during loading/discharging. Cargo will not be loaded in deep tanks, sidetanks, wing tanks, bunker spaces or any other confined spaces unreachable to the means of loading and discharging.
o- the ship has not carried mammalian meat, bone meal, fishmeal, asbestos cullet, or nuts for the past 12 months.
p- the ship has not been blacklisted/boycotted or arrested due to the ship's flag/ ownership/crew employment/age/past trading or for any other reason from local authorities and or unions in all countries designated under this Charter Party.
q- the ship is free of all encumbrances, arrests, and maritime liens.
r- the ship has no .... (insert countries which may affect the voyage) interests and is not black listed by Arab league countries.

s- the description of the ship as provided by the Owners is an accurate representation of the characteristics of the ship which can only be modified with the usual and acceptable shipping interpretation of the word 'about'. This clause shall override any inclusion of the words 'without guarantee' in the ship's description. The Owners are responsible and liable for any consequences/ delays/ losses/expenses that may arise when providing any misleading information on the ship's characteristics and on the terms agreed in this Charter Party.
t- the ship is ITF compliant or bona fide trade union's agreement or equivalent and will remain so throughout the duration of this charter and shall remain responsible for all consequences and/or delays if the ship is sailing under a flag of 'convenience'.
u- the ship, throughout the duration of the Charter Party will have on board and make available to the Charterers or their Agents or to the appointed surveyors, the capacity's plan, the deadweight and displacement scales, the hydrostatic curves and tables of displacement, the calibration scales for all ship's tanks and trimming correction tables as well as all sounding tables and all sounding tubes which will be properly maintained, in good condition and free from impediments. Plimsoll marks amidships and draft marks on port and starboard side, bow, amid ship and stern are always to be clearly cut and marked on shell platting and readable.

1.
Account Eagle Trading and Shipping Inc of British Virgin Islands
(Charterers advise their background)
Recent fixtures:
MV .... Fixed through .... Ph ...... Owners/brokers ph ....
MV .... Fixed through .... Ph .... Owners/brokers ph............
etc. etc.
2.
Sub Shippers'/Receivers' approval and stem which will be declarable at the latest 2 hours after the fixing of the main terms with the Charter Party details in order.
3.
the cargo stows about 44', but wog. No other cargo is allowed to be loaded.
4.
1 safe port -1/2 safe berth/s Constanza Romania always afloat, as loadport.
1 safe port-1/2 safe berth/s Tarragona Spain always afloat and/ or at chopt
1safe port-1/ 2 safe berth/s Valencia Spain always afloat, as discharge port/s.
If two discharge ports are used, the ship will discharge in rotation at the Owners' option and will be declared before the completion of the loading and the issuance of the Bill of Lading. The cargo quantity for each discharge port should be

in accordance with the ship's stress and the Master's approval.
If the second berth is used at all port/s, the shifting expenses, including officers'/ crew's overtime and consumed bunkers will be for the Owners' account, but time will continue to count as lay-time.
Owners are responsible for checking and satisfying themselves with all port/ berth restrictions and regulations at both ends.
Delete 'Sub Owners check the prevailing port expenses at all ports of contact' with the Owners' next counter offer.
5.
Please advise of ship's full itinerary/e.t.a. at load port. The ship's itinerary/e.t.a. will be part of and incorporated in the Charter Party.
6.
Freight USD 19.25 pmtn 1/1 free in/out spout/grabbed trimmed payable 90pct net of tax and bank charges into the Owners' nominated bank account within 5 banking days after the ship's sailing and signing/releasing of Congebill edition 1994 Bill of Lading marked as 'freight payable as per Charter Party', less the address commissions, less the estimated despatch at load port, less the Owners' contribution towards extra insurances and less any amount paid by the Charterers on the Owners' behalf at load port.
It is the Charterers' option the Bill of Lading to be marked as 'Freight Prepaid' in which case the original Bill of Lading will be kept in the Agents' custody and will be released to the Shippers upon the Charterers' confirmation of freight payment.
If a second discharge port is used, the Charterers will pay an additional freight of USD 1,50 pmtn on the entire cargo quantity.
After completion of the discharging operation, the Owners will courier/mail their original final freight statement/Time Sheets for all ports, supported by all the original load/discharge documents (N.O.R./S.O.F. etc.). The balance freight plus demurrages in both ends or minus despatch at discharge port/s or any amount due to the Owners or Charterers will be paid within 10 banking days after receipt of all documents.
The Owners are not entitled to claim for dead-freight unless a proper notice has been given and acknowledged by the Charterers/Shippers before sailing. However, and if there is a dispute, the Owners and/or Master have not the right to clause or delay the release of the presented Bills of Lading. If the Charterers are liable for dead freight, then the time allowed for the loading and discharging will be calculated based on the paid cargo quantity and not on the actual loaded cargo quantity.
If there is a dispute on the presented Time Sheets, the Charterers will pay the undisputed amount within 10 banking days together with the balance freight and the disputed amount will be settled amicably or will be referred to Arbitration.

The lay-time computation in all ports will be based only on the S.O.F. as issued in every port and duly signed by the Master and Agents. The Master will endeavor to have it also signed by the Shippers and Receivers at the load and discharge port, respectively.
Full freight deemed earned on signing the Bill of Lading discountless and non-returnable ship and or cargo lost or not lost.
7.
Loading rate/lay-time:
the cargo will be loaded free of risks and expenses to the ship at the rate of 4,000 metric tons daily or pro rata, per weather working day of 24 consecutive hours, Saturdays, Sundays and Holidays excluded even if used.
Time from Friday 17:00 hrs. until Monday 08:00 hrs. or from 17:00 hrs. on a day preceding a Holiday until 08:00 hrs. of the next working day, not to count as lay-time, even if used.
Valid Notice Of Readiness will be tendered in writing at berth, or, in case of congestion or for any other reason beyond the Owners'/ship's fault, the ship's berthing is not allowed upon arrival, from the anchorage or from the place where it will be instructed to wait within or out of the port's commercial limits, by any means of telecommunication (cable/ telex/e-mail/fax) to the Agents, during the official working hours from 08:00 hrs.- up to and including 17:00 hrs. from Monday to Friday and the lay-time will commence counting at 08:00 hrs. the next working day, whether in berth or not, whether in free pratique or not, whether customs are cleared or not, whether in port or not.
If the ship berths upon arrival, a valid N.O.R. will be tendered only after all preloading surveys including holds' inspections are fully passed by the Charterers' and/or Shippers' surveyors and/or by a national cargo bureau and/or local grain surveyors, and provided the ship is at the berth and the free pratique is granted.
In the case of congestion or for any other reason beyond the Owners'/ship's fault, the ship's berthing is not allowed upon arrival, then after the ship's berthing, lost time for the preloading surveys including lost time for holds' inspections and to grant free pratique, will not count as lay-time or time on demurrage.
The 'NOR is deemed to be tendered and accepted from the time of tendering and the lay-time will commence counting in accordance with the terms of the Charter Party.
Time used before the commencement of lay-time will not count as lay-time.
Discharging rate/lay-time:
the cargo will be discharged in one or two ports, free of risks and expenses to the ship, at the rate of 10,000 metric tons daily or pro rata, per weather working day of 24 consecutive hours, Saturdays, Sundays, and Holidays included.
A valid Notice Of Readiness at the first or sole discharge port will be tendered in writing at berth or, in case of congestion or for any other reason beyond the Own-

ers'/ ship's fault, the ship's berthing is not allowed upon arrival, from the anchorage or from the place where it will be instructed to wait within or out of the port's commercial limits, by any means of telecommunication (cable/ telex/e-mail/fax) to the Agents during the official working hours between 08:00 hrs.- up to and including 1700 hrs. from Monday to Sunday and the lay-time will commence counting at 08:00 hrs. of the next working day, whether in berth or not, whether in free pratique or not, whether customs are cleared or not, whether in port or not.
If the ship berths upon arrival, a valid N.O.R. shall be tendered only after all pre-discharging surveys/ inspections are fully passed by the Charterers' and/or Receivers' surveyors and/or by a national cargo's bureau and/or local grain surveyors and provided the ship is at berth and the free pratique is granted.
In the case of congestion or for any other reason beyond the Owners'/ship's fault, the ship's berthing is not allowed upon arrival, then, after the ship's berthing, the lost time for the pre-discharging surveys, including the time to grant free pratique, will not count as lay-time or time on demurrage.
The 'NOR is deemed to be tendered and accepted from the time of tendering and the lay-time will commence counting in accordance with the terms of the Charter Party.
Time used before the commencement of lay-time will not count as lay-time.
If a second discharge port is used, the NO.R. will be tendered in writing upon arrival at berth or, in case of congestion or for any other reason beyond the Owners'/ship's fault, the ship's berthing is not allowed upon arrival, from the anchorage or from the place where it will be instructed to wait within or out of the port's commercial limits as will be instructed to wait, by any means of telecommunication (cable/ telex/e-mail/fax) to the Agents, at any time day or night from Monday to Sunday. The lay-time will commence counting upon the first official resumption of work at the port, whether in berth or not, whether in free pratique or not, whether customs are cleared or not, whether in port or not. All time used for pre-discharging surveys, including free pratique, will count as lay-time unless the pre-discharging surveys, including free pratique, have been failed, then the time from failure until fully passing is not to count as lay-time or time on demurrage.
8.
Demurrage USD 13,000 per day, pro rata half despatch for all working time saved at both ends.
Lay-time to be non-reversible between the load/discharge port but to be reversible between the
discharging ports
9.
At no time during the draft survey (initial/intermediate or final) in all ports, should the liquids be shifted, taken in, or pumped out without prior advice to the

Charterers and all draft surveyors. If it is required, the Master is to forward to the Charterers/Agents, upon arrival at the loading and discharging port and before the commencement of the loading and discharging operation, a certificate indicating all the remaining ballast in the ship's tanks.
10.
a- first shifting time from the anchorage or standby berth or pilot station to the load or discharge berth will not count as lay-time, even if the ship is on demurrage.
b- time required for the initial/intermediate/final draft survey in all ports not to count as lay-time, even if the ship is on demurrage.
c- required time for the opening and closing of hatches will count as lay-time except for the first opening and last closing, where time will not count as lay-time, even if the ship is on demurrage.
d- national or state/local holidays as specified in the current bimco's calendar to apply only as holidays.
e- the ship will load with trim as the spout/grab can provide under the Charterers' time/expenses. Extra trimming required by the Master of the ship will be arranged and paid for by the Owners at their time/risk.
g- time used for the ship's shifting alongside quay by mooring lines in order to make all hatches available for the operation will count as lay-time. The shifting alongside will not be considered as berth shifting unless tugs are used in which case it will count as shifting. In all cases, the Officers' and Crews' expenses/overtime will be for the Owners' account.
l- the ship's Officers and Crew's overtime will be for the Owners' account.
n- if the loading and/or discharging operation is interrupted or extended due to the insufficiency of the ship's ballast/de-ballast pump capacity in relation to the loading/discharging capacity, then such time lost will not count as lay-time and extra expenses incurred will be for the Owners' account.
o- the Master of the ship will liaise with the stevedores to load/distribute the cargo in the cargo holds and discharge the ship within the safe operational limits and as per the SOLAS requirements. Time lost and expenses incurred due to interruption of the loading or discharging operation for stability reasons or for not adhering to the SOLAS provisions will be for the Owners' account.
p- the Owners are responsible and liable for all consequences, including, but not limited to, time lost and extra expenses incurred for unloading any overloaded cargo.
q- the ship will supply lights on board, when it is required, to lighten the deck's complete surface during the day and night, free of expenses to the Charterers.
r- all time lost and extra expenses incurred, due to the Owners' failure to provide the Agents with sufficient funds for covering the port expenses and other Owners' expenses incurred, before the ship's sailing from the port, will be for the Owners' account.

s- all time lost due to labourers boycotting the ship, because of its flag/ownership/registry/crew will not count as lay-time or time on demurrage and extra expense incurred will be for the Owners' account.
t- the holds' ladders must be in a good state of repair and in compliance with the safety at work and health act, otherwise, time lost will not count as lay-time on time on demurrages.
u- if the ship fails to pass the Port State Control inspection, time lost and expenses incurred in order to comply with recommendations or repairs made by the Authorities will not count as lay-time or time on demurrage.
11.
Extra insurance due to the ship's flag/ownership etc. will be in full for the Owners' account. Extra insurance due to the ship's age will be for the Charterers' account, but the Owners will contribute with a lump-sum amount of USD 10,000. -
12.
Any taxes/dues/wharfage dues/charges/duties calculated/levied on cargo including but not limited to the T3 tariff in Spain a/o on loaded cargo quantity a/o on Charter Party freight will be for the Charterers' account at both ends.
Any taxes/dues/wharfage dues/charges calculated/levied on ship/flag/ownership will be for the Owners' account at both ends.
The Owners will pay port expenses at both ends as per the port tariffs.
13.
The attending Agents will be selected by the Charterers at both ends and the Owners will appoint the same.
15.
In the case that the original Bills of Lading are not available at the discharge port, the Master/ Owners can allow the discharge/release of the entire cargo quantity without presentation of the original Bill of Lading in exchange for the Charterers' single L.O.I., as per the Owners' P&I Club's wording, signed only by the Charterers.
16.
Written notices of the ship's expected arrival at the load port will be tendered to the Charterers and Agents on fixing, then .../ .../.../... days approximate and .../.../... hours definite notices.
Owners'/Master's tendered 'approximate' notice, with the ship's estimated time of arrival at load port has changed by more than 1 day, then the Master/Owners are to keep the Charterers and attending Agents closely informed on the ship's exact position/full itinerary and cause of delay.
When the Master tenders the definite notices, he will state the expected time of arrival/draft on arrival, the anticipated draft on sailing and the expected loadable cargo quantity.
Owners/Master will remain responsible for the accuracy of the tendered 'defi-

nite' notices, which can be changed only by adverse weather conditions and/or unforeseen circumstances and/or delays in completion of cargo discharging operations at the current port, always beyond Owners'/Master's/ ship's fault. If has changed by more than 12 hours, then the Master/Owners are to keep the Charterers and attending Agents closely informed on the ship's exact position, full itinerary, and cause of delay.

If any willful misrepresentation of the definite notice is tendered by the Owners/ Master of the ship to the Charterers and/or Agents for the position and readiness of the ship, they will remain responsible and liable for any and all consequences/time lost/loss of turn for berthing until the berth becomes available again/additional costs including, but not limited to, the cost for cargo storage in the port, for the cancelled stevedoring shift etc.

On sailing from the load port, the Owners/Master will tender notices to the Charterers and Agents at the discharge port with the estimated time of arrival, the quantity of cargo loaded and the anticipated ship's draft on arrival.

.../.../... days approximate and .../.../... hours definite notices will be tendered thereafter.

The Charterers and Agents are to be kept by the Master/Owners continuously and fully advised of any change in the ship's position and its estimated time of arrival at the discharge port. When the Master tenders the definite notices, he will state the expected time of arrival and Owners/Master will remain responsible for the accuracy of the tendered 'definite' notices, which can be changed only by adverse weather conditions and/or unforeseen circumstances, always beyond Owners'/Master's/ ship's fault.

17.

The stevedores, although appointed by the Charterers/Shippers or their Agents at the loading port and by the Charterers/Receivers or their Agents at the discharging port respectively, will be deemed servants of the Owners and will be under the direction and control of the Master.

Claims for stevedore damage is to be settled directly between the Owners and stevedores, but the Charterers are to render all possible assistance and in case the Owners cannot settle with stevedores the Charterers to remain ultimately responsible.

19.

Holds' inspection:

The ship, upon arrival at the first or sole load port, must have all cargo holds and hatches free of loose rust flakes/rust scales and loose scales, free from salt and free from residues of previous cargo. It will also be clean/clear/fresh water washed and dried and free from insects and odours and in all respects ready and suitable for loading the Charterers' cargo, according to the rules and regulations prevailing in the port of loading.

If the ship fails the preloading surveys, including the holds' inspections, the Charterers and/or Shippers and/or Port Authorities have the right to order the ship to shift from the berth to anchorage or to a waiting place as deemed necessary for commercial and/or safety reasons, due to the ship's inability to pass the preloading surveys and inspections. All shifting expenses incurred, including tugs/pilots etc. to anchorage/waiting place and then back to berth will be for the Owners' account and the time lost will not count as lay-time or time on demurrage.
The Owners will take immediate corrective steps to expedite the readiness of the ship-cleaning, as fast as possible, either with the Crew of the ship, if it is permitted by the regulations/port labour unions, or with the shore labourers, in order to meet the surveyors' requirements at their cost/risk.
If the ship fails to pass re-inspection within 72 hours from the first inspection, the Charterers have the option to immediately declare the cancellation of this Charter Party, which will be considered null and void and no claim will be pursued, either by the Owners or Charterers or any third party. All in/out of the port expenses, including Agency fees, will be for the Owners' account.
Hatch covers' inspection:
The ship's hatch covers will be watertight and will be maintained in top condition. The Charterers/ Shippers may order, through their surveyors in the berth or in anchorage, for an ultrasonic test and/or a hose test or Chalk test on all hatch covers, but the Charterers' requirements will not exceed the ship's class rule. The costs of the survey will be for the Charterers' account, but the Master must give every facility to the Charterers and their surveyor, in order to conduct such tests and the lost time will count as lay-time or time on demurrage.
In the event that the ship has failed the hatch inspection, the Owners will immediately arrange for shore labourers, or, at their discretion, for the ship's crew, provided it is permitted by regulations/ port labour unions, to make repairs in the anchorage or in berth if allowed by the Shippers/Port Authorities. Time lost/ costs for this operation and extra expenses incurred, including the shift of the ship to anchorage and back to the berth will be for the Owners' account.
If the ship fails to pass re-inspection within 72 hours from the first inspection, the Charterers have the option to immediately declare the cancellation of this Charter Party which will be considered null and void and no claim will be pursued either by the Owners or Charterers or any third party. All in/out of the port expenses including Agency fees will be for the Owners' account.

OWNERS' COUNTER REPLY TO CHARTERERS' LAST:

The Owners could reply as follows with the untouched terms to have been accepted by them.

Thanks for Charterers last.
Owners A/E for reply within 30 minutes as follows:

Owners' and Managers' full style/contact numbers as follows:
-Owners: ......
-Managers: ......
-Disponent Owners: not applicable.

Comments:
The Owners reply on who the Owners/Managers are, except for the Disponent Owners, as the ship is trading with the head Owners.

The Owners continue:
3.
the cargo stows about 44' in the ship's holds.

Comments:
The Owners, in comparison to the Charterers' offer:
-accept the cargo to be transported as sole cargo
-accept the cargo stowage to be declared on an 'about' basis but reject 'wog'.

The only official party that could inform and agree with the Owners on the expected stowage factor are the Charterers, who are to be informed and agree with the Shippers.
The Owners cannot verify the same and if the Charterers insist, then they should have no other alternative but:
- to insist and to at least get the provision of 'about'.
- to drop the negotiations.
- take the risk by accepting the Charterers last and collect all possible information through the attending Agents at the load port.
- to propose and amend the offer on a lump sum basis.

The Owners continue:
4.
If a second berth is used at every load/discharge port/s, the shifting expenses are for the Charterers' account and the lost time will count as lay-time, but the Officers'/crew's overtime and consumed bunkers will be for the Owners' account.
The final sailing configuration is always subject to Master's stowage plan and satisfaction, which is in conformity with the ship's loading manual.
The ship should always be left in a seaworthy and safe stress/trim condition and with cargo on board safely stowed during its sailing and/or its shifting between

all ports, berths, anchorages and wharfs to the Master's satisfaction.
Upon completion of the loading and/or discharging and/or lightering operation at the first or subsequent port and/or berth/ anchorage/wharf or at any time it is requested by any party (Master, Port Authorities etc.) to bring the ship in a seaworthy and safe stress/trim condition, then all the expenses/ extra port expenses incurred will be entirely for the Charterers' account and the time lost will continue to count as lay-time or time on demurrage, even during ballasting/ de-ballasting the ship's tanks and until the Master is fully satisfied and the ship is ready to sail or shift.

Comments:
The Owners, in comparison to the Charterers' offer:
-accept the use of the second discharge port to be at the Charterers' option.
-accept the use of a second berth, but the shifting expenses to be for the Charterers' account.
-accept the Officers'/crew's overtime and consumed bunkers to be for the Owners' account.
-accept only the 'always afloat' term to be in effect
-accept the responsibility of the prevailing port restrictions and regulations in all ports of contact.
-accept to delete their reservation to check the port expenses in all ports.
-add the term that the ship is to be left in a seaworthy and safe stress/trim condition between the berths/ports.

The Owners continue:
5.
ship's itinerary....
e.t.a. at load port ...

Comments:
The Owners, in comparison to the Charterers' offer:
-inform the Charterers accordingly.
-accept the ship's itinerary/eta to be part of and incorporated in the Charter Party.

The Owners continue:
6.
Freight USD 20.50 pmtn 1/1 free in/out spout/grabbed trimmed. Freight is payable 95pct net of tax and bank charges into the Owners' nominated bank account within 3 banking days after completion of the loading and presentation/ signing/ releasing of Congebill edition 1994 Bill of Lading marked as 'freight pay-

able as per Charter Party', less the address commissions, less the Owners' contribution towards extra insurance and less any amount paid by the Charterers on the Owners' behalf at load port but made only after the Owners' written consent.
It is the Charterers' option the Bill of Lading to be marked as 'Freight Prepaid' in which case the original Bills of Lading will be kept in the Agents' custody or at the Owners' option by the ship's P&I Club's representatives and will be released to the Shippers upon the Charterers' bank confirmation directly to the Owners' office that 100pct of the freight, less the address commissions, less the Owners' contribution towards extra insurance and less any amount paid by the Charterers on the Owners' behalf at the load port, but made only after the Owners' written consent, has been irrevocably remitted with the same date value. If a second discharge port is used, the Charterers will pay an additional freight of USD 2,00 pmtn on the entire quantity of cargo, payable together with the initial freight.
The balance freight, plus demurrage or minus despatch in both ends or any amount due to the Owners or Charterers will be paid within 5 banking days after completion of the discharging operation and after the Owners' presentation of all relevant documents such as the N.O.R./S.O.F. and the Owners' Time Sheets from all ports, together with their final freight and demurrage invoice, which all are accepted even by fax. The original documents will be forwarded by mail/ courier to the Charterers' office as soon as possible.
The Owners are not entitled to claim for dead freight unless a proper notice has been given and acknowledged by the Charterers/Shippers before sailing. The Charterers guarantee that they will be contactable at any time it is required, and all time lost in waiting will be counted as lay-time.
If the Charterers are liable for dead freight, then the time allowed for the loading and discharging will be calculated based on the paid cargo quantity and not on the actual loaded cargo quantity.
If there is a dispute on the presented Time Sheets, the Charterers will pay the undisputed amount within 5 banking days together with the balance freight and the disputed amount will be settled amicably or will be referred to Arbitration.
The lay-time computation in all ports will be based only on the S.O.F. as issued in every port and duly signed by the Master and Agents, but the Master will endeavor, but without guarantee, to also have it signed by the Shippers and Receivers at the load and discharge port, respectively. Full freight is deemed earned upon shipment discountless and non-returnable ship and or cargo lost or not lost.

Comments:
The Owners, in comparison to the Charterers' offer:
-request USD 20,50 pmtn basis 1/1 (Charterers' offer was USD 19, 25 pmtn).
-accept the cargo trimming to be performed as grab and spout can provide.

-request the freight to be paid 95 percent within 3 banking days after the 'completion' of loading (the Charterers' offer was 90 percent within 5 banking days after the ship's 'sailing' from the port).
-accept the Charterers to have the right to deduct the address commission, the Owners' contribution on extra insurance and the amounts paid on the Owners' behalf, but only after the Owners' written consent.
-reject the Charterers' offer to deduct the estimated despatch at the load port
-accept the Bill of Lading to be marked as 'freight prepaid', however the freight will be paid 100 percent, minus some deductions.
-request to have the option for the Bills of Lading to also be kept by the ship's P&I Club representatives and to be released only when the freight payment is irrevocably remitted to the Owners' bank account with the same date value.
-request for the second discharge port extra USD 2,00 pmtn (Charterers' offer was USD 1,50 pmtn).
-request the balances due to the Owners or Charterers to be settled within 5 banking days after completion of discharge and presentation of the relevant documents which are acceptable even by fax.
-accept not to claim for 'dead freight', unless proper notice has been given and acknowledged by the Charterers/Shippers before sailing, however, they ask the Charterers to guarantee that they will be contactable at any time it is required and all lost time in waiting will be counted as lay-time.
-request, if there is a dispute, for the undisputed amount to be paid within 5 banking days (Charterers' offer was within 10 banking days).
- request the full freight to become due upon shipment, i.e., upon commencement of loading (Charterers' offer was 'on signing' the Bill of Lading).

The Owners continue:
7.
Loading rate/lay-time:
the cargo will be loaded free of risks and expenses to the ship, at the rate of 4,000 metric tons daily or pro rata per weather working day of 24 consecutive hours, Sundays and Holidays excluded unless used, in which case the actual time used to count as lay-time.
Time from Saturday noon until Monday 08:00 hrs. or from 17:00 hrs. on a day preceding a Holiday until 08:00 hrs. of the next working day, not to count as lay-time, unless used in which case the actual time used to count as lay-time.
A valid Notice Of Readiness will be tendered in writing at berth, or in case of congestion or for any other reason beyond the Owners'/ship's fault, the ship's berthing is not allowed upon arrival, from the anchorage or from the place where it will be instructed to wait within or out of the port's commercial limits, by any means of telecommunication (cable/ telex/e-mail/fax) to the Agents, during the

official working hours from 08:00 hrs. up to and including 17:00 hrs. from Monday to Friday and from 08:00 hrs. up to and including 12:00 hrs. on Saturday. The lay-time will commence counting at 14:00 hrs. if the Notice of Readiness is tendered before noon or at 08:00 hrs. the next working day if the Notice of Readiness is tendered in the afternoon, whether in berth or not, whether in free pratique or not, whether customs are cleared or not, whether in port or not.
Time used before the commencement of lay-time will count as lay-time.
Discharging rate/lay-time:
The lay-time will commence counting at 14:00 hrs. if the Notice of Readiness is tendered before noon or at 08:00 hrs. of the next working day, if the Notice of Readiness is tendered in the afternoon, whether in berth or not, whether in free pratique or not, whether customs are cleared or not, whether in port or not. Time used before the commencement of lay-time will count as lay-time.

Comments:
The Owners, in comparison to the Charterers' offer:
At the load port:
-accept the loading rate to be 4,000 mtns instead of 5,000 mtns as they had initially offered.
-reject the Charterers' term 'even if used' and insist on 'unless used.'
-request the exception period to be from Saturday noon to Monday 08:00 hrs. (Charterers' offer was from Friday 17:00 hrs. to Monday 08:00 hrs.).
-request the 08:00/14:00 hrs. term for commencement of lay-time to be in effect (Charterers' offer was the lay-time to commence counting at 08:00 hrs. the next working day).
-request the time used before the commencement of lay-time to count as lay-time (Charterers' offer was the lay-time not to count before the commencement of lay-time).
-accept that if the ship berths upon arrival, a valid N.O.R. will be tendered only after all preloading surveys, including holds' inspections, are fully passed and free pratique is granted.
-accept that in case of congestion and after the ship's berthing, the preloading surveys, including lost time for holds' inspections and to grant free pratique, are not to count as lay-time or time on demurrage.
At the discharge port:
-accept the discharging rate to be 10,000 mtns.
-request the 08:00 hrs./14:00 hrs. term for commencement of lay-time to be in effect (Charterers' offer was the lay-time to commence counting at 08:00 hrs. the next working day).
-accept that if the ship berths upon arrival, a valid N.O.R. will be tendered only after all pre-discharging surveys/inspections are fully passed.

-accept that, in case of congestion, and, after the ship's berthing, the pre-discharging surveys, including lost time to grant free pratique, are not to count as lay-time or time on demurrage.
-request the time used before the commencement of lay-time to count as lay-time.
-accept in full the Charterers' terms for the second discharge port.

The Owners continue:
10.
b- required time for the initial/intermediate/final draft survey in all ports will be equally shared between the Owners and Charterers. However, expenses incurred to hire a launch or tug in order for the draft survey to be carried out at the berth or in the anchorage will be for the Charterers' account and time lost will count as lay-time or time on demurrage.

Comments:
The Owners, in comparison to the Charterers' offer:
-request the necessary time lost for the draft survey to be shared equally between them.
-repeat the term which was already accepted by the Charterers on item no 9 that the expenses to hire a launch or tug as well as lost time will be for Charterers account.

The Owners continue:
11.
Extra insurance due to the ship's flag/ownership etc. will be for the Owners' account. Extra insurance due to the ship's age will be for the Charterers' account but the Owners will contribute with a maximum of USD 5,000 supported with original vouchers.

Comments:
The Owners, in comparison to the Charterers' offer:
-accept to pay all possible extra insurance, due to the ship's flag/ownership etc.
-can accept to contribute with a maximum of USD 5,000 for the extra insurance imposed on the ship's age (the Charterers' offer was to contribute with a lump-sum amount of USD 10,000), however, such an amount should be legally deducted when supported by original vouchers. If the supported vouchers are less than USD 5,000 then the Owners will pay for this presented amount.

The Owners continue:
15. The Charterers will endeavour for the original Bills of Lading to be available

on or before the ship's arrival to the discharge port. However, if original Bills of Lading are not available upon the ship's arrival at the discharge port, then the Owners/Master will allow the discharge of the cargo directly to the Receivers, in exchange for the Charterers' Letter of Indemnity (L.O.I.), as per the Owners' P&I Club's wording, but signed by the Charterers only. The L.O.I. will be issued on the Charterers' letterhead signed by a director. The Charterers will send a draft copy of the l.o.i. directly to the Owners' office for approval and the Original will be delivered to the Master's hands or to the Owners' premises, before the commencement of the discharging operation.
The Charterers will clearly advise on the name/full style/address of the nominated Receivers and will endeavour to hand over to the Master of the ship a statement on the Receivers' letterhead, stating that they are taking delivery of the cargo.
The Charterers remain responsible for the collection of the original Bills of Lading and the original l.o.i. will be returned to them only when the original Bills of Lading are handed over to the Master of the ship or sent to the Owners' premises directly, at a later stage.

Comments:
The Owners, in comparison to the Charterers' offer:
-reject the Charterers' term but accept to discharge without presentation of original Bills of Lading.
-accept to discharge but not to release the cargo (Charterers' offer was to discharge/release the cargo) unless it is delivered directly to the Receivers.
-request that the Charterers' director signs the l.o.i. which will be issued in their letterhead, as per the Owners' P&I Club's wording -request the original l.o.i. to be delivered to the Owners or Master before the commencement of the discharging operation.
-request that the Charterers be responsible for the collection of the original Bills of Lading and when they are presented to the Owners, they will take back their original l.o.i.

The Owners continue:
17.
The Stevedoring company will be appointed and paid by the Charterers, but will work under the supervision of the Master, if it is possible.
Charterers will be liable and responsible for any damage caused by the stevedores to the cargo or ship or its fittings at the ports of contact during this Voyage Charter, including hidden damage, which should be reported as soon as detected, but in any case, before the ship's sailing from the last discharge port.
If the ship suffers stevedore damage, the Master of the ship:

1. will report to the Charterers or their Agents in writing within 24 hours of occurrence or from the time of discovery, but in any case, before the ship sails from the port.
2. will hold the stevedores or the respective parties responsible and liable for any damage, in writing, with copies sent to the Charterers or their Agents. He will work with the Charterers or their Agents and will make every effort to reach a written acknowledgement from the parties responsible for causing the damage, unless the damage has in the meantime been repaired. If the Master is unable to obtain acceptance of liability and responsibility from the party who caused the damage, he is not allowed to sail from the port before coming into contact with the Charterers, allowing them to sort the matters out and the time lost will continue to count as lay-time or time on demurrage.
3. will try to get stevedores to repair the damage and try to settle the matter directly with them in the first stage and the time lost will continue to count as lay-time or time on demurrage.
For stevedore damage which:
a. affects the effectiveness of the ship and the safe operation and/or safety of the crew and/or affect the trading capabilities and/or seaworthiness and/or cargo worthiness and/or is subject to classification requirements, then such damage as will be ascertained by the ship's Class surveyor, who will be appointed by the Owners at the Charterers' cost, is to be repaired by the Charterers at their own expenses/risk, to the Master's and/or Class surveyor's entire satisfaction and the time lost will continue to count as lay-time or time on demurrage until completion of all repairs and the 'port clearance' is issued and delivered to the ship.
Charterers are to reimburse the Owners for the Class Surveyor's fees, once the Owners have submitted their final freight invoice, supported with the original vouchers.
The damage will be repaired at the place of occurrence, if it is possible and there are adequate facilities for permanent repairs, or otherwise, temporary repairs are to be made on the spot or as when detected, at the Charterers' risk and expenses and to the Master's and/or Class surveyor's satisfaction and the time lost will continue to count as lay-time or time on demurrage. The ship will sail or shift to the nearest shipyard, as chosen by the Charterers, with fuel consumed at sea or during shifting to be for the Charterers' account, which will be settled together with the Owners' final freight statement, in order to complete the repairs.
The time lost during the sailing or shifting to the shipyard and until completion of the works with the damage being repaired to the Master's and/or Class surveyor's entire satisfaction and the 'port clearance' is issued and delivered to the ship or, at the Owners' option, at equidistance from the discharge port, will count as lay-time or time on demurrage. Shipyard's expenses and/or port expenses, in-

cluding in and out tugs and/or pilots and/or agency fees etc., will be entirely for the Charterers' account.
b. does not affect the effectiveness of the ship and the safe operation and/or the safety of the crew and/or the trading capabilities and/or the seaworthiness or cargo worthiness and/or is subject to classification requirements and cannot be repaired by the stevedores, the Owners then undertake to accept the ship's sailing, in accordance with the terms of the Charter Party.
However, the said damage is to be assessed before the ship's departure from the discharge port and the costs to be agreed upon between the Master, Owners, Charterers and the respective survey company and to be paid together with the Owners' final freight statement. The survey company will be mutually appointed by the Owners and Charterers, if considered necessary, at the Charterers', Owners' equal expenses and will promptly inform the Charterers, Owners with the damage report, which will be issued before the ship sails, with a copy of all correspondence exchanged in this connection. The time lost will continue to count as lay-time or time on demurrage, until the 'port clearance' is issued and delivered to the ship.
Owners have the option of having the damage repaired at their next convenient time or at the ship's next Dry-Docking.
In the case that the Master, Owners, Charterers cannot reach an agreement on the cost with the respective survey company, the Charterers will then undertake to reimburse for the actual cost of the repairs, but not for the time lost, upon presentation of the original bills from the repairers or dockyard, that are consistent with the surveyors' report on the extent of such damage.
c. is small and has been repaired by the crew, the Charterers will reimburse the Owners for the cost on request, but not for the time lost and upon presentation of the Owners' repair statement, which is acceptable, even by fax or e-mail.
However, and in any case the Charterers are responsible for any directly related expenses resulting therefrom and incurred thereby and must pay compensation for any damage caused to the ship and/ or restore the ship to its initial condition under their own time and expenses, whether or not any payment has been made by the respective parties and/or stevedores to them.

Comments:
The Owners, in comparison to the Charterers' offer:
-reject the Charterers' term and repeat their last.

The Owners continue:
19.
Holds' inspection:
If part of the ship's holds have passed successfully and the loading operation has

commenced in the holds that have passed or the loading operation has commenced without all the pre-loading surveys to have been completed, the lay-time then will commence counting as per the Charter Party or will count as time on demurrage.
Hatch covers inspection:
If part of the ship's hatches have passed successfully and the loading operation has commenced in the hatches that have passed or the loading operation has commenced without all the pre-loading surveys to have been completed, the lay-time then will commence counting as per the Charter Party or will count as time on demurrage.
In the case of a dispute on the hatch inspection, an independent class approved surveyor will be mutually appointed and his opinion will be binding for all parties. Surveying fees will be paid by the party in default.

Comments:
The Owners, in comparison to the Charterers' offer:
-accept the Charterers' term in full.
-add that if part of the ship's holds or hatches are accepted then the lay-time will commence counting as per the Charter Party. Additionally, in the case of a dispute on the hatch inspection, then an independent class approved surveyor will be mutually appointed and his opinion will bind both parties.

TO SUMMARIZE THE OWNERS' COUNTER OFFER:

Thanks for Charterers last.
Owners A/E for reply within 30 minutes as follows:
Owners' and Managers' full style/contact numbers as follows:
-Owners: ......
-Managers: ......
-Disponent Owners: not applicable
3.
the cargo stows about 44' in the ship's holds.
4.
If a second berth is used at every load/discharge port/s, the shifting expenses are for the Charterers' account and the lost time will count as lay-time, but the Officers'/crew's overtime and consumed bunkers will be for the Owners' account.
The final sailing configuration is always subject to Master's stowage plan and satisfaction, which is in conformity with the ship's loading manual.
The ship should always be left in a seaworthy and safe stress/trim condition and with cargo on board safely stowed during its sailing and/or its shifting between all ports, berths, anchorages and wharfs to the Master's satisfaction.

Upon completion of the loading and/or discharging and/or lightering operation at the first or subsequent port and/or berth/ anchorage/wharf or at any time it is requested by any party (Master, Port Authorities etc.) to bring the ship in a seaworthy and safe stress/trim condition, then all the expenses/ extra port expenses incurred will be entirely for the Charterers' account and the time lost will continue to count as lay-time or time on demurrage, even during ballasting/ de-ballasting the ship's tanks and until the Master is fully satisfied and the ship is ready to sail or shift.

5.

ship's itinerary ....

e.t.a. at load port ....

6.

Freight USD 20.50 pmtn 1/1 free in/out spout/grabbed trimmed. Freight is payable 95pct net of tax and bank charges into the Owners' nominated bank account within 3 banking days after completion of the loading and presentation/ signing/ releasing of Congebill edition 1994 Bill of Lading marked as 'freight payable as per Charter Party', less the address commissions, less the Owners' contribution towards extra insurance and less any amount paid by the Charterers on the Owners' behalf at load port but made only after the Owners' written consent.

It is the Charterers' option the Bill of Lading to be marked as 'Freight Prepaid' in which case the original Bills of Lading will be kept in the Agents' custody or at the Owners' option by the ship's P&I Club's representatives and will be released to the Shippers upon the Charterers' bank confirmation directly to the Owners' office that 100pct of the freight, less the address commissions, less the Owners' contribution towards extra insurance and less any amount paid by the Charterers on the Owners' behalf at the load port, but made only after the Owners' written consent, has been irrevocably remitted with the same date value. If a second discharge port is used, the Charterers will pay an additional freight of USD 2,00 pmtn on the entire quantity of cargo, payable together with the initial freight.

The balance freight, plus demurrage or minus despatch in both ends or any amount due to the Owners or Charterers will be paid within 5 banking days after completion of the discharging operation and after the Owners' presentation of all relevant documents such as the N.O.R./S.O.F. and the Owners' Time Sheets from all ports, together with their final freight and demurrage invoice, which all are accepted even by fax. The original documents will be forwarded by mail/ courier to the Charterers' office, as soon as possible.

The Owners are not entitled to claim for dead freight unless proper notice has been given and acknowledged by the Charterers/Shippers before sailing. The Charterers guarantee that they will be contactable at any time it is required, and all time lost in waiting will be counted as lay-time.

If the Charterers are liable for dead freight, then the time allowed for the loading and discharging will be calculated based on the paid cargo quantity and not on the actual loaded cargo quantity.
If there is a dispute on the presented Time Sheets, the Charterers will pay the undisputed amount within 5 banking days together with the balance freight and the disputed amount will be settled amicably or will be referred to Arbitration.
The lay-time computation in all ports will be based only on the S.O.F. as issued in every port and duly signed by the Master and Agents, but the Master will endeavor, but without guarantee, to also have it signed by the Shippers and Receivers at the load and discharge port, respectively.
Full freight is deemed earned upon shipment discountless and non-returnable ship and or cargo lost or not lost.
7.
Loading rate/lay-time:
the cargo will be loaded free of risks and expenses to the ship, at the rate of 4,000 metric tons daily or pro rata per weather working day of 24 consecutive hours, Sundays and Holidays excluded unless used, in which case the actual time used to count as lay-time.
Time from Saturday noon until Monday 08:00 hrs. or from 17:00 hrs. on a day preceding a Holiday until 08:00 hrs. of the next working day, not to count as lay-time, unless used in which case the actual time used to count as lay-time.
A valid Notice Of Readiness will be tendered in writing at berth or, in case of congestion or for any other reason beyond the Owners'/ship's fault, the ship's berthing is not allowed upon arrival, from the anchorage or from the place where it will be instructed to wait within or out of the port's commercial limits, by any means of telecommunication (cable/ telex/e-mail/fax) to the Agents, during the official working hours from 08:00 hrs.- up to and including 17:00 hrs. from Monday to Friday and from 08:00 hrs.- up to and including 12:00 hrs. on Saturday.
The lay-time will commence counting at 14:00 hrs. if the Notice of Readiness is tendered before noon or at 08:00 hrs. the next working day if the Notice of Readiness is tendered in the afternoon, whether in berth or not, whether in free pratique or not, whether customs are cleared or not, whether in port or not.
Time used before the commencement of lay-time will count as lay-time.
Discharging rate/lay-time:
The lay-time will commence counting at 14:00 hrs. if the Notice of Readiness is tendered before noon or at 08:00 hrs. of the next working day, if the Notice of Readiness is tendered in the afternoon, whether in berth or not, whether in free pratique or not, whether customs are cleared or not, whether in port or not.
Time used before the commencement of lay-time will count as lay-time.
10.
b- required time for the initial/intermediate/final draft survey in all ports will

be equally shared between the Owners and Charterers. However, expenses incurred to hire a launch or tug in order for the draft survey to be carried out at the berth or in the anchorage will be for the Charterers' account and time lost will count as lay-time or time on demurrage.

11.

Extra insurance due to the ship's flag/ownership etc. will be for the Owners' account. Extra insurance due to the ship's age will be for the Charterers' account but the Owners will contribute with a maximum of USD 5,000 supported with original vouchers.

15.

The Charterers will endeavour for the original Bills of Lading to be available on or before the ship's arrival to the discharge port. However, if original Bills of Lading are not available upon the ship's arrival at the discharge port, then the Owners/Master will allow the discharge of the cargo directly to the Receivers, in exchange for the Charterers' Letter of Indemnity (L.O.I.), as per the Owners' P&I Club's wording, but signed by the Charterers only. The L.O.I. will be issued on the Charterers' letterhead signed by a director. The Charterers will send a draft copy of the l.o.i. directly to the Owners' office for approval and the Original will be delivered to the Master's hands or to the Owners' premises, before the commencement of the discharging operation.

The Charterers will clearly advise on the name/full style/address of the nominated Receivers and will endeavour to hand over to the Master of the ship a statement on the Receivers' letterhead, stating that they are taking delivery of the cargo.

The Charterers remain responsible for the collection of the original Bills of Lading and the original l.o.i. will be returned to them only when the original Bills of Lading are handed over to the Master of the ship or sent to the Owners' premises directly, at a later stage.

17.

The Stevedoring company will be appointed and paid by the Charterers, but will work under the supervision of the Master, if it is possible.

Charterers will be liable and responsible for any damage caused by the stevedores to the cargo or ship or its fittings at the ports of contact during this Voyage Charter, including hidden damage, which should be reported as soon as detected, but in any case, before the ship's sailing from the last discharge port.

If the ship suffers stevedore damage, the Master of the ship:

1. will report to the Charterers or their Agents in writing within 24 hours of occurrence or from the time of discovery, but in any case, before the ship sails from the port.

2. will hold the stevedores or the respective parties responsible and liable for any damage, in writing, with copies sent to the Charterers or their Agents. He will work with the Charterers or their Agents and will make every effort to reach a written acknowledgement from the parties responsible for causing the damage, unless the damage has in the meantime been repaired. If the Master is unable to obtain acceptance of liability and responsibility from the party who caused the damage, he is not allowed to sail from the port before coming into contact with the Charterers, allowing them to sort the matters out and the time lost will continue to count as lay-time or time on demurrage.

3. will try to get stevedores to repair the damage and try to settle the matter directly with them in the first stage and the time lost will continue to count as lay-time or time on demurrage.

For stevedore damage which:

a. affects the effectiveness of the ship and the safe operation and/or safety of the crew and/or affect the trading capabilities and/or seaworthiness and/or cargo worthiness and/or is subject to classification requirements, then such damage as will be ascertained by the ship's Class surveyor, who will be appointed by the Owners at the Charterers' cost, is to be repaired by the Charterers at their own expenses/risk, to the Master's and/or Class surveyor's entire satisfaction and the time lost will continue to count as lay-time or time on demurrage until completion of all repairs and the 'port clearance' is issued and delivered to the ship.

Charterers are to reimburse the Owners for the Class Surveyor's fees, once the Owners have submitted their final freight invoice, supported with the original vouchers.

The damage will be repaired at the place of occurrence, if it is possible and there are adequate facilities for permanent repairs, or otherwise, temporary repairs are to be made on the spot or as when detected, at the Charterers' risk and expenses and to the Master's and/or Class surveyor's satisfaction and the time lost will continue to count as lay-time or time on demurrage. The ship will sail or shift to the nearest shipyard, as chosen by the Charterers, with fuel consumed at sea or during shifting to be for the Charterers' account, which will be settled together with the Owners' final freight statement, in order to complete the repairs.

The time lost during the sailing or shifting to the shipyard and until completion of the works with the damage being repaired to the Master's and/or Class surveyor's entire satisfaction and the 'port clearance' is issued and delivered to the ship or, at the Owners' option, at equidistance from the discharge port, will count as lay-time or time on demurrage.

Shipyard's expenses and/or port expenses, including in and out tugs and/or pilots and/or agency fees etc., will be entirely for the Charterers' account.

b. does not affect the effectiveness of the ship and the safe operation and/or the safety of the crew and/or the trading capabilities and/or the seaworthiness or cargo worthiness and/or is subject to classification requirements and cannot be repaired by the stevedores, the Owners then undertake to accept the ship's sail-

ing, in accordance with the terms of the Charter Party.
However, the said damage is to be assessed before the ship's departure from the discharge port and the costs to be agreed upon between the Master, Owners, Charterers and the respective survey company and to be paid together with the Owners' final freight statement. The survey company will be mutually appointed by the Owners and Charterers, if considered necessary, at the Charterers', Owners' equal expenses and will promptly inform the Charterers, Owners with the damage report, which will be issued before the ship sails, with a copy of all correspondence exchanged in this connection. The time lost will continue to count as lay-time or time on demurrage, until the 'port clearance' is issued and delivered to the ship.
Owners have the option of having the damage repaired at their next convenient time or at the ship's next Dry-Docking.
In the case that the Master, Owners, Charterers cannot reach an agreement on the cost with the respective survey company, the Charterers will then undertake to reimburse for the actual cost of the repairs, but not for the time lost, upon presentation of the original bills from the repairers or dockyard, that are consistent with the surveyors' report on the extent of such damage.
c. is small and has been repaired by the crew, the Charterers will reimburse the Owners for the cost on request, but not for the time lost and upon presentation of the Owners' repair statement, which is acceptable, even by fax or e-mail.
However, and in any case the Charterers are responsible for any directly related expenses resulting therefrom and incurred thereby and must pay compensation for any damage caused to the ship and/or restore the ship to its initial condition under their own time and expenses, whether or not any payment has been made by the respective parties and/or stevedores to them.
19.
Holds' inspection:
If part of the ship's holds have passed successfully and the loading operation has commenced in the holds that have passed or the loading operation has commenced without all the pre-loading surveys to have been completed, the lay-time then will commence counting as per the Charter Party or will count as time on demurrage.
Hatch covers inspection:
If part of the ship's hatches have passed successfully and the loading operation has commenced in the hatches that have passed or the loading operation has commenced without all the pre-loading surveys to have been completed, the lay-time then will commence counting as per the Charter Party or will count as time on demurrage
In the case of a dispute on the hatch inspection, an independent class approved surveyor will be mutually appointed and his opinion will be binding for all par-

ties. Surveying fees will be paid by the party in default.

CHARTERERS' COUNTER REPLY TO OWNERS' LAST:

The Charterers could counter reply as follows:

Thanks Owners last, on which a/e for reply within 15 mnts as follows:

The Charterers continue:
4.
If a second berth is used in all ports, the shifting expenses, including officers'/ crew's overtime and consumed bunkers will be for the Owners' account, but the time will continue to count as lay-time.

Comments:
The Charterers, in comparison to the Owners' offer:
-actually, accept the Owners last, except that the shifting expenses for changing berth will be for the Owners' account.

The Charterers continue:
6.
Freight USD 20.00 pmtn 1/1 for fixing.

Comments:
The Charterers, in comparison to the Owners' offer:
-accept all as per the Owners', except for the freight rate which is given 'as fixing'.

The Charterers continue:
7.
Time used before the commencement of lay-time will not count as lay-time at load port.

Comments:
The Charterers, in comparison to the Owners' offer:
-accept all as per the Owners' last, except that the time used before the commencement of lay-time will not count as lay-time only at the load port.

The Charterers continue:
17.
The Owners guarantee that all vulnerable parts in the ship's holds will be prop-

erly protected under their time and expenses.
The Charterers will not to be liable for damage, when the Master has accepted a statement in writing from the stevedores and/or respective parties that the operations were carried out without any damage, or that the repairs were made to his satisfaction.

Comments:
The Charterers, in comparison to the Owners' offer:
-accept all as per the Owners' offer except that all vulnerable parts in the ship's holds will be properly protected and that the Charterers will not be liable for damage when the Master has accepted that the operations were carried out without any damage or that the repairs were made to his satisfaction.

TO SUMMARISE THE CHARTERERS' COUNTER OFFER:

Thanks Owners last, on which a/e for reply within 15 mnts as follows:
4.
If a second berth is used in all port/s, the shifting expenses, including officers'/ crew's overtime and consumed bunkers will be for the Owners' account, but the time will continue to count as lay-time.
6.
Freight USD 20.00 pmtn 1/1 for fixing.
7.
Time used before the commencement of lay-time will not count as lay-time at load port.
17.
The Owners guarantee that all vulnerable parts in the ship's holds will be properly protected under their time and expenses.
The Charterers will not to be liable for damage, when the Master has accepted a statement in writing from the stevedores and/or respective parties that the operations were carried out without any damage, or that the repairs were made to his satisfaction.

OWNERS' COUNTER REPLY TO CHARTERERS' LAST:

The Owners could counter reply as follows:
For reply within 15 mnts:
4.
If the second berth is used at every load/discharge port/s, the shifting expenses will be shared equally between the Owners and Charterers.

Comments:
The Owners, in comparison to the Charterers' offer:
-request the shifting expenses to be shared equally between them.

CHARTERERS' COUNTER REPLY TO OWNERS' LAST:

The Charterers could counter reply as follows:
Thanks Owners' last but Charterers repeat their last for reconfirmation only within 15 mnts.
or
Thanks Owners last but Charterers repeat their last for reply within 15 mnts basis take it or leave it.

Comments:
The Charterers refused the Owners' last and repeat their last for reconfirmation only. The Owners must decide if they will fix on Main Terms and further proceed to working the Charter Party details or if they will decline the biz and look for something else.

OWNERS' REPLY:

-Thanks Charterers last which is accepted. The ship is fixed on main terms with sub to Charter Party details based on the Charterers' executed proforma and sub to Shippers'/Receivers' approval and stem.

The Owners have accepted the Charterers last, therefore the ship is fixed on the Main Terms. The Charterers' broker will make the recapitulation which will be given to both parties as their guide on the terms that have been agreed. However, when complicated terms have been agreed, then will be given for their reconfirmation on the agreed terms.

THE RECAPITULATION COULD BE GIVEN AS FOLLOWS:

Ref Account Eagle Trading and Shipping Inc
40,000 mtns bulk wheat
Constanza/ Spain
MV TAXIARCHES
Please find here below recapitulation of main terms as agreed so far.
MV TAXIARCHES
....
....

....

Owners' and Managers' name/style/contact numbers as follows:
-Owners:
-Managers:
Owners guarantee that during the currency of this voyage:
a- the ship will not change ownership and/or class and will not be sold for scrap during or after this voyage without the Charterers' written consent.
b- the ship will proceed with the utmost despatch directly to the discharge port/s and will not deviate for any purpose whatsoever without the Charterers' consent, excluding bunkering en route and to the nearest bunkering station or for saving life.
c- the ship is fully covered by the P&I Club (as a member of an International Group of P&I Clubs) and has been covered for ship oil pollution risks and wreck removal risks and will remain so throughout the duration of this charter. In this respect, the Owners will furnish the Charterers with the P&I Club's written confirmation, if and whenever it is requested by them.
d- the ship is fully classed as Lloyd's 100A1 or equivalent with a classification society being a full member of the International Association of Classification Societies (IACS) throughout the duration of this charter. In this respect, the Owners will furnish the Charterers with the Classification society's written confirmation, if and whenever it is requested by them.
e- the ship's H&M insurance will be fully maintained and will not be changed. In this respect, the Owners will furnish the Charterers with the H&M underwriters' written confirmation, if and whenever it is requested by them.
f- the ship has no outstanding recommendations, restrictions or any other requirements of any nature in respect to the flag state, port state control, or classification inspections that may affect this voyage and the Charter Party.
g- the ship is suitable for entering and berthing at the specified ports, as well as loading/discharging the minimum contractual cargo quantity. The ship will comply with all applicable laws and regulations and is in all respects eligible for trading at the specified ports. The ship has all valid and up-to-date certificates and other documents on board required for this voyage.
h- the ship can open/close its hatches with its own means.
i- the ship has adequate mooring lines and will be tending them properly, whilst at the loading or discharging port.
j- the ship will give, free of expenses to the Charterers, the full use of its lighting on its deck and in its cargo holds, that will be adequate for all cargo operations.
k- the ship will comply with all applicable laws and regulations and is in all respects eligible for trading to the ports, places or countries specified in this Charter Party. The ship has all valid and up-to-date certificates and other documents on board that are required for this voyage.

l- the ship is fully fitted for the loading of grains according to the SOLAS 1974 regulations and any subsequent amendments thereto without requiring any shifting boards. The Owners guarantee that the ship is suitable for entering, berthing, and loading/discharging the contractual cargo quantity of wheat in bulk.
m- both, the ship and the company fully comply with ISPS/ISM requirements and will remain so throughout the duration of this charter.
n- the ship is sid/bc, steelfloored throughout, with hoppers port/starboard, with the engine/bridge aft and its holds/hatches/tank tops are clear and unobstructed without a centreline bulkhead, walkways, centrebeams and free of all/any obstructions which might incur difficult access to the cargo during loading/discharging. Cargo will not be loaded in deep tanks, sidetanks, wing tanks, bunker spaces, or any other confined spaces unreachable to the means of loading and discharging.
o- the ship has not carried mammalian meat, bone meal, fishmeal, asbestos cullet or nuts for the past 12 months.
p- the ship has not been blacklisted/boycotted or arrested due to the ship's flag/ownership/crew employment/age/past trading or for any other reason from local authorities and or unions in all countries designated under this Charter Party.
q- the ship is free of all encumbrances, arrests, and maritime liens.
r- the ship has no .... (insert countries which may affect the voyage) interests and is not black listed by Arab league countries.
s- the description of the ship as provided by the Owners is an accurate representation of the characteristics of the ship which can only be modified with the usual and acceptable shipping interpretation of the word 'about'. This clause shall override any inclusion of the words 'without guarantee' in the ship's description. The Owners are responsible and liable for any consequences/ delays/ losses/expenses that may arise when providing any misleading information on the ship's characteristics and on the terms agreed in this Charter Party.
t- the ship is ITF compliant or bona fide trade union's agreement or equivalent and will remain so throughout the duration of this charter and shall remain responsible for all consequences and/or delays if the ship is sailing under a flag of 'convenience'.
u- the ship, throughout the duration of the Charter Party will have on board and make available to the Charterers or their Agents or to the appointed surveyors, the capacity's plan, the deadweight and displacement scales, the hydrostatic curves and tables of displacement, the calibration scales for all ship's tanks and trimming correction tables as well as all sounding tables and all sounding tubes which will be properly maintained, in good condition and free from impediments. Plimsoll marks amidships and draft marks on port and starboard side, bow, amid ship and stern are always to be clearly cut and marked on shell platting and readable.

for:
1.
Account Eagle Trading and Shipping Inc of British Virgin Islands
(Charterers' background)
Recent fixtures:
MV .... Fixed through .... Ph ...... Owners/brokers ph: ....
MV .... Fixed through .... Ph .... Owners/brokers ph:............
Etc .etc.
(Charterers' recent fixtures)
2.
Sub Shippers'/Receivers' approval and stem which will be declarable at the latest 2 hours after the fixing of the main terms with the Charter Party details in order.
3.
cargo 40,000 metric tons bulk wheat, one grade, with 10pct moloo. The cargo stows about 44' in the ship's holds. No other cargo is allowed to be loaded.
4.
1 safe port -1/2 safe berth/s Constanza Romania always afloat, as loadport
1 safe port-1/2 safe berth/s Tarragona Spain always afloat and/or at chopt 1 safe port-1/ 2 safe berth/s Valencia Spain always afloat, as discharge port/s.
If two discharge ports are used, the ship will discharge in rotation at the Owners' option and will be declared before the completion of the loading and the issuance of the Bill of Lading. The cargo quantity for each discharge port should be in accordance with the ship's stress and the Master's approval.
If a second berth is used at all ports, the shifting expenses, including officers'/ crew's overtime and consumed bunkers will be for the Owners' account, but the time will continue to count as lay-time.
Owners are responsible for checking and satisfying themselves with all port/ berth restrictions and regulations at both ends.
The final sailing configuration is always subject to Master's stowage plan and satisfaction, which is in conformity with the ship's loading manual.
The ship should always be left in a seaworthy and safe stress/trim condition and with cargo on board safely stowed during its sailing and/or its shifting between all ports, berths, anchorages and wharfs to the Master's satisfaction.
Upon completion of the loading and/or discharging and/or lightering operation at the first or subsequent port and/or berth/ anchorage/wharf or at any time it is requested by any party (Master, Port Authorities etc.) to bring the ship in a seaworthy and safe stress/trim condition, then all the expenses/ extra port expenses incurred will be entirely for the Charterers' account and the time lost will continue to count as lay-time or time on demurrage, even during ballasting/ de-ballasting the ship's tanks and until the Master is fully satisfied and the ship

is ready to sail or shift.
5.
Laydays-Cancelling 20/25 January 2.... 24:00 hrs
ship's itinerary: ...
e.t.a. at load port: ...
6.
Freight USD 20.00 pmtn 1/1 free in/out spout/grabbed trimmed .Freight is payable 95pct net of tax and bank charges into the Owners' nominated bank account within 3 banking days after completion of the loading and presentation/ signing/ releasing of Congebill edition 1994 Bill of Lading marked as 'freight payable as per Charter Party', less the address commissions, less the Owners' contribution towards extra insurance and less any amount paid by the Charterers on the Owners' behalf at load port but made only after the Owners' written consent.
It is the Charterers' option the Bill of Lading to be marked as 'Freight Prepaid' in which case the original Bills of Lading will be kept in the Agents' custody or at the Owners' option by the ship's P&I Club's representatives and will be released to the Shippers upon the Charterers' bank confirmation directly to the Owners' office that 100pct of the freight, less the address commissions, less the Owners' contribution towards extra insurance and less any amount paid by the Charterers on the Owners' behalf at the load port but made only after the Owners' written consent, has been irrevocably remitted with the same date value.
If a second discharge port is used, the Charterers will pay an additional freight of USD 2,00 pmtn on the entire quantity of cargo, payable together with the initial freight.
The balance freight, plus demurrage or minus despatch in both ends or any amount due to the Owners or Charterers will be paid within 5 banking days after completion of the discharging operation and after the Owners' presentation of all relevant documents such as the N.O.R./S.O.F. and the Owners' Time Sheets from all ports, together with their final freight and demurrage invoice, which all are accepted even by fax. The original documents will be forwarded by mail/ courier to the Charterers' office as soon as possible.
The Owners are not entitled to claim for dead freight unless proper notice has been given and acknowledged by the Charterers/Shippers before sailing. The Charterers guarantee that they will be contactable at any time it is required, and all time lost in waiting will be counted as lay-time. However, and if there is a dispute, the Owners and/or Master have not the right to clause or delay the release of the presented Bills of Lading. If the Charterers are liable for dead freight, then the time allowed for the loading and discharging will be calculated based on the paid cargo quantity and not on the actual loaded cargo quantity.
If there is a dispute on the presented Time Sheets, the Charterers will pay the undisputed amount within 5 banking days together with the balance freight and

the disputed amount will be settled amicably or will be referred to Arbitration.
The lay-time computation in all ports will be based only on the S.O.F. as issued in every port and duly signed by the Master and Agents, but the Master will endeavor, but without guarantee, to also have it signed by the Shippers and Receivers at the load and discharge port, respectively.
Full freight is deemed earned upon shipment discountless and non-returnable ship and or cargo lost or not lost.
7.
Loading rate/lay-time:
the cargo will be loaded free of risks and expenses to the ship, at the rate of 4,000 metric tons daily or pro rata per weather working day of 24 consecutive hours, Sundays and Holidays excluded unless used, in which case the actual time used to count as lay-time.
Time from Saturday noon until Monday 08:00 hrs. or from 17:00 hrs. on a day preceding a Holiday until 08:00 hrs. of the next working day, not to count as lay-time, unless used in which case the actual time used to count as lay-time.
A valid Notice Of Readiness will be tendered in writing at berth, or in case of congestion or for any other reason beyond the Owners'/ship's fault, the ship's berthing is not allowed upon arrival, from the anchorage or from the place where it will be instructed to wait within or out of the port's commercial limits, by any means of telecommunication (cable/ telex/e-mail/fax) to the Agents, during the official working hours from 08:00 hrs. up to and including 17:00 hrs. from Monday to Friday and from 08:00 hrs. up to and including 12:00 hrs. on Saturday. The lay-time will commence counting at 14:00 hrs. if the Notice of Readiness is tendered before noon or at 08:00 hrs. the next working day if the Notice of Readiness is tendered in the afternoon, whether in berth or not, whether in free pratique or not, whether customs are cleared or not, whether in port or not.
If the ship berths upon arrival, a valid N.O.R. shall be tendered only after all preloading surveys including holds' inspections are fully passed by the Charterers' and/or Shippers' surveyors and/or by a national cargo bureau and/or local grain surveyors, and provided the ship is at the berth and the free pratique is granted.
In the case of congestion or for any other reason beyond the Owners'/ship's fault the ship's berthing is not allowed upon arrival, then after the ship's berthing, lost time for the preloading surveys including lost time for holds' inspections and to grant free pratique, will not count as lay-time or time on demurrage.
The 'NOR is deemed to be tendered and accepted from the time of tendering and the lay-time will commence counting in accordance with the terms of the Charter Party.
Time used before the commencement of lay-time will not count as lay-time.
Discharging rate/lay-time:

the cargo will be discharged in one or two ports, free of risks and expenses to the ship, at the rate of 10,000 metric tons daily or pro rata, per weather working day of 24 consecutive hours, Saturdays, Sundays, and Holidays included.
A valid Notice Of Readiness at the first or sole discharge port will be tendered in writing at berth, or, in case of congestion or for any other reason beyond the Owners'/ship's fault, the ship's berthing is not allowed upon arrival, from the anchorage or from the place where it will be instructed to wait within or out of the port's commercial limits, by any means of telecommunication (cable/ telex-/e-mail/fax) to the Agents during the official working hours between 08:00- up to and including 17:00 hrs. from Monday to Sunday and the lay-time will commence counting at 14:00 hrs. if the Notice of Readiness is tendered before noon or at 08:00 hrs. of the next working day, if the Notice of Readiness is tendered in the afternoon, whether in berth or not, whether in free pratique or not, whether customs are cleared or not, whether in port or not. If the ship berths upon arrival, a valid N.O.R. shall be tendered only after all pre-discharging surveys/inspections are fully passed by the Charterers' and/or Receivers' surveyors and/or by a national cargo's bureau and/ or local grain surveyors and provided the ship is at berth and the free pratique is granted. In the case of congestion, or for any other reason beyond the Owners'/ ship's fault, the ship's berthing is not allowed upon arrival, then, after the ship's berthing, the lost time for the pre-discharging surveys, including the time to grant free pratique, will not count as lay-time or time on demurrage.
The 'NOR is deemed to be tendered and accepted from the time of tendering and the lay-time will commence counting in accordance with the terms of the Charter Party.
Time used before the commencement of lay-time will count as lay-time. If a second discharge port is used, the NO.R. will be tendered in writing upon arrival at berth, or, in case of congestion or for any other reason beyond the Owners'/ ship's fault, the ship's berthing is not allowed upon arrival, from the anchorage or from the place where it will be instructed to wait within or out of the port's commercial limits as will be instructed to wait, by any means of telecommunication (cable/ telex/e-mail/fax) to the Agents, at any time day or night from Monday to Sunday. The lay-time will commence counting upon the first official resumption of work at the port, whether in berth or not, whether in free pratique or not, whether customs are cleared or not, whether in port or not.
All time used for pre-discharging surveys including free pratique will count as lay-time unless the pre-discharging surveys including free pratique have been failed, then the time from failure until full passing is not to count as lay-time or time on demurrage.
8.
Demurrages USD 13,000 per day, pro rata half despatch for all working time

saved at both ends.
Lay-time to be non-reversible between the load/discharge port but to be reversible between the discharging ports.
The demurrage claim will be settled between the Owners and Charterers and the Charterers will be solely responsible for paying the demurrage to the Owners.
The Laytime computation will be based on the signed Statement of Facts only
9.
Loaded and discharged cargo quantities will be ascertained by a joint draft survey which will be held between the Owners and Charterers/Shippers at the load port and between the Owners and Charterers/Receivers at the discharge port. Each party will appoint and pay for their surveyors. It is the Owners' option to appoint the ship's Master or Chief Officer as their surveyor. The appointed surveyors are to be allowed to have full access to the ship for the purpose of the draft survey.
One draft survey report will be issued which will be signed by all the surveyors involved. The draft survey will be carried out in berth in all ports, upon completion of the loading and discharging operation. Expenses incurred to hire a launch or tug in order for the draft survey to be carried out in berth or in the anchorage will be for the Charterers' account and time lost will count as lay-time or time on demurrage.
At the load and discharge port, the shore scale figures are not to be taken into consideration and the ship will load and discharge according to the findings of the carried out draft survey. The Owners/ Master/Ship will not be held responsible for any discrepancies which may arise between the draft survey and the shore figures and it is explicitly agreed that the Bill of Lading will be issued according to the draft survey findings. Delays incurred due to discrepancies between the draft survey and shore figures will count as lay-time or time on demurrage.
At no time during the draft survey (initial/intermediate or final) in all ports, should the liquids be shifted, taken in or pumped out without prior advice to the Charterers and all draft surveyors. If it is required, the Master is to forward to the Charterers/Agents, upon arrival at the loading and discharging port and before the commencement of the loading and discharging operation, a certificate indicating all the remaining ballast in the ship's tanks.
10.
during the contact of the ship with any port for operation under the terms of this Charter Party the:
a- first shifting time from the anchorage or standby berth or pilot station to the load or discharge berth will not count as lay-time, even if the ship is on demurrages.
b- time required for the initial/intermediate/final draft survey in all ports will

be shared equally between the Owners and Charterers. However, expenses incurred to hire a launch or tug in order for the draft survey to be carried out at the berth or in the anchorage will be for the Charterers' account and lost time will count as lay-time or time on demurrage.
c- time required for the opening and closing of hatches will count as lay-time except for the first opening and the last closing, where time will not count as lay-time, even if the ship is on demurrage.
d- national or state/local holidays as specified in the current bimco's calendar to apply only as holidays.
e- the ship will load with trim as the spout/grab can provide under the Charterers time/expenses. Extra trimming required by the Master of the ship will be arranged and paid for by the Owners at their time/risk.
f- time used for the shifting of shore cranes between the cargo holds will count as lay-time.
g- time used for the ship's shifting alongside quay by mooring lines in order to make all hatches available for the operation will count as lay-time. The shifting alongside will not be considered as berth shifting unless tugs are used in which case it will count as shifting. In all cases, the Officers' and Crews' expenses/overtime will be for the Owners' account.
h- time used for placing/removing/manoeuvring of bulldozers in the ship's cargo holds will count as lay-time.
i- time used for holds' or cargo fumigation before or after the completion of loading will count as lay-time.
j- time lost due to the Charterers, or Charterers' Agents or the terminal operator or the Shippers or Receivers or any Authority prohibiting, impeding, or restricting the loading or discharging operation, beyond the Owners'/ship's fault, will count as lay-time.
k- cargo documents including Bill/s of Lading will be presented to the Master within 2 hours after the completion of the operation in each port, but provided that the draft survey has been carried out. Upon expiration of those 2 hours and if all cargo documents including the Bill/s of Lading have not been presented, the lay-time will commence counting again, until all documents including the Bill/s of Lading have been presented accordingly. If the ship will be instructed to vacate the berth and shift to another waiting berth, then the shifting expenses will be for the Charterers' account and the time lost will count as lay-time or time on demurrage.
However, if the ship is already on demurrage then all time waiting for documents including Bill/s of Lading will count as lay-time.
l- at the loading and discharging port/s, any shifting beyond the number of shifting allowed by the Charter Party and if any overtime is ordered, the time and the expenses will be for the account of the party ordering the same, but if it is or-

dered by the Port Authorities, the expenses will be for the Charterers' account and the time lost will continue to count as lay-time or time on demurrage. The Ship's Officers and Crew's overtime will be for the Owners' account.
m- the ship will be left in a seaworthy and safe stress/trim condition to the Master's satisfaction during any shifting which will take place between the loading berths, as well as, between the discharging berths and between the discharging ports and the time lost will continue to count as lay-time or time on demurrage.
n- if the loading and/ or discharging operation is interrupted or extended due to the insufficiency of the ship's ballast/de-ballast pump capacity in relation to the loading/discharging capacity, then such time lost will not count as lay-time and extra expenses incurred will be for the Owners' account.
o- the Master of the ship will liaise with the stevedores to load/distribute the cargo in the cargo holds and discharge the ship within the safe operational limits and as per the SOLAS requirements. Lost time and expenses incurred due to interruption of the loading or discharging operation for stability reasons or for not adhering to the SOLAS provisions will be for the Owners' account.
p- the Owners are responsible and liable for all consequences, including, but not limited to, time lost and extra expenses incurred for unloading any overloaded cargo.
q- the ship will supply lights on board, when it is required, to lighten the deck's complete surface during the day and night, free of expenses to the Charterers.
r- all time lost and extra expenses incurred, due to the Owners' failure to provide the Agents with sufficient funds for covering the port expenses and other Owners' expenses incurred, before the ship's sailing from the port, will be for the Owners' account.
s- all time lost due to labourers boycotting the ship because of its flag/ownership/ registry/crew will not count as lay-time or time on demurrages and extra expenses incurred will be for the Owners' account.
t- the holds' ladders must be in a good state of repair and in compliance with the safety at work and health act, otherwise, time lost will not count as lay-time on time on demurrages.
u- if the ship fails to pass the Port State Control inspection, time lost and expenses incurred in order to comply with recommendations or repairs made by the Authorities will not count as lay-time or time on demurrage.
11.
Extra insurance due to the ship's flag/ownership etc. will be in full for the Owners' account. Extra insurance due to the ship's age will be for the Charterers' account but the Owners will contribute with a maximum of USD 5,000 supported with original vouchers.
12.
Any taxes/dues/wharfage dues/charges/duties calculated/levied on cargo in-

cluding but not limited to the T3 tariff in Spain a/o on loaded cargo quantity a/o on Charter Party freight will be for the Charterers' account at both ends.
Any taxes/dues/wharfage dues/charges calculated/levied on ship/flag/ ownership will be for the Owners' account at both ends.
The Owners will pay port expenses at both ends as per the port tariffs.
13.
The attending Agents will be selected by the Charterers at both ends and the Owners will appoint the same.
14.
Arbitration in London. English law will apply to the negotiations and to the eventual agreement.
15.
The Charterers will endeavour for the original Bills of Lading to be available on or before the ship's arrival to the discharge port. However, if original Bills of Lading are not available upon the ship's arrival at the discharge port, then the Owners/ Master will allow the discharge of the cargo directly to the Receivers, in exchange for the Charterers' Letter of Indemnity (L.O.I.), as per the Owners' P&I Club's wording, but signed by the Charterers only. The L.O.I. will be issued on the Charterers' letterhead signed by a director. The Charterers will send a draft copy of the l.o.i. directly to the Owners' office for approval and the Original will be delivered to the Master's hands or to the Owners' premises, before the commencement of the discharging operation.
The Charterers will clearly advise on the name/full style/address of the nominated Receivers and will endeavour to hand over to the Master of the ship a statement on the Receivers' letterhead, stating that they are taking delivery of the cargo.
The Charterers remain responsible for the collection of the original Bills of Lading and the original l.o.i. will be returned to them only when the original Bills of Lading are handed over to the Master of the ship or sent to the Owners' premises directly, at a later stage.
16.
Written notices of the ship's expected arrival at the load port will be tendered to the Charterers and Agents on fixing, then .../ .../.../... days approximate and .../.../... hours definite notices.
Owners'/Master's tendered 'approximate' notice, with the ship's estimated time of arrival at load port has changed by more than 1 day, then the Master/Owners are to keep the Charterers and attending Agents closely informed on the ship's exact position/full itinerary and cause of delay.
When the Master tenders the definite notices, he will state the expected time of arrival/draft on arrival, the anticipated draft on sailing and the expected loadable cargo quantity.

Owners/Master will remain responsible for the accuracy of the tendered 'definite' notices, which can be changed only by adverse weather conditions and/or unforeseen circumstances and/or delays in completion of cargo discharging operations at the current port, always beyond Owners'/Master's/ ship's fault. If has changed by more than 12 hours, then the Master/Owners are to keep the Charterers and attending Agents closely informed on the ship's exact position, full itinerary, and cause of delay.

If any willful misrepresentation of the definite notice is tendered by the Owners/ Master of the ship to the Charterers and/or Agents for the position and readiness of the ship, they will remain responsible and liable for any and all consequences/ time lost/loss of turn for berthing until the berth becomes available again/additional costs including, but not limited to, the cost for cargo storage in the port, for the cancelled stevedoring shift etc.

On sailing from the load port, the Owners/Master will tender notices to the Charterers and Agents at the discharge port with the estimated time of arrival, the quantity of cargo loaded and the anticipated ship's draft on arrival.

.../.../... days approximate and .../.../... hours definite notices will be tendered thereafter.

The Charterers and Agents are to be kept by the Master/Owners continuously and fully advised of any change in the ship's position and its estimated time of arrival at the discharge port. When the Master tenders the definite notices, he will state the expected time of arrival and Owners/Master will remain responsible for the accuracy of the tendered 'definite' notices, which can be changed only by adverse weather conditions and/or unforeseen circumstances, always beyond Owners'/ Master's/ ship's fault.

17.

The Stevedoring company will be appointed and paid by the Charterers, but will work under the supervision of the Master, if it is possible.

Charterers will be liable and responsible for any damage caused by the stevedores to the cargo or ship or its fittings at the ports of contact during this Voyage Charter, including hidden damage, which should be reported as soon as detected, but in any case, before the ship's sailing from the last discharge port.

If the ship suffers stevedore damage, the Master of the ship:

1. will report to the Charterers or their Agents in writing within 24 hours of occurrence or from the time of discovery, but in any case, before the ship sails from the port.

2. will hold the stevedores or the respective parties responsible and liable for any damage, in writing, with copies sent to the Charterers or their Agents. He will work with the Charterers or their Agents and will make every effort to reach a written acknowledgement from the parties responsible for causing the damage, unless the damage has in the meantime been repaired. If the Master is unable to obtain acceptance of liability and responsibility from the party who caused the damage, he is not allowed to sail from the port before coming into contact with the Charterers, allowing them to sort the matters out and the time lost will continue to count as lay-time or time on demurrage.

3. will try to get stevedores to repair the damage and try to settle the matter directly with them in the first stage and the time lost will continue to count as lay-time or time on demurrage.

For stevedore damage which:

a. affects the effectiveness of the ship and the safe operation and/or safety of the crew and/or affect the trading capabilities and/or seaworthiness and/or cargo worthiness and/or is subject to classification requirements, then such damage as will be ascertained by the ship's Class surveyor, who will be appointed by the Owners at the Charterers' cost, is to be repaired by the Charterers at their own expenses/risk, to the Master's and/or Class surveyor's entire satisfaction and the time lost will continue to count as lay-time or time on demurrage until completion of all repairs and the 'port clearance' is issued and delivered to the ship.

Charterers are to reimburse the Owners for the Class Surveyor's fees, once the Owners have submitted their final freight invoice, supported with the original vouchers. The damage will be repaired at the place of occurrence, if it is possible and there are adequate facilities for permanent repairs, or otherwise, temporary repairs are to be made on the spot or as when detected, at the Charterers' risk and expenses and to the Master's and/or Class surveyor's satisfaction and the time lost will continue to count as lay-time or time on demurrage. The ship will sail or shift to the nearest shipyard, as chosen by the Charterers, with fuel consumed at sea or during shifting to be for the Charterers' account, which will be settled together with the Owners' final freight statement, in order to complete the repairs.

The time lost during the sailing or shifting to the shipyard and until completion of the works with the damage being repaired to the Master's and/or Class surveyor's entire satisfaction and the 'port clearance' is issued and delivered to the ship or, at the Owners' option, at equidistance from the discharge port, will count as lay-time or time on demurrage.

Shipyard's expenses and/or port expenses, including in and out tugs and/or pilots and/or agency fees etc., will be entirely for the Charterers' account.

b. does not affect the effectiveness of the ship and the safe operation and/or the safety of the crew and/or the trading capabilities and/or the seaworthiness or cargo worthiness and/or is subject to classification requirements and cannot be repaired by the stevedores, the Owners then undertake to accept the ship's sailing, in accordance with the terms of the Charter Party.

However, the said damage is to be assessed before the ship's departure from the discharge port and the costs to be agreed upon between the Master, Owners, Charterers and the respective survey company and to be paid together with the Owners' final freight statement. The survey company will be mutually appointed by the Owners and Charterers, if considered necessary, at the Charterers', Owners' equal expenses and will promptly inform the Charterers, Owners with the damage report, which will be issued before the ship sails, with a copy of all correspondence exchanged in this connection. The time lost will continue to count as lay-time or time on demurrage, until the 'port clearance' is issued and delivered to the ship.

Owners have the option of having the damage repaired at their next convenient time or at the ship's next Dry-Docking.

In the case that the Master, Owners, Charterers cannot reach an agreement on the cost with the respective survey company, the Charterers will then undertake to reimburse for the actual cost of the repairs, but not for the time lost, upon presentation of the original bills from the repairers or dockyard, that are consistent with the surveyors' report on the extent of such damage.

c. is small and has been repaired by the crew, the Charterers will reimburse the Owners for the cost on request, but not for the time lost and upon presentation of the Owners' repair statement, which is acceptable, even by fax or e-mail.

However, and in any case the Charterers are responsible for any directly related expenses resulting therefrom and incurred thereby and must pay compensation for any damage caused to the ship and/or restore the ship to its initial condition under their own time and expenses, whether or not any payment has been made by the respective parties and/or stevedores to them.

The Owners guarantee that all vulnerable parts in the ship's holds will be properly protected under their time and expenses.

The Charterers will not to be liable for damage when the Master has accepted a statement in writing from the stevedores and/or respective parties that the operations were carried out without any damage, or that the repairs were made to his satisfaction.

18.

Sub further terms/details based on Synacomex 90 Charter Party, Charterers' executed proforma with 3,75pct ttl commission including 2,5pct address commission.

19.

Holds' inspection:

The ship, upon arrival at the first or sole load port, must have all cargo holds and hatches free of loose rust flakes/rust scales and loose scales, free from salt and free from residues of previous cargo. It will also be clean/clear/fresh water washed and dried and free from insects and odours and in all respects ready and suitable for loading the Charterers' cargo, according to the rules and regulations

prevailing at the port of loading.
If the ship fails the preloading surveys, including the holds' inspections, the Charterers and/or Shippers and/or Port Authorities have the right to order the ship to shift from the berth to anchorage or to a waiting place as deemed necessary for commercial and/or safety reasons, due to the ship's inability to pass the preloading surveys and inspections. All shifting expenses incurred, including tugs/pilots etc. to anchorage/waiting place and then back to berth will be for the Owners' account and the time lost will not count as lay-time or time on demurrage.
The Owners will take immediate corrective steps to expedite the readiness of the ship-cleaning, as fast as possible, either with the Crew of the ship, if it is permitted by the regulations/port labour unions, or with the shore labourers, in order to meet the surveyors' requirements at their cost/risk.
If the ship fails to pass re-inspection within 72 hours from the first inspection, the Charterers have the option to immediately declare the cancellation of this Charter Party, which will be considered null and void and no claim will be pursued, either by the Owners or Charterers or any third party. All in/out of the port expenses, including Agency fees, will be for the Owners' account.
If part of the ship's holds have passed successfully and the loading operation has commenced in the holds that have passed or the loading operation has commenced without all the pre-loading surveys to have been completed, the lay-time then will commence counting as per the Charter Party or will count as time on demurrage.
Hatch covers' inspection:
The ship's hatch covers will be watertight and will be maintained in top condition. The Charterers/ Shippers may order, through their surveyors in the berth or in anchorage, for an ultrasonic test and/or a hose test or Chalk test on all hatch covers, but the Charterers' requirements will not exceed the ship's class rule. The costs of the survey will be for the Charterers' account, but the Master must give every facility to the Charterers and their surveyor, in order to conduct such tests and the lost time will count as lay-time or time on demurrage.
In the event that the ship has failed the hatch inspection, the Owners will immediately arrange for shore labourers, or, at their discretion, for the ship's crew, provided it is permitted by regulations/ port labour unions, to make repairs in the anchorage or in berth if allowed by the Shippers/Port Authorities. Time lost/ costs for this operation and extra expenses incurred, including the shift of the ship to anchorage and back to the berth will be for the Owners' account.
If the ship fails to pass re-inspection within 72 hours from the first inspection, the Charterers have the option to immediately declare the cancellation of this Charter Party which will be considered null and void and no claim will be pursued either by the Owners or Charterers or any third party. All in/out of the port expenses including Agency fees will be for the Owners' account.

If part of the ship's hatches have passed successfully and the loading operation has commenced in the hatches that have passed or the loading operation has commenced without all the pre-loading surveys to have been completed, the lay-time then will commence counting as per the Charter Party or will count as time on demurrage.
In the case of a dispute on the hatch inspection, an independent class approved surveyor will be mutually appointed and his opinion will be binding for all parties. Surveying fees will be paid by the party in default.

PART III

CHAPTER I

FIXING ON TIME CHARTER

This is one of the three methods of fixing a ship, which applies to all types and sizes in the dry cargo sector.

The majority of fixtures on Time Charter are made on large sized ships, mostly of Handymax size and above. The number of fixtures on Time Charter for all sizes (except for coasters/smaller sizes) increases, when the shipping market has an upward trend and is showing significant improvement, in which case the Owners have the opportunity and the advantage to fix on Time Charter for a long period and in addition to achieve higher 'hire rates'.
The Owners' preference in this method of fixing is evident, as when fixing on a period with first class Charterers,
. mainly ensures the ship's trading and their investment, even if the market has thereupon dropped.
Many of the Voyage Charter's risks/costs/responsibilities lie on the Charterers, thus, the head Owners' profits that are actually based on Time Charter should in total be less of what they would be, if the ship was trading on Voyage Charter.
The Charterers should sustain the loss of time while at sea, or at port, because of bad weather, stormy seas, delays incurred in ports due to congestions, strikes, force majeure, as well as liabilities for cargo claims to third parties (should be apportioned in accordance with the Inter-Club New York Produce Exchange Agreement) and any other unforeseen circumstances resulting to delays beyond the ship's fault, which may arise during the voyage at sea and/or in ports.
In fact, Charterers, as regards to employment and decisions on the ship's trading, act as if they are the actual Owners, thus, when entering the open market

in search of suitable cargo to sub-Charter the ship, is referred to as
. 'disponent Owners'.

The ship remains at the head Owners' management/control and maintenance, who will organize and keep it safely manned, provide an efficient crew and will generally produce a seaworthy ship, ready and fitted in all respects to perform under the terms of the governing Time Charter Party. They should comply with all international and local regulations in the ports that will be contacted and should keep valid and up-to-date charts and original certificates on board. The ship should be kept in perfect condition and maintained the same for the entire duration of the Charter, for its obligations under the terms of the Charter Party to be fulfilled.

Fixing on Time Charter implies obligations and responsibilities upon Charterers to provide legitimate employment decisions and instructions for each trip and to direct the ship to proceed via safe port/s, safe berth/s, and safe anchorage/s. They will supply sufficient quantity of fuel required for the execution of the trip/s and will ascertain the ship's ability to approach the intended ports and ensure a safe docking/un-docking/departure.

Under the terms of the Time Charter Party, they will financially reimburse the Owners, if the ship suffers physical damage, due to insufficient draft at the port of operation or the berth/anchorage has been exposed to adverse weather conditions (unsafe berth/anchorage) or even for possible losses caused by the political situation in a port and generally, from any incorrect information that has been provided to the ship's Master by them or their Agents or representatives.

Charterers have the option to fix a ship/s and then to:

a- trade the same as disponent Owners for their own cargoes and/or in order to cover their Contract of Affreightment's (c.o.a.) commitments. They ensure a standard 'hire cost' for a period during which they must fulfil their trading obligations/contractual commitments with others, regardless of whether the market has thereupon dropped or arisen. They could additionally earn from the ship's quick despatch in the ports of call or arrange the same by ordering overtime etc., thus increasing their profits.

b- trade the same as disponent Owners in the open market for the best deal, primarily based on Voyage Charter.

They generally believe that they can achieve better numbers than the head Owners and will earn the difference between the standard 'hire rate' paid to the head Owners and the corresponding amount resulting after the completion of each voyage (Time Charter Equivalent rate). They could additionally earn from the ship's quick despatch in the ports of call or arrange the same by ordering overtime etc., thus increasing their profits.

c- sub-let out the entire or part of the Charter period to another Charterer for profitable /trading reasons (terms/conditions of the Sub-Charter should be on a 'back-to-back' basis with the head Charter Party). However, they remain responsible for the fulfilment of the terms of the head Time Charter Party (between them and the head Owners), even if the sub-Charterers have stepped down from their financial or other Time Charter Party obligations.

The Owners will deliver the ship to the Charterers' services:

- by receiving a specific amount as a daily 'hire rate' or pro rata for any part of the day, which the ship is on the Charterers' services.
- for a mutually agreed Time Charter period.
- at the point and dates, as will be mutually agreed upon.
- having the ship free of any trading commitments with others and in all respects clear to trade throughout the mutually agreed countries.
- having the holds cleaned/swept/dried and free of any cargo residues from its previous voyage, free of rust scales and loose rust and in all respect ready for loading the Charterers' intended cargo.
- capable to load the agreed type/grade of cargoes.
- having on board the agreed quantity of fuel on the time of delivery.

The Owners will not be affected:

- from loss of time, due to stress from weather/stoppages during the sea voyage and/or during the operations in the port/s, which is not the Owners' or ship's fault.
- from loss of time, due to force majeure/strikes in ports.
- from loss of time, due to congested port/s.
- from additional port costs that could be charged, due to unforeseen increased tariffs or costs that are in effect, due to the ship's prolonged stay in the port/s.
- from loss of time, due to stevedoring damage and until the damage is restored.
- from the process of the ship's refueling, which will be made by the Charterers for the execution of each trip and from its consequences (deviation/lost time supplying fuel to the ship/Agency fees etc.).
- from the entire liability for cargo claims made by third parties (however it should be adjusted in accordance with the Interclub Agreement clause).
- from the market's direction if it thereupon drops.

Owners undertake to pay for the costs, which include:

- the crew's wages and all related expenses, such as transportation/repatriation etc. and additional Agency fees or costs resulting therefrom the provisions/

stores/spare parts/dry-dock and special survey or other commitments to the ship's class/flag etc. and generally all costs that are
required to be spent for the ship's maintenance, including all additional Agency fees or costs resulting therefrom.
- the H&M (Hull and Machinery), P&I Club (Protection and indemnity).
- the basic War Risk insurance.
- the supply of lubricants required for the execution of the trip/period.
- the domestic consumption of fresh water.
- the quantity of fuel remaining in the ship's tanks at the time of redelivery
- the consular fees/taxes associated with the ship's flag/crew/ownership etc.
- the imposed fines/penalties and/or security required, because of Master's/Officers'/crew's illegal action.
- the proven overconsumption at sea/port, in violation of the Time Charter Party description clause (which is subject to prevailing weather conditions during the voyage).
- the address commissions and the nominal percentage (rate) of 1,25 percent to each broker involved in the ship's fixture, which is calculated on the gross amount of the due 'hire' and possible paid ballast bonus.
- the office expenses (staff wages/insurance/rental costs etc.).
- the installments on loans obtained from the banks for the purchase of the ship.

All above costs will be covered by the amount which has been achieved as a daily 'hire rate'.

<u>Charterers undertake to pay for the costs, which include:</u>

-the payment of the Time Charter Party daily 'hire rate' plus the possible ballast bonus.
-the quantity of fuel on board at the time of delivery.
-the supply of sufficient quantity of fuel required for the execution of the trip/s.
-the ordinary ship's port expenses in all ports of contact (tugs/ pilots/ wharfages/light dues etc.).
-the passage through the canals/straits. the taxes and/or freight/hire taxes and/or dues levied/calculated on cargo and/or on ship/flag etc.
-the extra insurances, as will be charged by the cargo insurances, due to the ship's age/flag/ownership etc.
-the extra insurances that will be charged to the ship by the H&M insurers for breaking the International Navigating Limits (if agreed to breaking the same).
-the additional War Risk insurances, because of trading to ports or areas which are designated as dangerous or likely to be dangerous or passing through des-

ignated piracy areas.
- the expenses that were incurred on behalf of the Charterers in the ports and at sea for communications/victualling/entertainment (abbreviated as 'c/v/e').
- the Owners' compensation for redelivery of the ship with unclean holds after completion of the Charter (in lieu of holds cleaning term, abbreviated as 'ilohc').
- compensation to Owners for the intermediate holds cleaning (the cleaning of holds by the ship's crew between the voyages).
- the fumigation required/carried out and related expenses/lodging expenses incurred, as a result of the ship's trading in infected areas or due to the type of the transported cargoes.
- the stevedoring damage occurred in the ship's holds/hatches/deck, during the progress of the operations.

Duties imposed on the Master/Owners, who undertake to fulfil the following obligations and are responsible under the terms of the Charter Party:

- to provide a seaworthy ship and to be in possession throughout the period of the Charter, of all necessary valid equipment, up-dated charts, and original certificates, complying with safety and health regulations, national and international regulations and all current requirements applicable to all allowed ports for contact or for passing through the canals/ straits and various straits.
- the ship to be delivered in such a condition and maintain the same for the entire duration of the Charter, ready to successfully pass all the 'Port State Control' inspections. The ship must be able to perform all the voyages fixed by the Charterers and be capable to trade/load/ discharge all allowable cargoes in all allowed countries, in accordance with the agreed Time Charter Party terms.
- to supervise the loading and discharging operations.
- to arrange and take any precautions required for the safety of the cargo/crew/ ship.
- to provide an efficient crew and keep the ship safely manned as per the requirements of the ship's flag.
- for cargo claims/shortages, in accordance with the 'Interclub agreement' terms.
- for the consequences/delays/damages/expenses which may arise, when they have provided any misleading information on the ship's characteristics and on the terms agreed in the Charter Party.
- to keep the ship fully operational during the execution of the Charter and to fulfil their obligations as per the agreed Time Charter Party terms.
- to execute the voyages with the utmost dispatch and keep the minimum described Charter Party speed and consumption.

Duties imposed on the Charterers, who undertake to fulfil the following obligations and are responsible under the terms of the Charter Party:

- to accept the delivery of the ship at the agreed point and range and redeliver it at the agreed point and range after completion of the Charter period.
- to fulfil their obligations if the ship has been sub-charted to other Charterers.
- to trade the ship only through the allowed countries/ranges and to provide for loading only the permitted by the Charter Party cargoes.
- to direct the ship to port/s which can be approached by the size of the ship and can safely berth/unberth/depart without being exposed to adverse weather conditions.
- to provide sufficient information and effective instructions to the Master with regard to the intended ports and the prevailing port restrictions and all the intended cargoes for loading.
- to pay regularly and punctually the agreed daily 'hire rate'.
- to redeliver the ship in the same condition as on delivery, after having repaired all damage incurred in the ship's holds/deck/hatches etc. during operations (ordinary wear and tear are excepted).
- to order an adequate quantity of fuel for the satisfactory execution of the intended voyage, in compliance with the described and agreed fuels specifications.
- to redeliver the ship with the same approximate quantity of fuel as on the time of delivery, otherwise as agreed.
- for the coverage of cargo claims/shortages, in accordance with the Interclub agreement terms.
- to release the arrested/detained ship, due to the actions, negligence or default of their Agents or their representatives or the contractors/Shippers or Consignees.

The ship remains on-hire for the time lost:

- at sea during its voyage, but subject to sea conditions exceeding the Beaufort wind scale 4 and state of sea Douglas 3 (sea and swell scale).
- due to adverse weather conditions during the operations in ports.
- due to force majeure or strikes in the ports of contact (stevedores/tugs/ pilots etc.).
- due to Charterers' failure to comply with the obligations of the Charter Party or for any delays due to Charterers' or their Agents'/servants' fault/s.
- due to oil or air pollution arising from Charterers' or Agents'/servants' negligence/fault.
- during the supply of fuel on board to run the trip/s or for redelivery purposes.

- due to arrest/detainment attributed to actions against Charterers or sub-Charterers or their Agents or their representatives or the Contractors/Shippers or Consignees.
- for repairs due to stevedore damage.
- for physical damage suffered due to insufficient draft or unsafe berths/ anchorages or due to the political situation in the port.
- for detention for quarantine due to Master's/Officers'/Crew's pestilence, epidemics and illness or failure to pass free pratique, due to the Charterers' previous trade in infested areas.

The ship will be placed off-hire for the time lost:

- if throughout the Time Charter period, it is not in possession of valid and up-to-date certificates on board the ship, complying with safety and health regulations and/or requirements in the countries which will be contacted.
- if the ship is not fitted to pass through the Suez/Panama Canals and various straits in violation to the Charter Party terms.
- arising due to its underperformance (poor speed during good weather conditions).
- for any deviation from the course of the voyage, due to breakdown of machinery, collision, stranding, fire, or accident of damage to the ship, dry-docking or periodical survey, sickness or accident to the Master-Officers-Crew, to disembark stowaways and for any other reason associated with the Owners'/ship's/Master's issues or fault.
- if the ship, due to action against the Owners, is arrested or detained in the port or while passing through canals etc. at the suit of any party having or supporting to have a claim against or any interest in the ship.
- arising due to smuggling from the Master/Officers/Crew.
- arising due to detention/quarantine of Master's/Officers'/Crew's pestilence, epidemics and illness or failure to pass free pratique due to crew members or for any other reason attributed to the ship's fault.
- due to breakdown, partial or otherwise, or non-conformity or disablement or malfunction at one or any/all winches/cranes or any other part of the ship's equipment preventing the ship's full operation.
- due to insufficient power provided to operate all winches/cranes at the same time during the operations in the ports, if it is required.
- due to cargo residues (upon delivery) in the holds from its previous voyage and/or rust scales and/or loose rust in cargo holds.
- if it is detained by the Port State Control inspection or by the International Transport Workers' Federation (ITF), preventing its sailing or interrupting its normal operation in the port.

- due to a strike of the crew and/or deficiency of officers/crew or stores.
- in the event of detention due to proven oil or air pollution, arisen by reason of the Owners' or Master's or crew's negligence/fault.
- if it is not eligible for bunkering at the places where the Charterers intend to and have agreed to supply the ship with fuel.
- arising from any other reason due to the Owners'/Master's fault/negligence or the ship's inability which could vitiate or create stoppages to its full operation.

The 'cargo orders' based on Time Charter will enter the market in accordance with the following modes:

1-Time Charter Trip Out or One Time Charter Trip (abbreviated as 'tct out' or 'one tct').

Time Charter Trip Out or One Time Charter Trip will be in effect, when the Charterers
- accept the delivery of the ship as per the contract terms,
- load from the intended port/s the intended cargo/oes,
- discharge to the intended port/s,
- redeliver the ship to the Owners as per the contract terms.

The offered and subsequently fixed ship should be a
. 'named' ship,
and under the terms of the Time Charter Party, the Owners do not have the right to substitute it with another or to sell it during the execution of the Trip without the prior consent of the Charterers.
The commitments of the ship's class/flag (dry dock/special and intermediate survey etc.) will not be passed, unless in the case of an emergency, as will be stipulated in the Charter Party.
The exact duration is unpredictable, and Charterers should try to avoid committing and agreeing to any specific duration of the trip (unless it is a very short trip). The Owners have the expertise required to calculate the approximate duration, otherwise they could agree to an escalation on the hire rate, if the ship is delayed for more than a specified maximum period.

As an example:
-Dwat 50/58,000 mtns
-Sid/max 15 years old/geared with min 30 mtns lifting capacity/ deck cargo essential
-delivery 1 sp China chopt (intention CJK)
-l/c 18/25 March 2....
-one time charter trip with intention steel prds/bagged cement/equipments/

containers etc
-dur about 60/70 days wog
-redelivery dop 1sp Dakar to Cape Town range
-comm 5 ttl incl 3,75 add comm.

The ship will execute only one loading leg and Charterers do not have the right to re-load the ship after completion of discharge at last port, within the range of West Africa/South Africa.
The delivery of the right ship in size will take place in China, basis on 'dop' of any port in the Charterers' option, but with the intention of CJK (located close to Shanghai), within the dates as specified in the 'cargo order'.
The exact type/grade of the cargoes intend for transport has not been declared and the 'cargo exclusion' clause should be discussed and mutually agreed upon.
The duration is 'about' but 'wog', in which case the Charterers have no obligations/liabilities as to the extent of the duration of the Charter but are just providing a guide to the Owners.

2-Time Charter Trip with 2 loading legs (abbreviated as '2ll') or more.

If there are two loading legs, it implies obligations of the Charterers
. to load and discharge the ship twice.
Upon completion of the first leg, Charterers will re-load other or similar cargo from the same or another country/port and discharge it elsewhere. Certainly, the ship will not be delivered back to the Owners, at the end of the first leg.
The interested and subsequently fixed ship should be a
. 'named' ship,
and under the terms of the Time Charter Party, the Owners do not have the right to substitute it with another or to sell it during the execution of the Trips.
The Charterers have the right to refuse the delivery, if the Owners have presented any other ship, even with better characteristics than the fixed one.
The commitments of the ship's class/flag (dry dock/special and intermediate survey etc.) will not be passed, unless in the case of an emergency, as will be stipulated in the Charter Party.
There is no commitment of the Charterers for the required time to execute the entire Charter or each leg, which could run between the allowed countries with irregular duration and transport of different cargoes. However, for fixed and regular schedules, an approximate duration could be advised to the Owners by the Charterers.

As an example:
-panamax

-Time Charter trip with 2 ll
-delivery North Brazil 20/30 March
-redelivery dop 1sp Skaw/Passero in chopt
-1st leg wit iron ore from Brazil to Argentina 2nd leg with agriprds/grainprds from Argentina to Skaw/passero
- 5 ttl comm.

Two loading legs will be executed by the Charterers. The first leg is iron ore cargo from Brazil to Argentina.
Upon completion, the ship will perform
. in direct continuation (as will be expressly stated in the Owners' offer)
the second leg with a cargo shipment of agricultural products and/or grain products from Argentina to the port/s which lie between the Skaw/Passero range in Europe. After completion of discharge, the ship will be redelivered to the Owners, at a specified point/area as per the contract terms. The schedule intentions are known to the Owners, so they can calculate the approximate duration of the Trip or they can ask to be informed by the Charterers.

Fixtures with a basis of 3 loading legs, imply obligations upon the Charterers
. to load and discharge the ship three times.

3-Time Charter Period (short or long period).

The Charterers' ability to keep the ship for a longer or shorter duration, without commitment as to the number of legs that it could execute will be expressed as:
a. short period,
which represents a Charter Party commitment of about 2 to about 5 months. The specific period should mutually be agreed upon.
During this short period:
- there is no limit to the number of trips that the ship will execute.
- irrespective of the executed trips, the ship will not be redelivered earlier or later than the allowed edges of the fixed period.
- the port/s of loading/discharging and the type/grade of cargo can be expressed only as 'intention', unless Charterers can guarantee the same, in which case the schedule/s intentions and cargoes could be limited to the Owners, so the arguments on the allowed trading and the cargoes for loading can be avoided.
- the prospective 'hire rate' could usually be higher than when fixing for a long period.

As an example:
-short period of min 2 months/max 4 mos at chopt

-Panamax max 20 years
-delivery port in Continent at chopt
-l/c 20/30 March
-trading within Atlantic/Med/Baltic
-redelivery dop 1sp or passing Skaw/Gib range
-5 ttl comm.

The ship that will be fixed, will be trading under the Charterers' services for a minimum of 2 months to a maximum of 4 months, however,
. the exact time of the ship's redelivery within the period allowed in terms of the Charter Party remains at the Charterers' option.
The Continent is the place of delivery and the ship will be trading only within the Atlantic/ Mediterranean /Baltic.
The interested and subsequently fixed ship should usually be a
. 'named' ship,
and under the terms of the Time Charter Party, the Owners do not have the right to substitute it with another or to sell it during the execution of the Charter.
The Charterers have the right to refuse the delivery, if the Owners have presented any other ship, even with better characteristics than the fixed one.
The commitments of the ship's class/flag (dry dock/special survey etc.) will not be passed, unless in the case of an emergency, as will be stipulated in the Charter Party.

b. long period,
which represents a Charter Party commitment of six months and above, where:
- there is no limit to the number of trips which the ship will execute,
- schedules' intentions and cargoes are usually unknown to the Owners, however, the same should be in compliance with the relevant Time Charter Party clauses, namely the 'cargo exclusions' and 'trade exclusions'.

-Charterers should search the market for Owners who are flexible on the trading and cargoes that are allowed.
- the ship will be trading around the world and the redelivery will take place on 'dlosps' in one port within the trading limits or 'passing' a specified point as will be mutually agreed upon.
- the 'hire' will usually be lower than when fixed in a short period.
- the Charterers, due to missing available cargoes (market is collapsed) or for commercial reasons could be authorised by the Charter Party terms, to instruct the ship to remain in an idle condition for a long period of time, or, even by a mutual agreement with the Owners, to lay-up the ship in a mutually agreed safe

place/anchorage etc., for the purpose of minimizing the operating costs and reducing the hire and loses accordingly.

As an example:
-Panamax max 13 years only
-1 year time charter period 1 month +/- chopt
-delivery Continent /Med
-trading ww
-redelivery dop ww
-3,75 ttl comm.

The Panamax ship that will be interested in the above-mentioned cargo order should not be more than 13 years old and will be fixed for one year on a Time Charter period, with one month more or less at the Charterers' option. It will be trading around the world, on the basis of redelivery on 'dlosps' at any port but within the trading limits.
The interested and subsequently fixed ship should be a
. 'named' ship,
but under the terms of the Charter Party of 'Sales/Substitution' clause the Owners might have the right to substitute it with another or to sell it during the execution of the Charter. However, and in any case, the Charterers' prior approval should be required.
The Owners, under the terms of the Time Charter Party of the 'Periodical Dry Dock/Special Survey/ Intermediate Survey' clause, should have the right during the execution of the Charter period, to pass the class and flag commitments. The ship will be placed off-hire during the time that is required, which could, however, be added on top of the agreed maximum period, as will be decided by Charterers.

A fixture based on a Time Charter Party in the dry cargo sector, is characterized by the following parties involved and terms, which, along with certain others, will apply and most of which could affect the 'hire rate':

1. The Owners and Charterers.

Owners/Managers:
they are the party who will provide for Time Charter a ship that is seaworthy/ suitable and in the right position for the needs of the Charterers.
Those who have a bad reputation in the market as unreliable, either because they do not meet their obligations under the terms of the Charter Party, or by the ship's maintenance which lead to the ship's detention in the ports of con-

tact, should be avoided to be fixed by the Charterers.
Charterers:
they are the party who will Charter (fix) a ship from the shipping market, for a specific period of time and will trade it for their contract cargoes or openly in the market for speculation reasons, in which case they will be called:
. Disponent Owners.
When fixing voyage cargoes, however, in order to cover their positions or for speculation reasons with or without presenting a 'named' ship, they will be called
. Operators.
Charterers who have had limited activity and have not been well established in the market, or who are not considered as 'first class', or who have previously given Owners a bad experience, may not be approved, as no one should take the risk of fixing a ship with someone, who might step down from full or partial fulfilment of their contractual obligations.
However, a fixture with Charterers with limited activity or who are unknown in the market could be saved if:
a. Charterers nominate as 'guarantors' another well-known or better-known reputable shipping company in the shipping community, which could be fully approved by the Owners.
A letter of guarantee, on the basis of the Owners' wording should be taken directly from the guarantors, before the final agreement is reached, confirming that they guarantee the Charterers' performance/payments. The guarantors should be requested to countersign the original Charter Party, showing that, apart from the guarantees that have been provided to the company, they additionally admit that they are also well aware of the existing Time Charter Party terms.
b. Charterers provide a revolving bank guarantee as will be issued by an absolutely first class bank, which roughly covers the entire 'hire' of the approximate duration and an approximate amount for port and other expenses, as well as the approximate fuel cost for the execution of the trip/s.

On Time Charter, the ship will be receiving legitimate employment instructions in compliance with the terms of the Charter Party,
. only from the Charterers.

2. The ship and Master's/officers'/crew's duties.

The Charterers will select a ship from the market,
- to either cover a single or certain specific trip, in which case the ship should meet the minimum characteristics required for the shipment/s,

or
-to fix it for a short/long period for trading freely in the market, in which case the Charterers should look for an attractive 'hire rate', along with the best description.

The cargo order for Time Charter will enter the market by the Charterers asking for a workable ship, by virtue of the:
a- general characteristic of the ship's size and ask, for example, for a 'panamax ship' or a 'handymax ship' etc.,
b- ship's specific size and ask, for example, a 'minimum 25,000 dwat' or '50/55,000 dwat' etc.,
c-minimum grain or bale capacity required for the trip and ask, for example, for 'sid/bc with minimum 3,000.000 cbft grain capacity' etc.

If the Charterers' attempt to sub-Charter the ship to others or to fix a cargo on Voyage Charter based on the named fixed ship, then, as 'disponent Owners', along with the ship's description and the Head Ownership, they could be subject to Letter of Credit terms and/or Head Charterers'/ Shippers'/ Receivers' approval.

I. Charterers' employment instructions.
Upon the ship's delivery to the Charterers' services, they must provide the Master of the ship, from time to time and for each voyage separately, with all requisite clear voyage instructions/ sailing directions and all useful cargo information, only in writing, by e-mail or fax and only in the English language. The Charterers should be obliged to make a response without delay to any clarifications sought by the Master, or will otherwise, under the terms of the Charter Party, be held responsible for the delays which might arise and/or expenses incurred, due to the Charterers' failure to respond.
The Master should follow instructions as if the ship was trading for the Owners' account and to give all proper notices or information that is required to be given to any party involved in the fixed voyage.
The Master should inform the Charterers daily as per given instructions, on the ship's navigation at sea, the movement in the ports, the progress of the operations and provide any other information which might be required by them. He must keep a full and correct deck and engine log of the voyages in the English language, to the Charterers' forms, as will be presented by them or in the forms as on board, showing, among others, the course of the ship, the running distance, the fuel and oil consumption, the prevailing weather conditions etc.

II. Ship's description.
For Trip Out Charter Parties and/or regular trades between certain countries,

where no special requirements are in effect and/or the approaches/berths are without port restrictions with regard to the size of the fixed ship, a "basic" Time Charter description could be enough to cover the Charterers' needs.
For long Time Charter periods, or in accordance with the Charterers' practice, which is applicable even for short trips, it may be asked from the Owners
. to fill in a standardized questionnaire
providing a wide and detailed description of the ship.
The Charterers' said questionnaire will be asked to be a part of and incorporated in the Charter Party. Depending on the prevailing conditions, there should always be easy access to the ship's detailed characteristics by the Charterers. Many claims have been made by Charterers for incorrect descriptions of the ship because the Owners have incorrectly judged the severity of this clause.

The 'basic' Time Charter description should include:
- the name and the ex-name/s.
Comments:
The current and the ex-name/s are required, enabling the Charterers to have access to the ship's history, before concluding the fixture.
- the summer dead weight all told (abbreviated as 'dwat').
Comments:
It represents the total and the maximum deadweight of the ship as designed by its yard, when trading within the summer zones, at the designated scantling draft. It includes all weights on board the ship, i.e., cargo/bunkers/stores/equipment etc.
- the draft based on summer salt water.
Comments:
The deepest draft (or draught) as designed by its yard (scantling draft) when the ship is fully loaded and trading within the summer zones. However, when trading within winter or tropical zones, the draft will fluctuate by decreasing or increasing its deadweight.
- the type of the ship.
Comments:
If it is a bulk carrier or tweendecker or multipurpose ship or logger or cement carrier etc.
- when it was built (date/month).
Comments:
It is the date and the month when the ship was delivered to the Owners for trading.
- the flag.
Comments:
The flag (the state), which follows the ship and inspects/certifies and issues all

documents referring to safety and pollution.
- the class.
Comments:
It is the classification society which follows/surveys/classifies the ship since the time of ordering the build and determines the operating condition when trading.
- the imo number.
It is the unique identification number as issued by the International Maritime Organization (IMO) and follows the ship, even if it changes its flag/ ownership/ management etc.
- the call sign.
Comments:
The identification number, which changes when the ship's flag is changed.
- the P&I club.
Comments:
The association which provides protection to its members against pollution claims, wreck removal, dock damage, seamen's/crew's risks, marine and cargo claims etc.
- the length over all (abbreviated as ' l.o.a'.).
Comments:
It is the ship's maximum hull length, i.e., the distance between the two extremes.
- the beam.
Comments:
It is the ship's maximum hull width.
- the length between perpendiculars (abbreviated as 'l.b.p.').
Comments:
It is the ship's maximum hull length between the forward and after the perpendicular.
- the depth moulded.
Comments:
It is the maximum distance from the top of the keel to the underside of the main deck and it is required when the ship contacts certain specific ports, for calculating part of the port expenses.
- the number of holds and type/number of hatches.
Comments:
It is the number of cargo holds as designed by the yard and the type and number of hatches which the ship can provide.
- the hatch dimensions.
Comments:
It is the length/width of the available hatches. An important point when load-

ing/ discharging with big grabs steel cargo/logs etc.
- the tonnes per inch and/or tonnes per centimetre (abbreviated as 'tpc/ tpi').
Comments:
It is required to calculate the amount of cargo loading per inch or per centimeter, mainly when trading in confined/river waters.
- the total grain capacity of the holds and its breakdown per hold.
Comments:
It is the maximum available space in the cargo holds (under deck) for commodities in bulk form, when loading cargoes with heavy or light stowage factors, which can spread themselves to all sides. The same is calculated in cubic feet or in cubic metres and is always larger than the bale capacity.
- the total bale capacity of the holds and its breakdown per hold.
Comments:
It is the maximum available space in cargo holds for commodities in form other than bulk and is calculated in cubic feet or in cubic metres. It is in effect mainly for smaller ships or for tweendecker/ multipurpose type ships, etc. which mainly carry cargoes that are bagged or in forms other than bulk.
- the gross tonnage/net tonnage (abbreviated as 'gt/nt').
Comments:
It is required in all ports around the word, for the assessment of the port expenses.
- Panama/Suez Canal gross tonnage/net tonnage.
Comments:
It is required for the calculation of expenses when the ship passes through the Suez or Panama Canal.
- the full cargo gear description.
Comments:
It is the description of the means of loading and/or discharging provided by the ship (geared ships).
- the speed and fuel consumption.
Comments:
The speed and fuel consumption which the ship can provide under specific weather/sea state conditions and under which conditions it can maintain the same.
- the fuel specifications.
Comments:
The specifications of the fuel for the correct supply by the Charterers.

The description of the ship, as offered by the Owners, must be an accurate representation of the characteristics of the ship, which cannot be altered after the ship's fixture. However, all ship's figures, under the terms of the Time Charter

Party, should be able to be modified with the usual and acceptable shipping interpretation of the word 'about'. In any other case the Owners should have guaranteed the same with accurate figures. The Owners are likely to be held fully responsible and liable for any consequences /delays /damages/expenses that may arise during the Charter, if they have offered misleading and inaccurate information on the ship's characteristics and have generally misdescribed its characteristics. In case of serious misdescription, the Charterers might be entitled with a notice to the Owners to terminate the Charter Party, or to even refuse the delivery of the ship.

The speed and consumption of a ship is one of the most important items for the Charterers, when making the "voyage calculation". Therefore, the word 'about' will protect the Owners in the case the ship fails to maintain its description,
. as it provides a negative variation of half a knot with the speed and a 5pct increase in the declared fuel consumption.

The provision 'all about and without guarantee', which could refer to the speed/ consumption, as well as, on the general description of the ship as will be provided, could cover the Owners or can minimize the responsibilities of the given data characteristics and would avoid claims for any possible incorrect description and/ or underperformance of the ship.
In fact, the majority of the Charterers only accept the provision of 'about' prevailing throughout the description of the ship, unless the 'wog' has been overlooked or they do not know the meaning of this phrase.
The brokers and Owners should exercise care when they describe and confirm certain specific characteristics of the ship during the progress of the negotiations. Simple and customary ship's description terms which could easily be confirmed by all, could incur additional obligations and responsibilities that could lead to losses.
A Self-trimming ship is that which, when loading full bulk cargo, by virtue of the design of its cargo holds, is in no need of extra trimming, during or after completion of the loading operation. However, this could not be confirmed for all types of bulk cargoes (such as pig iron, cement, gypsum, aggregates etc.),
. but only for free- flowing grain cargoes.
Owners should explicitly state, when confirming that the ship is self-trimming, that (as per Maritime Organizations' recommendation) it is
. 'STBC when loading of free-flowing grain cargoes only'.
The presentation by the Charterers or Receivers, of unreasonably large grabs, which could delay the discharging operation due to configuration of the holds, could give rise to claims against the Owners, because of the mistaken description that the ship is suitable for grab discharge. Owners should specifically state,

as a minimum act to protect the ship, that the ship
. *is suitable for a normal grab discharge operation, with normal size/type grabs and as far as a standard bulk carrier/ singledecker can be.*

III. Ship's certificates.
Throughout the duration of the Time Charter Party, the ship must be in possession of all necessary valid and up-to-date equipment and certificates that verify the size/type of the ship, its seaworthiness, its compliance with the health and safety regulations, national and international regulations and all current/ local requirements in all agreed ports/regions which the ship will contact, including valid certificates for passing through the Panama/Suez canals and several straits.
The Owners will remain responsible and liable for any delays/charges/consequences that may arise if the ship is unfit and not in possession of valid and up-to-date certificates before arrival in the port/s of contact or after inspection as will be conducted by the port Authorities and/or Port State Control inspectors. The Owners should take immediate corrective measures and all consequences, detention, consequential losses, fines, delays, damages, costs and expenses of claims and possible shifts which might be ordered, will be for the Owners' account.
The ship will also be placed off-hire if the stevedores, longshoremen or other port labourers are not allowed to work, by reason of the Master or Owners/ Owners' Agents failure to comply with such regulations or because the ship is not in possession of such valid and to date certificates and the cargo operations in the port have been interrupted or have interfered with the ship's unconditional operation.

IV. The quality of the supplied fuel.
The Charterers, when arranging and supplying fuel for the execution of the voyages, should be in accordance with the British standards. The supplied fuel should not include waste luboils and/or tar oil and/or inorganic acids and/ or nuclear waste and/or any other harmful substances and/or chemical waste or chemical polypropylene contaminations, which either jeopardize the safety of the ship or adversely affect the ship's performance or cause damage to the ship's engines or contribute to additional air pollution.

V. Holds'/hatches' cleanliness inspection on delivery.
The ship, on delivery or as will be agreed at the first or sole load port, will be inspected by the cargo surveyors for the readiness of its holds/hatches to load the Charterers' cargo.
The same will be contacted either by the Charterers' independent surveyors as

will be appointed and paid by them or by a national cargo bureau and/or a local grain surveyors' inspection at the port of call.
The ship must be in all respects ready for loading and:
-free of loose rust flakes/rust scales and loose scales,
-free from salt and free from residues of the previous transported cargo,
-washed with fresh water and dried,
-free from insects and odours.

The Charterers, under the terms of the Charter Party, may refuse the delivery of the ship, because of the dirty holds and in the case, they are not ready on or before the cancelling date, may become entitled from the terms of the Charter Party to cancel the Charter Party.

The ship will be placed off-hire if it fails its inspection and the Owners will be obliged to clean the holds/hatches without delay, either by using shore labourers or, if allowed by the port, with the ship's crew.
The Owners, under the terms of the Time Charter Party, will be held responsible for:
- the amount of fuel consumed during the off-hire period.
- the expenses incurred for the holds' cleaning.
- the necessary arrangement and supply of all chemicals, fresh water, consumable materials, tools such as water guns, pressure hoses, pumps, etc., when making the ship suitable for the cargo for loading.
- any directly related expenses incurred including stevedores' stand-by etc.
- for the shifting expenses to an anchorage or to waiting place if the Port Authorities instruct the ship to evacuate the berth.

Under the terms of the Charter Party the ship will again be on-hire when all holds/ hatches have successfully passed the re-inspection.
However, if the Master has objected to the rejection, there could be a provision in the Charter Party stating that both parties are to mutually appoint an independent surveyor, whose decision will be binding.
If two or more loading legs are to be executed, or a short/long period is fixed, where more than one voyage will in fact be executed, an
. 'intermediate holds' cleaning' clause
must be agreed and incorporated in the Charter Party.
The majority of the Time Charter Parties state that the Owners, against a fee, the crew is to render customary assistance in cleaning the ship's holds between cargoes and will keep the holds clean, as if trading the ship for their own account. However, the ship's crew are not regarded as skilled cleaning operatives and therefore there should be a limit on what cleaning can reasonably be effect-

ed by them. The removal of hard and rust/rust scale between the voyages is a major operation which cannot be performed by the crew, as such a cleaning is requiring the use of scrapping tools/rustybus machines etc. and cherry pickers for the access to the upper hopper areas/frames/corrugated bulkhead and the erection of staging. However, the crew can remove loose rust patches in accessible locations.

In fact, when intermediate hold/hatches cleaning is effected by the crew should be done without any guarantee whatsoever that the ship will pass any subsequent hold inspection and the ship/owners should not be in anyway liable if the ship's holds do not pass inspection for the next cargo. The crew/ Owners should not be kept liable for rust /rust scale in ship's holds between the executed voyages unless the Charterers have spelt out clearly in the Charter Party that the holds should remain in a specific standard cleaning under the Owners' risk/expenses. In addition, if Charterers accept the ship's holds on delivery

. without to reserve their rights of its condition,

then should be difficult to claim for damages when holds are failed during the execution of the balance Charter.

The Time Charterers should be obliged to provide and pay for all chemicals, fresh water, consumable materials, tools such as water guns, pressure hoses, pumps etc. necessary to be supplied in making the ship suitable for the cargo intended for loading.

If there is corrosion and/or pitting in the holds, because of the Charterers' previous trade of corrosive cargoes under this Time Charter Party, then the lost time and extra cleaning that could be required to be performed will be entirely for their account and under the terms of the Charter Party, they should bring the ship to its initial clean condition and before the redelivery takes place.

Dunnage, debris, lashing materials, which might have been stowed in the ship's holds and/or on its deck after completion of discharge, must be removed and discharged at the Charterers risk, by shore labourers, as arranged and paid by them, at the end of each voyage and before the redelivery takes place and/or the ship's sailing from the port.

VI. Officers'/crew's duties.

During the execution of the Time Charter, the Master/Officers and Crew of the ship will render their unconditional but regular/customary full assistance day and night and on overtime and will cooperate with all the parties involved, as they should normally perform when trading for the Owners' account, but subject to local regulation allowances. These procedures will contribute to the smooth execution of the Charter and to a fruitful cooperation between the Owners and Charterers.

However, all the additional work which may be required by the Time Charterers should be separately agreed upon during the negotiations of the Charter Party terms or even at the time of occurrence. The Charterers are to come, either directly or through their Agents, but always after the Owners' consent, in contact with the crew and agree on a compensation for their intended extra work. The Charterers are to benefit from such agreements, as the work from shore is costly and there would be loss of time. Such work could be the lashing/unlashing of the cargo on deck, the operation of the ship's cranes for loading and/or discharging purposes, the welding of padeyes on the deck or in the holds etc.

Some of the regular/customary Officers'/Crew's (subject to local shore regulations allowance) duties are:
- the readiness of the ship for immediate cargo operation on arrival at each port of call. The crew will open and close the hatches in connection with the loading and discharging operation.
- to supervise the loading/cargo stowage and discharging operation and everything related thereto and to insert any necessary remarks in the Mate's Receipts/Tally clerks/Statement of Facts.
- the bunkering operation of the ship.
- whenever required, the ballasting/de-ballasting of the ship's tanks during the loading and discharging operation, in order to comply with the port restrictions and/or the ship's stability.
- the docking/undocking/shifting between berths/warping alongside berths, whenever it is required by the Charterers/Agents/Port Authorities.
- the raising and lowering and the general preparation of the ship's cranes and/or gangways and/or grabs and/or grab devices, before the arrival at each port for the commencement of the loading and discharging operation without delay.
- to maintain sufficient and continuous electric power for the efficient work of the ship's equipment.
- attending to the deck cargo and its lashing, while the ship is at sea.

VII. Ship's cargo gear.

If the ship is described as a 'geared ship', the Charterers then have the right to load and/or discharge with the use of the ship's cargo gear as described and have at their disposal any related equipment on board the ship.
The ship's cargo gear and all other equipment must be kept in good working order/properly serviced and must comply with the regulations of the countries which the ship will contact and additionally, Owners should ensure that the ship is at all times in possession of valid and up-to-date certificates of efficiency, in compliance with such regulations, otherwise, the Owners must take immediate corrective measures. The Owners will bear the cost of all consequences and all consequential losses, delays, detention and/or fines, damages, costs, and ex-

penses of claims that may arise, due to their failure to comply with the port regulations.
In the case of breakdown or malfunction or disablement of the ship's cargo gear or even the supplied electrical power for working the cargo gear is not continuous or is insufficient, then the ship will be placed off-hire for all the actual time lost. The Owners will be asked to hire shore cranes in order to continue the operation and until this time, all extra expenses incurred including the time waiting and /or demurrage charges for trucks, barges, rail wagons and/or overtime charges for Shippers' and/or Receivers' and/or Charterers' employees/ servants and all other directly related expenses incurred, including to possible shifts, whether the affect is partial or not, will be for the Owners' account.

VIII. Ship's black list/Boycott.
The ship could be blacklisted or boycotted or even detained for various reasons by any organisation and/or governmental authorities and/or shore and/or port labourers and/or tugboats and/or linesmen and/or pilots and/or suppliers in the port of call, because of:
-its previous trade, leading to it being blacklisted by the country of its present call,
-its current trade, leading to it being blacklisted by other countries,
-its flag,
-its ownership and/or management,
-the nationality of its Master/officers/crew,
-the terms and conditions of employment of its Master/officers/crew,
-its class,
-its registry,
-the construction/cargo gear/fitting/other equipment.
Such boycotts and/or blacklistings could cause heavy delays and huge losses for the Owners. In the case that the ship remains detained or is boycotted for a long time, then the Charterers could have the option, under the terms of the Charter Party, to cancel the balance period and look for substitution.
Many Owners, when they negotiate to purchase a ship, ask from the sellers to confirm that the ship is not blacklisted from any port/country.
The Owners should make sure, through the terms of the Charter Party, that, if the political situation changed during the duration of the Charter and it is known that the ship will be blacklisted or be boycotted upon its contact to a port, then the Charterers must avoid ordering the ship to proceed to this port.

IX. Ship's deviation/put back/delays/off-hire.
During the voyage at sea, the ship might deviate from its normal route from one port to another, for various reasons, either attributed to the ship's/Mas-

ter's fault/ default or to avoid bad weather or for the purpose of saving life and/ or property or to assist or tow ship/s in distress.
The ship could deviate, because of:
-failure of machinery,
-collision,
-grounding,
-fire or any other accident or damage to the ship,
-dry-docking in case of emergency,
-illness or accident of the Master/Officers/Crew or any person on board except for persons travelling by request of the Charterers,
-stowaway or refugee,
-refusal of the Master/Officers/Crew to do their duties,
-any matters of the Owners or for the ship's maintenance or safety,
-bad weather or the ship driven into port or anchorage, due to stress of weather, -saving life and/or property and to assist or tow ship/s in distress
and for many other reasons.
The ship, during the execution of the Charter, could delay or be inoperative for various reasons and be placed off-hire, as has been already aforesaid. In general, if the ship is placed off-hire (provided that under the terms of the Time Charter Party, it has not been caused due to the act or omission or negligence or default of the Charterers and/or their Agents or Servants and/or their persons travelling on the ship and/or sub-Charterers and their Agents or Servants and/ or the Contractors and/or Shippers and/or Consignees and/or of the cargo carried and ports contacted under this Charter Party) and remained off-hire for a long, under the terms of the Charter Party then, the Charterers may become entitled to cancel the balance period.
This could happen when (subject to the Charter Party terms):
1- the ship is not in possession of all necessary valid and up-to-date equipment and certificates in compliance with health and safety regulations, national and international regulations and all current/local requirements, including valid certificates for passing the Panama/Suez canals and several straits.
2- the Master/Officers and Crew and/or any Owners' person travelling on the ship is captured, seized, detained, or arrested by any Authority or any legal process, during the period of the Charter, at the suit of any person having or purporting to have any claim against or interest in the ship and/or said Master/Officers and Crew and/or the person travelling on the ship.
3- because of its previous trade and/or because of its flag/ownership/class/registry/nationality and/or the terms and conditions, under which the officers and crew are employed and/or to the construction /cargo gear/fittings/other equipment and/or any other ship, under the same ownership/management and control:

- the ship is boycotted, picketed, blacklisted, delayed.
- the ship is not allowed to enter, load/discharge or depart from any port within the trading limits.
- the ship is rendered inoperative or immobilized or detained by Government and/or Authority and/or shore and/or port labourers and/or tugboats and/or linesmen and/or pilots and/or suppliers, or the ship cannot load or discharge at the intended port/place and is ordered to load or discharge in another place/ port.

4- the ship is inoperative from default and/or strike and/or deficiency of the Master/Officers/ crew or stores.

5- the ship is on fire or there is a breakdown/damage to hull or machinery and equipment or grounding or stranding or collision.

6- the ship is under repairs or other necessary measures in order to maintain its efficiency.

7- the ship is inoperative due to winches'/cranes' breakdown, partial or otherwise, or in non-conformity or disablement of one or any/all winches/cranes or any other part of the ship's equipment which prevents the ship's full operation or there is insufficient power to operate all winches/cranes at the same time, if it is required.

8- the ship failed to pass the free pratique due to pestilence, epidemics or illness of the Captain, Officers or Crew.

9- the ship is detained by the port state control.

10- the ship is detained due to proven oil pollution.

11- the ship is detained due to desertion of crew, or if the whole crew is not on board, when the ship is ready to sail.

12- the ship is unable to pass the holds'/hatch covers' inspection due to rust scales and/or loose rust in holds.

13- the ship is underperforming and generally does not comply with the agreed Charter Party description, or for any other reason including deviation, where the ship is not to the Charterers' full service and instructions.

If the ship is cancelled, the Owners should be obliged to refund the Charterers for any hire plus fuel paid in advance, or fuel supplied by the Charterers for their account and will be responsible for all extra expenses incurred, including, but not limited to, stevedores' standby. The redelivery of the ship obviously will take place without prejudice to any other right the Charterers may have on ship.

X. Ship's dry dock/special survey/intermediate survey.

During the execution of the Charter period the ship might have the obligation to pass its class/ flag commitments etc.

The Owners should have the right, under the terms of the Charter Party, to withdraw the ship from the Charterers' services for a reasonable period, in or-

der to pass its dry-dock/special survey/ intermediate survey commitments or to execute its annual or maintenance repairs.
Charterers should assist the Owners by fixing it to discharge, as close as, possible to the yard which has been previously selected by the Owners, in order to pass its surveys.
Logically, the ship will be placed off-hire on dlosps of the last discharge port and will again be on-hire, when all works have been completed and the ship is once more in the same position, as when it was placed off-hire, or at equidistance, as will be agreed upon.
All fuel consumed whilst the ship is off-hire should be for the Owners' account and the ship should be replenished and returned to the Charterers' services with approximately the same quantities of IFO and MGO on board, as when it was placed off-hire.
Charterers should reserve the right to add to the maximum period, any time that the ship was off hire during the passing of its surveys.

XI. Ship's lay-up condition.
The Charterers, for commercial reasons or due to a lack of tonnages in the area or due to a collapsed market, under the terms of the Charter Party, should request from the Owners, after giving proper notices, to lay-up the ship at a safe berth/ safe anchorage/safe place that is acceptable by the Owners and the ship's hull underwriters.
The Charterers should be kept responsible for all costs/time/expenses associated:
- with placing the ship in a lay-up condition, including the crew's expenses for repatriation (transportation, tickets etc.), indemnities payable to the crew and port expenses/lunch/ tugs/pilots/ Agency fees etc. incurred during the lay-up period.
- with placing the ship on reactivation upon request, including the crew's expenses for embarkation (transportation, tickets etc.)
- the cleaning of the ship's bottom/propeller and other parts from the marine growth, to the Master's satisfaction, before being entered again in the Charterers' services.
- with minor or serious damage/required repairs/stores etc. to the ship's hull or engine, which might be incurred during the time of lay-up, until reactivation and which are not covered by the ship's insurances.
After the ship is in a lay-up condition, the Owners should present the Charterers with a statement covering all possible estimated economic savings regarding operating costs, including signing-off of crew and insurance cover (but not on the insured ship's values). The hire should be mutually agreed to be adjusted between the Owners and Charterers, taking into consideration only the respec-

tive savings, as may be achieved and will continue to be paid to the Owners as per the relevant Charter Party payment clause.
Under the terms of the Charter Party, the Owners should not accept the ship being redelivered back to them in a lay-up condition or immediately after the reactivation.
For short periods there is not any reason for the Owners to accept such a clause, as the lay-up is not easily placed and may involve risk to the machine if it has been out of commission for a long period of time.

3. Delivery/Redelivery points.

The point of delivery and redelivery of the ship should be mutually agreed upon between the Owners and Charterers, during the negotiations of the Main Terms. This will determine, when the ship will be in commission/on-hire (commencing the period of the Time Charter) and when off-hire (ending the period of the Time Charter).

A ship is legally delivered when:
- presented by the Owners at the delivery point, free of cargo on board,
- presented with a quantity of fuel, in accordance with the Charter Party terms,
- presented within the Charter Party lay-days/cancelling date,
- its Master has tendered the delivery notice to the Charterers, and it is in fact ready, in all respects, to fulfil its obligations, under the terms and conditions of the governing Time Charter Party.

The normal delivery notice should contain:
the ship's name/call sign/by whom the ship is delivered (Owners or disponent Owners)/to whom the ship is delivered (Charterers)/point of delivery/date/local time of delivery and in UTC/fuel quantity on delivery.

A ship is legally redelivered when:
- it has received orders by the Charterers to proceed free of cargo on board to the redelivery point, as it is allowed from the Charter Party terms and has arrived at this point,
- it has executed the minimum agreed period,
- the Charterers have replenished the ship with fuel in accordance with the Charter Party terms,
- its Master has tendered the 'redelivery notice' to the Charterers, and it is free of stevedore damage that could affect its class and full operations.

The normal redelivery notice should contain:

the ship's name/call sign/by whom the ship is redelivered (the Charterers) and to whom the ship is delivered (Owners or disponent Owners)/point of redelivery/date/local time of redelivery and in UTC, fuel quantity on redelivery and the condition of holds upon redelivery (clean or unclean).

The lay-days/cancelling date for delivery and the time of delivery/redelivery could be expressed as follows:
-The ship will be delivered within the agreed lay-days, which usually take place from 00:01 hrs. of the first day of lay-days to 24:00 hrs. of the last day of lay-days (cancelling date). Under the terms of the Charter Party, for the purposes of delivery within the allowed lay-days,
. the local time will be in effect.
For example:
-lay/can 10/15 December 2.... 0001/24:00 hrs. For delivery purposes the local time will be in effect.
-The calculation of the hire, from the time of delivery until the time of redelivery will be made based on UTC time.
For example:
The ship delivered East Med on 1st March at 12:00 hrs. local time, however, the ship will be placed on-hire as per UTC time i.e., from 10:00 hrs. on 1st March.
-The provision of the delivery and redelivery term will be based on 'All Time Day or Night Saturdays Sundays Holidays Included' (abbreviated as 'atdnsshinc'), i.e., will take place all the time during the day or night, regardless of whether it is a Saturday/Sunday or Holiday.
For example:
-delivery on afsps Piraeus port, Greece atdnsshinc.

The delivery and redelivery of the ship at the agreed points could take place in accordance with the following key characteristics:

---

a)- On Arrival First Sea Pilot's Station (AFSPS) or on Arrival Pilot's Station (APS).

For delivery purposes:
The first pilot, who directs the ship from the open sea to the loading berth or to the entrance of a river channel is called a:
. 'Sea-Pilot'.
In river ports, the ship may need to use more than one pilot to reach the loading berth/place. The river-pilot/s that will board and direct the ship to the loading berth/place is called an:
. 'Inward Pilot/s'.

-On AFSPS (Arrival First Sea Pilot Station):
The ship will be delivered to the Charterers' services and commence the period of the Time Charter (on-hire), when the following conditions have been met:
- the ship has arrived at the specified port of delivery,
- it has reached the first sea pilot station, i.e., the point in the sea where the first sea-pilot will board the ship,
- the Master has tendered the delivery notice to the Charterers, whether the pilot embarked at the time or not.
For example, for delivery on an 'AFSPS':
- delivery on AFSPS Piraeus port, Greece, atdnsshinc
or
-delivery on APS Piraeus port, Greece, atdnsshinc:
On APS (Arrival Pilot Station), it basically has the same meaning as on AFSPS, however, in river ports a dispute may arise between the contracting parties as to which is the designated point of delivery, i.e., is it the first sea-pilot or the inward-pilots (river-pilots)?
In any case, the delivery point
. should be clearly defined on an AFSPS basis,
regardless of whether the loading port is in an area of the river or not.
The 'APS' could be used in conversations between the brokers for brevity's sake, but not when conducting firm negotiations.

For redelivery purposes:
Redelivery of the ship on an 'AFSPS' or 'APS' does not apply.

b)- On Dropping Last Outward Sea Pilot's Station (DLOSPS) or on Dropping Outward Pilot (DOP).
The expression on a 'DLOSPS' is mainly used for the purpose of redelivery at a port, however, it is also used for delivery purposes.

For delivery purposes:
The delivery of the ship for its next employment will take place at the point and time
. when the last sea-pilot of the last port of discharge of the previous voyage has disembarked from the ship.
Such delivery is possible, when the loading area is elsewhere and the ship will ballast in an empty cargo condition from the point in which it is delivered, to the port of loading.
When the same port will be used for loading, the AFSPS terms should be in effect.
For example, for delivery on a DLOSPS:

- delivery on DLOSPS Piraeus port, Greece, atdnsshinc.

For redelivery purposes:
-On DLOSPS (Dropping Last Outward Sea Pilot Station):
The majority of the Charter Parties provide the redelivery of the ship to take place on DLOSPS.
The Charter period will be terminated (off-hire) when the following conditions have been met:
- the ship has completed the discharging operation, the pilot has boarded and with tugs' assistance or not, it has been unmoored from the berth and directed to the open sea.
- the ship has reached the Last Sea Pilot Station, i.e., at the point at sea, where the last Sea-Pilot will disembark from it.
- the Master has tendered the redelivery notice to the Charterers.
For example, for redelivery on a 'DLOSPS':
- redelivery on DLOSPS Piraeus port, Greece, atdnsshinc
Or
-redelivery on DOP (Dropping Outward Pilot) Piraeus port, Greece, atdnsshinc:
Redelivery on DOP, it basically has the same meaning as on DLOSPS, however, in river ports a dispute may arise between the contracting parties as to which is the designated redelivery point, i.e., is it the last Inward pilot (river-pilot) or the last sea-pilot?
The redelivery point
. should be clearly defined on a DLOSPS basis in the Charter Party, regardless of whether the discharging port is in an area of the river or not.
The 'DOP' should be used in conversations between the brokers for brevity's sake, but not when conducting firm negotiations.

c)– On DLOSPS with Retroactive (shorten 'Retro') effect or passing a point with Retroactive effect.

For delivery purposes:
The delivery will take place at a point that the ship
. 'has already passed'
and the ship's route is apparently towards the Charterers' loading area.
The ship will be hired
- from the point and time the ship has departed (retroactive) from the last outward sea-pilot station of the last port of discharge
or
-from another point as will be specified, which the ship has already passed and is en route to the loading area.

For example, for delivery on a DLOSPS retro:
- delivery on DLOSPS Piraeus port, Greece, with retroactive effect, atdnsshinc.

For redelivery purposes:
Redelivery of the ship 'Retro' does not apply.

d)- 'Upon Passing' an agreed point / at 'Equidistant'.
There are some places/passes around the world, as defined by the contracting parties, as delivery and/or redelivery points. Such places are the 'passing Passero "passing Skaw' 'passing Gibraltar' 'passing Singapore "passing Durban' 'passing Muscat outbound' (abbreviated as 'pmo') etc.

For delivery purposes:
The delivery will take place:
. 'upon passing'
an agreed point as will be specified, which in any case is
. 'ahead'
of the ship's position and apparently en route to the place of loading.
For example, for delivery 'upon passing' for ships which are ballasting from East Med east of Cape Passero towards Gibraltar:
- delivery upon passing the Cape of Passero, Sicily, Italy westbound atdnsshinc.

For redelivery purposes:
The redelivery of the ship will take place
. 'upon passing'
a point as will be specified, which is apparently 'ahead' of the ship's position/ last discharge port and logically en route to a better loading area.
For example, for redelivery 'upon passing' for ships which have discharged at East Med, east of Cape Passero:
- redelivery upon passing the Cape of Passero, Sicily, Italy westbound atdnsshinc.
However, if the ship will discharge within the Adriatic, for example, and the redelivery point will be 'upon passing the Cape of Passero', the Owners could keep the option, during negotiations, of the ship being directed to the East Med or the Black Sea, in which case the agreement on redelivery could be expressed as follows:
- redelivery upon passing the Cape of Passero, Sicily, Italy westbound or at the Owners' option at equidistance atdnsshinc.
The ship will be redelivered in the same distance to the Cape of Passero, but eastwards.

e)- 'At the time and in position upon fixing'.

For delivery purposes:
The delivery will take place at the location and time the ship is
. 'upon fully fixing' (conclusion of fixture).
Obviously, the ship should be en route to a better loading area and the approximate position is to the Charterers' full knowledge.
The Master, when tendering the delivery notice to the Charterers, will include, along with the other standard information, the exact current position of the ship, by providing the longitude and latitude lines.
For example, for delivery 'at the time and in position upon fixing':
- delivery at the time and in position upon fully fixing, atdnsshinc.
The ship is under ballast condition to ECSA and its position this noon Monday 11th March 2..., was at longitude .... and latitude ......

For redelivery purposes:
Redelivery of the ship 'at the time and in position upon fixing' does not apply.

f)- 'When/ where ready'.

For delivery purposes:
The ship will be delivered to new Time Charterers:
. upon redelivery of the ship from the previous Charterers
or
. if on voyage, upon completion of the discharge of the previous cargo,
in the place/berth or anchorage where it was discharging.
Obviously, the loading will take place from the same port or even from the same berth/anchorage.
This case could be of interest to all.
For the Owners:
- if the ship is on Voyage Charter, it will save the outwards port expenses (tugs/pilots) and time that is required to shift from the berth to the last sea pilot.
- if the ship is on Time Charter Party and the previous Charterers agree to such redelivery, then both parties could share the saved outwards port expenses and time/fuel.
- the ship will be on-hire to new Time Charterers without delay
For the new Time Charterers:
- they will save the inwards port expenses (tugs/pilots) from the usual delivery point (AFSPS) to berth.
- they will save the shifting (lost time) from the first pilot to the berth and the fuel required to be consumed.

All above are subject to:
-the size of the ship,
-the type/grade of the last and next cargo,
-the new and previous hire rate.
When steel (non-oily)/palletized/project cargoes, etc. have been stowed and discharged from small capacity holds, the cleaning operation in the holds could easily be performed, or part of the holds could even be ready at the same time that they are emptied, while the ship is discharging from the rest of the holds. Difficulties could arise if the next cargo requires grain clean holds, however, for loading coals/minerals /aggregates /steel /limestone/bagged cargoes etc., even a careful sweep of the holds would be sufficient, in order to load the next cargo. All the operations are subject to the permission of the Port Authorities and their consent for the ship's occupation of the berth, even for a short time, for any purpose beyond the loading or discharging operation.
For example, for delivery 'when/where ready':
- delivery when/where ready at 1 safe berth or anchorage at Piraeus port, Greece, atdnsshinc.

For redelivery purposes:
The redelivery will take place:
- at the time of completion of discharge at berth/anchorage.
The Owners will bear the cost:
- for all outward port expenses (pilots/tugs),
- for the fuel consumed from the berth to open sea
and
- for the lost time for the shifting from the berth to open sea.
For example, for redelivery 'when/where ready':
- redelivery when/where ready at 1 safe berth or anchorage at Piraeus port, Greece, atdnsshinc.

4. Trading and Charter Party 'Trading Exclusions' clause.

The ship, under the terms of the Time Charter Party will be trading via safe port/s, safe berth/s, safe anchorage/s, always within the Institute Warranty Limits and always afloat.
Charterers are to direct the ship to the port/s, which can be safely approached and can supply sufficient number of tugs and pilots/lights for the safe anchoring/berthing /unberthing/departure, under the directions of the Port Authorities. The ship has to load and/or discharge the cargo from the anchorage/berth and sail, under normal conditions, without serious consequences due to stress from weather or any other reason attributed to a non-organized port.

The nominated port/s, anchorage/s, berth/s should, in all cases, be counted
. as safe,
throughout the duration of the ship's stay and this should be clearly incorporated in the Charter Party.
The ship, during the operations at the ports, should remain
. always afloat and without the risk of touching the bottom.
This term should be in effect
. throughout the Charter and should be clearly incorporated in the Charter Party. This provision should also be in effect, when passing through the canals/ straits/narrow waters etc.
All bids should include that:
- the port/s berth/s will remain safe from the time of the ship's arrival at the first sea pilot station until departure and the last sea pilot has left the ship.
- the ship will always lie afloat, throughout the period of the Charter.
However, for ships fixed for long period Charter, for worldwide trading or for ships fixed to operate in tidal ports/berths, the
. *'NAABSA' clause*
should be asked by Charterers to be in force.
Based on the *'NAABSA' (*'Not Always Afloat But Safely Aground) contractual term, the Charterers have the right to direct the ship to a tidal port/berth, where it can touch the bottom, but because the sea bottom is composed of soft mud or a sandy seabed, the ship may safely run aground during the low tide. The ship will not be instructed to operate or move during its aground.
However, in the case that the ship is directed for an NAABSA procedure at any places other than those located at East Coast of South America, then, after completion of the cargo operations or at the nearest next available port/place, the Charterers, under the terms of the Charter Party, under their time/expenses, should arrange to appoint a diving inspection, in order to certify whether any damage is caused, due to the ship's running aground.
The ship will not be instructed to navigate/trade in areas that will lead the ship to being uncovered by its Hull& Machinery insurance company, i.e., out of the International Navigating Limits (I.N.L.), formerly Institute Warranty Limits (I.W.L.) unless it is otherwise stated in the Charter Party. The insurance companies have designated certain areas, which the ships (mainly during the winter and in areas where there is ice) are not allowed to enter, before their written consent is achieved. If allowed to, the extra premium charge should usually be covered by the Charterers. Such areas are the northern Baltic Sea, North Canada etc.
The Charterers, during the negotiations, will try to achieve the maximum possible allowance trade, so that they can be more flexible in the open market, when it is required.
The Owners will specify the countries in which the ship, as well as the conditions

under which it is allowed to trade in a number of them.
The Master has the right to reject:
- the Charterers' employment instructions to reach an excluded area/port,
- the signing of Bills of Lading showing an excluded destination.
The trading allowance is directly linked with
. the ship's flag.
Ships hoisting a convention flag could be unworkable for the Charterers' intended trade, as these ships should avoid trading in countries, where the inspection of the ITF is strict.
Countries which may lead
. to a subsequent boycott of the ship by other countries
or countries where there is an embargo or economic sanctions from the International Organizations and/or areas with an increased risk of piracy or war/war like areas or are not ISPS compliance ports, could be excluded from the trading.

The Owners should generally and, subject to agreement, exclude areas/regions/ countries/ ports/ berths/anchorages
- that are not allowed by the country of the ship's registry,
- that are not allowed by the Owners' domicile,
- that are outside of the International Navigating Limits (I.N.L.),
- which may lead to a subsequent boycott by other countries,
- where the ITF inspection is strict,
-that restrict the ship's qualification to trade to USA/Australia/Canada and enforce costs in accordance with the laws of the relevant countries,
- that are subject to UN Security Council Resolutions and/or embargoes or sanctions of the NATO, EU, USA, or any other International Organization,
- that are ice blocked or ice threatened or have ice bound ports,
- that are areas or ports which are infested or where the world sanitary organisations have declared contagious diseases,
- that are at war or war like or piracy's zones or to trade into territorial waters of any of the countries or places that are described in the current War Trading Warranties, as adopted by the ship's underwriters,
- that are non ISPS compliant ports,
- where it is known that the ship will be boycotted by shore a/o port labourers a/o tugboats a/o pilots or by government a/o authorities by reason of the ship's flag/registry/class/ownership,
- that are affected during the high-risk season by the Asian Gypsy Moth
-where there are prevailing narrow waters and/or shallow harbours or tidal and/or ports in rivers or with bars,
and numerous others according to prevailing circumstances and the policy of the Owners' office.

5. Cargo exclusions.

The ship will be under the orders and directions of the Charterers regarding employment, who should trade the ship with cargoes as mutually agreed with the Owners and incorporated in the Charter Party.
The Charter Party should clearly stipulate, which cargoes are not allowed for loading or cargoes that can only be loaded under specific conditions.
If the Charterers present for loading a type/grade of cargo excluded from the Charter Party, the Master of the ship has the right
. to refuse the commencement of the loading operation
and instead to request an allowable cargo to be presented.
The Owners will face the same problems, if they refuse to load cargoes that are allowed by the Charter Party, but they have just omitted/forgotten to exclude the same during the negotiations.
The cargo quantities and the type/grade allowed for loading, which can safely be carried into the ship should be ensured by the Owners' relevant department, before the conclusion of the fixture.
For long period fixtures, even if the Charterers intend to trade only with its own regular cargoes, the 'cargo exclusion' clause could be discussed, even during the negotiations of the Main Terms. The Charterers will try to have the maximum number of allowable cargoes, enabling them to be more flexible/play better in the market, minimizing the risk of losing money.
However, for single trips, the intended cargo could be specified from the commencement of the negotiations or even from the moment when the biz was quoted in the market, thus enabling both parties to avoid arguments and long negotiations as to the cargoes allowed for trading.
For cargoes that are dirty or could expose the ship to danger or could cause damage to the holds etc., the owners can set a limit on the number of voyages allowed for transport of such cargoes and place specific conditions for carriage, which are considered necessary for the protection of the ship/crew.
All cargoes should be loaded/stowed/trimmed/discharged by the Charterers at their own time/risk/ expense and always in accordance with IMO/ IMSBC Code/ international regulations/ recommendations and local regulations, but to the extent that local regulations are more onerous/ restrictive than those under IMO/ IMSBC Code and/or regulations/recommendations of the countries through whose waters the ship will be navigated.
All solid bulk cargoes (solid are all cargoes except liquid or gas) and especially dangerous solid bulk cargoes should comply with the
. International Maritime Solid Bulk Cargoes Code (IMSBC Code).
The cargoes listed in the IMSBC Code have an official name (Bulk Cargo Shipping

Name-BCSN) and the sea transport documentations issued for the cargo must be identified with the BCSN sign.

The dry bulk cargoes are classified by the IMSBC Code into three groups:

Group A:
Solid bulk cargoes which may liquefy during the sea voyage, resulting in cargo shift and possible ship capsizing, are listed in this group. Such cargoes are nickel ore/mill scale/iron ore fines/ pyrites etc. Special care must be taken by the Master/Owners prior to commencement of the loading, in the areas where the cargo is exposed to rain without proper protection. The moisture content of the cargo must be measured/determined by the recognised international methods, which in any case must not be in excess of their Transportable Moisture Limit (TML). However, dry cargoes are not liable to liquefy.

Group B:
Solid bulk cargoes, which may pose a chemical hazard and could expose the ship and crew to danger, are listed in this group. These cargoes are dangerous, because when they come in contact with air or by interaction with water, they can come to be self-heating or emit flammable gases in dangerous quantities during the voyage or turn into radioactive materials or even contain toxic substances etc.
Dangerous goods are classified according to their particular characteristics (4.1/ 4.2/ 4.3/ 5.1/ 6.1/ 7/8/9), while certain others, when carried in bulk, are characterised as 'Materials Hazardous only in bulk' (MHB). Shipments will only be executed on ships that can provide the Appendix 'B' certificate, as will be issued by the ship's class, and can provide a CO2 firefighting system in their cargo holds.

Group C:
Solid bulk cargoes, which are simply not liable to liquefy and do not pose chemical hazards are listed in this group.

However, certain mineral cargoes are listed in both Group A and Group B, i.e. pose chemical hazards, and may liquefy during the sea voyage.

For all cargoes, whether they have been classified as dangerous or not, the Shippers are obliged before loading to provide the Master of the ship with a cargo declaration, as outlined in the IMSBC and the BCSN, as well a valid certificate with the information on the physical and chemical properties and the required conditions for the carriage.
Shippers/Stevedores will load/discharge the cargo strictly as per the agreed

ship's loading sequence/stowage plan/method of cargo operations and as per provisions of SOLAS chapter VI, regulation 7.3 (loading, unloading and stowage of solid bulk cargoes) and the Code of Practice for the Safe Loading and Unloading of Bulk Carriers (BLU Code).

The Charterers, during the execution of the Time Charter, could ask from the Master of the ship to present a stowage plan of the cargo, which is intended to be loaded. The same should be given as a 'preliminary stowage plan' and should state that:
- the cargo quantity, which will be loaded, and the holds' breakdown is subject to
. the actual stowage factor of cargo in the ship's holds.
. the ship's actual trim/stability/strength/stresses (hogging/sagging bending moments/ shear forces etc.).
. the available draft/salinity of the port.
. load line zones limit.

For cargo that is in the form of bags/pallets etc. the wording could be amended accordingly.

6. Duration.

The duration is the 'key' when fixing on Time Charter and affects the achievable 'hire rate'.
In an upward market, most of the Owners and especially those who are committed with 'new building' ships or whose companies are in the stock exchange, seek to fix on long periods, in order to ensure the ships' trading and look forward to their next investment.
The duration varies, depending on
- the Charterers' needs and according to their particular biz,
- the market's prediction and the Owners'/Charterers' expectations.
It begins with the Master, who, from the designated point of delivery
. sends the 'delivery notice' to the Charterers
by fax or e-mail or telex or cable, stating the delivery of the ship into their services.
It terminates when the Master, from the designated point of redelivery
. sends the 'redelivery notice' to the Charterers
by fax or e-mail or telex or cable, stating the delivery of the ship back to the Owners.

The ship, during the whole duration of the Charter:
- will remain under the Charterers' legitimate employment instructions and will

trade through countries/areas and load cargoes as will be fixed by the Charterers, but always in compliance with the relevant Charter Party clause.
- the Owners do not have the right to withdraw the ship without specific reason, which in any case should be in compliance with the Charter Party terms.

The contractual time of redelivery has given rise to many claims. The Charter party should clearly incorporate
- the minimum and maximum duration,
- a provision for the last voyage of the Charter,
- whether the off-hire periods incurred during the execution of the Charter will be added on top of the basic duration or should be an option.

When a premature redelivery takes place in violation of the Charter Party terms, the Owners should accept the redelivery, after having reserved their rights to claim for all and any losses or damages caused, as a result of this wrongful redelivery.
Owners should take steps to
. mitigate those losses, to the extent that it is possible to.
However, and in all cases, legal advices should be obtained, before any action is taken from the Owners.

-When trip out or loading legs are in effect, the exact duration is obviously unpredictable and the Charterers should try to avoid being committed to any specific duration for the Charter (unless it is a very short trip). An advice that could be given to the Owners on an 'about' basis, is to release Charterers from receiving claim for an earlier or later redelivery, only when it is followed by the subsequent introduction
. 'without guarantee' (wog).
As an example:
. duration about 30 days but wog.
That includes no responsibility for the Charterers, even if the Charter has been concluded much later or earlier. However, the Owners in order to avoid an unprofitable long duration and depends on the ship's position, could try to agree to an escalation of the hire rate, if the ship has been delayed for more than a specified maximum period.

-For a short or long period, the duration could be defined as:
a) from minimum to maximum basis, e.g., minimum 2 months/ maximum 4 months
No redelivery will take place earlier than 2 months and no later than 4 months.
b) on an 'about' basis, e.g., about 9 months to about 12 months

No redelivery will take place earlier than about 9 months and no later than about 12 months.
The Time Charter Parties usually define the word 'about' as 15 days more or less at the Charterers' options, otherwise, the tolerance of 10 days more/less should be considered, in accordance with the English Law.
c) With an option, e.g., about 3 months + about 3 months at chopt.
Before the expiry of the about 3 first month period, the Charterers will declare whether they will continue the fixture for another three months or will redeliver the ship to the Owners.

7. Hire/Ballast Bonus.

On Time Charter, the Owners will be compensated for the whole duration which the ship will be under the Charterers' services, with a daily 'hire rate' or pro rata for any part of a day.
The 'hire' commences from the time of delivery and the ship going on-hire, until the redelivery and the ship going off-hire.

For hire calculation purposes the UTC (Coordinated Universal Time) will be in effect, i.e., the ship, upon tendering the delivery notice to the Charterers, will include
. the time of delivery on local time and on UTC.
The UTC will be considered as the 'delivery time' for the calculation and payment of the hire. The same should be in effect upon the Master tendering the redelivery notice and the UTC time will be considered as the 'redelivery time'.

The amount which will be paid by the Charterers as 'hire' for the execution of the Charter, plus the possible ballast bonus will cover full or part of the Owners' below standard expenses:
- class and flag commitments (passing the special survey/dry-dock/ maintenance etc.).
- provisions, repairs, spares.
- crew's wages, transportations, repatriations etc.
- crew's/officers' overtime during the execution of the Charter.
- the basic ship's insurances (H&M, War Risk, P&I Club).
- lubricants and domestic consumption of fresh water.
- office expenses (staff wages/insurance/rental costs etc.).
- installments on loans obtained from the banks for the purchase of the ship.
- brokerage commissions.
- husbandry expenses incurred at the ports of contact.
and all other expenses that are necessary to be covered by the Owners for the

seaworthiness of the ship and for the successful execution of the Charter.

The Owners will have to decide if the achievable 'hire rate' is attractive and will proceed further with the conclusion of the fixture or will leave it for a better biz, taking into consideration:
- the resulting net or gross daily income.
- the total duration of the Charter.
- if the income is in line with the existing market.
- expectations of the ship's repositioning at the redelivery area.

The 'hire rate' is a basic term during the negotiations of the Main Terms and varies, depending on:
- type/size of the ship and its characteristics.
- type of Time Charter and of the extended time of period (duration).
- in which area/point the ship will be delivered/redelivered.
- through which countries and areas, the ship will trade.
- grade/type of cargoes to be transported.
- Owners' and Charterers' confidence in the existing market or market's tendency and expectations.

The crew's and/or officers' overtime which could be required to be provided out of their daily maximum normal time of work, are not to be covered by the Charterers and are to be included in the agreed 'hire rate'.
The 'hire rate' term should be given as:
- hire USD .... per day/pro rata including overtime (abbreviated as 'inclot')
or for short
- hire USD .... pd/pr inclot.

The standard procedure is the 'hire' to become due every
. 15 days in 'advance'.
Nevertheless, there might be an agreement that the first 15 days 'hire' are to be paid within 2/3 banking days after the ship's delivery, as the Charterers prefer to effect the payment after
. the ship's actual delivery to them and sometimes after the on-hire takes place. In this case, they ensure the delivery of a seaworthy ship and if the ship is in a convenient position and there is time (taking bunkers to execute the Charter etc.), they could carry out the on-hire survey and determine the exact quantity of fuel at the time of delivery, which is to be paid together with the first hire payment.
However, for all the subsequent due (15 days) hires, the Charterers' obligation is to make the remittances before the running 15 days have expired,

. i.e., in advance
and before the period of the next 15 days commences, even if it takes place during weekends/holidays, in which case the hire should be remitted on a working day before the commencement of a weekend/holiday.
For the last hire, which is usually much less than 15 days for payment, a day-by-day payment (very unusual/difficult to apply) could be agreed or every two/three/five days or so or even according to the information provided by the appointed agents on the expected date of completion of the discharging operation.
For very short trips, the first hire payment could be adjusted according to the expected maximum duration of the trip, i.e., if the expected period is 8 days or even 20 days, then the hire could be paid for the entire expected period and if remaining, any balance period could be payable day by day or so.

For ships in a ballast condition (empty of cargo en route to a loading area) and subject to the existing market, the hire could be accompanied by
. a ballast bonus.
It should be agreed on a lumpsum amount and will be negotiated together with the hire rate.
Owners should seek to cover:
- the cost of fuel from the last port of discharge, until reaching the requested point of delivery.
- the ship's running expenses.
In a good market, a decent ballast bonus could be achieved, which could cover the Owners' risk and the cost for possible delays/fuel overconsumption due to stress from weather or adverse currents, until the time of delivery of the ship plus a daily profit.
The ballast bonus is also achievable for ships opening near the area requested for delivery. The Owners take advantage of the lack of tonnages in the area and the Charterers accept to pay this amount, as they would have to pay for it in any case.
However, the Charterers would have
- ensured a safer ship's position,
- a chance to try to fix at a lower amount as ballast bonus,
- keep the daily hire rate at the side of the existing market.
The ballast bonus usually is payable together with the first hire payment,
. 'net' or 'gross'
of brokerage commissions and the same should be clearly stated during the negotiations.

In the case of a Charterers' failure to meet their financial obligations and under

the terms of the Charter Party, the Owners may become entitled
. to suspend
the fulfilment of any or all of their obligations under the Charter Party terms and/or
. withdraw
the ship from Charterers' services and/or
. have a lien on the cargo
and on all due to the Charterers' freight/demurrages/sub-freight/sub-hires etc. The Owners will seek to recover losses/damages caused therefrom. Legal advices should be obtained, before any action is taken from the Owners.
The main body of the NYPE 46 Charter Party incorporates a provision that, in the case of non-payment or late payment of hire, the Owners have the right to lien the cargo, freights etc., as well as to have the right to withdraw the ship from the Charterers' services. On these respective lines, the Owners may negotiate and insert certain additions or amendments necessary to protect their interests. In the case that there is no provision, then the Owners should seek legal advices for a complete
. lien/withdrawal clause,
which could be incorporated in the Charter Party accordingly.
Generally, if the hire failed to be timely paid, the Owners should have the right, under the terms of the Charter Party, to withdraw the ship from the Charterers' services, as well as, to have the right to lien the cargo and/or freights or sub-freights/demurrages etc. due to the Charterers from the sub-Charter Parties. The Owners might have the right to refuse to release the cargo to the consignees and to keep it at their possession, either on board the ship, or, after discharge, at the port's custody, until the Charterers fulfil their obligation, or otherwise, until they sell the cargo to others. The lien and the withdrawal will be directed from the Owners' law firm, as it is subject to the way in which the originally signed Bill of Lading was marked.
The provision
. for suspension
of the ship's performance of any or all of their obligations, provides the Owners with the privilege to exert pressure on the situation in order to get the due hires, without exercising their right to withdraw the ship. The suspension clause could be made by the Owners' law firm or their legal department.
Bimco has published a non-payment of hire clause for Time Charter Parties, which provides the Owners with the right to suspend the performance, until payment of due hire/s is made. The ship will remain on-hire and the suspension will be made without prejudice to any rights that the Owners may have under the Charter Party.
The notice for the ship's suspension of performance must be given after

. the grace period
of the respective clause has expired, provided that such a clause is also incorporated in the Charter Party.
A provision should be incorporated in the Charter Party, clearly stating that:
. any amount due as 'hire' should be paid by the Charterers to Owners, net of tax and bank charges.

8. Bunkers on delivery/redelivery (Bunkering Clause).

The Charterers, at the time of the ship's delivery, will buy the amount of fuel on board the ship, enabling them to commence the Charter. Subject to agreement during the negotiations, the Charterers are of the obligation to redeliver the ship with 'about' the same amount of fuel as on the time of delivery (the margin for 'about' is 5pct, however, will depend on the proximity of the place of the last bunker supply arranged by charterers to the point of redelivery), in which case the Owners will buy the remaining fuel onboard the ship. The fuel prices are mostly fixed at once and will remain the same for both transactions.
It is a basic term, as a ship can be considered unworkable due to the low quantity of fuel on board at the time of fixing (the ship is unworkable for a long ballast trip), or the opposite, because of the ship's full fuel capacity, in which case its cargo capacity is reduced making the ship unworkable for loading from ports located close to the delivery area.
Upon delivery, all the subsequent fuel supplies which will be ordered will be arranged only by the Charterers, through their bunker suppliers. The Charterers remain responsible for the payment and for the quality of the ordered and supplied fuel on the ship.
The fuel quality must strictly comply with
. the fuel specifications
that are given and included in the ship's description by the Owners, otherwise, huge claims may arise due to delivery of improper bunkers, resulting in damage caused to the ship's engine.
From each bunkering operation
. samples
should be taken from the ship's manifold, to verify the quality of the supplied fuel.
The supply of fuel to the ships constitutes the majority of pollution incidents. During the bunkering operation by barges, special attention and monitoring should be paid by the crew to the pollution and safety matters.
Disregarding bunkering and safety procedures in a ship's Safety Management System during bunker operations or crew which is not familiar with the ship and bunker procedures and generally ISM breaches by the ship could result to

accidental spillage. If the ship pollutes the pier/sea etc. the consequences are severe, and the ship could be arrested by the Port Authorities and sustain imposed fines and clean-up/costs etc.
Might also be Charterers' intention to supply bunker to the ship at the port by trucks. Based on the situation which occurred on many ships (pollution) the Owners office should send a technical circular and draw the crew's attention to the intended bunkering operation. The bunkering operation should be carried out only after careful inspection of the compatibility of the connectors provided by the bunker suppliers to the ship's bunkering manifold and only when the bunker delivery documents (delivery order) are completely prepared and acceptable. The ship's crew should attend during the whole bunkering operation, ready to prevent any accident.
The Charterers, for every bunkering operation arranged by them, under their time/expense, should appoint a well recognised independent surveyor, who should carry out a bunker survey before and after any bunkering operation takes place, in order to determine the quantity of fuel supplied to the ship.
The Charter Party should specify with
. what quantity of fuel and at what price the Owners will deliver the ship to the Charterers' services and with what quantity and price the Charterers will redeliver the ship to the Owners.
The fuel quantity on delivery will be ascertained
. by the on-hire
and on redelivery
. by the off-hire bunker survey.
If the Charterers, for any reason, avoid paying for the fuel ordered by them, the suppliers, if allowed and have the right of law, might move against the ship, as a minimum action to secure the payment. Thus some Masters try to enter remarks (if allowed by the suppliers), or put a stamp on the presented for signing 'shipping order', keeping the Owners harmless for the payment of such ordered by others quantity of fuels.

For a long Charter period, the Charterers should redeliver the ship,
. with approximately the same quantity of fuel as on the time of delivery.
The payment of the value of the bunkers on board the ship at the time of delivery will usually be made together with the first 15 days hire payment. However, the Charterers should have the right to deduct from the last hire payment or from the last sufficient hire payments, the value of the estimated redelivery bunkers.

For very short trips out, the ship could be delivered/redelivered
. with the quantity of fuel as on board,

i.e., without replenishment, but subject to the ship being able to provide enough bunkers to perform the trip and should additionally have enough bunkers, to contact the next bunkering station after redelivery. The Charterers should benefit, as they will save on the time required for replenishment, as well as, from any unexpected extra costs which may arise due to increased prices at the time of the ordering and of course from the fees of the attending Agents. The Charterers, with their first hire payment will pay for the approximate bunkers that are required to be consumed for the execution of the whole trip. Upon completion, the actual bunkers consumed will be determined with the off-hire bunker survey and will be settled in accordance with the issued and accepted final hire statement.

Nevertheless, for such short trips, the Charterers should guarantee

. the maximum duration

and remain responsible for replenishment, in the case that the duration, due to unexpected reasons (congestion/ strikes on the port/ bad weather etc.), overcomes the calculated and expected maximum duration.

Generally, for single trips there are several options to be agreed upon such as

- the Charterers to not buy the fuel and supply the ship with enough fuel to execute the trip,
- the Charterers to buy only part of the fuel on delivery and replenish the same before redelivery,

as well as numerous other options.

The Charterers, during the execution of the Charter and before its redelivery to the Owners, could ask from the Master of the ship to declare the amount of fuel necessary to be supplied, enabling the redelivery of quantities to meet the delivery quantities. A similar question could be made by the Charterers, for the amount of fuel required to execute a specific trip, enabling them to order sufficient fuel. In this case, any fuel quantity, which will be presented by the Master or Owners should be subject

-to the trip's final duration,

-to unexpected delays,

-to the ship's actual consumption, when manoeuvring at the ports/canals/ straits and/or in confined waters.

Information on any given bunker quantity to the Charterers should include the undermentioned reservation:

'The required quantity of fuel can be changed, subject to the trip's final duration and/or the ship's consumption, while it is maneuvering at the ports and/ or canals and/or straits and/or confined waters and subject to the final destination'.

9. On/Off hire survey.

The fuel quantity on delivery and redelivery as well as the ship's condition will be determined by the on-hire and off-hire survey. Unless it is a very short trip, where the Charterers could rely even on the declared speed/consumption of the Charter Party and on the Master's consumption report about the remaining bunkers on board, the Owners and Charterers should mutually agree
either
. upon delivery of the ship and at the place of delivery
or
. at the first opportunity during the steaming to load port
or
. upon arrival at the first/sole loading port,
that an on-hire survey be carried out to determine the quantity of fuel on board the ship, as well as the ship's hull/holds' condition for previous stevedoring damage
. that could later be transferred to the present Charterers.
A full report should be issued, and photos should be taken from the ship's holds, deck, and hatches, ascertaining the condition at the time of delivery.
The Owners and the Charterers will mutually agree
. either upon redelivery and at the place of redelivery
. or just before the completion of discharge at the last discharge port,
that an off-hire survey be carried out to determine the quantity of fuel remaining on board the ship, as well as the ship's hull/holds' condition. A full report should be issued, and photos should be taken of the ship's holds/deck/hatches, ascertaining the condition at the time of redelivery.
The survey can be conducted by a mutually acceptable surveyor/s and the costs could be shared equally between the Owners and Charterers, otherwise, each party will appoint and pay for their surveyor, but lost time remains equally shared. Sometimes the Owners propose to appoint the ship's Master/Chief Engineer to represent them, but in all cases the survey should be conducted jointly with a single report being issued by all parties, otherwise differences could be found among all issued reports.

10. Interclub Agreement.

On Voyage Charter, the Owners are liable for cargo claims as will be presented by third parties, however, on Time Charter and under the terms of the Charter Party,
. the liabilities for cargo claim to third parties
shall be apportioned between the Owners and the Time Charterers in accord-

ance with the rules of the Inter-Club New York Produce Exchange Agreement. The first agreement between the International Group of P&I Clubs was completed in 1970 and has thereafter been amended three times in 1984, in 1996 and more recently
. in September 2011.
The Inter-Club Agreement (abbreviated as 'ICA) clause can be used not only for the New York Produce Exchange form of Charter Parties or the Asbatime Charter parties, but could also be incorporated
. into other types of Time Charter Parties.
The ICA clause
. binds the contractual parties
and settles the liability claims depending on the cause of the cargo claim, avoiding the extra costs and lengthy litigations of the P&I Clubs involved in the fixture. For liabilities from cargo shortages, the Owners are entitled to recover 50 percent of the paid amount from the Charterers/ Charterers' P&I Club. However, shortages or damage due to cargo handling are 100 percent for the Charterers' account.

11. Weather routing.

The ship's performance during the execution of the Charter is one of the most important points for the Charterers and it is their success when making the voyage calculation for their fixed Voyage Charter.
The ship's actual speed/consumption and the validity of the agreed relevant clause could be a base for disputes/claims arising between the Owners and Charterers.
The method, which will be used and will be acceptable by both parties, to assess the ship's actual performance across the oceans during the currency of the Charter will be agreed when discussing this clause. In fact, the Charterers should not only rely on the route, which the Master will select and on evidence of weather conditions taken from the ship's deck logs. They will seek to fully protect their interests and ensure the fastest, safest, and most economical route, by appointing a recognised
. International Weather Bureau (Weather Routing Company or 'Ocean Route' as it is commonly called)
that will inform the Master of the expected winds, waves and prevailing currents on its route and recommend to (not impose on) him the fastest and safest route, from the point where the ship will commence the voyage up to its final destination.
The Master will choose the best route for his ship and will report the ship's daily noon position to them and will inform them (and document in the ship's log

book) of any changes in his route, which he deems necessary for the safety of the crew/ cargo, as well as, the reason for it.
At the end of the voyage at sea, the 'Ocean Route' will assess
. how fast and with what consumption the ship performed during the trip, from pilot (load port) to pilot (discharge port).
Because of this clause's validity, the ship's declared speed/consumption must be described on an
. "about" basis.
The Owners should be protected if the ship fails to maintain its description, as it provides
. *a negative speed variation of half a knot and a 5 percent increase in the declared fuel consumption.*
It is commonly accepted that the weather and the sea waves affect the performance of the ship, when
. the Beaufort Force 4 and Douglas Sea State 3 are exceeded
and the same should be clearly incorporated in the Charter Party.
When there are large discrepancies regarding the prevailing weather conditions between the log books and the weather reports that are issued by the Ocean Route, the Charter Party should spell out whose report will be binding for both parties, otherwise and upon referral to Arbitration, both should have the same value.
During the currency of the Charter Party and under the terms of the Charter Party, the ship's reduced speed, or delay, or any deviation made will not be taken into consideration and will not be considered as underperformance, if made due
-to Charterers instructions for their commercial reasons (subject to the Owners' consent),
-to trading in narrow or restricted waters with regard to the ship's size/draft,
-to the safety of ship/crew/cargo, because of weather conditions,
-to save lives/property or assist a ship in distress,
-to the ship's hull bottom becoming fouled by marine growth, but without prejudice to the 'bottom fouling clause',
-to a distance between the ports that is less than 10 hrs.,
-to navigation, which is not in an even keel condition,
-to poor visibility (fog etc.),
-when trading coastally and within 15 nautical miles from the coast,
-to the Charterers' supply of fuel not being in compliance with the fuel specifications provided by the Owners
or,
generally, for any other reason, which can affect the ship's performance and is beyond the Owners'/ship's/Master's fault/defect.

12. Bottom Fouling Clause.

When the ship trades, especially
. within warm tropical waters and remains in port for a certain period,
then the bottom/propeller of the ship may become heavily fouled with marine growth, thereby significantly reducing its sailing speed, and increasing its consumption at sea.
Owners must take care through the terms of the Charter Party and be well protected when the ship is trading within Tropical areas or generally in accordance with the trading directions by the Time Charterers.
In most of the Charter Parties, if the ship was complying with the Charterers' instructions to lie for a long period of time in such waters or in an idle condition, the obligations lie upon the Charterers to return the ship's hull to its original condition and redeliver the ship in a good order and condition with the costs of doing so and the time lost to be entirely for their account.
However, in the absence of such a clause in the Charter Party, then it falls within the Owners' general obligations to maintain the ship in an efficient condition, including keeping the hull clean and so expenses incurred would be regarded as ordinary expenses of trading the ship under a Time Charter. They will not have the right to pass the costs of the hull cleaning on to the Charterers and they will additionally bear the consequences of the ship's underperformance.
The ship's areas which are affected by the marine growth are the flat bottom, vertical sides, bilge keels, propeller, and rudder blade. The cleaning will be performed by organized diving companies with equipment/tools to remove the marine growth, which are in all cases to the Master's approval, in order to avoid damage to the paint of the ship's hull and damage to the propeller.
The most commonly used tools are the hand-scrapers or the more effective the airbrushes.
Sometimes, when it is possible to do so, the Charterers or Owners can instruct the ship
. to frequently spin around or change position in the anchorage, as a minimum action to prevent the bottom fouling.
It is a fact that after the cleaning, the marine growth will have been removed, but with the possible consequence
. of the removal/destruction of the ship's anti-fouling bottom protection paint, as had been applied during its last dry-dock.
The bottom will become
. sensitive to continuous fouling from marine growth and could direct the ship to dry-docking earlier than planned.
Therefore, the Charter Party should clearly state that the Charterers are to re-

main responsible for any further bottom cleaning that may be required and that will be arranged and carried out immediately under Charterers' time and expenses, regardless of the time the ship stayed in the port and/or anchorages/ waiting places.
It is a basic clause, when negotiating a Charter Period or even single trips and especially, when it is known that the ship will trade within tropical waters, or in areas where port congestion is common and foreseeable.
Tropical waters increase the ship's possibility of being fouled by marine growth, therefore, the number of days after which the ship might have been fouled by marine growth should be separately agreed upon in the Charter Party, when trading within tropical or non-tropical ports. If the ship is fouled by marine growth the Charterers are to employ a well organised and class approved diving company, under their own time and expenses, in order to determine the extent of any bottom fouling.
Should it be extensively fouled by marine growth, such as to affect the ship's Time Charter Party performance, the Charterers then should arrange for an underwater cleaning operation at the place of occurrence, by an organized diving company with effective equipment/tools in order to remove the marine growth. However, if there are no facilities and/or well organised/class approved diving companies with effective equipment/tools, then the ship should be directed, always on-hire, to a mutually appointed convenient place/port, after the sailing of the ship from the port, in order for the cleaning to be carried out and provided there is no cargo on board the ship.
Alternatively, and if the ship is on long period, then can be cleaned at the next available port/place.
In all cases and until a satisfactory cleaning has been conducted, the ship/Master/ Owners will not be held responsible for any reduction in the ship's declared Charter Party speed and/or an increased fuel consumption.

13. Communication/Victuals/Entertainment (C/V/E).

The monthly or pro rata
. lumpsum amount,
which will be paid to the Owners by the Charterers, in lieu of reimbursement for expenses incurred on behalf of the Charterers for communication, victualling, and entertainment during the ship's stay in the ports or at sea will be stipulated in the relevant clause of the Charter Party.
It lies within the Master's duties to tender, either by e-mail or cable or by fax, the number of notices as required by the Charterers and to make all the necessary phone calls to the relevant parties. He is to serve drinks/cigarettes, if deemed necessary, to pilots/customs officers/foremen etc. and to victual the

Charterers' representatives or to victual any important people as requested by the Charterers. All these costs are included in the agreed monthly amount which covers the period of a month or pro rata.

The agreement should be made based on one of the following three ways:
a. with a monthly or prorate for any part of the monthly coverage, which is widely used in the Charter Parties and is acceptable by both Charterers and Owners. The resulting monthly equal amount will be payable to the Owners together with the 15 days hire payment.
b. with a mutually agreed lumpsum amount for the entire duration of the Charter.
It is mainly applicable when the expected duration is less than one month, and it is usually payable at the end of the Charter.
c. payable at cost, in which case the Master should provide a statement of all expenses incurred on the Charterers' behalf. It should obviously be paid at the end of the Charter.

14. In Lieu Of Holds Cleaning (ILOHC).

The Charterers are to redeliver the ship in a condition as clean
. as on delivery.
After completion of discharge the Charterers must arrange and employ on board the ship shore workers to clean the ship's cargo holds compartments, prior to its redelivery to the Owners. That will take time and create heavy expenses.
Instead, they could agree with the Owners to compensate them and redeliver the ship with holds
. as left after the completion of the discharge of the last trip,
but in any case, without cargo residues, as will be stipulated in the Charter Party. The ship's crew will clean the holds, as the ship will now trade for the Owners' account. The paid amount is negotiable and varies depending on the size of the ship, the total number of holds which have been used etc.

The agreement should be made based on one of the following two ways:
- on a 'lumpsum' amount, which is widely used in the Charter Parties and it is acceptable by both Charterers and Owners. The agreed amount will be paid at the end of the Charter, regardless of the holds used by the Charterers.
- on a 'lumpsum' amount for each hold used by the Charterers. That is preferred by the Charterers, when they are to load the ship in alternative holds, therefore saving some money.

The inclusion or not to the agreed amount of the removal and discharge of the debris/dunnage and/or all other materials that have been used and were necessary, in order to secure and protect the cargo during the execution of the trip, is to be specified during the negotiations. Owners should take into consideration the fact that in some ports the discharge of such materials has a considerable cost, while in others they are not even allowed to be discharged. The Owners should agree that the same must be discharged from the ship, before its redelivery, entirely for the Charterers' own time/ risk and expenses. The ship should not sail from the port unless all such materials have been removed and discharged from the ship and any delays/ expenses/consequences and/or possible fines imposed by the Port Authorities, arising from the Charterers' inappropriate discharge of such materials or for their inappropriate delivery ashore, must be for the Charterers' account and under the terms of the Charter Party, the ship must always remain on-hire. It should also be agreed that in the case that the materials are not allowed to be discharged at the last port of discharge and always before the ship's redelivery, the ship should proceed to the most convenient anchorage or berth of another port or country and discharge the same in barges or ashore, under the Charterers' own time/risk/expenses. The ship should always remain on-hire, until it fully discharges the particular materials.

15. Commissions.

The Owners are the only party paying the nominal 1, 25 percent to each broker involved in the fixture, from the earned hire, plus gross ballast bonus, if it is agreed.
For off-hire periods, there is no commission.
The involvement of many brokers is not preferred by the Owners, as the total commission will not be kept to a minimum and will greatly affect the hire (heavy deductions) and will therefore decrease the daily income and profit of the ship.

The 'cargo orders', which are based on Time Charter could enter the market as follows:

=Time Charter Trip out or One Time Charter Trip
1.
Account ....
Handymax dwat about 45/55,000 mtns
Geared lifting capacity min 25 metric tons
Grabber preferred
Delivery East Coast India 10/15 Dec 2008
Time charter trip out via east coast India with bulk iron ore (fines)

duration about 25-35 days wog
Redelivery China
NYPE 46 C/P Charterers' proforma with 2,50 address commission past us.

At the top of the 'cargo order', the Charterers' name should be stated as a guide for the Owners. Many times, for commercial reasons, the name of the Charterers might not be disclosed and be replaced with the expression:
- 'fcc'
which means that the order is for the account of 'first class Charterers'.
The brokers involved inform the market that, although the name of the Charterers is not or cannot be disclosed in the open market for commercial reasons, they are dealing with a 'first class' name and the market shouldn't worry about anything serious.
However, this could include a wide range of responsibilities:
. as ensuring the market that it is dealing with A1 people.
In any case, the Charterers' name must be revealed upon the ship's commitment with them or upon their first counter offer. Charterers with little activity, which have not been well established in the market, should be avoided being qualified as 'fcc'.
The expression as:
. 'direct'
or even
. 'direct/close'
could be enough for brokers quoting the biz and it lies on the Owners to check the Charterers' credibility and financial standing in the market.

A. Which ships are workable for this trade?
The order states:
Handymax about 45/55,000 dwat
Geared lifting capacity min 25 metric tons/Grabber preferred

-all ships with a size of about dwat 45,000 mtns to about dwat 55,000 mtns, which are fitted with a minimum of 25 mtns lifting capacity are potentially workable, either opening at East Coast India or nearby and are willing to ballast the ship for delivery within the requested area.
-Charterers have not stated for any age restrictions, therefore overaged ships are also workable

B. Which ships have priority in working the biz?
The ideal ship could be:
-the larger size

-modern (not exceeding 15 years of age)
-opening within the loading area and able to meet the requested lay-days
-having grabs on board the ship.
Charterers will give the first chance to this ship.

Analyzing some positions in theory:
First example:
There are two ships at East Coast India both opening at around 10th December.
The first ship is:
-grabber
-size dwt 46,000 mtns
-built in 2... (not overaged)
And the other ship is:
-a non-grabber, but can provide sufficient gear according to the Charterers' needs
-similar size of dwat 47,000 mtns
-built in 2... (not overaged)
Both ships are modern, of almost similar size and have the same position. It is obvious that the first chance/first refusal will be given to the grabber ship.
Second example:
The first ship is:
-dwt 45,000 mtns
-built in 2... (not over aged)
-opening at East Coast India 10th December
-geared with lifting capacity of 30 mtns (above the required capacity)
-grabber
And the other ship is:
-dwat 55,000 mtns
-built in 19... (more than 15 years old)
-opening at Singapore but can be delivered as per Charterers' request and within the required lay-days
-geared with lifting capacity of 25 mtns
-without grabs.
Both ships are offering the same 'hire rate'.
Which of these is more workable?
The first ship (dwat 45,000 mtns):
-is modern, therefore no extra insurance is in effect and Charterers will save certain costs.
-can provide grabs. Charterers will save rental and other costs to accommodate the same to the ship's gear.
-opens near the place they intend to load. Apparently, the position is safer.

The second ship (dwat 55,000 mtns):
-is overaged. Charterers will pay additional cargo insurances.
-is actually far from the place of delivery and its e.t.a. to load port is always subject to weather conditions
-is without grabs, so the Charterers will bear the rental costs and other costs to accommodate the same to the ship's gear.
-the ship is larger, therefore consuming a larger quantity of fuel and has heavier port expenses that will be charged in both ends.
However, the larger 'cargo capacity' of about 10,000 mtns is just an extra cargo compared to the other smaller ship and this could be an advantage to this ship. A higher profit from the trade (selling a larger quantity) could be obtained, which could be sufficient for the Charterers not only to cover all the additional costs required for this ship, but also to earn extra money.
The Charterers will calculate and work out which ship is the most profitable for them.
Eventually everything is a matter of calculation.

C. Where is the delivery of the ship? / Do Charterers accept ballasters?
The order states:
Delivery East Coast India 10/15 December 2008.

The requested delivery point is the region on the Eastern Coast of India. Charterers will obviously not work/fix a ship open or willing to deliver away from this area. However, ships in ballast to the delivery area, without seeking compensation for the ballast, are potentially workable.

D. Where the ship will load from? / type of cargo to be loaded.
The order states:
Time Charter trip out via East Coast India with bulk iron ore (fines).

The ship will load from the region of the East Coast of India. The port is unknown but could be disclosed by the Charterers upon request. The cargo is iron ore fines.
Iron ore is one of the leading cargoes affecting the global market. It is loadable in all sizes and is required for the manufacture of steel.
The 'fines' type belongs to group 'A' and is a cargo that can liquefy during the trip if it is loaded in a wet condition. Care should be taken by the Master/Owners, before the commencement of loading, in areas where the cargo is exposed to rain without any protection. The moisture content of the cargo will be measured/determined by internationally recognised methods and must in no occasion exceed the Transportable Moisture Limit (TML). The ship's P&I Club should

be involved before and during the loading operation.

E. Where the ship will be redelivered to?
The order states:
Redelivery China.

Full China is the redelivery area, and it is a matter of negotiations, whether it should take place in southern or northern China. It is an ideal biz for ships scheduled to pass their special survey or dry dock in China. However, the biz could be ignored by many other Owners, who would prefer to check for alternative destinations, simply for the reason of the reposition.

F. What is the duration of the trip?
The order states:
duration about 25/35 days without guarantee.

The duration affects the hire rate. It is reasonable for each Owner to search for the longest possible duration, in order to avoid subsequent fixtures and in addition, to ensure a decent rate for a long duration.
The ship will be redelivered to the Owners after about 25/35 days, but the exact time of redelivery is subject to delays that the ship might encounter (congestions, bad weather, delays in piers etc.).

G. What is the type of Charter Party? / What is the commission?
The order states:
NYPE 46 C/P Charterers' proforma with 2, 50 address commission past us.

Charterers will use their own Charter Party proforma, which is based on the 'New York Produce Exchange' form, as amended in 1946.
The order states that the address commission is 2.50 percent, plus the brokers who will be involved.

2.
Account......
Any panamax size
Gearless /overaged ok
Delivery east Med /Black sea
Laycan : cargo ready –try vessel's dates
1 tct via B.Sea with dur about 45 days
Redelivery on dop full China pico
Nype 81 5 pct ttl com

A panamax sized ship is required, ranging from about dwt 60,000 mtns to about dwat 84,000 mtns.

Charterers have no problem in working a gearless and overaged ship.
Cargo is ready in port.

Analyzing the order:
A. Which ships are workable?
The order states:
Any panamax size
Gearless/overaged ok

All Panamax ships geared or gearless, even if they are overaged, are potentially workable. Charterers will start trading a ship, taking into consideration its position, its readiness to load, its characteristics/consumption, its size, its offered 'hire' etc.

B. Which ships have priority in working the biz?
A large/cheap ship with a good position, overrides all others to work the biz.

C. Where is the delivery of the ship? / Do Charterers accept ballasters?
The order states:
Delivery east Med/Black Sea
Lay-can: cargo ready –try vessel's dates.

Delivery will take place within the East Mediterranean to the Black Sea area. The East Med area in the Charter Parties is the extensive area east of Cape Passero (located in Sicily, Italy), with the Adriatic Sea to be included. It is considered as a discharging area for these types of ships and the Owners usually look for alternative areas to load, i.e., will either load from the Black Sea or will ballast to Gibraltar.
Obviously, the loading area for this cargo order is within the Black Sea and Charterers will try to fix a ship, as close as possible, to the Black Sea area. A mutually acceptable delivery point is the passage of the Canakkale straits.
The Charterers are searching for a spot ship (the cargo is ready), however, they can also discuss other, quite later dates, as will be proposed by the ship.

D. Where the ship will load from? / type of cargo to be loaded.
The order states:
1 tct via Black Sea.

The loading area is in the Black Sea. The Charterers have not disclosed the kind of cargo that they intend to load. The Owners should request the same or work the relevant 'cargo exclusions' clause.

E. Where the ship will be redelivered to?
The order states:
Redelivery on dop full China pico.

The ship will discharge in port/s in full China and will be redelivered upon dropping outward pilot at a "port in the Charterers' option" (abbreviated as "pico").

F. What is the duration of the trip?
The order states:
Duration about 45 days.

The duration is expected to be about 45 days.

G. What is the type of Charter Party? / What is the commission?
The order states:
Nype 81 5 pct ttl comm.

The Charter Party proforma will be based on the 'New York Produce Exchange' form, as amended in 1981.
The commission is 5 percent in total to the broker who quoted the biz. The address commission is included and the Owners have the right to request the analysis of the commission and how it has reached a total of 5 percent.

3.
Account......
Panamax OA OK to make Paranagua 1/10 January ....
One tct via ECSA with grain/grain products
Dlosp Skaw/Passero chopt passing PMO/Japan range atdnshinc
Dur bss APS about 40/65 days depends on redelivery
NYPE 46 C/P 5 pct ttl com

A. Which ships are workable?
The order states:
Panamax OA OK.

Any Panamax ship, i.e., with a size of about dwt 60,000 mtns to about dwt 84,000 mtns, overaged or not, is potentially workable.

B. Which ships have priority in working the biz?

The nearest Panamax ship to the load port, that is non-overaged with large cargo capacity will have priority in working this biz.

C. Where is the delivery of the ship and when? / Do Charterers accept balasters?

The order states:

to make Paranagua 1/10 January 2009.

Paranagua port is located in southern Brazil and it is a designated loading area for all sizes.

Ships in ballast are workable, if they can 'make' (arrive) at the loading port within the required lay-days. Ships coming in ballast from the Continent or from the Mediterranean can be fixed basis delivery 'passing Passero' or 'passing Gibraltar' and when coming from South Africa basis passing 'cape of Good Hope' or 'passing Durban'. If there is an unavailability of tonnages, Charterers may consider heavy ballasters, basis delivery 'Passing Muscat Outbound' ('PMO') or 'passing Singapore' or even delivery on DOP in a port in China. For the ships which are opening close to or at the Paranagua port, or for the Owners who accept to ballast the ship until the load port without hire, the delivery could be agreed on AFSPS at Paranagua port, with an agreed ballast bonus.

D. Where the ship will load from? / type of cargo to be loaded.

The order states:

One tct via ECSA with grain/grain products.

Charterers will load from the Eastern Coast of South America. The port of Paranagua will be used as sole or as first load port. The ship will load grains/grain products only.

E. Where the ship will be redelivered to?

The order states:

Dlosp Skaw/Passero chopt passing PMO/Japan range atdnshinc.

The cargo seems to be unsold. Charterers will investigate through their clients and will sell it to the most 'profitable' party for them.

The ship will be directed to discharge and be redelivered to either the Skaw/Passero range or the Passing Muscat Outbound/Japan range. A different hire rate will be in effect for each destination, because of its reposition.

F. What is the duration of the trip?

The order states:

Dur bss APS about 40/65 days depends on redelivery.

If it is Skaw/Passerro, the expected duration should be about 40/45 days basis delivery on APS, but if it is PMO/Japan, a longer duration should be expected (about 55/65 days).

G. What is the type of Charter Party? / What is the commission?
The order states:
NYPE 46 C/P 5 pct ttl com.

Charterers will use the 'New York Produce Exchange' form Charter Party as amended in 1946 and the existing total commission (including the address commission plus the brokerage) is 5 percent.

-Time Charter Trip with 2 loading legs or even more declared loading legs:
1.
Account.......
40/50,000 dwt max 28 years
Gearless ok
Dely continent ppt/onwards
Tc with 2 ll
Redely Atlantic
3,75 ttl com

A. Which ships are workable?
The order states:
40/50,000 dwt max 28 years
Gearless workable

Ships with about dwat 40,000 mtns to about dwat 50,000 mtns, geared or gearless are both workable, if their age is no more than 28 years.

B. Which ships have priority in working the biz?
Non-overaged ships and ships with a larger size will have a priority in this biz. C. Where is the delivery of the ship and when? / Do Charterers accept ballasters?
The order states:
Dely continent ppt/onwards

The ship will be delivered to the Continent area and will promptly load, i.e., a few days after the cargo order entered the market or later.
The Continent is a discharging area for many ships, there are therefore not many chances for ballasters.

D. Where the ship will load from? / type of cargo to be loaded.
The order states:
Tc with 2 ll.

The Charterers will execute two loading legs. At the first leg, it is obvious that the Charterers will load from the area of delivery (Continent) and discharge in an area where the next cargo is already secured, or it is not difficult to find and fix one.

E. Where the ship will be redelivered to?
The order states:
Redely Atlantic.

At the second leg, the Charterers will load or search to load from the area of discharging the first leg. The Owners can negotiate and exclude certain places, which are not favourable for the ship's next employment or for safety reasons or for any other reason the Owners are reluctant to trade there.

F. What is the duration of the trip?
No duration is given, but the Owners could ask the same of the Charterers.

G. What is the type of Charter Party? / What is the commission?
The order states:
3,75 ttl com.

The Charterers have not declared the type of Charter Party, and the commission is 3,75 percent in total to the broker who is quoting the biz, including the address commission.

2.
Account....
Handymax max 15 years old –minimum 25 mtns cranes
Delivery to make EC Aussie 10/20 March
Tct with 1/3 ll chopt – try 3+3 mos tc period
Redelivery Aden/Japan range
Nype 81 Charterers' proforma with only brokerage commissions

A. Which ships are workable?
The order states:
Handymax max 15 years old –minimum 25 mtns cranes.

All Handymax ships which are modern and are fitted with a minimum of 25 mtns cranes are potentially workable.

B. Which ships have priority in working the biz?
The cheapest modern/well geared ships will have priority in working the biz. C. Where is the delivery of the ship and when? / Do Charterers accept ballasters? The order states:
Delivery to make EC Aussie 10/20 March.

The Charterers ask for a ship which can make delivery at the East Coast of Australia within the required dates. Ships in ballast from south/southeast Asia/China range are potentially workable.

D. Where the ship will load from? / type of cargo to be loaded.
The ship will load from port/s located in the East Coast of Australia. The Charterers have not declared what cargo they intend to load.

E. Where the ship will be redelivered to?
The order states:
Redelivery Aden/Japan range.

Charterers request a wide range for discharge. Owners may negotiate and limit the redelivery area.

F. What is the duration of the trip?
The order states:
Tct with 1/3 ll chopt – try 3+3 mos tc period.

The Charterers can fix a ship for only one loading leg, but with an option to perform another one or another two trips. If the Owners are reluctant on such a way of work, the Charterers could try to work for a short period of 3+3 months.

G. What is the type of Charter Party? / What is the commission?
The order states:
Nype 81 Charterers' proforma with only brokerage commissions.

Charterers will use the NYPE Charter Party as amended in 1981.
As per the order there is no address commission, in which case only the commissions of the brokers involved will be in effect.

-Time Charter short period
1.
Account......
35/50,000 sdbc geared max 20 years
Abt 3 or 5 months chopt tc long period
Delivery Skaw/ Baltic
Laycan 20/30 December
Redelivery full Atlantic
5pct add com pus.

A. Which ships are workable?

The order states:
35/50,000 sdbc geared max 20 years.

Any handy/handymax ship with about dwat 35,000 mtns to about dwat 50.000 mtns is potentially workable if its age is no more than 28 years.
The Charterers are asking for a geared ship, without indicating the required minimum lifting capacity.

B. Which ships have priority in working the biz?

Charterers are flexible on the quantity that they will load, but the largest ship located near the loading area (within the Baltic area) should be the most workable. Charterers when take a ship opening within Baltic will also save the cost for pilots requiring for passing the Great Belts.

C. Where is the delivery of the ship and when? / Do Charterers accept ballasters?

The order states:
Delivery Skaw/ Baltic.
Laycan 20/30 December.

Ballasters coming from the region of the Continent should be workable, however, the Owners must accept that the delivery takes place on the entrance of the Baltic Sea, at Skaw and not before. Skaw is a cape that belongs to Denmark and is the most northern point of the country. It is a designated point for delivery/redelivery purposes for ships trading or going to trade within the Baltic Sea.

D. Where the ship will load from? / type of cargo to be loaded.

The loading place is within the Baltic Sea. The cargo is unknown and the 'cargo exclusions' and the 'trade exclusions' clauses will be discussed during the progress of the negotiations.

E. Where the ship will be redelivered to?

The order states:

Redelivery full Atlantic.

The redelivery will take place within the Atlantic. The Owners could negotiate for the usual redelivery areas to apply, such as the Boston/Houston range, including the Caribbean Sea or the Recife/Buenos Aires range or the Skaw/Gibraltar range or the Skaw/Dakar range.

F. What is the duration of the trip?

The order states:

Abt 3 or 5 months chopt tc long period.

The Charterers are searching for a ship for short period of about 3 to 5 months, with the exact duration at their option. A longer period could be discussed, if the ship's characteristics and the proposed 'hire rate' is attractive and satisfies the Charterers.

G. What is the type of Charter Party? / What is the commission?

The order states:

5pct add com pus.

The type of Charter Party which will be used is not disclosed and an address commission of 5 percent is in effect plus the brokerage commissions.

2.
25/30000 dwt sid/bc oaok gearless ok
Dely SW PASS 10/20 March
4/6 months tc period
Redely Spore/Japan or Skaw /Passero or New York/Baires range
can try tct out redely east Med/Black Sea -dur about 50/60 days
1st trip grains from USG to east Med /Black sea
Nype 46 C/P with 3,75 ttl comm incl 2,5 add

A. Which ships are workable?

The order states:

25/30000 dwt sid/bc oaok gearless ok.

All ships from about dwt 25.000 mtns to about dwt 30.000 mtns, of any age, geared or not, are workable for the biz.

B. Which ships have priority in working the biz?

Usually the youngest/better positioned/cheaper ship will have priority in working the biz.

C. Where is the delivery of the ship and when? / Do Charterers accept ballasters?

The order states:

Dely SW PASS 10/20 March.

Delivery of the ship will take place in the Southwest Pass, a designated point of delivery located at the entrance of the Mississippi Delta at Louisiana of the U.S.A.

Charterers' required lay can is for 10/20 March. All the right sized ships in ballast, which could meet the Charterers' delivery point, are potentially workable.

D. Where the ship will load from? / type of cargo to be loaded.

The order states:

1st trip grains from USG to east Med/Black sea.

The Charterers indicate that the first trip will be grains, loading from the United States Gulf to the East Mediterranean/Black sea area.

The 'cargo exclusions' and 'trade exclusions' clauses in the Charter Party will determine the allowed places for trading and the allowed cargoes for loading.

E. Where the ship will be redelivered to?

The order states:

Redely Spore/Japan or Skaw /Passero or New York/Baires range
can try tct out redely east Med/Black Sea
-dur about 50/60 days.

A wide range of redelivery is requested, as the Charterers, after completing the first leg, will speculate on the market for the best deal.

Alternatively, the Charterers could discuss only a trip out, with redelivery in the East Med/Black sea area.

Owners should decide what they prefer and will offer to the Charterers accordingly.

F. What is the duration of the trip?

The order states:

4/6 months tc period
can try tct out redely east Med/Black Sea -dur about 50/60 days.

A duration of 4/6 months (the exact duration is always at the Charterers' option), but for a single trip, about 50/60 days.

G. What is the type of Charter Party? / What is the commission?
The order states:
Nype 46 C/P with 3,75 ttl comm incl 2,5 add.

The 'New York Produce Exchange' Charter Party as amended in 1946 will be used. The address commission is 2,5 percent + 1,25 percent brokerage commission to the Charterers' broker.

-Time Charter period
1.
LME upto Kamsarmax maximum 10 years
January delivery
Del Skaw/Passero/West Africa range
Min 3-max 5 years chopt Time Charter Period
Redely tba
Nype 46 Charterers proforma with 3.75 pct ttl com.

A. Which ships are workable?
The order states:
LME up to Kamsarmax 10 years.

Large – Modern – Economical (LME) Panamax sized ships up to Kamsarmax are workable. The LME are the ships with a dwat above 70,000 mtns. The Kamsarmax size can provide a dwat of about 82,000 mtns at the maximum length over all (loa) of 229 mtrs, which allows it to load at Kamsar port in Guinea and maximize the cargo capacity in accordance with the prevailing restrictions there.

B. Which ships have priority in working the biz?
The Kamsarmax sized ships have the priority.

C. Where is the delivery of the ship and when? / Do Charterers accept ballasters?
The order states:
January delivery
Del Skaw/Passero/West Africa range.

The delivery of the ship will take place within the Skaw/Passero range or West Africa and the Charterers will accept all ships giving lay-days within January. During negotiations, the lay-can will be narrowed, enabling Charterers to arrange for their program and especially for the ship's first voyage.

D. Where the ship will load from? / type of cargo to be loaded.
The first cargo could be minerals from Kamsar port or grains from ECSA.

E. Where the ship will be redelivered to?
The order states:
Redely tba.

The Charterers and the Owners will discuss, during the progress of negotiations, various places for the ship's redelivery.

F. What is the duration of the trip?
The order states:
Min 3-max 5 years chopt Time Charter Period.

The duration of the period is a minimum of 3 up to a maximum of 5 years. The exact duration will be at the Charterers' option (chopt).

G. What is the type of Charter Party? / What is the commission?
The order states:
NYPE 46 Charterers' proforma with 3,75pct ttl com.

Charterers will work on a New York Produce Exchange form as amended in 1946 and they will propose to the Owners their own proforma for consideration. The commission is 3,75 percent in total including the address commission.

2.
Handymax/Supramax OA ok
Min 4x30 mtns cranes + 10 cbm Grabs
Delivery P.G./Singapore range
Feb/March dates
12/18 months or try 2 years +/- one month
For W/W trading including Indian Coastal biz with iron ore and coal
Redely W/W tba
5 ttl including 3,75 add comm.

A. Which ships are workable?
The order states:
Handymax/Supramax non OA
Min 4x30 mtns cranes + 10 cbm Grabs.

Modern ships, of handymax or supramax size are workable, if they can provide a minimum of 4 cranes with a minimum lifting capacity of 30 mtns and grabs with a minimum 10 cbm capacity.

B. Which ships have priority in working the biz?

The cheapest starting ship, in a good position, according to the Charterers' needs, has priority in working the biz.

C. Where is the delivery of the ship and when? / Do Charterers accept ballasters?

The order states:
Delivery P.G./Singapore range
Feb/March dates.

Charterers can accept to work a ship opening within the range from the Persian Gulf up to Singapore.
It can work for any date of delivery, but it should be within the months of February and March.

D. Where the ship will load from? / type of cargo to be loaded.

The order states:
For W/W trading including Indian Coastal biz with iron ore and coal.

Charterers need a ship for trading around the world, but it seems that they are actually looking for Indian Coastal trading.

E. Where the ship will be redelivered to?

The order states:
Redely W/W tba.

The Charterers will discuss for Worldwide redelivery and in areas which will be mutually agreed upon.

F. What is the duration of the trip?

The order states:
12/18 months or try 2 years +/- one month.

The Charterers request 12/18 months but could try for a period of two years with one more or less month. All depend on the proposed 'hire rate' and the characteristics of the interested and presented ship.

G. What is the type of C/P? / What is the commission?

The order states:

5 ttl including 3,75 add comm.

Charterers have not declared the type of Charter Party. The commission is 5 percent total to the broker who is quoting the biz, including 3,75 percent as address commission.

PART III

CHAPTER II

HOW TO MAKE AN OFFER FOR A TIME CHARTER

The forms of bids on main terms for Time Charter are not as complicated compared to those that are based on Voyage Charter. Many important terms could be left and discussed during the negotiations of the Charter Party details.
Owners and Charterers should take care to protect their interests during the negotiations of both Main Terms and Charter Party details and should only conclude the fixture after having obtained a full coverage and once having agreed to all and clear terms. Disputes should be limited to the extent that an amicable solution is possible, or else, unclear/confused terms may lead to the disputes being referred to Arbitration, with considerable costs and unpredictable awards for the contracting parties.

Nothing is taken for granted, fix it with crystal clear terms, so as to prevent the contracting parties from referring to arbitration.

The interest of the Owners for an 'order' could be expressed with one of the following three forms (all in writing):
- As a 'firm bid' or 'firm offer'.

The Owners will make an 'offer' for the Charterers' consideration, provided that the biz is 'firm' and the Charterers are in fact seriously looking to fix a ship either for covering their Voyage cargo/es, or for speculating on the open market for a suitable biz after fixing the ship.
If the ship is workable for the biz and 'free' to work in the market (there is no commitment to any other biz), the 'firm offer' will be passed on and the negotiations on the presented terms will be in progress. 'Firm offers' and 'counter re-

plies' keep each other (Owners and Charterers) committed for a specific period of time. It is not fair for the Owners to work on two 'orders' for the same ship at the same time or for the Charterers to work two ships for the same biz.
- As an 'indication'.
For 'orders', which are not yet 'firm' or the ship is not 'free' to work any biz on a 'firm' basis, the Owners will only make an 'indication'. There is no contractual or ethical commitment/obligation, but the Charterers simply take a note of the Owners' intention for the biz at this time. The 'hire rate' may change when the Owners make a bid at a later stage, as it is subject to the market's direction and the availability of other firm biz in the area.
- As a 'firm indication'.
There are no commitments, but it just shows that the Owners are ready to offer for their biz, in accordance with the presented terms and to alter the same into a 'bid', upon the Charterers' readiness to work or request.
The 'indication' and 'firm indication' should include:
-the basic Time Charter description of the ship.
-the delivery point.
-the dates for the ship's delivery.
-the trade and duration.
-the redelivery point.
-the hire rate.
-the type of Charter Party and the existing total brokerage commission.

The firm negotiations and the subsequent fixture will be concluded in two parts:
a). the Main Terms.
The Owners' bid with its negotiated terms represents the Main Terms. The conclusion of the Main Terms is the agreement that is reached on certain basic terms, which are actually based on a standard motive, with amendments only depending on the intended trade.
b). the Charter Party terms.
Following the agreement on the Main Terms, the parties involved and depending on the law, which is in effect during negotiations, or, according to shipping ethics, will be obliged or will be required to further develop, through the negotiation of the Charter Party proforma and its additional terms and conditions, referred to as 'details'.
Mainly the Charterers, or, if it is agreed, the Owners will produce their Charter Party pro forma (usually a previously executed one which should be based on a standard type of Charter Party) to the Owners or Charterers respectively, for their consideration and for their likely comments.
The Charterers will state, from the time that the 'cargo order' circulated in the market, or, at the latest, from the beginning of the negotiations, the type of the

Charter Party that will be in use.

The Charter Party proforma based on a standard type of Charter Party consists of two parts:
1. the Main Body.
The Main Body is a standard wording as adopted or issued or recommended by various shipping organizations and has been amended from time to time. These terms set the limit of obligations and responsibilities of each contracting party. However, all printing terms are negotiable, and the Charterers/Owners make their deletions/amendments/ additions in accordance with the prevailing trading conditions and circumstances.
2. the Rider or Additional Clauses.
The Rider or Additional clauses cover any omissions of the Main Body or add new regulations, as well as other specific/special terms that are in effect to the trip/s and/or to protect the Owners' and/ or Charterers' interests. Many important terms allocate liabilities and responsibilities for specific risks for each contracting party and try to make each case "crystal clear' without doubts or room of any disagreement, which might otherwise lead to extra costs and lengthy litigations.
Until conclusion of the agreement and subject to the prevailing law during negotiations, each party, in the case of a disagreement or for other serious reasons, has the right to withdraw its interest on the biz/ship.
Particularly, under the terms of the English law and until the additional completion of successful negotiations on the Charter Party 'details', there is no contract, and each party could 'escape' from the negotiations. However, this will be contrary to shipping ethics and can cost the Owners or Charterers, their reputation. Under the terms of the American law, the consent of 'subject to the details' on the Main Terms, binds both parties to work the same and they must make an attempt to conclude the negotiations.
The writing abbreviations that are used during the negotiations, including those on rider/additional clauses, do not exist when the Charter Party will be issued. Whole words must be entered, which in this case would minimize disputes for any misunderstandings due to abbreviations.

Before the parties interested enter into negotiations and eventually come to the conclusion of the agreement (final fixture), the following procedures should be followed:

1. The Charterers should
- enter the market with their 'order' through their brokers,
- search for a ship which can meet the minimum requirement of their intended

shipment/s and their needs,
- search for a ship which can be delivered within the required dates.
This is the first move before any further moves take place.

2. The ship should
- be in the right position,
- agree to perform on a Time Charter,
- agree to be committed with these Charterers, for this particular biz for a period of time.
The ship, on delivery, must be free of any commitments from its previous biz and clean/ready to fulfil the ordinary and specific obligations of the governing Charter Party.

3. The offer, pursuant with the customary procedure, should be made by the Owners to the Charterers and the negotiations on the Main Terms will be in progress and should hopefully be concluded.
The 'offer' will be given for the Charterers' consideration and subsequent acceptance, or for their 'counter offer' reply within a specified period, after which, it cannot be revoked or asked to be modified by the Owners.
The Charterers' clean acceptance, depending on the law applicable to the negotiations, could bind the Owners and Charterers to develop further by working the Charter Party details.
The reply to the Owners' offer (as well as all subsequent replies from both parties), always within a specified period, is expressed
. as 'counter offer',
which should be given on an Accept/Except basis, i.e., the negotiating parties respond just with the terms they want to amend/alter. All other terms are to be considered as 'accepted' and any reference back to them for re-negotiation should be considered
. as 'back-trade',
which might result in the termination of the negotiations.

The negotiations on the Main Terms and the Owners'/Charterers' commitment could be terminated:
-if the Charterers did not reply at all to the Owners' offer for various reasons.
-if the Charterers decline the Owners' offer and send a binding offer for the Owners' acceptance/ counter offer that is declined by the Owners.
-if any party replies out of the specified period, throughout the negotiations.
-if any party fails to clean accept the offer or the counteroffer as given from the other party, throughout the negotiations.
It would be unfair and contrary to the ethics of shipping and would cost the

Owners' and Charterers' reputation, if the ship has been dually committed on a 'firm' basis or the Charterers have been working two ships simultaneously or they drop the negotiations for another cheaper ship or the Owners drop it for a better biz, without giving to the other party the
. 'first refusal'.

4. When the negotiations are concluded (with Charter Party details in order), the Charterers must confirm the final fixture within the deadline, for lifting the subjects, as it will be agreed during the negotiations of the Main Terms.
- 'Subjects' or as shorten called 'subs',
are the Charterers' duly kept reservations to conclude the fixture. They will present the ship, as the case may be, to the Shippers/Sellers and/or to the Consignees/ Buyers for approval of its characteristics and will additionally arrange the readiness of the cargo (stem= subject to enough merchandise).
However, if the ship is fixed by the Charterers for speculative reasons, without having fixed any specific voyage cargo, after which they will work on the open market with forward or prompt delivery, for a short or long period, they could request some time only for the
. Board of Directors' (abbreviated as 'b.o.d.') approval'.
The respective company's directors will make a decision, on whether the fixed ship has prospective chances to be profitable, according to its position/agreed 'hire rate'/ship's characteristics and the existing and prospective shipping market.

<u>If the Owners are to bid, they must be aware of:</u>

a- the Charterers' name.
At the top of the cargo order, the Charterers' name should be revealed as a guide for the Owners'.
However, many times for commercial reasons, the name of the Charterers might not be disclosed and be replaced with the expression:
. 'fcc',
which means that the 'order' is for the account of "First Class Charterers".
The involved brokers inform the market that although the name of the Charterers does not have or cannot be disclosed in the open market for commercial or other reasons, they are dealing with a first-class name and the market should not worry about anything serious. However, this could include a wide range of responsibilities for the brokers
. as ensuring the market that they are dealing with A1 people, who, if eventually step down from their obligations, could cause them problems.
In any case, the Charterers' name must be revealed upon the ship's commit-

ment with them or upon their first counter offer.
Charterers with little activity, which have not been well established in the market, should be avoided being qualified as 'fcc'.
The expression:
. 'direct' or even 'direct/close'
could be enough for brokers quoting the biz and it lies on the Owners to check the Charterers' credibility and their financial standing in the market.

b- the size of the required ship.
The available cargo quantity and its stowage factor, the form of cargo and the port restrictions will basically determine the size and type of the required ship.
Usually, the Charterers enter the market as we have already aforementioned by virtue of the:
1-general characteristic of the ship's size and ask, for example, for a 'panamax ship' or a 'handymax ship' etc.
2-ship's specific size and ask, for example, a 'minimum 25,000 dwat' or '50/55,000 dwat' etc.
3-minimum grain or bale capacity required for the trip and ask, for example, for 'sid/bc with minimum 3,000.000 cbft grain capacity' etc.

c-the intended cargo and its form of transportation.
The cargo will determine if a specific type of ship is required, i.e., a logger for logs cargo or a multipurpose type for general cargoes or a specific ship which can carry dangerous goods etc.

d- the point of delivery and redelivery.
The point of delivery and redelivery, which could both have an effect and influence on the required 'hire rate', as well as, on the possible 'ballast bonus'.
The ship could be declined by the Charterers when the steaming time (distance) from the ship's last discharge port to the delivery point is long and make the Owners' to offer with a 'hire rate' unattractive to the Charterers or the Owners to refuse to work the biz, even if an attractive 'hire rate' is achievable, but because of the redelivery point, which could be in a place where the ship will suffer to find its next employment, the Owners should prefer to look for other alternatives.

e- the duration.
The duration is the 'key' when fixing on Time Charter and in an upward market the Owners should search for long periods, as they will mainly ensure the ship's trading and their investment, even if the market has thereupon dropped.

f- the type of Charter Party and the existing total brokerage commissions.

In short, before the Owners send their offer, they should have knowledge of:
- the Charterers' name,
- the size and type of the required ship/any age or any other ownership/flag restrictions applying, which could affect the ship's workability,
- the point of delivery,
- the point of redelivery,
- the intended cargo/es for the trip out or even for short periods,
- the duration of the Charter, or otherwise the intended trip,
- the type of Charter Party that will work and the existing total brokerage commission.

PREPARING TO OFFER.

The 'cargo order' has been circulated as follows:

Account Eagle Trading and Shipping Inc
Handymax dwat about 45/55,000 mtns-oa ok
Geared lifting capacity min 25 metric tons
Grabber preferred
Delivery East Coast India 10/15 Dec 2....
Time charter trip out via east coast India to Far East with bulk agriprds
Redelivery Far East (intention: dop 1sp China chopt)
duration about 25-35 days wog
NYPE 46 C/P Charterers' proforma with 2,50 address commission past us.

The market and the Owners, who might be interested in this particular biz, have a full picture of how this biz stands. The Owners can send an 'offer' to the Charterers, through the preferred brokerage channel from those who spread the biz in the market, or otherwise, through the channel that first called them and committed the ship for this particular biz.

THE OFFER.

The Owners' offer should basically include:
-the basic Time Charter description of the ship,
-the Charterers' name, otherwise relevant reservations should be kept for the Charterers' approval by the Owners,
-when the 'subjects' will be lifted, in order for the ship to be considered fully fixed,

-the point at which the ship will be delivered,
-the delivery dates,
-how the ship will trade,
-the point at which the ship will be redelivered,
-the 'trade exclusions' clause, unless the trade has been declared by the Charterers,
-the 'cargo exclusions' clause, unless the cargo/es have been declared by the Charterers,
-the 'hire rate' per day or pro rata for any part of the day which the ship will remain into the Charterers' services,
-the bunkering clause (the quantity of fuel on delivery/redelivery),
-the cost for Communications/Victualing/Entertainment (C/V/E),
-the amount for the Owners' compensation in lieu of holds' cleaning (abbreviated as 'ilohc') and if the ship executes more than one leg, then the 'intermediate holds' cleaning' clause,
-the law and arbitration forum that will apply to negotiations and to the eventual fixture,
-the type of Charter Party that will be the executed proforma and the existing total brokerage commission.

The Owners' offer could start as follows:

The name of the Charterers is Messrs. 'Eagle Trading and Shipping Inc'.
The name of the ship is 'MV FOTINI.'
At the top of the message, the Owners should state to which 'order' (biz) they are referring to.

Ref Account Eagle Trading and Shipping Inc
Handymax dwt 45/55,000 mtns
Dely East Coast India – Redelivery Far East (intention: dop 1sp China chopt)
cargo bulk agriprds
MV FOTINI

Reference is made to the above mentioned 'order'.
The Owners are in the position to firm bid for reply here within 1 hour as follows:
or
The Owners offer you firm for 1 hour reply as follows:
(or in any other form, which clearly states that the ship is being offered and committed with them for a specific period of time.)

The Owners continue:
"
MV FOTINI
....
....
"

Comments:
The ship's description was given as the case may be, and it is applicable when working on a single or short/long Time Charter period.
The description of the ship as provided by the Owners must be
. an accurate representation of the characteristics of the ship,
which can only be modified with the usual and acceptable shipping interpretation of the word 'about'. The Owners should be kept responsible and liable under the terms of the Charter Party for any consequences/delays/damages/expenses that may arise, if they have provided any misleading information on the ship's characteristics or because they have incorrectly judged the severity of this clause.

1. Account (Charterers' name).
The order states:
Account Eagle Trading and Shipping Inc.

Comments:
The Charterers' name is revealed as a guide for the Owners; however, they are unknown to them.

The offer could be given as follows:
For:
1.
Account Messrs. Eagle Trading and Shipping Inc. Sub Charterers approval by the Owners. Please advise of the Charterers' full background, where they are based and recent fixtures made by them, for the Owners' approval.

Comments:
Only the Charterers' name does not commercially cover the Owners.
The Charterers must provide:
. their full background, which includes, among others, the company's history / when it was established as well as their activities,
. the place where the company is based,
. a list of recent fixtures with other Owners. The Charterers must provide the name of the ships/ the owning or managing companies/through which brokers

they have fixed and their contact numbers.
The Owners should be able to check through the Owners or brokers, however, the Charterers will be approved only when the Owners have been fully satisfied on their reliability and performance.
. Until the Charterers have been approved, they will remain subject to the Owners' approval during the negotiations.
In case of dissatisfaction, the Owners have the right
. to drop the negotiations at any time, as no one should take the risk of fixing with someone who might step down from full or partial fulfilment of their contractual obligations.

The Owners continue:
2. Subjects.

The offer could be given as follows:
2.
All subs will be lifted upon fully fixing with Charter Party details in order.

Comments:
The Owners state that all subjects (abbreviated as 'subs'), which would possibly apply for this fixture, should be lifted
. upon the full agreement on the Main Terms and the Charter Party details.
The Charterers may counter reply based on a different time which should comply with the actual time that is required to present the ship to the parties involved and make the cargo ready, with all relevant approvals, or, if they fixed for speculative reasons, for the b.o.d. approval.

The Owners continue:
3. Delivery point.
The order states:
Delivery East Coast India

Comments:
The delivery of the ship will take place at the Eastern Coast of India. The intended port of loading is unknown to the market.
The actual position of the ship along with the prevailing market in general and particularly in the loading area, will determine which of the following three ways the Owners will choose and give as their offer:

a) on 'DLOSPS'
at the ship's last discharge port.

In this case the ship will sail 'on-hire' to reach the port of loading.
It is applicable when:
-the ship will open free of cargo in a port within the delivery area, where such delivery could be easily accepted by the Charterers
-the ship will open in a port outside the delivery area, however, due to the market's tendency and a lack of tonnages in the area, the Charterers could consider and eventually accept such a delivery.

If the ship will open at the port of Madras (located in E.C.India and within the requested delivery area), then the delivery term could be given as follows:
3.
Delivery on dlosps Madras port, East Coast India atdnsshinc.

Comments:
The ship will be delivered:
-on Dropping Last Outward Sea Pilot Station at Madras port.
The ship will tender the delivery notice to the Charterers and will sail 'on-hire', within the agreed lay-days, at the point where the last sea pilot of Madras port will disembark from the ship and provided that it is in all respects ready to trade/load the intended cargo under the terms of the Charter Party. The Charterers will direct the ship to the port of loading and apart from the payment of the hire will also bear the cost for any time lost beyond the ship's fault and for all fuel consumed from the point of delivery until the arrival at the load port.
The location of the port of delivery must be stated (notably 'East Coast India') in the given offers, mainly preventing misunderstandings
. with other synonymous ports,
but also, for identifying the country of loading.
Brokers should try to make all terms clear, so that both sides will have eventually agreed on crystal clear terms, as nothing should be taken for granted.

b) on 'AFSPS'
at the port of loading within the East coast of India as specified by the Charterers during the negotiations or even after the fixture.
The Owners will direct the ship to the delivery point
. and will bear the cost of lost time and fuel consumed from the point which the ship left until arrival.

The delivery term could be given as follows:
when the port is unknown
3.
Delivery on afsps 1sp within the East Coast India atdnsshinc. Port of delivery at chopt.

Or when the port is specified
3.
Delivery on afsps .... (insert the name of the port) port in East Coast India atdnsshinc.

Comments:
The ship will be delivered and:
-will be placed on-hire and the Charter will commence, upon the ship's Arrival at the First Sea Pilot Station at the port of loading and the tendering of the delivery notice.

c) on PASSING the farthermost point
when entering the delivery range, i.e., when the ship is positioned elsewhere, for example within the West Coast of India and has sailed to reach the range of delivery, the requested delivery point will be
. upon passing the farthermost port of the delivery range at the side on which the ship is approaching the range.
In this particular case, the Tuticorin port is the most known farthermost port in the East Coast of India for ships coming in ballast from the West Coast of India. The Charterers will direct the ship to the port of loading and apart from the payment of the hire, will also bear the cost of any lost time beyond the ship's fault and of all fuel consumed from the point of delivery until the arrival at the load port.

The delivery term could be given as follows:
3.
Delivery on passing Tuticorin port northbound in East Coast India atdnsshinc.

Comments:
All ways of delivery will take place on a basis of 'All Time Day or Night, Saturdays, Sundays and Holidays Included' (abbreviated as 'atdnsshinc'). The Master, when the ship is ready and in compliance with the terms of the Charter Party, will tender the delivery notice via e-mail or fax to the designated parties,
. at any time during day or night, including weekends and holidays.
The Time Charter will commence, and the first hire will run and will be paid as per the Charter Party terms.

The Owners continue:
4. Lay-can.
The order states:
10/15 Dec 2....

Comments:
Lay-days/Cancelling (abbreviated as 'l/c') is the span of dates, within which the ship, which will be in all respects ready to trade, must tender the delivery notice to the Charterers. The tendering of the notice of delivery outside of the agreed lay-days/cancelling is invalid and if it is on the late side could lead to the cancelation of the Charter Party

The offer could be given as follows:
4.
Lay-days 10th December 2... 00:01 hrs. – Cancelling 15th December 2... 24:00 hrs. For lay-days/cancelling the local time will be in effect
Or shorten:
l/c 10/15 Dec 2.... 00:01/24:00 hrs. - for l/c the local time will be in effect.

Comments:
The Owners have offered:
-that the delivery of the ship will take place between 10th and 15th December 2....
Obviously, the Charterers will accept it.
Even though the ship's readiness for delivery is within different dates, the ship is of full interest due to its characteristics or due to the low 'hire rate' offered, so the Charterers could apply and check for the possibility of expanding the lay-days for an earlier or later loading, depending on the case.
The exact delivery time within the agreed l/c is
. at the Owners' option, particularly from 00:01 hrs. of the 10th December until 24:00 hrs. of the 15th of December.
For the l/c
. the local time should be in effect and the same must be inserted in the bid and incorporated in the issued Charter Party.

The Owners continue:
5. Trading.
The order states:
Time charter trip out via East Coast India to Far East with bulk agriprds.

Comments:
The Charterers will direct the ship through port/s, which can be approached by the size/draft of the ship and can safely berth/unberth/depart under the Port Authorities' directions.
The ports/berths/anchorages should provide competent pilots, number of tugs,

lighthouses etc. so that the ship can load and discharge the cargo and sail under normal conditions without serious consequences, due to stress from weather or any other reason attributed to a non-organized port.
The nominated port/s berth/s should in all cases be considered 'safe' throughout the duration that the ship is in the port/berth and this should be clearly stated in the Owners' offer and incorporated in the issued Charter Party.
The Charterers should direct the ship to navigate/trade/approach the ports/berths where it will remain for loading/discharging, always afloat, and will not be at risk of touching the bottom. These terms should be in effect throughout the Charter, including when crossing canals/straits/narrow waters etc.
For tidal ports/berths, the 'NAABSA' (Not Always Afloat But Safely Aground) term should be agreed upon. The ship might touch the bottom, but, due to the soft muddy or sandy seabed, will run aground safely.

The offer could be given as follows:
5.
Time Charter Trip out (or 'One Time Charter trip') via East Coast India to China, via safe port/s, safe berth/s, always within I.N.L. with cargo of bulk agricultural products.
The port/s berth/s will remain safe from the time of the ship's arrival at the first sea pilot station until departure and the last sea pilot has left the ship.
The ship will always lie afloat throughout the period of the Charter.

Comments:
The Owners have offered:
- that the Charterers will perform a single Time Charter Trip. The ship will load only once and after completion of discharge will be redelivered to the Owners to the designated point.
-that the ship will be instructed to trade only via safe port/s, via safe berth/berths and will remain safe during its entire stay in the ports/berths.
-that the ship will trade always afloat during the entire trip.
-that the ship will always trade within I.N.L. (International Navigating Limits).
The Charterers do not have the right to direct and trade the ship in areas that will lead the ship to be uncovered by its Hull&Machinery insurance company.
-the ship will load from East Coast India to China with bulk agricultural products, as cargo.

The Owners continue:
6. Duration of the trip.
The order states:
duration about 25-35 days wog.

Comments:
The duration is the 'key' when fixing on Time Charter and affects the achievable 'hire rate'.
In an upward market, most of the Owners and especially those who are committed with 'new building' ships or their company is in the stock market exchange, search to fix for long periods, in order to ensure the ships' trading and look forward to their next investment.

The offer could be given as follows:
6.
Charterers guarantee a minimum duration of 30 days to about 35 days.

Comments:
The Owners have offered:
-Charterers to guarantee a minimum of 30 days as duration.
The Owners seek to ensure the minimum duration, otherwise, if the biz is shorter, they could lose their interest in it. However, it is risky for Charterers to accept it, because if, for any reason, they exceed their initial voyage calculations and the ship completes earlier, the Owners, depending on the market, could claim for losses against them.
-the Charterers have the right to trade the ship for up to 'about' 35 days. The Charterers should be covered for a slightly late redelivery more than the 35 days, however, the Owners, depending on the market, could claim for losses against them, if the ship is heavily delayed.

The Owners continue:
7. Redelivery of the ship.
The order states:
Redelivery Far East (intention: dop 1sp China chopt).

Comments:
The Charterers will redeliver the ship when Dropping the Outward Pilot in a safe port (the port is at the Charterers' option) in the Far East, with China as their intention.
The Owners could focus only on the China discharge/redelivery and should decide if such a repositioning of the ship in full China is of interest or could otherwise offer that the basis ship opens in an area, which could assist it in its next employment.

On a basis of discharge and therefore redelivery in full China, the offer could be

given as follows:
7.
Redelivery on dlosps 1 safe port full China atdnsshinc. Redelivery port at Charterers' option.
Or shorten
Redelivery on dlosps 1sp full China atdnsshinc. Redelivery port at chopt.

Comments:
The Owners have offered:
-that the Charterers have the right to discharge the ship and redeliver it on a dlosps to a safe port, which could be located in any place in China.
or,
if the Owners reject full China as the redelivery area, the offer could be given as follows:
7.
Redelivery on dlosps 1safe port not northern of Shanghai port or passing southbound Shanghai port atdnsshinc. Redelivery port at the Charterers' option.

Comments:
The Charterers have the right to discharge the ship south of the Shanghai port and redeliver it on dlosps to a safe port or to even discharge north of the Shanghai port, however, under their own time/expenses, they will
. ballast the ship
empty of cargo, from the discharge port until they pass the Shanghai port for redelivery purposes.

If the discharge port is nominated by the Charterers, the offer could be given as follows:
7.
Redelivery on dlosps .... port, China atdnsshinc.

Comments:
In all cases the redelivery point in a port should be clearly defined on a DLOSPS basis and not as DOP, regardless of whether the discharging port is in a river area or not.
The Master will tender the redelivery notice by e-mail or fax to the Charterers
. at any time
during day or night, including weekends and holidays and the Time Charter will be terminated from that tendered time.

The Owners continue:

8. Hire/Hire payment.

The 'hire rate' will cover each calendar day that the ship will be in the Charterers' services or pro rata for any part of a day and should include the Master's/Officers'/Crew's overtime costs, i.e., the Charterers will not pay extra fees if the ship's officers/crew are required to provide their customary services out of their daily maximum normal time of work.

Usually, under the terms of the Charter Party, the hire is due every 15 days in advance and for the last days it is payable day by day as it becomes due or as has been agreed upon. For hire calculation purposes the UTC (Coordinated Universal Time) will be in effect, i.e., the ship, upon tendering the delivery notice to the Charterers, will include

. the time of delivery on local time and on UTC.

The UTC will be considered as the 'delivery time' for the calculation and payment of the hire. The same should be in effect upon the Master tendering the redelivery notice and the UTC time will be considered as the 'redelivery time'.

The offer could be given as follows:

8.

Hire USD 25,000 per day/pro rata including overtime. Hire is payable every 15 days always in advance net of tax and bank charges.

For hire calculation purposes the UTC time will be in effect.

Or, in short:

Hire USD 25,000 pdpr inclot payable every 15 days always in advance net of tax and bank charges.

For hire calculation purposes the UTC time will be in effect.

Comments:

The Owners have offered:

-a 'hire rate' per day or pro rata for part of a day that the ship would be in the Charterers' services.

-that the hire is due for payment every 15 days always in advance net of tax and bank charges.

-that for hire calculation purposes the UTC time will be in effect.

-that the hire includes the crew's overtime.

The Owners continue:

9. Trading exclusions.

The order states:

- *Time charter trip out via east coast India to Far East.*

Comments:
The ship will execute only one loading leg via east coast India to the Far East. The Charterers do not have the right to re-load the ship after completion of discharge and the ship should be redelivered to the Owners in accordance with the terms of the Charter Party. The Owners must choose between two options. To offer on the basis that
- the ship will only load from the Eastern Coast of India and discharge only in China
or
-to work the 'trading exclusions' term and exclude countries around the world in which the ship cannot trade.
The Charterers' intention for discharge in China is known from the beginning and the Owners/ Charterers have a good opportunity to work on the first option which, if it is accepted, will save time and arguments for both sides, as to the countries which will be excluded.

The offer could be given as follows:
First option:
9.
Charterers guarantee that the ship will load only from the East Coast India and will discharge only in China.
Second option:
9.
The ship will not trade/load/discharge in ....

The Owners continue:
10. Cargo exclusions.
The order states:
-Time charter trip out via east coast India to Far East with bulk agriprds.

Comments:
The Owners must choose between two options.
To offer on the basis that
- the ship will only load bulk harmless agricultural products
or
-will work the 'trading exclusions' term and exclude cargoes that the ship cannot load for various reasons.
The Charterers' intention for loading bulk agricultural products is known from the beginning and the Owners/Charterers have a good opportunity to work on the first option which, if it is accepted, will save time and arguments for both sides as to the cargoes which will be excluded.

The offer could be given as follows:
First option:
10.
The ship will only load bulk agricultural products and no other cargo will be loaded on board the ship. The Charterers confirm that the cargo is harmless, non-hazardous and belongs to group C of the International Maritime Solid Bulk Cargoes Code (abbreviated as 'IMSBC Code')
Second option:
10.
The ship is not allowed to load the following cargoes: ....

The Owners continue:
11. Bunkering.
The consumed quantity of fuel, from the time of the ship's delivery until the time of its redelivery, is for the Charterers' account. However, for a short/long period the Charterers, in order to commence trading, will buy the fuel on board the ship upon delivery, at an agreed price.
At the end of the Charter the Charterers will sell the fuel inside the ship's tanks back to the Owners, at an agreed price (mostly on about same quantities as on delivery and at the same price).
For single trips there are several options to be agreed upon such as:
- the Charterers to not buy the fuel and supply the ship with enough fuel to execute the trip
- the Charterers to buy only part of the fuel on delivery and replenish the same before redelivery
- the Charterers to buy the required fuel quantity for the trip, but not to replenish the same and redeliver the ship with fuel 'as on board',
as well as numerous other options.
The payment of the value of the bunkers on board the ship at the time of delivery will usually be made together with the first 15 days hire payment.

The offer could be given as follows:
11.
The ship will be delivered with about .... mtns VLSFO and about .... mtns LSMGO. The ship will be redelivered with the same quality of fuel and with about the same quantity of fuel as on delivery.
The value of the fuel on delivery is payable together with the first 15-day hire payment.
The fuel prices on delivery and redelivery to be mutually agreed on. Same prices will be in effect on delivery and redelivery.

The Master's or Owners' information of the quantity of fuel that is required to be supplied in order to execute the trip or for redelivery purposes, is subject to trip's final duration and/or ship's consumption while manoeuvring in ports and/ or confined waters and subject to final destination.

Comments:
The Owners have offered with:
- what quantity of fuel the ship will be delivered with (the exact fuel quantity will be determined with the carried out on-hire bunker survey).
- what quantity of fuel the ship will be redelivered with (the exact fuel quantity will be determined with the carried out off-hire bunker survey).
-that the value of the delivered fuel will be due and be paid together with the first 15-day hire payment.
-that Owners and Charterers should mutually agree on the price with which the Charterers will buy the fuel from the Owners on delivery and the price with which they will sell it back to the Owners on redelivery. Both sides need some time to check on the prevailing prices through their suppliers.
-that any declaration by the Master or Owners on the required bunkers for redelivery or in order to execute a voyage is subject to final duration, ship's consumption while manoeuvring in ports and in confined waters and subject to final destination.

The Owners continue:
12. Communication/Victualing/Entertainment (c/v/e).
All costs incurred for communications made by the Master on the Charterers' behalf, to various involved parties in accordance with the terms of the Charter Party or in accordance with the Charterers' instructions during the execution of the Charter/trip, including any victualing/ entertainment that will be provided to pilots/stevedores and/or to the Charterers' visitors/ representatives etc. should be covered by this term.

The agreement should be made based on one of the following three ways:
a- with a monthly or prorate for any part of the monthly coverage, which is widely used in the Charter Parties and is acceptable by both Charterers and Owners. The respective monthly amount is payable together with the hire payment.
b- with a mutually agreed lumpsum amount for the entire duration of the Charter.
It is mainly applicable when the expected duration is less than one month, and it is usually payable together with the first hire payment.
c- payable at cost, in which case the Master will provide a statement for all ex-

penses incurred on the Charterers' behalf at the end of the Charter for a single trip, or, for a period, at the end of each voyage. For a single trip, it is obviously payable at the end of the Charter.

The offer could be given as follows:
12.
The Charterers, for the whole duration of this Charter period, will compensate the Owners with a lump sum amount of USD .... per month or pro rata for any part of the month for communications/ victuals/entertainment. The respective monthly amount is payable to the Owners together with the hire payments.
or
12.
The Charterers will compensate the Owners with a lump sum amount of USD .... for the entire duration of the Charter for communications/victuals/entertainment. The respective amount is payable together with the first hire payment.
or
12.
C/V/E is payable by the Charterers at cost at the end of the Charter and in accordance with the presented statement/bills or the Master's statement of expenses incurred on the Charterers' behalf.

For a period: *C/V/E is payable by the Charterers at cost at the end of each voyage.*

Comments:
The Owners have offered:
-the amount that will be compensated for the costs incurred and made on the Charterers' behalf.

The Owners continue:
<u>13. In Lieu Of Holds' Cleaning Clause (abbreviated as 'ilohc').</u>
The Charterers are of the contractual obligation to redeliver the ship's holds in a condition such as on the time of delivery. However, through this term they will keep the option of paying an amount to the Owners as compensation, in lieu of holds' cleaning. In fact, this amount will cover the crew's work and loss of time in order to clean the dirty holds. It is quite difficult for the Charterers to organize the holds' cleaning and this term is necessary to be discussed when working on the Time Charter.
The agreement should be made based on one of the following two ways:
- on a lumpsum amount, which is widely used in the Charter Parties and is acceptable by both Charterers and Owners. The agreed amount will be paid at the

end of the Charter, regardless of the holds used by the Charterers.
- on a lumpsum amount for each hold used by the Charterers. That is preferred by the Charterers when they load the ship in alternative holds and will be paid at the end of the Charter.

As an example:
A ship has 7 holds, but the Charterers have used only 4 holds (heavy cargo loaded in alternative holds)
-based on the first option the agreed amount for ilohc was a lumpsum amount of USD 7,000.
The Charterers will pay this amount regardless of the holds used for loading.
-based on the second option the agreed amount for ilohc was a lumpsum amount per hold used for loading of USD 1,000, so the Charterers will obviously save money, as they will pay only for the used holds i.e., usd 4,000.

The offer could be given as follows:
13.
The ship will be redelivered to the Owners with its holds in the same condition as on delivery, normal wear and tear excepted.
However, the Charterers will have the option of redelivering the ship with unclean holds, by paying a lumpsum amount of USD .... The amount due will be paid together with the Charterers' final hire statement.
or
However, the Charterers will have the option of redelivering the ship with unclean holds, by paying a lumpsum amount of USD .... per hold used for loading. The amount due will be paid together with the Charterers' final hire statement.
The ship in any case will be redelivered to the Owners free of cargo residues in its cargo holds.
Debris and/or dunnage and/or lashing or any other materials that have been used to secure and/or protect the cargo during its sea voyage will be discharged from the ship before its redelivery, at the Charterers' entire time, risk and expenses. The ship will not sail from the port unless all such materials have been removed and discharged from the ship and any delays, expenses, consequences and/or possible fines imposed by the Port Authorities, arising from the Charterers' inappropriate discharge of the materials or inappropriate delivery ashore, will be for the Charterers' account and the ship will remain always on-hire.
In the case that the materials are not allowed to be discharged at the last port of discharge, then, before the ship's redelivery, the ship will proceed to the most convenient anchorage, berth of another port or country in order to discharge the same in barges or ashore, under the Charterers' time, risk and expenses. The ship will remain always on-hire, until having fully discharged the respective ma-

terials and the ship is redelivered on dlosps or at the Owners' option at equidistance from the actual point of redelivery.

Comments:
The Owners have offered:
-the amount that will be compensated if the holds have been redelivered dirty
-the ship in any case must be redelivered free of cargo residues.
-that the debris/dunnage/lashing, or any other materials used to secure/protect the cargo during its sea voyage will be discharged from the ship before its redelivery, entirely at the Charterers' time/risk/expenses.

The Owners continue:
14. Law and Arbitration.
This term will specify that the disputes arising from the Charter Party will be resolved by Arbitration. The place of arbitration and the prevailing law during negotiations and eventual agreement should be discussed and agreed upon from the negotiations on the Main Terms.
The main arbitration venues are
. in London with English Law to apply
and
. in New York with U.S. Law to apply.
Bimco has added the option for two more arbitration venues
. in Singapore with Singapore or English law to apply.
. in Hong Kong with Hong Kong or English law to apply.

The offer could be given as follows:
14.
Arbitration in London. English law will apply to negotiations and to the eventual agreement.
Sub to agree to a full clause.

Comments:
The Owners have offered:
-the Arbitration forum and law that will apply to negotiations and eventual agreement
-the full Arbitration clause will be discussed and agreed upon when working on the Charter Party details.

The Owners continue:
15. Charter Party and Commission.
The order states:

NYPE 46 C/P Charterers' proforma with 2, 50 address commission past us.

The offer could be given as follows:
15.
Sub further terms/details which will be based on NYPE 46 C/P at the Charterers' executed proforma with 3,75 percent total at your end including 2,5 percent address commission.

Comments:
The Owners have offered:
- the Owners accept the Charterers' proforma to be based on the New York produce Exchange 46 form.
- the proforma to be on an executed one, i.e. on a proforma which has already been worked by others and, hopefully, many unfavourable terms will have been amended or deleted by them.
-the Charterers' Charter Party proforma is subject to the Owners' acceptance and it is possible that the Owners will revert with their comments on the terms or even add other terms.
-that the existing brokerage commission is 3,75 percent in total.

TO SUMMARIZE THE OWNERS' OFFER:

Ref Account Eagle Trading and Shipping Inc
Handymax dwt 45/55,000 mtns
Dely East Coast India – Redelivery Far East (intention: dop 1sp China chopt)
cargo bulk agriprds
MV FOTINI

Reference is made to the above mentioned 'order'.
The Owners are in the position to firm bid for reply here within 1 hour as follows:
or
The Owners offer you firm for 1 hour reply as follows:
(or in any other form, which clearly states that the ship is being offered and committed with them for a specific period of time.)

MV FOTINI
....
....

For:

1.

Account Messrs. Eagle Trading and Shipping Inc. Sub Charterers approval by the Owners. Please advise of the Charterers' full background, where they are based and recent fixtures made by them, for the Owners' approval.

2.

All subs will be lifted upon fully fixing with Charter Party details in order.

3.

Delivery on dlosps Madras port East Coast India atdnsshinc

or

Delivery on afsps 1sp within the East Coast India atdnsshinc. Port of delivery at chopt.

or

Delivery on afsps .... (insert the name of the port) port in the East Coast India atdnsshinc.

or.

Delivery on Passing Tuticorin port northbound in East Coast India atdnsshinc

4.

l/c 10/15 Dec 2.... 00:01/24:00 hrs. - for l/c the local time will be in effect.

5.

Time Charter Trip out (Or 'One Time Charter trip') via East Coast India to China, via safe port/s, safe berth/s, always within I.N.L. with cargo of bulk agricultural products.

The port/s berth/s will remain safe from the time of the ship's arrival at the first sea pilot station until departure and the last sea pilot has left the ship.

The ship will always lie afloat throughout the period of the Charter.

6.

Charterers guarantee a minimum duration of 30 days to about 35 days.

7.

Redelivery on dlosps 1sp full China atdnsshinc. Redelivery port at chopt

or

Redelivery on dlosps 1safe port not northern of Shanghai port or passing southbound Shanghai port atdnsshinc. Redelivey port at the Charterers' option.

or

Redelivery on dlosps .... port China atdnsshinc.

8.

Hire USD 25,000 pdpr inclot payable every 15 days always in advance net of tax and bank charges.

For hire calculation purposes the UTC time will be in effect.

9.

Charterers guarantee that the ship will load only from the East Coast India and will discharge only in China.

or

The ship shall not trade/load/discharge in ....

10.

The ship will load only bulk agricultural products and no other cargo will be loaded on board the ship. The Charterers confirm that the cargo is harmless, non-hazardous and belongs to group C of the International Maritime Solid Bulk Cargoes Code (abbreviated as 'IMSBC Code')

or

The ship is not allowed to load the following cargoes: ...

11.

The ship will be delivered with about .... mtns VLSFO and about .... mtns LSMGO. The ship will be redelivered with the same quality of fuel and with about the same quantity of fuel as on delivery.

The value of the fuel on delivery is payable together with the first 15-day hire payment.

The fuel prices on delivery and redelivery to be mutually agreed on. Same prices will be in effect on delivery and redelivery.

The Master's or Owners' information of the quantity of fuel that is required to be supplied in order to execute the trip or for redelivery purposes, is subject to trip's final duration and/or ship's consumption while manoeuvring in ports and/or confined waters and subject to final destination.

12.

The Charterers, for the whole duration of this Charter period will compensate the

Owners with a lump sum amount of USD .... per month or pro rata for any part of the month for communications/ victuals/entertainment. The respective monthly amount is payable to the Owners together with the hire payments.

or

12.

The Charterers will compensate the Owners with a lump sum amount of USD .... for the entire duration of the Charter for communications/victuals/entertainment. The respective amount is payable together with the first hire payment.

or

12.

C/V/E is payable by the Charterers at cost at the end of the Charter and in accordance with the presented statement/bills or the Master's statement of expenses incurred on the Charterers' behalf.

For a period: *C/V/E is payable by the Charterers at cost at the end of each voyage.*

13.

The ship will be redelivered to the Owners with its holds in the same condition as

on delivery, normal wear and tear excepted.
However, the Charterers will have the option of redelivering the ship with unclean holds, by paying a lumpsum amount of USD .... The amount due will be paid together with the Charterers' final hire statement.
or
However, the Charterers will have the option of redelivering the ship with unclean holds, by paying a lumpsum amount of USD .... per hold used for loading. The amount due will be paid together with the Charterers' final hire statement.
The ship in any case will be redelivered to the Owners free of cargo residues in its cargo holds.
Debris and/or dunnage and/or lashing or any other materials that have been used to secure and/or protect the cargo during its sea voyage will be discharged from the ship before its redelivery, at the Charterers' entire time, risk and expenses. The ship will not sail from the port unless all such materials have been removed and discharged from the ship and any delays, expenses, consequences and/or possible fines imposed by the Port Authorities, arising from the Charterers' inappropriate discharge of the materials or inappropriate delivery ashore, will be for the Charterers' account and the ship will remain always on-hire.
In the case that the materials are not allowed to be discharged at the last port of discharge, then, before the ship's redelivery, the ship will proceed to the most convenient anchorage, berth of another port or country in order to discharge the same in barges or ashore, under the Charterers' time, risk and expenses. The ship will remain always on-hire, until having fully discharged the respective materials and the ship is redelivered on dlosps or at the Owners' option at equidistance from the actual point of redelivery.
14.
Arbitration in London. English law will apply to negotiations and to the eventual agreement.
Sub to agree to a full clause.
15.
Sub further terms/details which will be based on NYPE 46 C/P at the Charterers' executed proforma with 3,75 percent total at your end including 2,5 percent address commission.

CHARTERERS' COUNTER REPLY TO OWNERS' LAST:

The Charterers:
either
- will agree to work on the Owners' offer, making a clean acceptance or counter offer on an accept/except basis to the terms on which they do not agree
or

- will decline the Owners' offer and forward a full re-offer for the Owners' acceptance or counter reply.
On the basis that the Charterers have accepted to work on the received offer, a counter offer on an Accept/Except basis could be given as follows (terms not touched, have been fully accepted by the Charterers):

Ref MV FOTINI
Account Eagle Trading and Shipping Inc
East coast India/China

Thanks for the Owners' offer on which the Charterers' counter offer on an A/E basis, for reply latest at 14:00 hrs. London time today, as follows:
or
Charterers a/e for 30 mnts reply as follows:
(or by any other form stating that the Charterers' counter offer is on an accept/except basis to the Owners' offer and that they are committed on the ship for a specific time.)

Comments:
The Owners should respond
-timely within the period that is allowed by the Charterers,
-or will ask for an extension, as they need some additional time to reply

The Charterers continue:
-MV FOTINI
As described by the Owners.
-Please advise of cycle speed of cranes (hoisting speed-lowering speed-luffing time-slewing speed).
-Please advise of outreach distance of cranes.

Comments:
The ship is geared, and the Charterers seek some additional information on the operation of the cranes.
The technical department of the Owners will take up the matter and will inform the parties accordingly.

The Charterers continue:
-Please advise of the full Charter Party chain including the name/style/contact numbers of the Head Owners/Managers and Disponent Owners.
-Please advise of contact details of the ship and the Master's name.
-Please advise of ship's Constants/FW

Comments:
The Charterers request:
-the full Charter Party chain to be advised, i.e., all parties involved should be revealed along with the Owners' next counter offer, including the name/style and contact numbers of the head Owners/ Managers and possible Disponent Owners.
-the ship's contact details and the Master's name should be stated enabling the Charterers to contact him after conclusion of the fixture.
-the Constants (the standard weights on board the ship) and the capacity of the Fresh Water, which, along with the fuel on board, will determine the ship's 'cargo capacity', i.e., the actual loadable quantity on delivery, for the Charterers to calculate the same themselves whenever it is required.

The Charterers continue:
Owners guarantee that during the currency of this Charter:
a- the ship will not change ownership and/or class and will not be sold for scrap during this Charter without the Charterers' written consent.
b- the ship is fully covered by the P&I Club (as a member of an International Group of P&I Clubs) and has been covered for ship oil pollution risks and wreck removal risks and will remain so throughout the duration of this charter. In this respect, the Owners will furnish the Charterers with the P&I Club's written confirmation, if and whenever it is requested by them.
c- the ship is fully classed as Lloyd's 100A1 or equivalent with a classification society being a full member of the International Association of Classification Societies (IACS) throughout the duration of this charter. In this respect, the Owners will furnish the Charterers with the Classification society's written confirmation, if and whenever it is requested by them.
d- the ship has no outstanding recommendations, restrictions or any other requirements of any nature in respect to flag state, port state control, or classification inspections that may affect this voyage and the Charter Party.
e- the ship's H&M insurance will be fully maintained and will not be changed. In this respect, the Owners will furnish the Charterers with the H&M underwriters' written confirmation, if and whenever it is requested by them.
f- the ship will comply with all applicable laws and regulations and is in all respects eligible for trading to the ports, places or countries allowed in this Charter Party. The ship has all valid and up-to-date certificates and other documents on board required for this Charter.
g- the ship is fully fitted for the loading of grains according to the SOLAS 1974 regulations and any subsequent amendments thereto without requiring any shifting boards. The ship can open/close its hatches with its own means.

h- the ship and the company are fully in accordance with ISPS/ISM requirements and will remain so throughout the duration of this charter.
i- the ship is sid/bc, steelfloored throughout, with hoppers port/starboard, with the engine/bridge aft and its holds/hatches/tank tops are clear and unobstructed without a centreline bulkhead, walkways, centrebeams and free of all/any obstructions which might incur difficult access to the allowed cargoes during the loading/discharging. The cargo will not be loaded in deep tanks, sidetanks, wing tanks, bunker spaces, or any other confined spaces unreachable to the means of loading and discharging.
j- the ship has not carried mammalian meat, bone meal, fishmeal, asbestos cullet, or nuts for the past 12 months.
k- the ship has not been blacklisted/boycotted or arrested due to the ship's flag/ ownership/crew employment/age/past trading or for any other reason, from local authorities and or unions in all allowed countries under this Charter Party.
l- the ship is free of all encumbrances, arrests, and maritime liens.
m-the ship has no .... interests (insert countries which may affect the voyage) and is not black listed by Arab league countries.
n- the description of the ship as provided by the Owners is an accurate representation of the characteristics of the ship which can only be modified with the usual and acceptable shipping interpretation of the word 'about'. This clause will override any inclusion of the words 'without guarantee' in the ship's description. The Owners are responsible and liable for any consequences/ delays/loss/expenses that may arise when providing any misleading information on the ship's characteristics and on the terms agreed in this Charter Party.
o- that the ship is ITF's compliant or bona fide trade union's agreement or equivalent and shall remain so throughout the duration of this Charter and will remain responsible for all consequences and/or delays if the ship is sailing under a flag of 'convenience'.
p- the ship will, throughout the duration of the Charter Party, have on board and make available to the Charterers or their agents or to the appointed surveyors, the capacity plan, the deadweight and displacement scales, the hydrostatic curves and tables of displacement, the calibration scales for all ship's tanks and trimming correction tables as well as all sounding tables and all sounding tubes which will be properly maintained, in good condition and free from impediments. Plimsoll marks amidships and draft marks on port and starboard side, bow, amid ship and stern are always to be clearly cut/marked on shell platting and readable.

Comments:
The Owners should guarantee certain important points which are necessary for making the ship workable for the shipment. In terms which the ship does not

comply with, the Owners should clearly indicate to the contrary, otherwise, in the case that the ship is fixed, claims could arise against them with huge consequences.

The Charterers continue:
For:
1. - Account Eagle Trading and Shipping Inc. of British Virgin Islands
(Charterers advise their background)
Recent fixtures:
MV .... Fixed through .... Ph. ...... Owners/brokers ph ....
MV .... Fixed through .... Ph. .... Owners/brokers ph............
etc. etc.

Comments:
The Charterers provide their full background/full name etc. Some recent fixtures have also been provided along with the involved brokers that will allow the Owners to make a quick search of the Charterers' reliability/performance.

The Charterers continue:
2. Sub shippers'/receivers' approval and stem will be declarable latest 1 working day after the fixing of main terms.

Comments:
The Charterers, in comparison to the Owners' offer:
-reject the Owners' proposal as to the time of lifting of subjects (The Owners' offer was that all subs should be lifted upon fully fixing with Charter Party details in order)
The fixture is subject:
- to readiness of the cargo (stem= subject to enough merchandise)
- to the Shippers' and Receivers' approval to the ship's characteristics/Owners etc.
Both will be declarable within 1 working day after completion of negotiations on the Main Terms.
However, if a ship is in a spot 'waiting for orders' and the biz is not so attractive, it is risky for the Owners to accept it. If it is a Friday, the Owners should wait until the following Monday and if the subs eventually fail, then they will have lost another three days. Some Owners, depending on the market, could drop the negotiations, letting the ship sail empty of cargo to a better area for loading, or even commence working another biz which is even worst but with quicker 'subs'.

The Charterers continue:
5.
Time Charter Trip out (Or 'One Time Charter trip') via East Coast India to China, via safe port/s, safe berth/s, safe anchorages always within I.N.L with cargo of bulk agricultural products.
The port/s, berth/s, anchorage/s will remain safe from the time of the ship's arrival at the first sea pilot station until departure and the last sea pilot has left the ship.

Comments:
The Charterers, in comparison to the Owners' offer:
-accept the ship to lie always afloat and to trade through safe ports/berths however,
-they add that the ship will also trade through 'safe anchorages'.
The Charterers have the right to operate the ship either for loading or discharging or take bunkers through 'safe anchorage/s' where the ship will obviously not be exposed to risk. However, the Owners should protect their interests and should agree to a protective clause for the ship's operation in the anchorages.

The Charterers continue:
6.
Duration about 25-35 days wog.

Comments:
The Charterers, in comparison to the Owners' offer:
-reject to guarantee any minimum duration (The Owners' offer was for the Charterers to guarantee a minimum duration of 30 days to about 35 days).
-they counter reply that the duration should be about 25-35 days without guarantee.
The Charterers will not be kept liable for the ship's possible earlier or later redelivery.

The Charterers continue:
8.
Hire USD 22,000 pd/pr/inclot and lubricants.

Comments:
The Charterers, in comparison to the Owners' offer:
-accept the Owners' last except the 'hire rate' which should be at USD 22,000 (the Owners' offer was at USD 25,000). However, they add that the cost of lubricants will be consumed during the charter should be included in the hire.

The Charterers continue:

11.

Any minor difference on the bunkers' value on delivery and redelivery will be settled together with the final settlement with the exact figures.

The Charterers have the right to deduct the value of the bunkers on redelivery, from the last 15-day hire payment, but if it is not sufficient it will be extended until the Charterers deduct the full value of the bunkers.

Prices for VLSFO USD ...... and for LSMGO USD ......to be in effect for delivery and redelivery.

Comments:

The Charterers, in comparison to the Owners' offer:

-accept the delivery fuel quantity as given by the Owners.

-accept to redeliver the ship with bunkers with the same qualities and with about the same quantities.

-accept to pay the delivery bunkers along with the payment of the first hire.

-accept same bunker prices to be in effect for delivery/redelivery.

-accept the Owners reservations when they declare the bunkers required to be supplied.

-add that the minor differences on the bunkers' value on delivery/redelivery are to be settled together with the final settlement with exact figures.

-ask to have the right to deduct the value of the bunkers on redelivery from the last sufficient hire/s -they provide bunker prices for the Owners' consideration.

The Charterers continue:

12.

This payment includes incidental expenses made on the Charterers' behalf, such as cigarettes, drinks and petty expenses incurred by the Master, Officers to the Charterers' representatives and/or pilots and/or custom officers and/or foremen etc. It also includes the cost of radio telegrams, telexes, fax, and telephone communications made by the Master, Officers to the Charterers or their Agents or servants, in direct performance of this Charter Party.

Meals to tally clerks and/or stevedores and/or foremen etc. as well other expenses which have been previously approved by the Charterers and incurred by the Master, Officers on the Charterers' behalf, will be paid at cost and on presentation even by fax of the supporting vouchers and/or the Master's statement of expenses.

Comments:

The Charterers, in comparison to the Owners' offer:

-specifically, state what the agreed cost for c/v/e includes.
-request the respective meals to tally clerks/stevedores/foremen to be paid at cost and on presentation of supporting vouchers.

The Charterers continue:
15.
Sub further terms/details on NYPE 46 C/P at Charterers' executed proforma based on MV.... C/P dated ...

Comments:
The Charterers, in comparison to the Owners' offer:
-accept to work an executed Charter Party proforma which should be based on MV... with Charter Party dated....

<u>TO SUMMARISE THE CHARTERERS' COUNTER OFFER:</u>

Ref MV FOTINI
Account Eagle Trading and Shipping Inc
East coast India/China

Thanks for the Owners' offer on which the Charterers' counter offer on an A/E basis, for reply latest at 14:00 hrs. London time today, as follows:
or
Charterers a/e for 30 mnts reply as follows:
-MV FOTINI- As described by the Owners.

-Please advise of cycle speed of cranes (hoisting speed-lowering speed-luffing time-slewing speed).
-Please advise of outreach distance of cranes.
-Please advise of the full Charter Party chain including the name/style/contact numbers of the Head Owners/Managers and Disponent Owners.
-Please advise of contact details of the ship and the Master's name
-Please advise of ship's Constants/FW.
Owners guarantee that during the currency of this Charter:
a- the ship will not change ownership and/or class and will not be sold for scrap during this Charter without the Charterers' written consent.
b- the ship is fully covered by the P&I Club (as a member of an International Group of P&I Clubs) and has been covered for ship oil pollution risks and wreck removal risks and will remain so throughout the duration of this charter. In this respect, the Owners will furnish the Charterers with the P&I Club's written confirmation, if and whenever it is requested by them.

c- the ship is fully classed as Lloyd's 100A1 or equivalent with a classification society being a full member of the International Association of Classification Societies (IACS) throughout the duration of this charter. In this respect, the Owners will furnish the Charterers with the Classification society's written confirmation, if and whenever it is requested by them.
d- the ship has no outstanding recommendations, restrictions or any other requirements of any nature in respect to flag state, port state control, or classification inspections that may affect this voyage and the Charter Party.
e- the ship's H&M insurance will be fully maintained and will not be changed. In this respect, the Owners will furnish the Charterers with the H&M underwriters' written confirmation, if and whenever it is requested by them.
f- the ship will comply with all applicable laws and regulations and is in all respects eligible for trading to the ports, places or countries allowed in this Charter Party. The ship has all valid and up-to-date certificates and other documents on board required for this Charter.
g- the ship is fully fitted for the loading of grains according to the SOLAS 1974 regulations and any subsequent amendments thereto without requiring any shifting boards. The ship can open/close its hatches with its own means.
h- the ship and the company are fully in accordance with ISPS/ISM requirements and will remain so throughout the duration of this charter.
i- the ship is sid/bc, steelfloored throughout, with hoppers port/starboard, with the engine/bridge aft and its holds/hatches/tank tops are clear and unobstructed without a centreline bulkhead, walkways, centrebeams and free of all/any obstructions which might incur difficult access to the allowed cargoes during the loading/discharging. The cargo will not be loaded in deep tanks, sidetanks, wing tanks, bunker spaces, or any other confined spaces unreachable to the means of loading and discharging.
j- the ship has not carried mammalian meat, bone meal, fishmeal, asbestos cullet or nuts for the past 12 months.
k- the ship has not been blacklisted/boycotted or arrested due to the ship's flag/ownership/crew employment/age/past trading or for any other reason, from local authorities and or unions in all allowed countries under this Charter Party.
l- the ship is free of all encumbrances, arrests, and maritime liens.
m- the ship has no .... interests (insert countries which may affect the voyage) and is not black listed by Arab league countries.
n- the description of the ship as provided by the Owners is an accurate representation of the characteristics of the ship which can only be modified with the usual and acceptable shipping interpretation of the word 'about'. This clause will override any inclusion of the words 'without guarantee' in the ship's description. The Owners are responsible and liable for any consequences/ delays/loss/expenses that may arise when providing any misleading information on the ship's

characteristics and on the terms agreed in this Charter Party.
o- that the ship is ITF's compliant or bona fide trade union's agreement or equivalent and shall remain so throughout the duration of this Charter and will remain responsible for all consequences and/or delays if the ship is sailing under a flag of 'convenience'.
p- the ship will, throughout the duration of the Charter Party, have on board and make available to the Charterers or their agents or to the appointed surveyors, the capacity plan, the deadweight and displacement scales, the hydrostatic curves and tables of displacement, the calibration scales for all ship's tanks and trimming correction tables as well as all sounding tables and all sounding tubes which will be properly maintained, in good condition and free from impediments. Plimsoll marks amidships and draft marks on port and starboard side, bow, amid ship and stern are always to be clearly cut/marked on shell platting and readable.

For:
1. - Account Eagle Trading and Shipping Inc. of British Virgin Islands
(Charterers advise their background)
Recent fixtures:
MV .... Fixed through .... Ph ...... Owners/brokers ph. ....
MV .... Fixed through .... Ph .... Owners/brokers ph.............
etc. etc.
2. Sub shippers'/receivers' approval and stem will be declarable at the latest 1 working day after the fixing of main terms.
5.
Time Charter Trip out (Or 'One Time Charter trip') via East Coast India to China, via safe port/s, safe berth/s, safe anchorages always within I.N.L. with cargo of bulk agricultural products. The port/s, berth/s, anchorage/s will remain safe from the time of the ship's arrival at the first sea pilot station until departure and the last sea pilot has left the ship.
6.
Duration about 25-35 days wog.
8.
Hire USD 22,000 pd/pr/inclot and lubricants.
11.
Any minor difference on the bunkers' value on delivery and redelivery will be settled together with the final settlement with the exact figures.
The Charterers have the right to deduct the value of the bunkers on redelivery, from the last 15-day hire payment, but if it is not sufficient it will be extended until the Charterers deduct the full value of the bunkers.
Prices for VLSFO USD ...... and for LSMGO USD ......to be in effect for delivery and redelivery.

12.
This payment includes incidental expenses made on the Charterers' behalf, such as cigarettes, drinks and petty expenses incurred by the Master, Officers to the Charterers' representatives and/or pilots and/or custom officers and/or foremen etc. It also includes the cost of radio telegrams, telexes, fax and telephone communications made by the Master,Officers to the Charterers or their Agents or servants, in direct performance of this Charter Party.
Meals to tally clerks and/or stevedores and/or foremen etc. as well other expenses which have been previously approved by the Charterers and incurred by the Master, Officers on the Charterers' behalf, will be paid at cost and on presentation even by fax of the supporting vouchers and/or the Master's statement of expenses.
15.
Sub further terms/details on NYPE 46 C/P at the Charterers' executed proforma based on MV.... C/P dated ....

OWNERS' COUNTER REPLY TO CHARTERERS' LAST:

The Owners could reply as follows, with untouched terms to have been accepted by them.

Thanks Charterers' last. Owners A/E for reply within 30 minutes as follows:
MV FOTINI
Cranes: ....

Comments:
The required information about the operation of the cranes was given to the Charterers.

The Owners continue:
Owners: ....
Managers: ....
ship's contact details....
Master's name: ....
Constants: ...... mtns
Fw: .... mtns

Comments:
The Owners reply to the Charterers' request.
The ship is under the employment control of the head Owners/Managers, therefore 'disponent Owners' is not applicable.

The Owners continue:

5.

If anchorage/s is used, the Bimco's current 'Ship to Ship Transfer Clause for Time Charter Parties' will be in effect. Sub to agree to a full clause.

Comments:

The Owners, in comparison to the Charterers' offer:

-accept the ship to trade via safe anchorages, however, the Bimco's current 'Ship to Ship Transfer Clause for Time Charter Parties' to apply. The ship will be protected from possible damage that may occur during the operation in the anchorage.

-request to reserve the right to propose a full clause (additional terms except for the Bimco's ones) which should obviously be discussed during the work on the Charter Party details.

The Owners continue:

6.

Charterers guarantee a minimum duration of 30 days, to about 35 days.

The Owners, in comparison to the Charterers' offer:

Comments:

-Owners reject the Charterers' counter offer and repeat their last term about the duration.

The Owners continue:

8.

Hire rate USD 24,500 pd/pr.

Comments:

The Owners, in comparison to the Charterers' offer:

-request as 'hire rate' USD 24,500 pd/pr/inclot (the Charterers' offer was for USD 22,000).

The Owners continue:

15.

Please fax the Charterers' executed Charter Party proforma for the Owners' possible comments.

Comments:

The Owners, in comparison to the Charterers' offer:

-request the Charterers' proforma to be sent for study and possible comments.

TO SUMMARISE THE OWNERS' COUNTER OFFER:

Thanks Charterers' last. Owners A/E for reply within 30 minutes as follows:
MV FOTINI
Cranes: ....
....
The Owners continue:
Owners: ....
Managers: ....
ship's contact details....
Master's name: ....
Constants: ...... mtns
Fw: .... mtns
5.
If anchorage/s is used, the Bimco's current 'Ship to Ship Transfer Clause for Time Charter Parties' will be in effect. Sub to agree to a full clause.
6.
Charterers guarantee a minimum duration of 30 days, to about 35 days.
8.
Hire rate USD 24,500 pd/pt
15.
Please fax the Charterers' executed Charter Party proforma for the Owners' possible comments.

CHARTERERS' COUNTER REPLY TO OWNERS' LAST:

Charterers A/E for reply within 30 minutes as follows:
8.
Hire USD 23,500. -

The Charterers, in comparison to the Owners' offer:
-request for a 'hire rate' of USD 23,500 (the Owners' offer was at USD 24,500-).

6. Charterers guarantee a minimum duration of 30 days to about 35 days but wog.

The Charterers, in comparison to the Owners' offer:
-guarantee only the minimum duration which could obviously satisfy the Owners, as they secure a minimum period for which the ship will be on Charter with

them.
-insist for the maximum duration to remain without guarantee.

OWNERS' COUNTER REPLY TO CHARTERERS' LAST:

Owners A/E for reply within 15 minutes as follows:
8. Hire USD 24,000. -

Comments:
The Owners, in comparison to the Charterers' offer:
-request as 'hire rate' USD 24,000 (the Charterers' offer was at USD 23,500. -).
The Owners need faster negotiations, as only one point remains pending.

CHARTERERS' COUNTER REPLY TO OWNERS' LAST:

Charterers A/E for reply within 15 minutes as follows.
8. Hire USD 23,750. -

Comments:
The Charterers, in comparison to the Owners' offer:
-request as hire rate USD 23,750 (the Owners' offer was at USD 24,000. -).

OWNERS' COUNTER REPLY TO CHARTERERS' LAST:

The Owners thank the Charterers for their last, which is accepted and will be reverting shortly with possible comments on Charter Party details. Please make a recapitulation (abbreviated as 'recap') *of Main Terms that have been agreed so far.*

Comments:
The Owners have accepted the Charterers last, therefore the ship is fixed on the Main Terms. The Charterers' broker will make the recapitulation which will be given to both parties as their guide on the terms that have been agreed. However, when complicated terms have been agreed, then will be given for their reconfirmation on the agreed terms.

THE RECAP COULD BE GIVEN AS FOLLOWS:

Ref Account Eagle Trading and Shipping Inc
Handymax dwt 45/55,000 mtns
Dely East Coast India – Redelivery Far East (intention: dop 1sp China chopt)

cargo bulk agriprds
MV FOTINI

Recap of main terms agreed:
MV FOTINI
....
....
Cranes: ....
....
Owners: ....
Managers: ....
ship's contact details....
Master's name: ....
Constants: ...... mtns
Fw: .... mtns

Owners guarantee that during the currency of this Charter:
a- the ship will not change ownership and/or class and will not be sold for scrap during this Charter without the Charterers' written consent.
b- the ship is fully covered by the P&I Club (as a member of an International Group of P&I Clubs) and has been covered for ship oil pollution risks and wreck removal risks and will remain so throughout the duration of this charter. In this respect, the Owners will furnish the Charterers with the P&I Club's written confirmation, if and whenever it is requested by them.
c- the ship is fully classed as Lloyd's 100A1 or equivalent with a classification society being a full member of the International Association of Classification Societies (IACS) throughout the duration of this charter. In this respect, the Owners will furnish the Charterers with the Classification society's written confirmation, if and whenever it is requested by them.
d- the ship has no outstanding recommendations, restrictions or any other requirements of any nature in respect to flag state, port state control, or classification inspections that may affect this voyage and the Charter Party.
e- the ship's H&M insurance will be fully maintained and will not be changed. In this respect, the Owners will furnish the Charterers with the H&M underwriters' written confirmation, if and whenever it is requested by them.
f- the ship will comply with all applicable laws and regulations and is in all respects eligible for trading to the ports, places or countries allowed in this Charter Party. The ship has all valid and up-to-date certificates and other documents on board required for this Charter.
g- the ship is fully fitted for the loading of grains according to the SOLAS 1974

regulations and any subsequent amendments thereto without requiring any shifting boards. The ship can open/close its hatches with its own means.
h- the ship and the company are fully in accordance with ISPS/ISM requirements and will remain so throughout the duration of this charter.
i- the ship is sid/bc, steelfloored throughout, with hoppers port/starboard, with the engine/bridge aft and its holds/hatches/tank tops are clear and unobstructed without a centreline bulkhead, walkways, centrebeams and free of all/any obstructions which might incur difficult access to the allowed cargoes during the loading/discharging. The cargo will not be loaded in deep tanks, sidetanks, wing tanks, bunker spaces, or any other confined spaces unreachable to the means of loading and discharging.
j- the ship has not carried mammalian meat, bone meal, fishmeal, asbestos cullet or nuts for the past 12 months.
k- the ship has not been blacklisted/boycotted or arrested due to the ship's flag/ownership/crew employment/age/past trading or for any other reason, from local authorities and or unions in all allowed countries under this Charter Party.
l- the ship is free of all encumbrances, arrests, and maritime liens.
m- the ship has no .... interests (insert countries which may affect the voyage) and is not black listed by Arab league countries.
n- the description of the ship as provided by the Owners is an accurate representation of the characteristics of the ship which can only be modified with the usual and acceptable shipping interpretation of the word 'about'. This clause will override any inclusion of the words 'without guarantee' in the ship's description. The Owners are responsible and liable for any consequences/ delays/loss/expenses that may arise when providing any misleading information on the ship's characteristics and on the terms agreed in this Charter Party.
o- that the ship is ITF's compliant or bona fide trade union's agreement or equivalent and shall remain so throughout the duration of this Charter and will remain responsible for all consequences and/or delays if the ship is sailing under a flag of 'convenience'.
p- the ship will, throughout the duration of the Charter Party, have on board and make available to the Charterers or their agents or to the appointed surveyors, the capacity plan, the deadweight and displacement scales, the hydrostatic curves and tables of displacement, the calibration scales for all ship's tanks and trimming correction tables as well as all sounding tables and all sounding tubes which will be properly maintained, in good condition and free from impediments. Plimsoll marks amidships and draft marks on port and starboard side, bow, amid ship and stern are always to be clearly cut/marked on shell platting and readable.

For:
1. - Account Eagle Trading and Shipping Inc. of British Virgin Islands

(Charterers advise their background)
Recent fixtures:
MV .... Fixed through .... Ph. ...... Owners/brokers ph. ....
MV .... Fixed through .... Ph. .... Owners/brokers ph. ....
Etc. etc.
2. Sub shippers'/receivers' approval and stem will be declarable at the latest 1 working day after the fixing of main terms.
3.
Delivery on dlosps Madras port East Coast India atdnsshinc
or
Delivery on afsps 1sp within the East Coast India atdnsshinc. Port of delivery at chopt.
or
Delivery on afsps .... (insert the name of the port) port in the East Coast India atdnsshinc.
or
Delivery on Passing Tuticorin port northbound in East Coast India atdnsshinc
4.
l/c 10/15 Dec 2.... 00:01/24:00 hrs. - for l/c the local time will be in effect.
5.
Time Charter Trip out (Or 'One Time Charter trip') via East Coast India to China, via safe port/s, safe berth/s, safe anchorages always within I.N.L. with cargo of bulk agricultural products.
The port/s, berth/s, anchorage/s will remain safe from the time of the ship's arrival at the first sea pilot station until departure and the last sea pilot has left the ship.
If anchorage/s are used, the Bimco's current 'Ship to Ship Transfer Clause for Time Charter Parties' will be in effect. Sub to agree to a full clause.
The ship will always lie afloat throughout the period of the Charter.
6.
Charterers guarantee a minimum duration of 30 days to about 35 days but wog.
7.
Redelivery on dlosps 1sp full China atdnsshinc. Redelivery port at chopt
or
Redelivery on dlosps 1safe port not northern of Shanghai port or passing southbound Shanghai port atdnsshinc. Redelivey port at the Charterers' option.
or
Redelivery on dlosps ...... port China atdnsshinc.
8.
Hire USD 23,750 pdpr inclot and lubricants, payable every 15 days always in advance net of tax and bank charges.

For hire calculation purposes the UTC time will be in effect.

9.

Charterers guarantee that the ship will load only from the East Coast India and will discharge only in China.

Or

The ship shall not trade/load/discharge in ....

10.

The ship will load only bulk agricultural products and no other cargo will be loaded on board the ship. The Charterers confirm that the cargo is harmless, non-hazardous and belongs to group C of the International Maritime Solid Bulk Cargoes Code (abbreviated as 'IMSBC Code')

or

The ship is not allowed to load the following cargoes: ...

11.

The ship will be delivered with about .... mtns VLSFO and about .... mtns LSMGO. The ship will be redelivered with the same quality of fuel and with about the same quantity of fuel as on delivery.

Any minor difference on the bunkers' value on delivery and redelivery will be settled together with the final settlement with the exact figures.

The value of the fuel on delivery is payable together with the first 15-day hire payment.

Prices for VLSFO USD ...... and for LSMGO USD ......to be in effect for delivery and redelivery.

The Charterers have the right to deduct the value of the bunkers on redelivery, from the last 15-day hire payment, but if it is not sufficient it will be extended until the Charterers deduct the full value of the bunkers.

The Master's or Owners' information of the quantity of fuel that is required to be supplied in order to execute the trip or for redelivery purposes, is subject to trip's final duration and/or ship's consumption while manoeuvring in ports and/or confined waters and subject to final destination.

12.

The Charterers, for the whole duration of this Charter period will compensate the Owners with a lump sum amount of USD .... per month or pro rata for any part of the month for communications/ victuals/entertainment. The respective monthly amount is payable to the Owners together with the hire payments.

or

The Charterers will compensate the Owners with a lump sum amount of USD .... for the entire duration of the Charter for communications/victuals/ entertainment.

The respective amount is payable together with the first hire payment.

or

C/V/E is payable by the Charterers at cost at the end of the Charter and in accordance with the presented statement/bills or the Master's statement of expenses incurred on the Charterers' behalf.
This payment includes incidental expenses made on the Charterers' behalf, such as cigarettes, drinks and petty expenses incurred by the Master, Officers to the Charterers' representatives and/or pilots and/or custom officers and/or foremen etc. It also includes the cost of radio telegrams, telexes, fax and telephone communications made by the Master, Officers to the Charterers or their Agents or servants, in direct performance of this Charter Party.
Meals to tally clerks and/or stevedores and/or foremen etc. as well other expenses which have been previously approved by the Charterers and incurred by the Master, Officers on the Charterers' behalf, will be paid at cost and on presentation even by fax of the supporting vouchers and/or the Master's statement of expenses.
13.
The ship will be redelivered to the Owners with its holds in the same condition as on delivery, normal wear and tear excepted.
However, the Charterers will have the option of redelivering the ship with unclean holds, by paying a lumpsum amount of USD .... The amount due will be paid together with the Charterers' final hire statement.
or
However, the Charterers will have the option of redelivering the ship with unclean holds, by paying a lumpsum amount of USD .... per used hold. The amount due will be paid together with the Charterers' final hire statement.
The ship in any case will be redelivered to the Owners free of cargo residues in its cargo holds.
Debris and/or dunnage and/or lashing or any other materials that have been used to secure and/or protect the cargo during its sea voyage will be discharged from the ship before its redelivery, at the Charterers' entire time, risk and expenses. The ship will not sail from the port unless all such materials have been removed and discharged from the ship and any delays, expenses, consequences and/or possible fines imposed by the Port Authorities, arising from the Charterers' inappropriate discharge of the materials or inappropriate delivery ashore, will be for the Charterers' account and the ship will remain always on-hire.
In the case that the materials are not allowed to be discharged at the last port of discharge, then, before the ship's redelivery, the ship will proceed to the most convenient anchorage, berth of another port or country in order to discharge the same in barges or ashore, under the Charterers' time, risk and expenses. The ship will remain always on-hire, until having fully discharged the respective materials and the ship is redelivered on dlosps or at the Owners' option at equidistance from the actual point of redelivery.

14.
Arbitration in London. English law will apply to negotiations and to the eventual agreement.
Sub to agree to a full clause.
15.
Sub further terms/details on NYPE 46 C/P at Charterers' executed proforma based on MV.... C/P dated .... with 3,75 percent total commission including 2,5 percent address commission.

2ND OFFER.

The order circulated in the market as follows:
Account Messrs. Eagle Trading and Shipping Inc of British Virgin Islands
Panamax OA OK to make Paranagua 1/10 January 2....
One tct via ECSA with bulk corn
Dlosp Skaw/Gib, chopt Passero/EMed
NYPE C/P 46 with 5 pct ttl com

From the data given:
-a Panamax ship is required without any age restriction (oa ok).
-the ship will load from the East Coast South America, which consists of the countries of Brazil /Argentina/Uruguay. The delivery could take place in the area of loading or elsewhere but provided the ship can reach the Paranagua port within the lay-days required for loading.
-the redelivery will take place at a port located between the Skaw and Gibraltar range. Skaw is a cape that belongs to Denmark and is the country's northernmost point. The rock of Gibraltar is located at the entrance to the Mediterranean Sea.
-the Charterers keep the option of the ship discharging between the range of Passero, which is a cape located in Sicily, Italy and the Eastern Mediterranean and be redelivered to a port after dropping the last outward sea pilot station.
-the cargo is bulk corn.
-Nype Charter Party will be used with 5 percent total as brokerage commission.

Taking into consideration the fact that:
-the name of the Charterers is 'Messrs. 'Eagle Trading and Shipping Inc.' and that they are well known in the market as 'first class'.
-the interested ship is the 'MV FOTINI', which is a Panamax ship open in Antwerp on mid-December.
Then the Owners' offer could be given as follows:
Ref Account Messrs. Eagle Trading and Shipping Inc

Pnmx to make Paranagua 1/10 Jan.
Tct via ECSA to Skaw/Gib, chopt Passero/EMed,
MV FOTINI

The Owners firm bid for 2 hours reply as follows:

Comments:
The ship will be 'committed' with the particular Charterers, for the particular biz, for a maximum period of two hours from the time which it was sent by the Owners or by the Owners' brokers.

The Owners continue:
MV FOTINI
....
....

Comments:
The ship's description was given, as the case maybe and it is applicable when working on a single trip.

The Owners continue:
For:
1.
Account Messrs. Eagle Trading and Shipping Inc. of British Virgin Islands
Sub to receive Charterers' last three fixtures.

Comments:
The Owners have offered:
-that even though the Charterers are known in the market, they must advise of their last three fixtures, enabling the Owners to check and re-confirm the Charterers' reliability/performance and that nothing has changed with them.

The Owners continue:
2.
All subs will be lifted upon fixing the main terms.

Comments:
The Owners have offered that all subjects will be lifted when the Main Terms are concluded and before working on the Charter Party details. The Charterers should reply if it is feasible, or otherwise, they will revert with the time required

to lift the respective subjects.

The Owners continue:
3.
Delivery on dlosps Antwerp port, Belgium atdnsshinc.

Comments:
The Owners have offered:
-that the ship will be delivered on 'Dropping Last Outward Sea Pilot Station' at the port of Antwerp. The Charterers, upon delivery, will direct the ship 'on-hire' to the loading area and the consumed fuel/delays due to stress from weather will be for their account.

The Owners continue:
4.
14/19 December 2....
00:01-2400 hrs. For lay-can the local time to be in effect.

Comments:
The Owners have offered:
- the dates within which the ship can be ready for delivery.
The distance from Antwerp to the Paranagua port in Brazil is about 5,544 Nautical Miles and with an average speed of 13 knots, the ship should arrive in early January (5,544 miles: 13 knots: 24 hrs. = 17,76 days).
The given dates for delivery at Antwerp are encouraging that weather permitting and with unforeseen circumstances excepted (abbreviated as 'uce'), the ship can safely arrive at the load port within the Charterers' lay-days requested for loading.

The Owners continue:
5.
Time Charter Trip out (Or 'One Time Charter trip') via East Coast South America to Continent or Med, via safe port/s, safe berth/s, always within I.N.L. with cargo of bulk corn.
The port/s, berth/s, anchorage/s will remain safe from the time of the ship's arrival at the first sea pilot station until departure and the last sea pilot has left the ship.
The ship will always lie afloat throughout the period of the Charter.

Comments:
The Owners have offered:

-that the ship will trade always via safe port/s, safe berth/s, always afloat and always within the International Navigating Limits -that the cargo is bulk corn.

The Owners continue:
6.
Redelivery on dlosps 1 safe port Skaw Denmark/Gibraltar range or 1 safe port Passero Italy/East Med atdnsshinc. Discharging range and port at the Charterers' option.

Comments:
The Owners have offered:
-that the redelivery will take place in a safe port within the Skaw/Gibraltar range or within Passero/East Med.
-Port and discharging ranges are at the Charterers' option.

The Owners continue:
7.
Hire USD 12,000 per day/pro rata including overtime. The hire is payable every 15 days always in advance net of tax and bank charges.
For hire calculation the UTC time to be in effect.

Comments:
The Owners have offered that:
-the 'hire rate' includes crew's/officers' overtime.
-the hire is payable every 15 days in advance.
-the delivery/redelivery time will be based on UTC hours.

The Owners continue:
8.
Charterers guarantee that the ship will load only from the East Coast South America and if it loads in Argentina and/or Uruguay, not above the San Lorenzo and Nueva Palmira ports, respectively. The Charterers guarantee that the ship will discharge only within the Gibraltar/ Skaw range, including the United Kingdom and Ireland or, at the Charterers' option, Passero Italy/Black sea range including the full Adriatic Sea, but excluding .... (Owners may insert the countries they want to exclude from the trading)
Comments:
The Owners have offered:
-that the ship will load only from the East Coast South America and if it loads from Argentina, not to trade above the San Lorenzo port and if it loads from Uruguay not above the Nueva Palmira port. That is requested because the pre-

vailing drafts for ports located above San Lorenzo and Nueva Palmira are limited, making the navigation more dangerous.
-the ranges in which the ship can discharge and the countries within which these ranges are excluded.

The Owners continue:
9.
The ship will load only bulk corn and no other cargo is allowed to be loaded.

Comments:
The Owners have offered that:
- the ship will load only bulk corn and no other cargo is allowed.

The Owners continue:
10.
Duration about 60-70 days.

Comments:
The Owners have offered that:
-the duration will be about 60/70 days from the date of delivery until its redelivery back to the Owners.

The Owners continue:
11.
The ship will be delivered with about .... mtns VLSFO and about .... mtns LSMGO.
The ship will be redelivered with the same fuel qualities and with about the same quantity of fuel as on delivery.
The value of the fuel on delivery is payable together with the 1st hire.
Fuel prices on delivery and redelivery for the VLSFO USD ... and for the LSMGO USD ....
The Master's or Owners' declaration of the quantity of fuel that is required to be supplied in order to execute the trip or for redelivery purposes, is subject to final duration and/or ship's consumption while manoeuvring in ports and/or confined waters and subject to final destination.
Sub to agree to a full bunker clause.

Comments:
The Owners have offered:
-the quantity of bunkers with which the ship will be delivered and redelivered and their prices.
-that a full clause should be agreed.

The Owners continue:
12.
C/V/E USD 1400. Per month /pro rata.

Comments:
The Owners have offered:
-a monthly rate or prorate which is required to cover the communications /victualing/entertainment expenses made on the Charterers' behalf.

The Owners continue:
13.
The ship will be redelivered to the Owners with its holds in the same condition as on delivery, normal wear and tear excepted.
However, the Charterers will have the option of redelivering the ship with unclean holds, by paying a lumpsum amount of USD 6,000-. The amount due will be paid together with the Charterers' final hire statement.

Comments:
The Owners have offered:
-with a lumpsum amount payable in lieu of holds cleaning.

The Owners continue:
14.
Arbitration in London. English law will apply to negotiations and to the eventual agreement.
Sub to agree to a full clause.

Comments:
The Owners have offered:
- the Arbitration forum which will apply and under which law.

The Owners continue:
15.
Sub further terms/details will be based on NYPE C/P 46 at the Owners' executed proforma with 5.00 percent total your end. Please provide breakdown of the commission.

Comments:
The Owners have offered:
-that the Charter Party details will be worked on an Owners' executed profor-

ma. -that they want to know the address commission and the number of the brokers involved.

TO SUMMARIZE THE OWNERS' OFFER:

Ref Account Messrs. Eagle Trading and Shipping Inc
Pnmx to make Paranagua 1/10 Jan.
Tct via ECSA to Skaw/Gib, chopt Passero/EMed,
MV FOTINI

The Owners firm bid for 2 hours reply as follows:
MV FOTINI
....
....
For:
1.
Account Messrs. Eagle Trading and Shipping Inc. of British Virgin Islands
Sub to receive Charterers' last three fixtures.
2.
All subs will be lifted upon fixing the main terms.
3.
Delivery on dlosps Antwerp port, Belgium atdnsshinc
4.
14/19 December 2....
00:01-2400 hrs. For lay-can the local time to be in effect.
5.
Time Charter Trip out (Or 'One Time Charter trip') via East Coast South America to Continent or Med, via safe port/s, safe berth/s, always within I.N.L. with cargo of bulk corn.
The port/s, berth/s, anchorage/s will remain safe from the time of the ship's arrival at the first sea pilot station until departure and the last sea pilot has left the ship.
The ship will always lie afloat throughout the period of the Charter.
6.
Redelivery on dlosps 1 safe port Skaw Denmark/Gibraltar range or 1 safe port Passero, Italy/East Med atdnsshinc. Discharging range and port at the Charterers' option.
7.
Hire USD 12,000 per day/pro rata including overtime. The hire is payable every 15 days always in advance net of tax and bank charges.

For hire calculation the UTC time to be in effect.
8.
Charterers guarantee that the ship will load only from the East Coast South America and if it loads in Argentina and/or Uruguay, not above the San Lorenzo and Nueva Palmira ports, respectively. The Charterers guarantee that the ship will discharge only within the Gibraltar/ Skaw range including the United Kingdom and Ireland or, at the Charterers' option, Passero, Italy/Black sea range including the full Adriatic Sea but excluding ....
9.
The ship will load only bulk corn and no other cargo is allowed to be loaded.
10.
Duration about 60/70 days.
11.
The ship will be delivered with about .... mtns VLSFO and about .... mtns LSMGO. The ship will be redelivered with the same fuel qualities and with about the same quantity of fuel as on delivery.
The value of the fuel on delivery is payable together with the 1st hire.
Fuel prices on delivery and redelivery for the VLSFO USD ... and for the LSMGO USD ....
The Master's or Owners' declaration of the quantity of fuel that is required to be supplied in order to execute the trip or for redelivery purposes, is subject to final duration and/or ship's consumption while manoeuvring in ports and/or confined waters and subject to final destination.
Sub to agree to a full bunker clause.
12.
C/V/E USD 1400. Per month /pro rata.
13.
The ship will be redelivered to the Owners with its holds in the same condition as on delivery, normal wear and tear excepted.
However, the Charterers will have the option of redelivering the ship with unclean holds, by paying a lumpsum amount of USD 6,000-. The amount due will be paid together with the Charterers' final hire statement.
14.
Arbitration in London. English law will apply to negotiations and to the eventual agreement.
Sub to agree to a full clause.
15.
Sub further terms/details will be based on NYPE C/P 46 at the Owners' executed proforma with 5.00 percent total your end. Please provide breakdown of the commission.

CHARTERERS' COUNTER REPLY TO OWNERS' LAST:

Charterers:
Either
- will agree to work on the Owners' offer and will make a clean acceptance or a counter offer on an accept/except basis on the terms which they do not agree
Or
- will decline the Owners' offer and forward a full re-offer for the Owners' acceptance or counter reply.

If the Charterers decline the Owners' offer, as they prefer to start trading the ship on a basis of their form of Main Terms, the RE-OFFER could be given as follows:

Ref Account Messrs. Eagle Trading and Shipping Inc
Pnmx to make Paranagua 1/10 Jan.
MV FOTINI

Thanks for the Owners' offer which the Charterers respectfully decline and firm re offer for reply latest within 1 hour as follows:
MV FOTINI
....
....
-The description of the ship as provided by the Owners is an accurate representation of the characteristics of the ship, which can only be modified with the usual shipping interpretation of the word 'about'. This clause supersedes any and all adding of the words 'without guarantee' in the description of the ship.

Comments:
The Charterers have offered:
- that they accept the ship's description as provided by the Owners and clarify that the given description is accurate and can only be modified by the word 'about'. Any expression 'without guarantee' on the ship's description is without any validity.

The Charterers continue:
Owners guarantee that the:
-ship is a single deck bulk carrier and that it was originally constructed as a bulk carrier.
-ship has clear unobstructed main holds.
-ship does not have any horizontal frames and or beams within its holds -ship has no reefer space or special cargo locker's space.

-cargo will be loaded in the main holds only.
-ship is steel floored throughout.
-ship has its engine, bridge aft.
-ship's crew and officers will be ITF approved or equivalent acceptable to the ITF
-ship is ISPS fitted.
Owners guarantee that during the duration of this Charter Party:
-the ship will not change ownership or class without the Charterers' written consent.
-the ship's hull and machinery insurance will be fully maintained and will not be changed.
-the ship is fully P&I Club covered, which will be maintained and will not be changed.
Owners confirm:
- the ship has no outstanding recommendations, restrictions or any other requirements of any nature in respect to flag state, port state control or classification inspection.
- the ship will comply with regulations and requirements at all ports of call and has valid certificates on board applicable to those ports, countries where the ship will trade under this Charter.
The Owners will fax copies of the certificates of the undermentioned ship which are valid for the duration of the Charter Party:
- DOC
-SMC
-CLASS-ISPS
-ISSC/ITC/P&I CLUB

Comments:
The Charterers have offered:
-and request certain guarantees to be confirmed to by the Owners. The Owners would reply on which of the above the ship does not comply with,
-and request copies faxed to them of certain important certificates that the ship must have valid.

The Charterers continue:
For:
1.
Account Messrs. Eagle Trading and Shipping Inc. of British Virgin Islands
Charterers' last three fixtures as follows:
Mv ...... Fixed through ......
Mv....... Fixed through......
Mv ......Fixed through....

Comments:
The Charterers have:
-provided their full name/where they are based and the last three fixtures as per the Owners' initial request.

The Charterers continue:
2.
Subject to Shippers' and Receivers' approval declarable 1 hr. after fixing the main terms.

Comments:
The Charterers have offered that:
-they will declare if the subs are lifted or not within 1 hour after fixing the main terms.

The Charterers continue:
3.
delivery on aps Paranagua port, Brazil atdnsshinc.

Comments:
The Charterers have offered that:
-the delivery of the ship will take place upon Arrival at the Pilot Station of the Paranagua port 'All Time Day and Night Saturdays Sundays and Holidays Included'.

The Charterers continue:
4.
lay-can 00:01 hours 1st January 2.... - 24:00 hours 10th January 2...
Please advise of the ship's full itinerary and the Agents' name at the last discharge port/last 3 cargoes/e.t.a. Paranagua.

Comments:
The Charterers have offered:
- the requested lay-days and cancelling date,
-that the Owners, when counter offering to the Charterers, should advise them of:
-the ship's full itinerary (place of discharge/when completed etc.),
-the Agents' name at the last discharge port, enabling them to check if the position of the ship is safe and if all is going smoothly during the discharging operation,

-the last three loaded cargoes, which will determine the condition and cleanliness of the cargo holds,
-the estimated time of arrival (e.t.a.) at the load port.

The Charterers continue:
5.
TCT out via ECSA to Continent/Baltic/Med/Black Sea or chopt to PMO/Japan range including PG.
The ship will trade always afloat except at ECSA where the NAABSA clause will be in effect via sp(s)/sb(s)/sa(s), always within I.N.L. with any cargo of bulk grains/ grain products/ agriproducts.

Comments:
The Charterers have offered:
-that they need the ship for a trip out.
-contrary to the 'order' as quoted in the market, that they need the option of the ship also discharging at the Middle East/Far East. The Owners may reject such an option, but if the Charterers insist the negotiations could be terminated
-that the ship will trade 'always afloat' except at the East Coast of South America where it is customary for the ships to load Not Always Afloat But Safely Aground (NAABSA).
-that the ship will trade via safe ports and safe berths/safe anchorages and always within the International Navigating Limits.
-contrary to the 'order', the cargo could not only be bulk corn but any bulk grains/ grain products/ agriproducts.

The Charterers continue:
6.
Redelivery on dop 1 safe port or passing the Skaw/Passero range including the UK and Ireland or chopt on dop 1 safe port or passing the Muscat outbound/Japan range. Redelivery port at the Charterers' option. The ship will be redelivered atdnsshinc.

Comments:
The Charterers have offered:
-that they will redeliver the ship on dropping the outward pilot in 1 safe port or passing the given ranges. Obviously, the Charterers have the option of discharging in a port outside of the redelivery range and after completion to ballast the ship within the given ranges for redelivery purposes.

The Charterers continue:

7.
Duration if redelivery at Skaw/Passero about 35/45 days wog - if redelivery at PMO/Japan about 60/70 days wog.

Comments:
The Charterers have offered:
-that the duration is subject to the redelivery range and it is given without guarantee.

The Charterers continue:
8.
Hire USD 11,500 pdprinclot and lubricants, plus gross ballast bonus of USD 400,000 for redelivery at Skaw/Passero.
Hire USD 16,000 pdprinclot plus gross ballast bonus of USD 400,000 for redelivery at PMO/ Japan range.
The hire is payable every 15 days always in advance net of tax and bank charges, however, the first 15-day hire is payable 3 banking days after the ship's delivery and the receipt of the Owner's first hire invoice.

Comments:
The Charterers have offered:
-two different hire rates, depending on the region of the redelivery of the ship, plus a gross ballast bonus. When 'gross ballast bonus' applies, the brokers involved in the fixture are also entitled to a commission on this amount, otherwise it would be expressed as 'net ballast bonus'.
-that the first 'hire' will be paid by the Charterers within 3 banking days after the ships' delivery and after having received the first hire invoice by the Owners. Thereafter all 'hires' are payable in advance.
-that the hire includes the crew's overtime and the cost of the lubricants which will be consumed during the charter.

However, the Charterers have not stated:
-when they will declare the discharging range for the hire purposes
-when the ballast bonus will be paid to the Owners.

The Charterers continue:
9.
The bunkers on delivery will be about VLSFO ... mtns and about LSMGO .... mtns and the ship will be redelivered with about the same quantities.
Any minor difference on the bunkers' value on delivery and redelivery will be settled together with the final settlement with the exact figures.

The value of the bunkers on delivery is payable together with the first 15-day hire payment.
Owners guarantee that the ship is eligible to be supplied with bunkers in any country within the trading limits that have been agreed under this Charter Party.
Charterers have the right to deduct the value of the bunkers on redelivery, from the last 15-day hire payment, but if it is not sufficient it will be extended until the Charterers deduct the full value of the bunkers.
The Bunkers' prices on delivery and redelivery will be the same, USD .... per metric ton for the VLSFO and USD ... per metric ton for the LSMGO.
The Charterers have the right to bunker the ship before the delivery, as long as that the same does not interfere with the Owners' operation and is permitted by the current Charterers.

Comments:
The Charterers have offered:
-the quantity of fuel with which the ship will be delivered and redelivered and their prices.
-when the value of the bunkers is payable on delivery/redelivery.
-to have the option to bunker the ship for their account, before the delivery takes place and obviously, at the last port before the ship's redelivery from its previous trade. This operation should be made, provided that it does not interfere with the ship's operation. The Owners, before agreeing, must obtain the approval of the current Charterers (unless the ship is on a voyage basis). A provision should be made that all extra expenses or lost time which might be incurred and any potential liabilities for pollution will be for the Charterers' account. The Charterers will benefit by the time required for the ship's additional bunkering to perform their trip and/or the cheaper prices that are achievable in this port.

The Charterers continue:
10.
On delivery the ship must be suitable in every respect for loading and safe carriage of any grains, grain products and agricultural products in bulk.
Owners guarantee that the ship can fully load any grain, grain products and agricultural products in bulk without using shifting boards or any other fittings. The ship must have on board, at any time during the duration of this Charter, complete Grain Trimming Tables and valid documents and certificates for grain trade issued by the ship's classification society and any other national authority on the basis of the SOLAS 1974 regulations and any subsequent thereto.

Comments:
The Charterers have offered that:

-the ship must be suitable for loading the Charterers' intended cargo without using shifting boards or any other fittings.
-all documents and respective certificates for grain trade must be valid.

The Charterers continue:
11.
The ship will load from the ECSA and will discharge in the Continent/Baltic/Med/ Black sea, excluding ...., or at chopt at full PG/Japan range, including Pakistan/full India/full South East Asia/Philippines but excluding ....

Comments:
The Charterers have offered:
-and declare where the ship will load and discharge from. The 'trade exclusion' clause will not be discussed.

The Charterers continue:
12.
It is expressly agreed that:
1.The ship on delivery:
the ship on delivery or, at the latest, on arrival or berthing at the first or sole load port will have all cargo holds and hatch covers clean:
-free of rust, loose rust flakes, rust scales, and loose scales,
-free from salt and free from residues of previous cargo/oes,
-fresh water washed and dried,
-free from insects and odours
and in all respects ready for loading the Charterers' intended cargo, according to the rules and regulations prevailing at the port of loading and to the satisfaction of the independent surveyors, as have been selected by the Charterers and/ or Shippers, under their time and expenses and/or to the satisfaction of a national cargo bureau and/or local grain surveyors' inspection.
2. Cargo holds' and/or hatch covers' failure on delivery or at each loading port:
if, the ship fails from the cargo holds and/or hatch covers survey inspection, then the ship will be placed fully off-hire, from the time of failure until such time as all the rejected cargo holds and/or hatch covers have successfully fully passed inspection and are ready for loading.
The Owners are to arrange without delay for the cargo holds' and/or hatch covers' cleaning, at their time and expenses with the ship's Crew, if allowed by the port's regulations and/or labour unions, or, otherwise with shore labourers, in order to meet the surveyors' requirements.
The Owners are responsible for:
-the amount of the fuel consumed during the off-hire period.

-the expenses incurred for the purpose of holds' and/or hatch covers' cleaning.
-all chemicals, fresh water, consumable materials, tools such as water guns, pressure hoses, pumps, mechanical de-rusting equipment, rustybus machines etc. necessary to be arranged and supplied to the ship, in making the same suitable for the intended for loading cargo.
-any directly related expenses incurred, but supported with original vouchers, including, but not limited to, stevedores stand-by, but up to a maximum of one shift.
If the Port Authorities and/or Shippers, order the ship to evacuate the berth and shift to anchorage or to change waiting place for the purpose of cargo holds' and/or hatch covers cleaning, then all the delays incurred and extra expenses incurred, including, but not limited to, tugs and/or pilots and/or Agency fees etc., until the ship is back at the berth, or in its initial position, will be entirely for the Owners' account.

Comments:
The Charterers have offered:
-how the ship must be presented for loading. The ship should comply with this clause, or otherwise it will be placed off-hire and all expenses incurred will be for their account.

The Charterers continue:
13.
The Charterers will pay the Owners a lumpsum of USD 1200 per month or pro rata for communications/victualing/entertainment.

Comments:
The Charterers have offered:
-the amount with which the Owners will be compensated for the c/v/e expenses incurred and made on the Charterers' behalf.

The Charterers continue:
14.
The ship will be redelivered to the Owners with its holds in the same condition as on delivery, normal wear and tear excepted.
However, the Charterers will have the option of redelivering the ship with unclean holds, by paying a lumpsum amount of USD 5,000.-. The amount due will be paid together with the Charterers' final hire statement.

Comments:
The Charterers have offered:

-the rate with which the Owners will be compensated in the case that the ship is redelivered with unclean holds.

The Charterers continue:
15.
A joint on-hire bunker and condition survey will be held by an independent surveyor, formally agreed on by both parties, at the port of delivery or, if mutually agreed, at the most convenient and proximate time/place/port to the delivery time/place/port.
Time actually lost due to this survey will be for the Owners' account, but the costs of the survey will be equally shared between the Owners and Charterers.
A joint off-hire bunker and condition survey will be held by an independent surveyor formally agreed by both parties, at the port of redelivery or, if mutually agreed, at the most convenient and proximate time/place/port to the redelivery time/place/ port.
Time actually lost due to this survey will be for the Charterers' account, but the costs of the survey will be equally shared between the Owners and Charterers.

Comments:
The Charterers have offered that:
-the 'on' and 'off-hire survey' will be carried out by an independent surveyor, who will determine the bunkers' quantity on delivery and redelivery and the condition of the ship. Lost time and expenses are shared equally between Charterers and Owners.

The Charterers continue:
16.
The Charterers will make every endeavour to ensure that the original Bill of Lading is available at the discharge port on or before the ship's arrival to discharge. However, if the original Bills of Lading are not available to discharge, then the Owners/Master will permit the discharge of the cargo without delay in exchange for the Charterers' single L.O.I., as per the Owners' P&I Club's wording, signed only by the Charterers.

Comments:
The Charterers have offered:
-the ship to discharge with a letter of indemnity (l.o.i.) in exchange for delivery of the cargo without production of the original Bill of Lading.
The acceptance of the l.o.i. is the Owners' commercial decision, which could lead to the ship falling outside of the P&I Club's rules/cover. The l.o.i must be properly worded and should only be accepted when offered by absolute first

class Charterers, otherwise, the Owners must insist the l.o.i. to be also counter-signed by a first-class bank.

The Charterers continue:
17.
Arbitration in London with the current Bimco's 'Dispute Resolution Clause' to apply. English Law will apply to negotiations and to the eventual fixture.

Comments:
The Charterers have offered:
-which arbitration forum and which law will apply to the negotiations and to the eventual fixture.
-that the current Bimco's 'Dispute Resolution Clause' to apply.

The Charterers continue:
18.
All negotiations and eventual fixture will be kept pnc and will not be disclosed to others.

Comments:
The Charterers have offered:
-that the negotiations and eventual fixture shall be kept private and confidential (abbreviated as 'pnc'). The fixture must not to be disclosed and reported to any other party which is not involved in this fixture.

The Charterers continue:
19.
Sub details that are based on NYPE 46, at the Charterers' executed proforma with 5 percent total commission including 3,75 percent address com.

Comments:
The Charterers have offered:
-that the Charter Party details will be negotiated on the Charterers' executed proforma based on the NYPE 46 standard form.

TO SUMMARIZE THE CHARTERERS' RE-OFFER:

Ref Account Messrs. Eagle Trading and Shipping Inc
Pnmx to make Paranagua 1/10 Jan.
MV FOTINI

Thanks for the Owners' offer which the Charterers respectfully decline and firm re offer for reply latest within 1 hour as follows:

MV FOTINI
....
....
-The description of the ship as provided by the Owners is an accurate representation of the characteristics of the ship, which can only be modified with the usual shipping interpretation of the word 'about'. This clause supersedes any and all adding of the words 'without guarantee' in the description of the ship.
Owners guarantee that the:
-ship is a single deck bulk carrier and that it was originally constructed as a bulk carrier.
-ship has clear unobstructed main holds.
-ship does not have any horizontal frames and or beams within its holds -ship has no reefer space or special cargo locker's space.
-cargo will be loaded in the main holds only.
-ship is steel floored throughout -ship has its engine, bridge aft.
-ship's crew and officers will be ITF approved or equivalent acceptable to the ITF
-ship is ISPS fitted.
Owners guarantee that during the duration of this Charter Party:
-the ship will not change ownership or class without the Charterers' written consent
-the ship's hull and machinery insurance will be fully maintained and will not be changed.
-the ship is fully P&I Club covered, which will be maintained and will not be changed.
Owners confirm:
-that the ship has no outstanding recommendations, restrictions or any other requirements of any nature in respect to flag state, port state control or classification inspection.
-that the ship will comply with regulations and requirements at all ports of call and has valid certificates on board applicable to those ports, countries where the ship will trade under this Charter.
The Owners will fax copies of the certificates of the undermentioned ship which are valid for the duration of the Charter Party:
- DOC
-SMC
-CLASS
-ISPS
-ISSC/ITC/P&I CLUB

For:
1.
Account Messrs. Eagle Trading and Shipping Inc. of British Virgin Islands
Charterers' last three fixtures as follows:
Mv ...... Fixed through ......
Mv....... Fixed through......
Mv ......Fixed through....
2.
Subject to Shippers' and Receivers' approval declarable 1 hr after fixing the main terms.
3.
delivery on aps Paranagua port, Brazil atdnsshinc
4.
lay-can 00:01 hours 1st January 2.... - 24:00 hours 10th January 2...
Please advise of the ship's full itinerary and the Agents' name at the last discharge port/last 3 cargoes/e.t.a. Paranagua
5.
TCT out via ECSA to Continent/Baltic/Med/Black Sea or chopt to PMO/Japan range including PG.
The ship will trade always afloat except at ECSA where the NAABSA clause will be in effect via sp(s)/sb(s)/sa(s), always within I.N.L. with any cargo of bulk grains/ grain products/ agriproducts.
6.
Redelivery on dop 1 safe port or passing the Skaw/Passero range including the UK and Ireland or chopt on dop 1 safe port or passing the Muscat outbound/Japan range. Redelivery port at the Charterers' option. The ship will be redelivered atdnsshinc.
7.
Duration if redelivery at Skaw/Passero about 35/45 days wog- if redelivery at PMO/Japan about 60/70 days wog.
8.
Hire USD 11,500 pdprinclot and lubricants, plus gross ballast bonus of USD 400,000 for redelivery at Skaw/Passero.
Hire USD 16,000 pdprinclot plus gross ballast bonus of USD 400,000 for redelivery at PMO/ Japan range.
The hire is payable every 15 days always in advance net of tax and bank charges, however, the first 15-day hire is payable 3 banking days after the ship's delivery and the receipt of the Owner's first hire invoice.
9.
The bunkers on delivery will be about VLSFO ... mtns and about LSMGO .... mtns

and the ship will be redelivered with about the same quantities.
Any minor difference on the bunkers' value on delivery and redelivery will be settled together with the final settlement with the exact figures.
The value of the bunkers on delivery is payable together with the first 15-day hire payment.
Owners guarantee that the ship is eligible to be supplied with bunkers in any country within the trading limits that have been agreed under this Charter Party.
Charterers have the right to deduct the value of the bunkers on redelivery, from the last 15-day hire payment, but if it is not sufficient it will be extended until the Charterers deduct the full value of the bunkers.
The Bunkers' prices on delivery and redelivery will be the same, USD .... per metric ton for the VLSFO and USD ... per metric ton for the LSMGO.
The Charterers have the right to bunker the ship before the delivery, as long as that the same does not interfere with the Owners' operation and is permitted by the current Charterers.
10.
On delivery the ship must be suitable in every respect for loading and safe carriage of any grains, grain products and agricultural products in bulk.
Owners guarantee that the ship can fully load any grain, grain products and agricultural products in bulk without using shifting boards or any other fittings. The ship must have on board, at any time during the duration of this Charter, complete Grain Trimming Tables and valid documents and certificates for grain trade issued by the ship's classification society and any other national authority on the basis of the SOLAS 1974 regulations and any subsequent thereto.
11.
The ship will load from the ECSA and will discharge in the Continent/Baltic/Med/ Black sea, excluding ...., or at chopt at full PG/Japan range, including Pakistan/full India/full South East Asia/Philippines, but excluding ....
12.
It is expressly agreed that:
1.The ship on delivery:
the ship on delivery or, at the latest, on arrival or berthing at the first or sole load port will have all cargo holds and hatch covers clean:
-free of rust, loose rust flakes, rust scales, and loose scales,
-free from salt and free from residues of previous cargo/oes,
-fresh water washed and dried,
-free from insects and odours,
and in all respects ready for loading the Charterers' intended cargo, according to the rules and regulations prevailing at the port of loading and to the satisfaction of the independent surveyors, as have been selected by the Charterers and/ or Shippers, under their time and expenses and/or to the satisfaction of a nation-

al cargo bureau and/or local grain surveyors' inspection.
2. Cargo holds' and/or hatch covers' failure on delivery or at each loading port: if, the ship fails from the cargo holds and/or hatch covers survey inspection, then the ship will be placed fully off-hire, from the time of failure until such time as all the rejected cargo holds and/or hatch covers have successfully fully passed inspection and are ready for loading.
The Owners are to arrange without delay for the cargo holds' and/or hatch covers' cleaning, at their time and expenses with the ship's Crew, if allowed by the port's regulations and/or labour unions, or, otherwise with shore labourers, in order to meet the surveyors' requirements.
The Owners are responsible for:
-the amount of the fuel consumed during the off-hire period.
-the expenses incurred for the purpose of holds' and/or hatch covers' cleaning.
-all chemicals, fresh water, consumable materials, tools such as water guns, pressure hoses, pumps, mechanical de-rusting equipment, rustybus machines etc. necessary to be arranged and supplied to the ship, in making the same suitable for the intended for loading cargo.
-any directly related expenses incurred, but supported with original vouchers, including, but not limited to, stevedores stand-by, but up to a maximum of one shift.
If the Port Authorities and/or Shippers, order the ship to evacuate the berth and shift to anchorage or to change waiting place for the purpose of cargo holds' and/ or hatch covers cleaning, then all the delays incurred and extra expenses incurred, including, but not limited to, tugs and/or pilots and/or Agency fees etc., until the ship is back at the berth, or in its initial position, will be entirely for the Owners' account.
13.
The Charterers will pay the Owners a lumpsum of USD 1200 per month or pro rata for communications/victualing /entertainment.
14.
The ship will be redelivered to the Owners with its holds in the same condition as on delivery, normal wear and tear excepted.
However, the Charterers will have the option of redelivering the ship with unclean holds, by paying a lumpsum amount of USD 5,000.-. The amount due will be paid together with the Charterers' final hire statement.
15.
A joint on-hire bunker and condition survey will be held by an independent surveyor, formally agreed on by both parties, at the port of delivery or, if mutually agreed, at the most convenient and proximate time/place/port to the delivery time/place/port.
Time actually lost due to this survey will be for the Owners' account, but the

costs of the survey will be equally shared between the Owners and Charterers.
A joint off-hire bunker and condition survey will be held by an independent surveyor formally agreed by both parties, at the port of redelivery or, if mutually agreed, at the most convenient and proximate time/place/port to the redelivery time/ place/port.
Time actually lost due to this survey will be for the Charterers' account, but the costs of the survey will be equally shared between the Owners and Charterers.
16.
The Charterers will make every endeavour to ensure that the original Bill of Lading is available at the discharge port on or before the ship's arrival to discharge. However, if the original Bills of Lading are not available to discharge, then the Owners/ Master will permit the discharge of the cargo without delay in exchange for the Charterers' single L.O.I., as per the Owners' P&I Club's wording, signed only by the Charterers.
17.
Arbitration in London with the current Bimco's 'Dispute Resolution Clause' to apply.
English Law will apply to negotiations and to the eventual fixture.
18.
All negotiations and eventual fixture will be kept pnc and will not be disclosed to others.
19.
Sub details that are based on NYPE 46, at the Charterers' executed proforma with 5 percent total commission including 3,75 percent address com.

OWNERS' COUNTER REPLY TO CHARTERERS' LAST:

Mv Fotini/ Eagle Trading and Shipping Inc

The Owners thank the Charterers for their offer, on which a/e for reply within 1 hrs. as follows:

Comments:
The Owners have offered that:
-they grant one hour for the Charterers to counter reply.

The Owners continue:
3.
Delivery on dlosps Antwerp port, Belgium atdnsshinc

Comments:

The Owners, in comparison to the Charterers' offer:
-reject the Charterers' delivery place and offer the ship to be delivered on dlosps Antwerp port.

The Owners continue:
4.
L/C 14/19 December .... 00:01-2400 hrs. For lay-can the local time to be in effect ship's full itinerary as follows: ....
Agents: ....
last 3 cargoes (starting from the last) as follows: ....
eta Paranagua .... agw wp uce. (all going well - weather permitting - unforeseen circumstances excepted)

Comments:
The Owners, in comparison to the Charterers' offer:
-request for lay-can according to the ship's readiness at the place of delivery.
-reply to the Charterers' requested information.

The Owners continue:
5.
TCT out via ECSA to Continent/Med/Black Sea or chopt to PMO/Japan range including PG.
The ship will trade via sp(s)/sb(s)/sa(s), always within I.N.L. with cargo of bulk corn. The port/s, berth/s, anchorage/s will remain safe from the time of the ship's arrival at the first sea pilot station until departure and the last sea pilot has left the ship.
If anchorage/s is used, the Bimco's current 'Ship to Ship Transfer Clause for Time Charter Parties' will be in effect. Sub to agree to a full clause.
The ship will always lie afloat throughout the period of the Charter except at ECSA where the NAABSA clause will apply.
NAABSA Clause:
Charterers have the right to load or discharge the ship at any ports Up River, but not north of San Lorenzo timbues area and Nueva Palmira or in Brazilian port/s not north of Vitoria port (Vitoria to be excluded), provided that it is customary for ships of similar dimensions and size to navigate there, where they can safely lie aground on the soft muddy sea-bed, but always at the Charterers' risk and subject to the permissible and prevailing draft declared by the Port Authorities and/ or pilots.
If the ship lies aground and the Master suspects that the hull is damaged, then the Charterers are to employ, under their time and expenses, an inspection of the ship's flat bottom, including, but not limited to, the vertical sides, bilge keels,

propeller and rudder blade, by a well-organized and ship's class approved diving company, to determine the extent of any damage. The inspection will be carried out under the appointed by the Owners' ship's class surveyor attendance, but at the Charterers' cost.
Should the ship's bottom and underwater parts, propeller, rudder be extensively damaged, affecting the ship's seaworthiness and class, the ship will be directed immediately and always on-hire to a mutually appointed convenient place or port or yard to carry out the repairs at the Charterers' time and cost and to the class surveyor's entire satisfaction. The Shipyard's expenses and port expenses, including in and out tugs, pilots, agency fees etc., will be entirely for the Charterers' account.
Time lost for unloading the cargo at the load port, if it is required, and any expenses incurred will be for the Charterers' account.
Damage not affecting the ship's seaworthiness and class will be repaired at the Charterers' cost, but not for the time lost, at the next scheduled dry docking, normal wear and tear excepted.
If, after the ship's underwater inspection, no damage is found, either minor or not, then all the directly related expenses incurred such as the diving company's expenses, ship's class surveyor's fees etc. will be shared equally between the Owners and Charterers, but the ship will remain always on-hire.

Comments:
The Owners, in comparison to the Charterers' offer:
-accept the Charterers to have the option the ship to also discharge at the PMO/Japan range.
-accept the ship to also trade via safe anchorages, however, the Bimco's current 'Ship to Ship Transfer Clause for Time Charter Parties' will apply. The ship will be protected from possible damage that may be occurred during the operation in the anchorage.
-request the ship to remain safe during its entire stay in the ports/berths/anchorages.
-request the NAABSA clause to apply at ECSA.

The Owners continue:
6.
Redelivery on dropping last outward sea pilot station.

Comments:
The Owners, in comparison to the Charterers' offer:
-accept the Charterers' redelivery places but request the redelivery to take place on dlosps (the Charterers' offer was on dop).

The Owners continue:
7.
Duration if redelivery at Skaw/Passero about 60/70 days wog.
If redelivery at PMO/Japan about 85/95 days wog.

Comments:
The Owners, in comparison to the Charterers' offer:
-calculate the duration basis delivery at Antwerp port and they accept to be 'wog'.

The Owners continue:
8.
Hire USD 12,000. - for redelivery at Skaw/Passero range.
Hire USD 17,000. - for redelivery at PMO/Japan range.
Charterers will declare the discharging range before the payment of the first hire. The hire is payable every 15 days always in advance net of tax and bank charges, however, the first 15-day hire is payable 1 banking day after the ship's delivery and the receipt of the hire invoice from the Owners.

Comments:
The Owners, in comparison to the Charterers' offer:
-request a different 'hire rate' in accordance with the ship's destination.
-request the Charterers to declare the discharging range before the delivery takes place.
-request the first 15 days hire to be paid within 1 banking day after the ship's delivery.
-the hire is payable net of tax and bank charges.

The Owners continue:
9.
The Charterers, before the ship is delivered to their services and at the last port of discharge of the ship's previous trade, have the option to supply the ship with fuel for their account, as long as that the same does not interfere with the full operation of the ship and is allowed (written approval) by the current Charterers and/ or after the Owners' consent. The Charterers are responsible for any loss of time which will result in the ship being placed off-hire, arising from any reason due to the Charterers' supply of fuel to the ship before its delivery to them. The Charterers will reimburse the Owners, as per the provided off-hire statement from the previous Charterers, for all time lost and fuel consumed during the off-hire period. Any expenses, losses, damages and/or consequences of any nature

relating directly to the bunkering of the ship, including damages from pollution and fines imposed by any party, will be for the Charterers' account, liability, and responsibility.
The Owners have the option of bunkering the ship en route to its last port of call or at the last port of call before its redelivery for their account and after the Charterers' consent, as long as that the same does not interfere with the full operation of the ship or with the effect of not loading the full intended cargo quantity. Time lost, because of the bunkering for the Owners' account should be counted as off-hire.
The Charterers, for every bunkering operation arranged by them, under their time and expenses, will appoint a well recognised independent Surveyor, who will carry out a bunker survey before and after any bunkering operation takes place, in order to determine the quantity of fuel supplied to the ship.
Information on any given quantity of fuel to the Charterers is subject to the trip's final duration, the ship's consumption while manoeuvring in ports and in confined waters and subject to final destination.

Comments:
The Owners, in comparison to the Charterers' offer:
-accept the Charterers to have the option to bunker the ship before delivery but under certain specific conditions.
-request to have the option to bunker the ship for their account en route to the last port of call or at the last port of call before its redelivery.
-request that the Charterers, when supplying bunkers to the ship, appoint a Surveyor to carry out a bunker survey, under their time/expense, before and after any bunkering operation takes place.
-that any declaration by the Master or Owners on the required bunkers for redelivery or to execute a voyage is subject to final duration, ship's consumption while maneuvering in ports and in confined waters and subject to final destination.

The Owners continue:
10.
The ship will load only bulk corn. On delivery the ship must be suitable in every respect for loading and safe carriage of corn in bulk.
Owners guarantee that the ship can fully load corn in bulk without using shifting boards or any other fittings. The ship must have on board, at any time during the duration of this Charter, complete Grain Trimming Tables and valid documents and certificates for grain trade issued by the ship's classification society and any other national authority on the basis of SOLAS 1974 Regulations and any subsequent thereto.

Comments:
The Owners, in comparison to the Charterers' offer:
-accept the Charterers' clause, however, request the ship to load only bulk corn.

The Owners continue:
11.
Charterers guarantee that the ship will load only from the East Coast South America and if it loads in Argentina and/or Uruguay, not above the San Lorenzo and Nueva Palmira ports, respectively.
Charterers guarantee that the vessel will discharge only in full Mediterranean, including the Black sea area, but excluding .... or at the Charterers' option at Gibraltar/Skaw range, including the United Kingdom and Ireland.
Charterers have the option the ship to discharge at PG/Japan range, including Pakistan /full India/full South East Asia/Philippines, but excluding ....

Comments:
The Owners, in comparison to the Charterers' offer:
-exclude to load north of the San Lorenzo and Nueva Palmira ports.
-exclude the Baltic as a discharging area.
-in addition, exclude some countries for discharge.

The Owners continue:
12.
-If more than one load port is going to be used for completion of the loading, then a cargo holds' and hatch covers' cleanliness inspection will take place only in the cargo holds and hatch covers indented for loading, at each port of contact. -if the loading operation partly commenced in the number of cargo holds and hatch covers accepted, then the time will count pro rata in relation to the ships' total cargo holds. However, if the ship commenced its full load operation which the port can provide in the approved cargo holds and hatch covers, then the time will count in full. If the approved cargo holds and hatch covers are fully loaded or the Master stopped the loading for reasons of stability and the rest of the cargo holds and/or hatch covers remain unacceptable, then the ship will be placed fully off-hire, from the time of the interruption, until such time as all the rejected cargo holds and/or hatch covers have successfully fully passed re-inspection and are ready for loading.
-if the ship's cargo holds and/or hatch covers have been partially or totally rejected and the Master has objected to the rejection, another well recognised independent surveyor then will be mutually appointed by the Charterers and Owners and his decision will be binding. The expenses incurred including the survey-

or's fees will be for the account of the party in fault. Until such time as the independent surveyor's decision is given, the ship will remain on-hire.

Comments:
The Owners, in comparison to the Charterers' offer:
-accept the Charterers offer but add that if part of the holds have been accepted and the loading operation has commenced in these holds then the time shall count pro rata in relation to the ship's total holds, but if it has commenced its full load operation then the time will count in full.
- if the Master has objected to the rejection, the decision of another independent surveyor will bind.

The Owners continue:
14.
USD 6,000 lumpsum ilohc
The ship's holds will be redelivered free of cargo residues.

Comments:
The Owners, in comparison to the Charterers' offer:
-request USD 6,000. - lumpsum in lieu of holds' cleaning (Charterers offer was at USD 5,000. -).
-request the holds to be redelivered free of cargo residues.

The Owners continue:
16.
The l.o.i. will be issued in the Charterers' letterhead and will be signed by a director of the company. The Original L.O.I. will be delivered to the Master's hands or to the Owners' premises, before the commencement of the discharging operation.

Comments:
The Owners, in comparison to the Charterers' offer:
-accept to discharge with an l.o.i. signed by the Charterers,
-request the l.o.i. to be issued in the Charterers' letterhead,
-request the l.o.i. to be signed by a director of the company and delivered to the Master or directly to the Owners before the commencement of the discharging operation.

<u>TO SUMMARIZE THE OWNERS' COUNTER REPLY:</u>

Mv Fotini/ <u>*Eagle Trading and Shipping Inc.*</u>

The Owners thank the Charterers for their offer, on which a/e for reply within 1 hr. as follows:

3.
Delivery on dlosps Antwerp port, Belgium atdnsshinc
4.
L/C 14/19 December .... 00:01-2400 hrs. For lay-can the local time to be in effect ship's full itinerary as follows: ....
Agents: ....
last 3 cargoes (starting from the last) as follows: ....
eta Paranagua .... agw wp uce
5.
TCT out via ECSA to Continent/Med/Black Sea or chopt to PMO/Japan range including PG.
The ship will trade via sp(s)/sb(s)/sa(s), always within I.N.L. with cargo of bulk corn.
The port/s, berth/s, anchorage/s will remain safe from the time of the ship's arrival at the first sea pilot station until departure and the last sea pilot has left the ship.
If anchorage/s is used, the Bimco's current 'Ship to Ship Transfer Clause for Time Charter Parties' will be in effect. Sub to agree to a full clause.
The ship will always lie afloat throughout the period of the Charter except at ECSA where the NAABSA clause will apply.
NAABSA Clause:
Charterers have the right to load or discharge the ship at any ports Up River, but not north of San Lorenzo timbues area and Nueva Palmira or in Brazilian port/s not north of Vitoria port (Vitoria to be excluded), provided that it is customary for ships of similar dimensions and size to navigate there, where they can safely lie aground on the soft muddy sea-bed, but always at the Charterers' risk and subject to the permissible and prevailing draft declared by the Port Authorities and/ or pilots.
If the ship lies aground and the Master suspects that the hull is damaged, then the Charterers are to employ, under their time and expenses, an inspection of the ship's flat bottom, including, but not limited to, the vertical sides, bilge keels, propeller and rudder blade, by a well-organized and ship's class approved diving company, to determine the extent of any damage. The inspection will be carried out under the appointed by the Owners' ship's class surveyor attendance, but at the Charterers' cost.
Should the ship's bottom and underwater parts, propeller, rudder be extensively damaged, affecting the ship's seaworthiness and class, the ship will be direct-

ed immediately and always on-hire to a mutually appointed convenient place or port or yard to carry out the repairs at the Charterers' time and cost and to the class surveyor's entire satisfaction. The Shipyard's expenses and port expenses, including in and out tugs, pilots, agency fees etc., will be entirely for the Charterers' account.

Time lost for unloading the cargo at the load port, if it is required, and any expenses incurred will be for the Charterers' account.

Damage not affecting the ship's seaworthiness and class will be repaired at the Charterers' cost, but not for the time lost, at the next scheduled dry docking, normal wear and tear excepted.

If, after the ship's underwater inspection, no damage is found, either minor or not, then all the directly related expenses incurred such as the diving company's expenses, ship's class surveyor's fees etc. will be shared equally between the Owners and Charterers, but the ship will remain always on-hire.

6.

Redelivery on dropping last outward sea pilot station.

7.

Duration if redelivery at Skaw/Passero about 60/70 days wog.

If redelivery at PMO/Japan about 85/95 days wog.

8.

Hire USD 12,000. - for redelivery at Skaw/Passero range.

Hire USD 17,000. - for redelivery at PMO/Japan range.

Charterers will declare the discharging range before the payment of the first hire.

The hire is payable every 15 days always in advance net of tax and bank charges, however, the first 15-day hire is payable 1 banking day after the ship's delivery and the receipt of the hire invoice from the Owners.

9.

The Charterers, before the ship is delivered to their services and at the last port of discharge of the ship's previous trade, have the option to supply the ship with fuel for their account, as long as that the same does not interfere with the full operation of the ship and is allowed (written approval) by the current Charterers and/ or after the Owners' consent. The Charterers are responsible for any loss of time which will result in the ship being placed off-hire, arising from any reason due to the Charterers' supply of fuel to the ship before its delivery to them. The Charterers will reimburse the Owners, as per the provided off-hire statement from the previous Charterers, for all time lost and fuel consumed during the off-hire period. Any expenses, losses, damages and/or consequences of any nature relating directly to the bunkering of the ship, including damages from pollution and fines imposed by any party, will be for the Charterers' account, liability, and responsibility.

The Owners have the option of bunkering the ship en route to its last port of call or at the last port of call before its redelivery for their account and after the Charterers' consent, as long as that the same does not interfere with the full operation of the ship or with the effect of not loading the full intended cargo quantity. Time lost, because of the bunkering for the Owners' account should be counted as off-hire.

The Charterers, for every bunkering operation arranged by them, under their time and expenses, will appoint a well recognised independent Surveyor, who will carry out a bunker survey before and after any bunkering operation takes place, in order to determine the quantity of fuel supplied to the ship.

Information on any given quantity of fuel to the Charterers is subject to the trip's final duration, the ship's consumption while manoeuvring in ports and in confined waters and subject to final destination.

10.

The ship will load only bulk corn. On delivery the ship must be suitable in every respect for loading and safe carriage of corn in bulk.

The Owners guarantee that the ship can fully load corn in bulk without using shifting boards or any other fittings. The ship must have on board, at any time during the duration of this Charter, complete Grain Trimming Tables and valid documents and certificates for grain trade issued by the ship's classification society and any other national authority on the basis of SOLAS 1974 Regulations and any subsequent thereto.

11.

Charterers guarantee that the ship will load only from the East Coast South America and if it loads in Argentina and/or Uruguay, not above the San Lorenzo and Nueva Palmira ports, respectively.

Charterers guarantee that the vessel will discharge only in full Mediterranean, including the Black sea area, but excluding .... or at the Charterers' option at Gibraltar/Skaw range, including the United Kingdom and Ireland.

Charterers have the option the ship to discharge at PG/Japan range, including Pakistan /full India/full South East Asia/Philippines, but excluding ....

12.

-If more than one load port is going to be used for completion of the loading, then a cargo holds' and hatch covers' cleanliness inspection will take place only in the cargo holds and hatch covers indented for loading, at each port of contact. -if the loading operation partly commenced in the number of cargo holds and hatch covers accepted, then the time will count pro rata in relation to the ships' total cargo holds. However, if the ship commenced its full load operation which the port can provide in the approved cargo holds and hatch covers, then the time will count in full. If the approved cargo holds and hatch covers are fully loaded or the Master stopped the loading for reasons of stability and the rest

of the cargo holds and/or hatch covers remain unacceptable, then the ship will be placed fully off-hire, from the time of the interruption, until such time as all the rejected cargo holds and/or hatch covers have successfully fully passed re-inspection and are ready for loading.
-if the ship's cargo holds and/or hatch covers have been partially or totally rejected and the Master has objected to the rejection, another well recognised independent surveyor then will be mutually appointed by the Charterers and Owners and his decision will be binding. The expenses incurred including the surveyor's fees will be for the account of the party in fault. Until such time as the independent surveyor's decision is given, the ship will remain on-hire.
14.
USD 6,000 lumpsum ilohc.
The ship's holds will be redelivered free of cargo residues.
16.
The l.o.i. will be issued in the Charterers' letterhead and will be signed by a director of the company. The Original L.O.I. will be delivered to the Master's hands or to the Owners' premises, before the commencement of the discharging operation.

CHARTERERS' COUNTER REPLY TO OWNERS' LAST:

Mv Fotini Eagle Trading and Shipping Inc.

Thanks for the Owners' counter on which a/e for reply within 30 mnts as follows:

Comments:
The Charterers' offer is valid for 30 mnts reply.

The Charterers continue:
8.
Hire USD 11,000. - for redelivery at Skaw/Passero.
Hire USD 16,000. - for redelivery at PMO/Japan range.
The hire is payable every 15 days always in advance, however the first 15-day hire is payable 2 banking days after the ship's delivery and the receipt of the hire invoice from the Owners.

Comments:
The Charterers, in comparison to the Owners' offer:
-have accepted delivery in Antwerp port and offer as 'hire rate' USD 11,000.
-for Skaw/Passero (Owners offer was at USD 12,000) and USD 16,000. - for PMO/Japan range (the Owners' offer was at USD 17,000. -).

-accept to declare the discharging range before the payment of the first hire.
-request the first 15 days hire to be paid within 2 banking days after the ship's delivery (Owners' offer was within 1 banking day).

OWNERS' COUNTER REPLY TO CHARTERERS' LAST:

Mv Fotini/ *Eagle Trading and Shipping Inc.*

a/e for reply within 15 mnts, basis take it or leave it:

8.
Hire USD 11,750. - for redelivery at Skaw/Passero.
Hire usd 16,750. - for redelivery at PMO/Japan range.

Comments:
The Owners' counter is based on 'take it or leave it', i.e., the Charterers will either accept it or will drop the negotiations.

CHARTERERS' COUNTER REPLY TO OWNERS' LAST:

Mv Fotini Eagle Trading and Shipping Inc.

Thanks for the Owners' last, but the Charterers regretfully cannot accept it.
The Owners and Charterers are free to work other biz/ship.

Comments:
The negotiations failed and both sides are free to look for other alternatives. Nevertheless, the brokers involved in the negotiations should not give the case up and should exert all their influence to bring the two sides back to negotiations and try to bridge the difference.
In similar cases, usually the brokers propose to both parties to
. 'split the difference',
i.e., to fix at, say, USD 11,400 for Skaw /Passero and USD 16,400 for PMO/Japan. Both parties will lose some money, but the fixture is saved.
If both parties finally accepted to return to negotiations and have agreed to split the difference at a 'hire rate' for Skaw/Passero at USD 11,400 and USD 16,400 for PMO/Japan, then the ship is fixed on the Main Terms.
The Charterers' broker will make the recapitulation which will be given to both parties as their guide on the terms that have been agreed. However, when complicated terms have been agreed, then will be given for their reconfirmation on the agreed terms.

THE RECAP (RECAPITULATION) COULD BE GIVEN AS FOLLOWS:

Ref Account Messrs Eagle Trading and Shipping Inc
Pnmx to make Paranagua 1/10 Jan.
*TCT out via ECSA to Continent/Med/Black Sea or chopt to PMO/Japan range including PG MV FOTINI_*

Recap of main terms agreed:
MV FOTINI
....
....
-The description of the ship as provided by the Owners is an accurate representation of the characteristics of the ship, which can only be modified with the usual shipping interpretation of the word 'about'. This clause supersedes any and all adding of the words 'without guarantee' in the description of the ship.
Owners guarantee that the:
-ship is a single deck bulk carrier and that it was originally constructed as a bulk carrier.
-ship has clear unobstructed main holds.
-ship does not have any horizontal frames and or beams within its holds.
-ship has no reefer space or special cargo locker's space.
-cargo will be loaded in the main holds only.
-ship is steel floored throughout -ship has its engine, bridge aft.
-ship's crew and officers will be ITF approved or equivalent acceptable to the ITF.
-ship is ISPS fitted.
Owners guarantee that during the duration of this Charter Party:
-the ship will not change ownership or class without the Charterers' written consent.
-the ship's hull and machinery insurance will be fully maintained and will not be changed.
-the ship is fully P&I Club covered, which will be maintained and will not be changed.
Owners confirm:
-that the ship has no outstanding recommendations, restrictions or any other requirements of any nature in respect to flag state, port state control or classification inspection.
-that the ship will comply with regulations and requirements at all ports of call and has valid certificates on board applicable to those ports, countries where the ship will trade under this Charter.
The Owners will fax copies of the certificates of the undermentioned ship which

are valid for the duration of the Charter Party:
- DOC
-SMC
-CLASS
-ISPS
-ISSC/ITC/P&I CLUB

For:
1.
Account Messrs. Eagle Trading and Shipping Inc. of British Virgin Islands.
Charterers' last three fixtures as follows:
Mv ...... Fixed through ......
Mv....... Fixed through......
Mv ......Fixed through....
2.
Subject to Shippers' and Receivers' approval declarable 1 hr after fixing the main terms.
3.
Delivery on dlosps Antwerp port, Belgium atdnsshinc.
4.
L/C 14/19 December .... 00:01-2400 hrs. For lay-can the local time to be in effect ship's full itinerary as follows: ....
Agents: ....
last 3 cargoes (starting from the last) as follows: ....
eta Paranagua .... agw wp uce
5.
TCT out via ECSA to Continent/Med/Black Sea or chopt to PMO/Japan range including PG.
The ship will trade via sp(s)/sb(s)/sa(s), always within I.N.L. with cargo of bulk corn.
The port/s, berth/s, anchorage/s will remain safe from the time of the ship's arrival at the first sea pilot station until departure and the last sea pilot has left the ship.
If anchorage/s is used, the Bimco's current 'Ship to Ship Transfer Clause for Time Charter Parties' will be in effect. Sub to agree to a full clause.
The ship will always lie afloat throughout the period of the Charter except at EC-SA where the NAABSA clause will apply.
NAABSA Clause:
Charterers have the right to load or discharge the ship at any ports Up River, but not north of San Lorenzo timbues area and Nueva Palmira or in Brazilian port/s not north of Vitoria port (Vitoria to be excluded), provided that it is customary

for ships of similar dimensions and size to navigate there, where they can safely lie aground on the soft muddy sea-bed, but always at the Charterers' risk and subject to the permissible and prevailing draft declared by the Port Authorities and/ or pilots.
If the ship lies aground and the Master suspects that the hull is damaged, then the Charterers are to employ, under their time and expenses, an inspection of the ship's flat bottom, including, but not limited to, the vertical sides, bilge keels, propeller and rudder blade, by a well-organized and ship's class approved diving company, to determine the extent of any damage. The inspection will be carried out under the appointed by the Owners' ship's class surveyor attendance, but at the Charterers' cost.
Should the ship's bottom and underwater parts, propeller, rudder be extensively damaged, affecting the ship's seaworthiness and class, the ship will be directed immediately and always on-hire to a mutually appointed convenient place or port or yard to carry out the repairs at the Charterers' time and cost and to the class surveyor's entire satisfaction. The Shipyard's expenses and port expenses, including in and out tugs, pilots, agency fees etc., will be entirely for the Charterers' account.
Time lost for unloading the cargo at the load port, if it is required, and any expenses incurred will be for the Charterers' account.
Damage not affecting the ship's seaworthiness and class will be repaired at the Charterers' cost, but not for the time lost, at the next scheduled dry docking, normal wear and tear excepted.
If, after the ship's underwater inspection, no damage is found, either minor or not, then all the directly related expenses incurred such as the diving company's expenses, ship's class surveyor's fees etc. will be shared equally between the Owners and Charterers, but the ship will remain always on-hire.
6.
Redelivery on dlosps 1 safe port or passing Skaw/Passero range including the UK and Ireland or chopt on dlosps 1 safe port or passing Muscat outbound/Japan range. Redelivery port at the Charterers' option. The ship will be redelivered atdnsshinc.
7.
Duration if redelivery at Skaw/Passero about 60/70 days wog- if redelivery at PMO/Japan about 85/95 days wog.
8.
Hire USD 11,400 pdprinclot and lubricants for redelivery at Skaw/Passero.
Hire USD 16,400 pdprinclot and lubricants for redelivery at PMO/Japan range.
Charterers will declare the discharging range before the payment of the first hire.
The hire is payable every 15 days always in advance net of tax and bank charg-

es, however the first 15-day hire is payable 2 banking days after the ship's delivery and the receipt of the hire invoice from the Owners.
9.
The bunkers on delivery will be about VLSFO ... mtns and about LSMGO .... mtns and the ship will be redelivered with about the same quantities.
Any minor difference on the bunkers' value on delivery and redelivery will be settled together with the final settlement with the exact figures.
The value of the bunkers on delivery is payable together with the first 15-day hire payment.
Owners guarantee that the ship is eligible to be supplied with bunkers in any country within the trading limits that have been agreed under this Charter Party.
Charterers have the right to deduct the value of the bunkers on redelivery, from the last 15-day hire payment, but if it is not sufficient it will be extended until the Charterers deduct the full value of the bunkers.
The Bunkers' prices on delivery and redelivery will be the same, USD .... per metric ton for the VLSFO and USD ... per metric ton for the LSMGO.
The Charterers, before the ship is delivered to their services and at the last port of discharge of the ship's previous trade, have the option to supply the ship with fuel for their account, as long as that the same does not interfere with the full operation of the ship and is allowed (written approval) by the current Charterers and/ or after the Owners' consent. The Charterers are responsible for any loss of time which will result in the ship being placed off-hire, arising from any reason due to the Charterers' supply of fuel to the ship before its delivery to them. The Charterers will reimburse the Owners, as per the provided off-hire statement from the previous Charterers, for all time lost and fuel consumed during the off-hire period. Any expenses, losses, damages and/or consequences of any nature relating directly to the bunkering of the ship, including damages from pollution and fines imposed by any party, will be for the Charterers' account, liability, and responsibility.
The Owners have the option of bunkering the ship en route to its last port of call or at the last port of call before its redelivery for their account and after the Charterers' consent, as long as that the same does not interfere with the full operation of the ship or with the effect of not loading the full intended cargo quantity. Time lost, because of the bunkering for the Owners' account should be counted as off-hire.
The Charterers, for every bunkering operation arranged by them, under their time and expenses, will appoint a well recognised independent Surveyor, who will carry out a bunker survey before and after any bunkering operation takes place, in order to determine the quantity of fuel supplied to the ship.
Information on any given quantity of fuel to the Charterers is subject to the trip's final duration, the ship's consumption while manoeuvring in ports and in con-

fined waters and subject to final destination.

10.

The ship will load only bulk corn. On delivery the ship must be suitable in every respect for loading and safe carriage of corn in bulk. Owners guarantee that the ship can fully load corn in bulk without using shifting boards or any other fittings. The ship must have on board, at any time during the duration of this Charter, complete Grain Trimming Tables and valid documents and certificates for grain trade issued by the ship's classification society and any other national authority on the basis of SOLAS 1974 Regulations and any subsequent thereto.

11.

Charterers guarantee that the ship will load only from the East Coast South America and if it loads in Argentina and/or Uruguay, not above the San Lorenzo and Nueva Palmira ports, respectively.

Charterers guarantee that the vessel will discharge only in full Mediterranean, including the Black sea area, but excluding .... or at the Charterers' option at Gibraltar/Skaw range, including the United Kingdom and Ireland.

Charterers have the option the ship to discharge at PG/Japan range, including Pakistan /full India/full South East Asia/Philippines, but excluding ....

12.

It is expressly agreed that:

1.The ship on delivery:

the ship on delivery or, at the latest, on arrival or berthing at the first or sole load port will have all cargo holds and hatch covers clean:

-free of rust, loose rust flakes, rust scales, and loose scales.

-free from salt and free from residues of previous cargo/oes.

-fresh water washed and dried.

-free from insects and odours.

and in all respects ready for loading the Charterers' intended cargo, according to the rules and regulations prevailing at the port of loading and to the satisfaction of the independent surveyors, as have been selected by the Charterers and/or Shippers, under their time and expenses and/or to the satisfaction of a national cargo bureau and/or local grain surveyors' inspection.

2. Cargo holds' and/or hatch covers' failure on delivery or at each loading port: if, the ship fails from the cargo holds and/or hatch covers survey inspection, then the ship will be placed fully off-hire, from the time of failure until such time as all the rejected cargo holds and/or hatch covers have successfully fully passed inspection and are ready for loading.

The Owners are to arrange without delay for the cargo holds' and/or hatch covers' cleaning, at their time and expenses with the ship's Crew, if allowed by the port's regulations and/or labour unions, or, otherwise with shore labourers, in order to meet the surveyors' requirements.

The Owners are responsible for:
-the amount of the fuel consumed during the off-hire period
-the expenses incurred for the purpose of holds' and/or hatch covers' cleaning
-all chemicals, fresh water, consumable materials, tools such as water guns, pressure hoses, pumps, mechanical de-rusting equipment, rustybus machines etc. necessary to be arranged and supplied to the ship, in making the same suitable for the intended for loading cargo.
-any directly related expenses incurred, but supported with original vouchers, including, but not limited to, stevedores stand-by, but up to a maximum of one shift If the Port Authorities and/or Shippers, order the ship to evacuate the berth and shift to anchorage or to change waiting place for the purpose of cargo holds' and/or hatch covers cleaning, then all the delays incurred and extra expenses incurred, including, but not limited to, tugs and/or pilots and/or Agency fees etc., until the ship is back at the berth, or in its initial position, will be entirely for the Owners' account.
-If more than one load port is going to be used for completion of the loading, then a cargo holds' and hatch covers' cleanliness inspection will take place only in the cargo holds and hatch covers indented for loading, at each port of contact. -if the loading operation partly commenced in the number of cargo holds and hatch covers accepted, then the time will count pro rata in relation to the ships' total cargo holds. However, if the ship commenced its full load operation which the port can provide in the approved cargo holds and hatch covers, then the time will count in full. If the approved cargo holds and hatch covers are fully loaded or the Master stopped the loading for reasons of stability and the rest of the cargo holds and/or hatch covers remain unacceptable, then the ship will be placed fully off-hire, from the time of the interruption, until such time as all the rejected cargo holds and/or hatch covers have successfully fully passed re-inspection and are ready for loading.
-if the ship's cargo holds and/or hatch covers have been partially or totally rejected and the Master has objected to the rejection, another well recognised independent surveyor then will be mutually appointed by the Charterers and Owners and his decision will be binding. The expenses incurred including the surveyor's fees will be for the account of the party in fault. Until such time as the independent surveyor's decision is given, the ship will remain on-hire.
13.
The Charterers will pay the Owners a lumpsum of USD 1,200 per month or pro rata for communications/victualing /entertainment.
14.
The ship will be redelivered to the Owners with its holds in the same condition as on delivery, normal wear and tear excepted.
However, the Charterers will have the option of redelivering the ship with un-

clean holds, by paying a lumpsum amount of USD 6,000-. The amount due will be paid together with the Charterers' final hire statement. The ship's holds will be redelivered free of cargo residues.
15.
A joint on-hire bunker and condition survey will be held by an independent surveyor, formally agreed on by both parties, at the port of delivery or, if mutually agreed, at the most convenient and proximate time/place/port to the delivery time/place/port.
Time actually lost due to this survey will be for the Owners' account, but the costs of the survey will be equally shared between the Owners and Charterers.
A joint off-hire bunker and condition survey will be held by an independent surveyor formally agreed by both parties, at the port of redelivery or, if mutually agreed, at the most convenient and proximate time/place/port to the redelivery time/ place/port.
Time actually lost due to this survey will be for the Charterers' account, but the costs of the survey will be equally shared between the Owners and Charterers.
16.
The Charterers will make every endeavour to ensure that the original Bill of Lading is available at the discharge port on or before the ship's arrival to discharge. However, if the original Bills of Lading are not available to discharge, then the Owners/ Master will permit the discharge of the cargo without delay in exchange for the Charterers' single L.O.I., as per the Owners' P&I Club's wording, signed only by the Charterers.
The l.o.i. will be issued in the Charterers' letterhead and will be signed by a director of the company. The Original L.O.I. will be delivered to the Master's hands or to the Owners' premises, before the commencement of the discharging operation.
17.
Arbitration in London with the current Bimco's 'Dispute Resolution Clause' to apply.
English Law will apply to negotiations and to the eventual fixture.
18.
All negotiations and eventual fixture will be kept pnc and will not be disclosed to others.
19.
Sub details that are based on NYPE 46, at the Charterers' executed proforma with 5 percent total commission including 3,75 percent address com.

PART III

CHAPTER III

CHARTER PARTY ADDITIONAL (OR RIDER) CLAUSES FOR TIME CHARTER

When the Owners and Charterers have finally reached an agreement on the Main Terms, they will further develop by negotiating the Charter Party Main Body and Additional Terms, commonly known as the 'details'.

The Charterers or, if mutually agreed, the Owners, will present their Charter Party pro forma (usually an executed one and based on a standard type of Charter Party) to the Owners or Charterers respectively, for their consideration and likely comments.

The Charter Party pro forma consists of two parts:

1. the Main Body.

The Main Body of a standard type of Charter Party is a standard wording as adopted or issued or recommended by various shipping organizations and amended from time to time.

These terms set the limit of obligations and responsibilities of each contracting party. However, all printing terms are negotiable, and the Charterers/Owners can make their deletions/amendments/ additions in accordance with the prevailing trading conditions and circumstances.

2. the Rider or Additional Clauses.

The Rider or Additional clauses cover any omissions of the main body or add new regulations as well as other specific/special terms that are in effect to the intended trip/s and/or to protect the Owners' and/or Charterers' interests.

The 'details', when working on the clauses of a Charter Party proforma for Time Charter, relegate the importance of the negotiations. The brokers involved in the negotiations will many times deem that they will conclude the fixture after having agreed on the Main Terms and, in any case, it is usually quite difficult to fail on the 'details'. However, the 'details' still comprise of many important terms that allocate liabilities and responsibilities for specific risks to each contracting party and should be seriously considered by all parties involved in the fixture.

In a few words:
the Owners and/or Charterers/Operators should exercise care and through the 'details', each party must try to make each case/term/clause "crystal clear', without doubts or room for any disagreement, otherwise the same might lead to extra costs and lengthy litigations.

The negotiations on the incorporated clauses will be in progress on an 'accept/ except' basis and both parties will work on a joint consideration and acceptance of all terms, by replying within
. a reasonable period of time (there is usually no time limit).
Nevertheless, until conclusion of the negotiations, in the case of a disagreement, each party has the right to withdraw its interest from the biz/ship. Here the shipbroker's experience is a necessity in order to save the fixture.
The presented Charter Party pro forma is usually a previously executed Charter Party with others and the same should be requested by the Owners from the time of presentation of their initial 'offer'. Otherwise, the Charterers should have the right to present their general not previously negotiated pro forma, obviously with terms entirely favourable to them. These terms should have not been approved and/or altered/amended by others, thus must be discussed at length, and might give rise to difficulties in concluding the agreement. The previously executed one is much preferred by all, as some of the unfavourable terms for the Owners should already have been negotiated and, logically, have been altered or deleted.
Some Charterers could also accept to work on an executed pro forma, as will be presented by the Owners. The Charterers will study it and they will respond with their possible comments, but, in any case, this way of working brings the final agreement much closer than the others.
In any case, the Charter Party executed proforma which will be in use is a bargaining point during the negotiations of the Main Terms.
Depending on the pressure of the market, the Owners or Charterers could show increased tolerance and lenience for certain terms during the negotiations on the Charter Party details, enabling the fixture to incorporate clear terms and to eventually be concluded. However, the Letter of Indemnity (l.o.i.), the type of Bill of Lading which will be issued for the Voyage and the way in which it will be presented/ marked, the double banking clause, the intermediate holds clean-

ing, the Interclub agreement, the bottom fouling clause etc., are some of the important terms which, if not taken seriously, could create huge losses for both parties.
The writing abbreviations that are used during the negotiations, including those on rider/ additional clauses, do not exist when the Charter Party will be issued. Whole words must be entered, which in this case would minimize disputes for any misunderstandings due to abbreviations.

For the terms/conditions of the Charter Party nothing is taken for granted, fix it with crystal clear terms so as to prevent the contracting parties from referring to arbitration.

The presentation of the undermentioned contents - Disclaimer notice:

For a Charter Party, which could be negotiated mainly for a long period, the Main Body could be worked as below and could contain, among others, the undermentioned additional clauses. A precise copy of certain selected previously executed Charter Parties clauses is given, but the presentation of the undermentioned contents is without any liability and responsibility, as well as the recommendations/ comments/information as will be provided here below. The contents, as well as the recommendations/comments/information contained here below, may not be accurate at times and presented without any responsibility and liability for the use of the same, and for any errors or omissions. The recommended full clauses can be altered/amended in accordance with the prevailing negotiations/conditions and it is up to the Owners', Charterers', or any other parties' choice on whether they will adopt the same or not. Many of the additional terms could also be used for a Voyage Charter, with slight amendments.

Time Charter

GOVERNMENT FORM

Approved by the New York Produce Exchange

November 6th, 1913 - Amended October 20th, 1921; August 6th, 1931; October 3rd, 1946

1. This Charter Party, made and concluded in .... day of ....19 ....

RECOMMENDATIONS:

INSERT THE PLACE AND THE DATE OF FIXTURE.

2. Between ....

RECOMMENDATIONS:

INSERT THE FULL STYLE OF THE SHIP OWNING COMPANY.

3. Owners of the good Steamship/Motorship .... of......................

RECOMMENDATIONS:

DELETE THE WORD 'steamship' AND AFTER 'motorship' INSERT HE NAME OF THE SHIP DELETE THE WORD 'of' AND INSERT 'see ship's description clause....'.

4. of ......tons gross register, and .... tons net register, having engines of ....indicated horse power

RECOMMENDATIONS:

INSERT THE GROSS TONNAGE (GT) AND THE NET TONNAGE (NT) OF THE SHIP.
DELETE THE WORDS 'engines of... indicated horse power'
OR
DELETE THE ENTIRE LINE.

5. and with hull, machinery and equipment in a thoroughly efficient state, and classed...........................

RECOMMENDATIONS:

INSERT THE CLASS OF THE SHIP
OR
DELETE THE ENTIRE LINE.

6. at ....of about ....cubic feet bale capacity, and about............. tons of 2240 lbs.

RECOMMENDATIONS:

FOR CARGOES IN BULK INSERT THE CUBIC FEET GRAINS OF THE SHIP, DELETE THE WORD 'bale' AND INSERT 'grain'.
FOR CARGOES IN ANY OTHE FORM EXCEPT BULK, THE WORD 'bale' REMAINS AND INSERT THE SHIP'S CUBIC FEET BALES.
AFTER 'capacity' INSERT 'in main holds' OR 'in clear/unobstructed main holds'
AFTER THE WORDS 'and about', INSERT THE SHIP'S SUMMER DWAT AND DELETE THE WORDS 'of 2240 lbs.'
OR
DELETE THE ENTIRE LINE.

7. deadweight capacity (cargo and bunkers, including fresh water and stores not exceeding one and one half percent of
ship's deadweight capacity, 418

RECOMMENDATIONS:

DELETE 'capacity' INSERT 'all told ....' AND DELETE 'cargo and bunkers .... deadweight capacity,'

OR

DELETE THE ENTIRE LINE.

8. allowing a minimum of fifty tons) on a draft of ....feet ....inches on .... Summer freeboard, inclusive of permanent bunkers,

RECOMMENDATIONS:

DELETE THE WORDS 'allowing a minimum of fifty tons'. AFTER THE WORDS 'on a draft of' INSERT THE SHIP'S DRAFT EITHER IN FEET AND
INCHES OR IN METERS, IN WHICH CASE DELETE THE WORDS 'feet .... . inches' AND INSERT 'meters'.

THE SHIP'S DWAT /DRAFT IS ALWAYS GIVEN IN SUMMER SALT WATER, THEREFORE, BEFORE THE WORD 'Summer' INSERT THE WORDS 'salt water'

OR

DELETE THE ENTIRE LINE.

9. which are of the capacity of about ....tons of fuel, and capable of steaming, fully laden, under good weather

10. conditions about .... knots on a consumption of about .... tons of best Welsh coal best grade fuel oil best grade Diesel oil,

RECOMMENDATIONS:

AFTER 'of the capacity of about' INSERT THE CAPACITY OF THE BUNKERS
AFTER 'condition' INSERT 'throughout the currency of this Charter Party'
AFTER BOTH WORDS 'about' INSERT THE SERVICE SPEED AND THE CONSUMPTION RESPECTIVELY AND DELETE THE REST OF THE WORDS 'of best Welsh coal.......'

OR

DELETE BOTH LINES AND INSERT 'see ship's description clause....'.

11. now ....

RECOMMENDATIONS:

INSERT THE SHIP'S ACTUAL POSITION ON THE DATE OF FIXTURE, ALTHOUGH THE WORD 'trading' IS USUALLY USED.

12. .... and ....Charterers of City of ....

RECOMMENDATIONS:

AFTER 'and' INSERT THE FULL STYLE OF THE CHARTERERS AND AFTER 'of city of' THE PLACE OF THEIR BIZ.

13. Witnesseth, that the said Owners agree to let, and the said Charterers

agree to hire the said vessel, from the time of delivery, for
14. about……………………………………………………………………………………………
15. within below mentioned trading limits.
RECOMMENDATIONS:
INSERT THE AGREED CHARTER PERIOD OR THE AGREED WORDING.
FOR EXAMPLE, 'one time charter trip with one loading leg……etc.'
THE WORD 'about' WILL REMAIN OR NOT DEPENDING ON THE AGREEMENT FOR THE TIME OF PERIOD.
16. Charterers to have liberty to sublet the vessel for all or any part of the time covered by this Charter, but Charterers remaining responsible for
17. the fulfillment of this Charter Party.
RECOMMENDATIONS:
AFTER 'sublet the vessel' INSERT 'under the same terms and conditions'
AFTER THE WORD 'fulfillment' INSERT '/payment of hires/performance/and settle all claims with Owners, whether the Charterers have settled with their Sub-Charterers or not. The ship not to be sub-chartered to companies based in …. /…../….. The ship not to
be sub-chartered to companies which may lead to the ship's subsequent boycotting. The Charterers will advise the Owners in advance of the intended company. Acceptance of delivery by Charterers will not constitute any waiver of the Owners' obligations hereunder.'
18. Vessel to be placed at the disposal of the Charterers, at ……………………………………………
19. ……………………………………………
RECOMMENDATIONS:
INSERT THE DELIVERY PORT/AREA AND THE POINT OF DELIVERY. FOR EXAMPLE. 'on afsps' OR 'passing….' ETC.
IF THE DELIVERY IS PASSING CHANAKALE IN TURKEY THEN INSERT 'any delays in reaching the delivery point due to traffic/congestion in entering Dardanelles to be for the Charterers' account'.
20. in such dock or at such wharf or place (where she may safely lie, always afloat, at all times of tide, except as otherwise provided in clause No 6) as
21. the Charterers may direct. If such dock, wharf or place be not available time to count as provided for in clause No 5.
RECOMMENDATIONS:
COMPLETELY DELETE THE LINES.
21. …………Vessel on her delivery to be
RECOMMENDATIONS:
DELETE 'delivery' INSERT 'arrival first loading port',
OR
REMAINS AS PRINTED AND INSERT 'or latest upon arrival at first loading port'.

22. ready to receive cargo with clean-swept holds and tight, staunch, strong and in every way fitted for service, having water ballast, winches and

RECOMMENDATIONS:

AFTER 'ready' INSERT 'in all respects'.

AFTER 'receive' INSERT 'any permissible'.

AFTER 'with clean-swept holds' INSERT 'see holds' cleanliness inspection on delivery Clause ....' AND DELETE THE REST OF THE LINE.

OTHERWISE:

AFTER 'clean-swept' INSERT 'free of smell, fresh water washed and dried'.

AFTER 'holds' INSERT 'free from loose rust flakes, scale and residues of previous cargoes'.

AFTER 'every way fitted for the service' INSERT 'and for Charterers' cargoes/ trade and will be so maintained whilst under this Charter party and is construed and equipped in accordance with SOLAS 1974 Regulations and any subsequent amendments and complies with such regulations at all times under the present Charter'.

23. donkey boiler with sufficient steam power, or if not equipped with donkey boiler, then other power sufficient to run all the winches at one and the same

RECOMMENDATIONS:

IF THE SHIP IS FITTED WITH CRANES THEN DELETE 'winches' AND INSERT 'cranes'.

IF THE SHIP IS GEARLESS THEN DELETE THE ENTIRE LINE.

24. time (and with full complement of officers, seamen, engineers and firemen

for a vessel of her tonnage), to be employed, in carrying lawful merchan-

RECOMMENDATIONS:

DELETE 'seamen, engineers and firemen' INSERT 'and crew according to regulations'.

AFTER 'in carrying' INSERT 'only'.

AFTER 'lawful' INSERT 'harmless'.

25. dise, including petroleum or its products , in proper containers, excluding.....................................

RECOMMENDATIONS:

DELETE 'including petroleum or its products, in proper containers,'.

AFTER THE WORD 'excluding 'INSERT 'see Cargo Exclusions Clause....'.

26. (vessel is not to be employed in the carriage of Live Stock, but Charterers are to have the privilege of shipping a small number on deck at their risk,

27. all necessary fitting and other equipments to be for account of Charterers) in such lawful trades, between safe port and/or ports in British North

28. America, and/or United States of America, and/or West Indies, and/or Central America, and/or Caribbean Sea, and/or Gulf of Mexico, and/or

29. Mexico, and/or South America ....
.. and/or Europe
30. and/or Africa , and/or Asia , and/or Australia, and/or Tasmania , and/or New Zealand, but excluding Magdalene River, River St. Lawrence between
31. October 31st and May 15th , Hudson Bay and all unsafe ports . also excluding when out of season, White Sea, Black Sea and the Baltic,
RECOMMENDATIONS:
COMPLETELY DELETE THE LINES.
32. ....
RECOMMENDATIONS:
INSERT 'see Trading Exclusions Clause....'.
33. ....
34. ....
35. as the Charterers or their Agents shall direct, on the following conditions:
RECOMMENDATIONS:
DELETE THE WORD 'or their Agents' OTHERWISE THE CHARTERERS SHOULD CONFIRM IN WRITING THAT THE MASTER WILL ALSO FOLLOW THE CHARTERERS' AGENTS' INSTRUCTIONS.
36. 1.That the Owners shall provide and pay for all provisions, wages and consular shipping and discharging fees of the Crew; shall pay for the
RECOMMENDATIONS:
AFTER 'provisions' INSERT 'stores, lubricants'.
AFTER 'wages' INSERT 'including officer's and crew's overtime/boatage for their own business/immigration/repatriation, also consular fees pertaining to ship's nationality or flag, non-compulsory garbage removal and non-compulsory gangway watchmen and all other charges relating to the Master, Officers and crew, fines imposed by the authorities on crew and/or ship, unless caused by neglect or default on the part of the Charterers or their agents/servants'.
37. insurance of the vessel, also for all the cabin, deck, engine-room and other necessary stores, including boiler water and maintain her class and keep
38. the vessel in a thoroughly efficient state in hull, machinery and equipment for and during the service.
RECOMMENDATIONS:
AFTER 'necessary stores' INSERT 'including towing and mooring line'.
DELETE 'boiler water' INSERT 'domestic consumption of fresh water'.
AFTER 'hull' INSERT 'cargo holds' AND IF THERE IS CARGO ON DECK, INSERT 'cargo spaces'.
AFTER 'service' INSERT 'with all valid certificates and fittings necessary to comply with current requirements at all ports of call and canals, failing which the Owners shall be responsible for all lost time and for the directly related expenses incurred thereby.'

ALSO, THE FOLLOWING COULD BE ADDED 'Owners will pay to the Charterers USD 500 per month in respect of fuel consumed for the ship's domestic purposes'.

39. 2. That the Charterers shall provide and pay for all the fuel, except as otherwise agreed. Port Charges,, Pilotages, Agencies, Commissions,

40. Consular Charges (except those pertaining to the Crew), and all other usual expenses except those before stated, but when the vessel puts into

RECOMMENDATIONS:

AFTER 'that the Charterers' INSERT 'whilst on hire'.

AFTER 'pilotages' INSERT 'compulsory or customary or recommended by a governmental Authority or internationally recognized body, including Pilotages for bunkering purposes in Singapore and for pilotages used for Passing Torres Strait/ Barrier Reef pilots/ Japan sea inland pilots but Owners to co-operate and minimize pilotage at Japan whenever it is possible and weather permitting, Bosporus and Chanakale pilots, Grate Belt/and Sound pilots (or passing Skaw/Great Belts to the last pilot station at Gedser), Magellan Straits pilots but the Master will endeavor his best not to use non-compulsory pilots provided weather/traffic/visibility/sea conditions permitting and pilotages which the Master deems necessary for maneuvering due to bad weather and/or shallow/narrow waters, as long as it is understood that the Master's request is not to be unreasonable for ships of similar size and type to do so'

OR

AFTER 'pilotages' INSERT 'compulsory, otherwise, for non-compulsory pilotage, on occasion subject to the Charterers' approval'.

AFTER 'Agencies' INSERT 'except those levied for the Owners' business, ports disbursements, Canal tolls, deck Helmsmen, tuggage including special hawswer in ports where it is required, any and all taxes/dues/wharfages calculated or levied on cargo/ship/voyage freight/charter hire/flag/ownership, extra insurance on cargo and ship's age, compulsory garbage removal, boatage on Charterers business'.

AFTER 'to the crew 'INSERT 'or 'flag/ownership'.

41. a port for causes for which vessel is responsible, then all such charges incurred shall be paid by the Owners. Fumigations ordered because of

RECOMMENDATIONS:

AFTER 'vessel' INSERT 'and/or Owners'.

42. illness of the crew to be for Owners account. Fumigations ordered because of cargoes carried or ports visited while vessel is employed under this

43. charter to be for Charterers account. All other fumigations to be for Charterers account after vessel has been on charter for a continuous period

44. of six months or more.

RECOMMENDATIONS:

COMPLETELY DELETE THE LINES AND INSERT 'see. Fumigation clause......',
OR
AFTER illness of the crew' INSERT 'on delivery or at first load port, or cargoes carried before to delivery'.
AFTER 'under this charter' INSERT 'and all directly related expenses including lodging expenses'.
AFTER 'to be for Charterers account.' INSERT 'Fumigations will be carried out in accordance with IMO/SOLAS regulations and recommendations'.
THE CHARTERERS USUALLY DELETE 'All other fumigations ......of six months or more'.
AT THE END ADD: 'Owners to provide and keep on board valid deratisation certificate or deratting exemption certificate throughout the Time Charter period.
45. Charterers are to provide necessary dunnage and shifting boards, also any extra fittings requisite for a special trade or unusual cargo, but
46. Owners to allow them the use of any dunnage and shifting boards already aboard vessel. Charterers to have the privilege of using shifting boards
47. for dunnage, they making good any damage thereto.
RECOMMENDATIONS:
LINE 45 AFTER 'necessary dunnage 'INSERT '/materials for separation/ lashing/ securing'.
AFTER 'use of any dunnage' INSERT '/lashing equipment'.
LINE 46 AFTER 'the use of any dunnage' INSERT 'and lashing materials but in case of any damage, the Charterers are to replace the same to the Owners before redelivery ',
OR
'and lashing materials but in case of any damage, the Charterers are to return such material in the same good order and condition as delivered, ordinary wear and tear excepted'.
48. 3.That the Charterers at the port of delivery, and the Owners, at the port of re delivery, shall take over and pay for all fuel remai ning on
49. board the vessel at the current prices in the respective ports, the vessel to be delivered with not less than......... ....tons and not more than
50. ....tons and to be re delivered with not less than .... tons and not more than .... tons
RECOMMENDATIONS:
COMPLETELY DELETE THE LINES AND INSERT 'see Bunkering Clause ...'.
51. 4. That the Charterers shall pay for the use and hire of the said Vessel at the rate of ....
52. United States Currency, per ton on vessel's total deadweight carrying capacity ,including bunkers and
RECOMMENDATIONS:

INSERT THE AGREED HIRE RATE.
DELETE THE ENTIRE LINE 52.

53. stores, on .... summer freeboard, per Calendar Month, commencing on and from the day of her delivery, as aforesaid, and at

RECOMMENDATIONS:

DELETE 'stores, on .... summer freeboard, per Calendar Month,'.
DELETE 'the day' AND INSERT 'the time'.

54. and after the same rate for any part of a month; hire to continue until the hour of the day of her re-delivery in like good order and condition, ordinary
55. wear and tear excepted, to the Owners (unless lost), at..........

RECOMMENDATIONS:

DELETE 'a month' INSERT 'a day'.
DELETE 'hour of the day' INSERT 'time'.
AFTER 'wear and tear excepted' INSERT 'but stevedore damage not to be considered fair wear and tear'.
AFTER 'to the Owners (unless lost) at ......' INSERT THE DURATION AND THE RE-DELIVERY POINT OF THE SHIP.

56. unless otherwise mutually agreed. Charterers are to give Owners not less than .... days
57. notice of vessels expected date of re-delivery, and probable port .

RECOMMENDATIONS:

COMPLETELY DELETE THE LINES AND INSERT 'the Charterers are to give the Owners no less than.../.../.... approximate redelivery notices and probable port and 10/5/3/2/1 definite notices including definite port',
OR
'see Charterers' notices on redelivery cl....'.

58. 5. Payment of said hire to be made in New York in cash in United States Currency, semi monthly in advance, and for the last half month or

RECOMMENDATIONS:

AFTER 'payment of said hire' INSERT 'discountless except for where otherwise specified in this Charter Party'.
DELETE 'New York' AND INSERT THE PLACE WHERE THE OWNERS' BANK IS BASED
OR
INSERT 'to the Owners' designated bank account'.
DELETE 'United States Currency' AND INSERT 'USD'.
DELETE 'semi month' INSERT 'every 15 days always in advance'.
DELETE 'half month' INSERT '15 days'.

59. part of same the approximate amount of hire , and should same not cover the actual time, hire is to be paid for the balance day by day ,as it becomes
60. due, if so required by Owners, unless bank guarantee or deposit is made

by the Charterers, otherwise failing the punctual and regular payment of the
61. hire, or bank guarantee, or on any breach of this Charter Party, the Owners shall be at liberty to withdraw the vessel from the service of the Char-
62. terers, without prejudice to any claim they (the Owners) may otherwise have on the Charterers.
Time to count from 7 a.m. on the working day
63. following that on which written notice of readiness has been given to Charterers or their Agents before 4 p.m. but if required by Charterers, they
64. to have the privilege of using vessel at once, such time used to count as hire.

RECOMMENDATIONS:

DELETE FROM 'Time to count......until ...4 p.m.' AND INSERT 'Time to count as from the ship's delivery, all time day/night /Saturdays/ Sundays/Holidays included.'.

DELETE 'at once' INSERT 'earlier, if available and ready'.

65.Cash for vessel's ordinary disbursements at any port may be advanced as required by the Captain , by the Charterers or their Agents, subj ect
66. to 2 1/2% commission and such advances shall be deducted from the hire. The Charterers, however, shall in no way be responsible for the application
67. of such advances.

RECOMMENDATIONS:

DELETE 'Captain' AND INSERT 'Owners'

OR

REMAINS AS PRINTED AND AFTER 'Captain' INSERT 'but subject to the Owners' written approval'

OR

'but the Charterers are not to advance any cash to the Master on behalf of the Owners, unless mutually agreed'.

AT THE END ADD 'and in case the Owners' outlays are disputed, the Owners are to settle disputed items with the Agents involved directly'.

68. 6. That the cargo or cargoes be laden and/or discharged in any dock or at any wharf or place, that Charterers or their Agents may

RECOMMENDATIONS:

DELETE 'in any dock or at any wharf or place' INSERT 'in any safe berth'.

IF CHARTERERS ADD 'and in any safe anchorage' THEN INSERT 'The Bimco's current 'ship to ship transfer clause for time charter parties' to apply'.

69. direct, provided the vessel can safely lie always afloat at any time of tide, except at such place where it is customary for similar size vessels to safely
70. lie aground.

RECOMMENDATIONS:

DELETE 'such places' INSERT 'customary ports in Argentina/Uruguay/Brazil not

North of Vitoria (Vitoria to be excluded) and provided there is a soft, muddy sea-bed. The vessel is not to be asked to maneuver or shift when safely aground',
OR
'The Charterers have the option at customary ports in Argentina/Uruguay/Brazil not North of Vitoria (Vitoria to be excluded), and, provided there is a soft muddy sea-bed, for the ship to safely lie aground where ships of similar size and draft are accustomed to lie safely aground in the above mentioned areas. In case the ship is directed for an NAABSA procedure at any place other than those aforementioned, then, after completion of each loading and/or discharging operation or at the nearest next available port/place, the Charterers, under their time/expenses, will appoint a diving inspection in order to certify whether any damage is caused due to the ship's running aground.
OR
'see Trading Exclusions NAABSA clause....'
71. 7. That the whole reach of the Vessel's Hold, Decks, and usual places of loading (not more than she can reasonably stow and carry), also
72. accommodations for Supercargo, if carried, shall be at the Charterers' disposal, reserving only proper and sufficient space for Ship's officers, crew,
73.tackle, apparel, furniture, provisions, stores and fuel. Charterers have the privilege of passengers as far as accommodations allow, Charterers
74. paying Owners .... Per day per passenger for accommodations and meals. However , it is agreed that in case any fines or extra expenses are
75. incurred in the consequences of the carriage of passengers, Charterers are to bear such risk and expense.
<u>RECOMMENDATIONS:</u>
DELETE 'decks' INSERT 'no deck cargo'.
OR
'see Deck Cargo clause....'.
DELETE 'charterers have the privilege of passengers.... until the end of the line.
75. ' INSERT 'no passengers except for the Port Captain or the Charterers' representative in the port are allowed to board the ship only'
76. 8. That the Captain shall prosecute his voyages with the utmost despatch , and shall render all customary assistance with ship's crew and
<u>RECOMMENDATIONS:</u>
AFTER 'utmost dispatch' INSERT 'subject to weather and sea conditions permitting'.
77. boats. The Captain (although appointed by the Owners), shall be under the orders and directions of the Charterers as regards employment and
78. agency, and Charterers are to load, stow, and trim the cargo at their expense under the supervision of the Captain, who is to sign Bills of Lading for
79.cargo as presented, in conformity with Mate's receipts or Tally Clerk's re-

ceipts.

RECOMMENDATIONS:

AFTER 'agency' INSERT 'and all communication/instruction to be given to the Master in clear English without abbreviations'.

DELETE THE WORD 'and' LOCATED BETWEEN THE WORDS 'stow' AND 'trim'.

AND AFTER 'trim' INSERT 'discharge, lash, unlash, secure, unsecure, chock, unchock, tally'.

DELETE 'as presented'.

BEFORE 'conformity' INSERT 'always in strict'.

DELETE 'or' INSERT 'and/or '.

80. 9. That if the Charterers shall have reason to be dissatisfied with the conduct of the Captain, Officers, or Engineers, the Owners shal on

81.receiving particulars of the complaint, investigate the same, and, if necessary, make a change in the Appointments

RECOMMENDATIONS:

AT THE END ADD: 'The provision of this clause does not affect the Charterers' right to advance any claim or require arbitration of any dispute regarding the conduct of the Master in prosecution of the voyages pursuant to the Charter Party terms and in carrying out the legitimate orders and directions as regards employment of the Charterers.'.

82. 10. That the Charterers shall have permission to appoint a Supercargo, who shall accompany the vessel and see that voyages are prosecuted

83. with the utmost despatch. He is to be furnished with free accommodation, and same fare as provided for Captain's table. Charterers paying at the

RECOMMENDATIONS:

COMPLETELY DELETE THE LINES AND INSERT 'no Supercargo is allowed'

OTHERWISE

DELETE 'who shall accompany the vessel and see that voyages are prosecuted with the utmost dispatch' AND INSERT 'to monitor the loading or discharging in the ports'.

AFTER 'free accommodation' INSERT 'if available'.

AFTER 'for Captain's table.' INSERT 'Supercargo and his office are obliged before boarding the ship to sign the owners' P&I Club's letter 'free pass agreement' and original letters to be handed over to the Master on boarding.'.

84. rate of $ 1.00 per day. Owners to victual Pilots and Customs Officers, and also, when authorized by Charterers or their Agents, to victual Tally

85. Clerks, Stevedore's Foreman, etc., Charterers paying at the current rate per meal, for all such victualling.

RECOMMENDATIONS:

DELETE '$ 1,00' INSERT '$ 15'.

AT THE END INSERT 'C/V/E charges do not include the above victualing'.

86. 11. That the Charterers shall furnish the Captain from time to time with all requisite instructions and sailing directions, in writing , and the
87. Captain shall keep a full and correct Log of the voyage or voyages, which are to be patent to the Charterers or their Agents, and furnish the Char
88. terers, their Agents or Supercargo, when required, with a true copy of daily Logs, showing the course of the vessel and distance run and the con-
89. sumption of fuel.

RECOMMENDATIONS:

AFTER 'in writing' INSERT 'in English language'.

AFTER 'true copy of daily Logs, INSERT 'abstracts for deck and engine in English Language'.

AT THE END ADD 'as well as revolutions of main engine and velocity and direction of wind and sea, all in English language.

90. 12. That the Captain shall use diligence in caring for the ventilation of the cargo.

RECOMMENDATIONS:

AT THE END INSERT 'The ship can provide only natural ventilation. The Charterers are to give specific instructions to the Master for any ventilation requirement out of the ordinary',

OR

AT THE END INSERT 'The ship will be held responsible for any damage to the cargo caused by inadequate ventilation or caused by water through ventilators'.

91. 13. That the Charterers shall have the option of continuing this charter for a further period of ....
92. ....
93. on giving written notice thereof to the Owners and their Agents ....days previous to the expiration of the first named term, or any declaredo ption.

RECOMMENDATIONS:

COMPLETELY DELETE THE LINES.

94. 14.That if required by Charterers, time not to commence before .... and should vessel
95.not have given written notice of readiness on or before .... But not later than 4 p.m. Charterers or
96. their Agents to have the option of canceling this Charter at any time not later than the day of vessel's readiness.

RECOMMENDATIONS:

AFTER 'time not to commence before' INSERT '(the first lay-day) 00:01 hrs'.

AFTER 'on or before' INSERT '(the cancelling date) 24:00 hrs'.

DELETE 'but not later than 4 p.m.'

AT THE END ADD 'Should it become evident that for unforeseen reasons beyond the Owners' control, the ship will miss its canceling date, the Owners are to notify the Charterers immediately stating the reason for the delay, the ship's actual position, e.t.a. and the new canceling date which is required. Within two working days of receipt of such notice, the Charterers will either grant an extension as was required or will exercise their option to cancel the ship.

97. 15. That in the event of the loss of time from deficiency of men or stores, fire, breakdown or damages to hull, machinery or equipment ,

98. grounding, detention by average accidents to ship or cargo, drydocking for the purpose of examination or painting bottom, or by any other cause

RECOMMENDATIONS:

AFTER 'loss of time' INSERT 'including the loss of time for deviation'.

AFTER 'from' INSERT 'default and/or sickness, accident or'.

AFTER 'deficiency of' INSERT 'Owners'.

AFTER 'grounding' INSERT 'unless such damage/breakdown/grounding caused by fault or wrong/unclear information given by the Charterers and/or the Charterers' agents or servants,'.

AFTER 'average detention' INSERT 'for which the Owners are found to be responsible by virtue of the terms of the Charter Party'.

AFTER 'to ship or cargo' INSERT 'unless resulting from inherent vice, quality or defect of the cargo'.

DELETE 'drydocking for the purpose of examination or painting of the bottom'.

AFTER 'by any other cause' INSERT 'attributable to the ship and/or crew unless it is the Charterers' or the Charterers' agents or Shippers'/Receivers' default or for any other reason beyond the ship's fault defect whatsoever'.

99. preventing the full working of the vessel, the payment of hire shall cease for the time thereby lost and if upon the voyage the speed be reduced by

RECOMMENDATIONS:

DELETE 'working' INSERT 'use including the Master's/Officers'/crew's strike'.

AFTER 'of the vessel' INSERT 'to the Charterers'.

AFTER 'shall cease' INSERT 'and fuel consumed to be for the Owners' account'.

AFTER 'shall cease for the' INSERT 'actual'.

AFTER 'cease for the time thereby lost' INSERT 'until the ship has returned to the same or equivalent position and resumed its voyage'

OR

'until the ship is again in full efficiency and resumes its service from a position no less favorable to the Charterers than that at which such loss of time commenced'.

100. defect in or breakdown of any part of her hull, machinery or equipment, the time so lost , and the cost of any extra fuel consumed in consequence

101. thereof, and all extra expenses shall be deducted from the hire.

RECOMMENDATIONS:

AFTER 'the time' INSERT 'actual'.

AFTER 'so lost' INSERT "until the ship has returned to the same or equivalent position'.
AFTER 'all extra' INSERT 'directly related undisputed'
AFTER 'expenses' INSERT 'supported by vouchers'.
AT THE END ADD 'Charterers shall give credit for fuel/diesel oil or any other expenses saved in consequence of such reduction in speed or off-hire'.

102. 16. That should the Vessel be lost, money paid in advance and not earned (reckoning from the date of loss or being last heard of) shall be
103. returned to the Charterers at once. The act of God, enemies, fire, restraint of of Princes, Rulers and People, and all dangers and accidents of the Seas,
104. Rivers, Machinery, Boilers and Steam Navigation, and errors of Navigation throughout this Charter Party, always mutually excepted.

RECOMMENDATIONS:

AFTER 'and not earned 'INSERT 'and any overpaid hire or excess deposit'.
AFTER 'errors' INSERT 'and negligence'.

105. The vessel shall have the liberty to sail with or without pilots, to tow and to be towed, to assist vessels in distress, and to deviate for the
106. purpose of saving life and property.

RECOMMENDATIONS:

AT THE END ADD 'see Deviation-Put Back clause....'.

107. 17. That should any dispute arise between Owners and the Charterers, the matter in dispute shall be referred to three persons at New York,
108. one to be appointed by each of the parties hereto, and the third by the two so chosen. Their decision or that of any two of them, shall be final, and for
109. the purpose of enforcing any award, this agreement may be made a rule of the Court. The arbitrators shall be commercial men.

RECOMMENDATIONS:

DELETE FROM 'to three persons at New York....... until......The arbitrators shall be commercial men' AND INSERT 'in London.
This Charter Party shall be governed by and construed in accordance with English Law. See Arbitration clause....'.

110. 18. That the Owners shall have a lien upon all cargoes, and all sub-freights for any amounts due under this Charter, including General Aver-

RECOMMENDATIONS:

AFTER 'sub-freights' INSERT 'freights/sub-hires/dead freights/demurrages/sub-demurrages/times for detention and all and any other amounts due to the Charterers under this Charter Party'.
DELETE 'freights for any amounts due under this Charter'.

111. age contributions, and the Charterers to have a lien on the Ship for all monies paid in advance and not earned, and any overpaid hire or excess
112. deposit to be returned at once .Charterers will not suffer, nor permit to be

continued, any lien or encumbrance incurred by them or their agents, which
113. might have priority over the title and interest of the owners in the vessel.

<u>RECOMMENDATIONS:</u>

AFTER 'or their agents' INSERT 'or servants'.

AT THE END ADD 'The Charterers undertake that, during the period of this Charter Party, they will not procure any supplies or necessities or service, including any port expenses and bunkers, on the credit of the Owners/ship or in the Owners' account.'.

114. 19. That all derelicts and salvage shall be for Owners' and Charterers' equal benefit after deducting Owners' and Charterers' expenses and
115. Crew's proportion . General Average shall be adjusted, stated and settled according to Rules 1 to 15, inclusive, 17 to 22, inclusive, and Rule F of
116. York-Antwerp Rules 1924,at such port or place in the United States as may be selected by the carrier, and as to matters not provided for by these
117. Rules, according to the laws and usages at the port of New York. In such adjustment disbursements in foreign currencies shall be exchanged into
118. United States money at the rate prevailing on the dates made and allowances for damage to cargo claimed in foreign currency shall be converted at
119. the rate prevailing on the last day of discharge at the port or place of final discharge of such damaged cargo from the ship. Average agreement or
120. bond and such additional security, as may be required by the carrier, must be furnished before delivery of the goods. Such cash deposit as the carrier
121. or his agents may deem sufficient as additional security for the contribution of the goods and for any salvage and special charges thereon, shall, if
122. required, be made by the goods, shippers, consignees or owners of the goods to the carrier before delivery. Such
deposit shall, at the option of the
123. carrier, be payable in United States money and be remitted to the adjuster. When so remitted the deposit shall be held in a special account at the
124. place of adjustment in the name of the adjuster pending settlement of the General Average and refunds or credit balances, if any, shall be paid in
125. United States money.

<u>RECOMMENDATIONS:</u>

DELETE 'to Rules 1 to 15, inclusive, 17 to 22, inclusive and Rule F of'.

AFTER 'York-Antwerp Rules' INSERT '1994 in London with English Law to apply' (for Owners who explicitly want to exclude the Rules 2004 they may add: "application of the York-Antwerp Rules 2004 are explicitly excluded").

DELETE FROM '1924 at such port……. until……. shall be paid in United States money'.

AT THE END ADD 'Hire and fuel not to contribute to the General Average'

OR
'Hire and fuels to contribute to the General Average',
OR
'Hire and fuels not to contribute to the General Average'.

126. In the event of accident, danger, damage, or disaster, before or after commencement of the voyage resulting from any cause whatsoever,
127. whether due to negligence or not, for which, or for the consequence of which, the carrier is not responsible, by statute, contract, or otherwise, the
128. goods, the shipper and the consignee, jointly and severally, shall contribute with the carrier in general average to the payment of any sacrifices,
129. losses, or expenses of a general average nature that may be made or incurred, and shall pay salvage and special charges incurred in respect of the
130. goods. If a salving ship is owned or operated by the carrier, salvage shall be paid for as fully and in the same manner as if such salving ship or
131. ships belonged to strangers.

RECOMMENDATIONS:

COMPLETELY DELETE THE LINES.

132. Provisions as to General Average in accordance with the above are to be included in all bills of lading issued hereunder.

RECOMMENDATIONS:

AFTER 'in accordance with the' INSERT 'York-Antwerp Rules 1994 (for Owners who explicitly want to exclude the Rules 2004 they may add: "application of the York-Antwerp Rules 2004 is explicitly excluded") in London with English Law to apply'.

DELETE THE WORD 'above'.

133. 20. Fuel used by the vessel while off hire, also cooking, condensing water, or for grates and stoves to be agreed to as to quantity, and the
134. cost of replacing same, to be allowed by Owners.

RECOMMENDATIONS:

DELETE 'also for cooking, condensing water, or for grates and stoves to be agreed to as to quantity'.

AT THE END ADD 'no deductions shall be made for domestic consumption'.

135. 21. That as the vessel may be from time to time employed in tropical waters during the term of this Charter, Vessel is to be dockeda t a
136. convenient place, bottom cleaned and painted whenever Charterers and Captain think necessary, at least once every six months, reckoning from
137. time of last painting, and payment of the hire to be suspended until she is again in proper state for the service.
138. ....
139. ....

RECOMMENDATIONS:

COMPLETELY DELETE THE LINES AND INSERT 'see Bottom Fouling Clause....'.

140. 22. Owners shall maintain the gear of the ship as fitted, providing gear (for all derricks) capable of handling lifts up to three tons, also
141. providing ropes, falls, slings and blocks. If vessel is fitted with derricks capable of handling heavier lifts, Owners are to provide necessary gear for
142. same, otherwise equipment and gear for heavier lifts shall be for Charterers' account. Owners also to provide on the vessel lanterns and oil for
143. night work , and vessel to give use of electric light when so fitted, but any additional lights over those on board to be at Charterers' expense. The
144. Charterers to have the use of any gear on board the vessel.

RECOMMENDATIONS:

COMPLETELY DELETE THE LINES

OR, FOR GEARED VESSELS, INSERT 'See Cargo gear clause....'.

AT THE END ADD 'The Owners are to provide sufficient light for night work in all holds and the deck simultaneously, as on board and to maintain the same in an efficient working order. Additional lights over those on board shall be for the Charterers' account'.

145. 23.Vessel to work night and day, if required by Charterers, and all winches to be at Charterers' disposal during loadingand discharging,
146. steamer to provide one winchman per hatch to work winches day and night, as required, Charterers agreeing to pay officers, engineers, winchmen,
147. deck hands and donkeymen for overtime work done in accordance with the working hours and rates stated in the ship's articles. If the rules of the
148. port, or labor unions, prevent crew from driving winches, shore Winchmen to be paid by Charterers. In the event of a disabled winch or winches, or
149.insufficient power to operate winches, Owners to pay for shore engine, or engines, in lieu thereof, if required, and pay a ny loss of time occasioned
150. thereby.

RECOMMENDATIONS:

DELETE FROM 'and all winches to be...until....... Of time occasioned thereby'

OR FOR GEARED SHIPS INSERT 'See Cargo Gear clause...'.

151. 24. It is also mutually agreed that this Charter is subject to all terms and provisions of and all the exemptions from liability contained
152. in the Act of Congress of the United States approved on the 13th day of February, 1893, and entitled "An Act relating to Navigation of Vessels
153. etc." in respect of all cargo shipped under this charter to or from the United States of America. It is further subject to the following clauses ,both
154. of which are to be included in all bills of lading issued hereunder :
155. U.S.A. Clause Paramount
156. This bill of lading shall have effect subject to the provisions of the Car-

riage of Goods by Sea Act of the United States, approved April
157. 16, 1936, which shall be deemed to be incorporated herein, and nothing herein contained shall be deemed a surrender by the carrier of
158. any of its rights or immunities or an increase of any of its responsibilities or liabilities under said Act.
If any term of this bill of lading
159. be repugnant to said Act to any extent, such term shall be void to the extent, but no further .
160. Both to Blame Collision Clause
161. If the ship comes into collision with another ship as a result of the negligence of the other ship and any act, neglect or default of the
162. Master, mariner, pilot or the servants of the Carrier in the navigation or in the management of the ship, the owners of the goods carried
163. hereunder will indemnify the Carrier against all loss or liability to the other or non carrying ship or her owners in so far as such less
164. or liability represents loss of, or any claim whatsoever of the owners of said goods, paid or payable by the other non
165. carrying ship or her owners to the owners of said goods and set off, recouped or recovered by the other or non carrying ship or her
166. owner as part of their claim against the carrying ship or carrier.

<u>RECOMMENDATIONS:</u>

COMPLETELY DELETE THE LINES AND INSERT 'See Protective/Standard clause....'.

167. 25 The vessel shall not be required to enter any ice bound port, or any port where lights or light ships have been or are about to be with
168. drawn by reason of ice, or where there is risk that in the ordinary course of things the vessel will not be able on account of ice to safely enter the
169. port or to get out after having completed loading or discharging.

<u>RECOMMENDATIONS:</u>

AFTER 'the vessel' INSERT always to trade in ice free ports and'.

AFTER 'not be required to' INSERT 'force ice or to follow ice breakers, not to'.

AT THE END ADD 'See Trading Exclusions clause...'

OR

WHEN THE OWNERS AGREE TO NAVIGATE IN ICE AT THE END ADD 'The Charterers confirm that the ship will not navigate in ice without following a convoy and icebreaker at short distance and the Charterers are to guarantee that the agents will arrange for the ship to be the second or third ship of the convoy.'.

170. 26. Nothing herein stated is to be construed as a demise of the vessel to the Time Charterers. The owners to remain responsible for the
171. navigation of the vessel, insurance, crew, and all other matters, same as when trading for their own account.

<u>RECOMMENDATIONS:</u>

AFTER 'navigation of the vessel' INSERT 'seaworthiness, acts of pilots and tug-boats and linesmen. See also Pilots-Tugs acts clause…'.

AFTER 'and all other matters' INSERT 'except for Time Charterers' responsibilities as stipulated in this charter'.

172. 27. A commission of .... per cent is payable by the Vessel and Owners to

173. ....

174. on hire earned and paid under this Charter, and also upon any continuation or extension of this Charter.

RECOMMENDATIONS:

AFTER 'on hire' INSERT 'and ballast bonus'.

175. 28. An address commission of 2 ½ per cent payable to .... on the hire earned and paid under this Charter.

RECOMMENDATIONS:

AFTER 'on the hire' INSERT 'and ballast bonus'.

Additional Clauses Nos …., both inclusive and the questionnaire, as attached hereto, are deemed to be fully incorporated in and to form an integral part of this charter party.

THE OWNERS: THE CHARTERERS:

ADDITIONAL CHARTER PARTY CLAUSES -TABLE OF CONTENTS

38. Last voyage.
39. Gangway watchmen.
40. War and extra war risk insurance.
41. Requisition.
42. Oil pollution.
43. Marine insurance /Return of premium.
44. Exports/Imports/Government approval of cargo.
45. Fumigation.
46. Hire/Hire payment/ Owners' bank details.
47. Final settlement.
48. Delays/Off-hire.
49. Periodical Dry Dock/Special Survey/Intermediate Survey. Survey/Dry- Dock in case of emergency.
50. Draft survey/Plans.
51. Lay-up condition.
52. Sales /Substitution.
53. Communications/Victualing/Entertainment (C/V/E)
54. Charterers' house flag/Markings/Colours.
55. Slow/Fast steaming.
56. Master's and Crew's assistance/Services/Duties.
57. Superficial inspection/Port Captain/Purplefinder.
58. Deck/Hatch covers loading.
59. Arbitration/Mediation.
60. Cargo gear.
61. Pilots/ Tugs acts.
62. Smuggling-Fines.
63. In Lieu Of Holds' Cleaning (ILOHC).
64. Instructions to the Charterers' Servants/ Charterers' procurement.
65. Winter navigation
66. Drug and alcohol guidelines.
67. ISPS compliant ports.
68. ISM.
69. Confidential fixture.
70. Supersede.

If there are titles of the clauses in the Charter Party, the following must be incorporated in the Charter Party:

'The interpretation of the clauses will not be affected by the titles of the clauses, which are provided for reference purposes only.'

1. Ship's description.

The ship's description usually constitutes the first clause when working and eventually issuing the Charter Party additional clauses (or rider clauses). It stipulates the description of the ship, as described by the Owners and has been accepted by the Charterers, since the commencements of the negotiations of the Main Terms. Actually, for a trip out and/or for a regular trade between two countries, where special requirements are not in effect and/or without port restrictions with regard to the size of the fixed ship, the Charterers could be satisfied with a "basic" Time Charter description clause.
For a long Time Charter periods, or in accordance with the Charterers' practice which is applicable even for short trips, the Owners may be asked
. to fill in a standardized questionnaire
which provides a wide and detailed description of a ship. The Charterers' said questionnaire is considered to be a part of and is incorporated in the Charter Party. Depends on the prevailing conditions, there should be an easy and at any time access to the ship's detailed characteristics by the Charterers. Charterers have many times made claims for incorrect descriptions of a ship, because the Owners incorrectly judged the severity of this clause or when fill in the Charterers' questionnaires.
The 'basic' description should mainly include:
. the name and the ex-name/s.
. the dead weight all told and the draft on summer salt water.
. the type of the ship-when it was built (date/month)-the flag.
. the class-the imo number-the call sign.
. the P&I club.
. the length over all-the beam-the length between perpendiculars (abbreviated as 'l.b.p.')-the depth moulded.
. the number of holds and type/number of hatches-the hatch dimensions.
. the tonnes per inch and/or tonnes per centimetre (abbreviated as 'tpi/tpc').
. the total grain capacity of the holds and its breakdown per hold.
. the total bale capacity of the holds and its breakdown per hold.
. the gross tonnage/net tonnage (abbreviated as 'gt/nt').
. Panama/Suez Canal gross tonnage/net tonnage.
. the full gear description.
. the speed and consumption.
. the fuel specifications.
The ship's description, as provided by the Owners, should include the undermentioned key characteristics, which should be taken into consideration by the Charterers:
a. the word 'about' which prevails on the presented figures and the validity of

the same:
The description of the ship, as provided by the Owners, must be an accurate representation of the characteristics of the ship. The Owners are likely to be held fully responsible and liable for any consequences/delays/losses/expenses that may arise if they have offered misleading and inaccurate information on the ship's characteristics and have generally presented misdescribed characteristics.
In case of serious misdescription, the Charterers have the right to terminate the Charter Party or even refuse the delivery of the ship.
However, all of the ship's figures should be able to be modified with the usual and acceptable shipping interpretation of the word 'about', otherwise, and in the case of omission of this word, the Owners would have guaranteed the same as accurate figures, which could lead to huge claims and losses for the Owners.
Taking into account that the speed and consumption of the ship is one of the most important items for the Charterers when making the "voyage calculation", the word 'about' could provide protection to the Owners in the case that the ship fails to maintain its Charter Party speed description.
The word 'about' provides to the Owners the benefit
. of a negative speed variation of half a knot
and
. a 5 percent increase in the declared fuel consumption.
The provision of 'all about and without guarantee', which could refer to the speed/ consumption, as well as on the general provided description of the ship could cover the Owners or can minimize the responsibilities for the presented data characteristics and prevent potential claims for any possible incorrect description and/ or underperformance of the ship. In fact, the majority of the Charterers only accept the provision of "about' prevailing throughout the description of the ship, unless the 'wog' has been overlooked by them, or they do not know the meaning of this phrase.
b. state of wind (Beaufort) and the sea state (Douglas):
The ship's declared speed and consumption could be affected when:
-trading in narrow waters.
-navigating with unfavorable currents.
-the state of wind exceeds Beaufort 4, and the Douglas sea state exceeds 3.
The conditions under which the speed and consumption would remain unchanged should be clearly incorporated in the terms of the Charter Party, otherwise, the ship is contractually obligated to keep the declared speed and consumption under any adverse conditions that the ship and crew may encounter during the sea voyage.
c. provisions of SOLAS for the safe loading:
Some major export terminals, in order to speed up the loading of bulk coal/iron

ore and reduce congestions, exercise pressure on the Master of the ship to arrive in the port with as little ballast as possible and load without taking into account the safe operational limits (loading sequences).

This results in the ship's safety being at risk, with potential damage to its structural strength.

The IMO (International Maritime Organization) issued a circular on 'Additional Considerations for the Safe Loading of Bulk Carriers', which recognize that it is a SOLAS requirement for the Master and the terminal to agree, prior to the commencement of loading, on a loading plan which prevents over-stressing of the ship's structure. In fact, the Master of the ship is responsible for the safety of the ship and the Owners must take the relevant precautionary measures.

Even though it is mandatory for the ship to load/discharge strictly as per the provisions of SOLAS, the respective protected provision should be inserted and be a part of the Owners' standard Time Charter ship's description.

d. provisions of SOLAS XII/14:

As per the IMO Maritime Safety Committee meeting, they have agreed upon new requirements (SOLAS XII/14) to apply when the bulk carriers have reached 10 years of age and are to carry heavy cargo with a density above 1,780kg/m3 in alternate loading conditions when fully loaded.

It should be part of the Owners' standard Time Charter ship's description that the ship will load as per the provisions of SOLAS XII/14.

e. the supply of fuel:

The Charterers, when arranging and supplying fuel, should be in accordance with the British standards. The supplied fuel must not include waste luboils/inorganic acids/tar oil/ nuclear waste or any other harmful substances or chemical wastes or chemical polypropylene contaminations, which either jeopardize the safety of the ship or adversely affect the performance of the ship or cause damage to the ship's engine or contribute to additional air pollution.

f. the maximum sulphur:

The Charterers must supply fuel of such specifications and grades as to permit the ship to, at all times, meet the maximum sulphur content requirements of any emission control zone when the ship is navigating and contacting ports within that zone.

g. off-set:

If the ship, during the trip, fails to maintain the described speed or fuel consumption for various reasons, it could be agreed that any saving in fuel consumption will be off-set against any reduction in speed and vise versa.

h. Bimco's clauses:

Certain published Bimco's clauses, could be incorporated into the ship's standard Time Charter description.

A Panamax gearless ship for Time Charter could be described as follows (basic description):

Date of build/name of builder/place of construction
Class .... - Flag: .... IMO .... Call Sign ....
Type of the ship (Bc-Panamax-Handy etc)
Summer Dwat / Draft 76,015 mt on 14.035 mtrs
Tropical Dwat / Draft 77,997.6 mt on 14.327 mtrs
Winter Dwat / Draft 74,036 mt on 13.743 mtrs
TPC summer 67.88
GT / NT 40,002 / 26,101
Suez GRT / NRT 41786.66/38350.49
Panama NRT 33,069
LOA 225.00 mtrs
L.B.P. 217,00 mtrs
Breath molded 32,26 mtrs
Depth molded 19,30 mtrs
7/7 Ho/Ha - Side rolling hatch covers.
Total Grain Capacity 91,356.6 cbm
Hatch sizes: No.1 15.57 mtrs x 12.80 mtrs
Nos 2-7) 17.30 mtrs x 15.40 mtrs
Ballast about 14,00 knots on about 30,00 mtns VLSFO380RMG
Laden about 13,00 knots on about 30,50 mtns VLSFO380RMG
ECO: Ballast about 12,50 knots on about 24,00 mtns VLSFO380RMG
Laden about 12,00 knots on about 25,00 mtns VLSFO380RMG
In port/berth/idle/working about 1,60 mtns VLSFO380RMG + about 1,60 mtns LSMGO (Boiler+A/E)
When charterers supply bunkers shall be according to British standards for VLS-FO380RMG max 0,5% sulphur as per ISO 8217/2017 or any subsequent amendment -LSMGO as per ISO 8217/2017 DMA 0,1%.
Charterers to ensure that their suppliers will comply fully with Annex VI regs 14 and 18 of Marpol 73/78 and resolution Mepc 96(47) or any subsequent amendments thereof.
The vessel will load/discharge strictly as per provisions of SOLAS chapter VI, regulation 7.3 and the Code of Practice for the Safe Loading and
Unloading of Bulk Carriers (BLU Code).
Ventilation natural
Vessel entitles to consume LSMGO for main engine/generator engines when maneuvering in confined/narrow waters/channels /canals/rivers when sailing in bad weather/rough sea or if below 10 knots, entering/leaving ports, shift of generator engine, exchanging ballast water at sea and while ballasting/ debal-

lasting operations and in general restricted to navigation/maneuvering area with or without pilots to master's discretion.
The described speed of the vessel is always subject to smooth sea and good weather periods of 24 consecutive hours running from noon to noon, which are defined as no adverse currents and tidal stream, wind not exceeding BF 4 a/o Douglas Sea State 3 (defined as a total combined sea and swell height not exceeding 1.25mtrs).
The speed is on average basis taken over the entire charter period including ballast voyage/s provided clean hull, clean bunker, even keel excluding canals, narrows restricted waters and poor visibility passage.
All the above figures on about basis.

The Owners could also add:
-A representative sample, taken at ship's manifold, of fuel supplied during the currency of this Charter Party will be analysed for quality by the .... on Owners' behalf and Charterers are to inform their suppliers accordingly.
- Any savings in consumption must be off-set against any reduction in speed, any saving in time must be off-set against any excess consumption,
any saving in VLSFO must be off set against increased LSMGO and vice versa.
-the ship's performance will be calculated in accordance with the English Law and favourable currents not to be considered in the calculations.
Time loss and over-consumption calculation should not apply to entire voyage, but performance claims to limit to good weather period/s only.

2. Owners' notices on delivery / Charterers' notices on redelivery.

This clause stipulates the number of notices that the Owners/Master should tender before the delivery and the number of notices that the Charterers should tender to the Owners before the redelivery.
-For the purpose of delivery:
It is directly connected with the Charterers' readiness for the implementation of their obligations, either under the terms of the fixed Time Charter Party and the circulation of the position of the ship in the market for speculative reasons, or of the fixed Voyage Charter Party (sub-Charter Party) and the information of the interested parties.
The notices will be tendered initially as
. 'approximate' and then finally as 'definite' notices.
Sometimes the Charterers also request notices
. 'on fixing',
i.e., to be tendered on conclusion of all negotiations and lifting of subjects. By the 'on fixing' or the 'approximate' notices, the ship's position is still uncertain

and may change at any time. However, the 'definite' notices, that are usually less than three days before the expected date of the ship's readiness at the point of delivery, declare that the ship's position is clear and if any delay occurs it will be due to unforeseen circumstances (continued bad weather conditions, accidents etc.).
The 'definite' notices
. increase the responsibilities of the party tendering the same.
-For the purpose of redelivery:
It allows the Owners to timely search the market for the ship's next employment. Usually, when the Charterers tender the approximate notices, they must declare the last discharge port and the point of redelivery.
Reservations should be kept when tendering the notices and especially the 'definite' notices, for any unforeseen circumstances which may arise and incur delays to the ship's delivery/redelivery. If they have not been given correctly or omitted by the party responsible, the other party may claim
. for losses suffered, due to breach of this clause.

The "Owners' Notices on Delivery/Charterers' Notices on Redelivery" clause, which, among others, should include:
-the approximate and definite notices for the ship's delivery.
-the approximate and definite notices for the ship's redelivery and the time of declaration of the last discharge port/point of redelivery.

An executed clause that has been run in a Charter Party is as follows (precise copy):
Owners/Master to advise Charterers approximate .../.../.../.... / days of vessel's readiness to deliver under this Charter Party and to keep Charterers informed of any changes immediately they occur.
Comments:
The Owners are to tender some approximate notices, as well as to keep the Charterers informed of any changes in the ship's position. There is no obligation for the tendering of any definite notices.

A complete clause for Time Charter period, which refers to the "Owners' notices on delivery/Charterers' notices on redelivery" could be worded along the following lines:

Owners and/or Master are to tender on fixing and then .../.../.../... days of approximate notices of the expected date and time of delivery of the ship to the Charterers' services.
Owners' and/or Master's tendered 'approximate' notice is given in good faith and in accordance with the currently available information. The ship's estimated time

of arrival at the delivery point is subject to completion of cargo discharging operations at the current port, adverse weather conditions and unforeseen circumstances, always beyond Owners'/Master's/ship's fault. If the tendered 'approximate' notice has changed by more than 12 hours, then the Master/Owners are to keep the Charterers and/or their Agents closely informed on the ship's exact position/ full itinerary and cause of delay.
Owners and/or Master are to tender.../.../...days of definite notices of the expected date and time of delivery of the ship to the Charterers' services. Owners/ship will remain responsible for the accuracy of the tendered 'definite' notices, which can be changed only by adverse weather conditions and/or unforeseen circumstances, always beyond Owners'/ Master's/ship's fault, default.
Charterers are to tender .../.../.../... days of approximate notices of the expected date and time of redelivery of the ship to the Owners. The Charterers, upon tendering the approximate notice of .... days must declare the last discharge port and the point of redelivery. Charterers' tendered 'approximate' notice is given in good faith and in accordance with the currently available information. The ship's estimated time of arrival at the redelivery point is subject to completion of cargo discharging operations at the last discharge port, adverse weather conditions and unforeseen circumstances, always beyond Charterers' fault. If the tendered 'approximate' notice has changed by more than 12 hours, then the Charterers are to keep the Owners closely informed on the ship's exact position/full itinerary and cause of delay.
Charterers are to tender.../.../...days of definite notices of the expected date and time of redelivery of the ship to the Owners. Charterers will remain responsible for the accuracy of the tendered 'definite' notices, which can be changed only by adverse weather conditions and/or unforeseen circumstances, always beyond Charterers' fault, default.

3. Time on delivery/redelivery.

The agreed lay-can and the time at the point of delivery and redelivery will be based

. on local time.

However, when computing the period for the hire payment, the local time will be converted in accordance with the Coordinated Universal Time (abbreviated as 'UTC'), which was formerly known as Greenwich Mean Time (abbreviated as 'GMT').

A complete clause for Time Charter period, which refers to the "Time on delivery/ redelivery" could be worded along the following lines:

For the time of the ship's delivery to the Charterers' services and redelivery to the

Owners, the local time is to apply. For computing the payment of the hires, the UTC time is to apply.

4. On and Off hire survey.

This clause defines when and at whose cost the on-hire and off-hire survey will be conducted. It is an essential clause that is incorporated in all Charter Parties, except for very short trips, where the Charterers could rely on the Charter Party declared speed/consumption and on the Master's consumption report about the remaining bunkers on board on completion of the Charter.
The Owners and Charterers will mutually agree on an on-hire survey to be carried out
. either at the point of delivery
. or at the first opportunity during the steaming time to the load port
. or on arrival at the first/sole loading port,
in order to determine the quantity of fuel on board the ship and the ship's hull/ holds' condition for previous stevedoring damage, that could be later transferred to the present Charterers. A full report should be issued, and photos should be taken from the ship's holds/deck/hatches, ascertaining the condition at the time of delivery.
Owners and Charterers will mutually agree on an off-hire survey to be carried out
. either at the point of redelivery
. or just before the completion of the discharge at the last discharge port,
in order to determine the quantity of fuel remaining on board and the ship's hull/ holds' condition. A full report should be issued, and photos should be taken from the ship's holds/deck and hatches, ascertaining the condition at the time of redelivery.
The survey can be carried out by a mutually acceptable surveyor and the costs could be shared equally between the Owners and Charterers, or otherwise each party can appoint and pay for their surveyor, but the lost time should be shared equally. Sometimes, the Owners propose to appoint the ship's Master/Chief Engineer to represent them, but in all cases the survey must be conducted jointly, and a single report is to be issued and accepted by all parties, otherwise differences could be found among all issued reports.

The "On-Off hire survey" clause, which, among others, should include:
-when and at which place the on-hire survey will be carried out.
-when and at which place the off-hire survey will be carried out.
-who will carry out the survey.
-who will bear the cost and the lost time of the carried out survey.

-that the Owners and Charterers have the right to take samples for laboratory analysis to determine the quality of the delivery fuel.

An executed clause that has been run in a Charter Party is as follows (precise copy):
At first port of loading a joint on-hire survey to be carried out in Owners time, unless same simultaneously carried out during vessel's loading or vessel is idle, to ascertain vessel's bunkers on board at the time of delivery as well as the condition of vessel's cargo compartments.
At the port of redelivery, a joint of-hire survey to be carried out in Charterers time to ascertain vessel's bunkers on board and vessel's condition at the time of her redelivery. Costs for both surveys to be shared equally between Owners and Charterers. Owners have the option to appoint Master/Chief Engineer to act as Owners surveyor, in which case Owners will not contribute.
Comments:
A joint on-hire survey will be carried out at the Owners' time, unless the ship is loading during this survey and provided that it does not affect the loading operation or the ship is in an idle condition, in which case no time will be deducted. The time for the off-hire survey at the discharge port will be for the Charterers' account. The cost of the surveyor will be shared equally between the Owners/Charterers, but the Owners have the option of appointing the Master/Chief engineer to act as the Owners' surveyor, in which case the Owners will not contribute to the cost of the appointed surveyor by the Charterers.

Another clause reads as follows (precise copy):
A joint on/off hire condition/bunker survey to be held at load/discharge port prior or concurrent with loading/discharging operation and no time shall be deducted. Each party to appoint their surveyor and pay him. Owners' right to appoint Master and/or Chief Engineer as Owners' surveyors.
Comments:
A joint on/off hire survey will be carried out, which will be conducted before or concurrently with the loading or discharging operation. No time will be deducted and each party will appoint and pay for their surveyor. However, the Owners keep the right to appoint the Master and/or the ship's Chief engineer to represent them.

A complete clause for Time Charter period, which refers to the "On-Off hire survey" could be worded along the following lines:

It is expressly agreed that:
1. on-hire survey.

Each party will appoint a surveyor and pay him for a joint on-hire survey, which will be conducted either at the port of delivery or at the first load port, or, otherwise if mutually agreed, at a more convenient time and place.
Owners have the right to appoint the Master and/or the Chief Engineer to act as the Owners' surveyors.
The survey will determine the actual quantity of fuel on board at the time of delivery and the condition of the ship for previous stevedoring damage.
Photos will be taken of the ship's deck, hatch covers and all cargo compartments.
The on-hire survey will be conducted before or concurrently with the loading operation and, if it interferes with the loading operation, the actual loss of time will be for the Owners' account.
Whenever it is required, the Master will endeavour to bring the ship into an even trim, if the same is possible and circumstances allow it, in order to ensure the correct bunker soundings.
The ship's Agents are to provide the Master and/or Owners with a copy of the single issued on-hire report, as soon as the same is ready.
In the case of a disagreement between the Owners' and Charterers' surveyors, then the on-hire survey will be conducted by a qualified independent surveyor, as will mutually be appointed by the Owners and Charterers and his report will be binding on both parties. Costs incurred and actual lost time will be shared equally between the Owners and Charterers.
If be required, the Charterers and Owners have the option of appointing their own surveyors, under their time and expenses, to take fuel samples for laboratory analysis, in order to determine the quality of the delivery fuel.
2. off-hire survey.
Each party will appoint a surveyor and pay him for a joint off-hire survey, which will be conducted either at the port of redelivery, or, if mutually agreed, at a more convenient time and place, but always within the redelivery area and, in any case, before the redelivery of the ship.
Owners have the right to appoint the Master and/or the Chief Engineer to act as the Owners' surveyors.
The survey will determine the actual quantity of fuel on board at the time of redelivery and the condition of the ship for stevedoring damage. Photos will be taken of the ship's deck and hatch covers and all cargo compartments.
Whenever it is required, the Master will endeavour to bring the ship into an even trim, if the same is possible and circumstances allow it, in order to ensure the correct bunker soundings.
Any time lost due to the off-hire survey will be for the Charterers' account.
The ship's Agents are to provide the Master and/or Owners with a copy of the single issued off-hire report, as soon as the same is ready.
In the case of a disagreement between the Owners' and Charterers' surveyors,

then the off-hire survey will be conducted by a qualified independent surveyor, as will mutually be appointed by the Owners and Charterers and his report will be binding on both parties. Costs incurred and actual lost time will be shared equally between the Owners and Charterers.

-If a single surveyor will carry out the on and off- hire survey, it could be agreed as follows:

It is expressly agreed that:
1. on-hire survey.
A joint on-hire survey will be conducted by an independent surveyor formally agreed by both parties either at the port of delivery or at the first load port, or, otherwise if mutually agreed, at a more convenient time and place.
The cost of the survey will be shared equally between the Owners and Charterers.
The survey will determine the actual quantity of fuel on board at the time of delivery and the condition of the ship for previous stevedoring damage.
Photos will be taken of the ship's deck, hatch covers and all cargo compartments.
The on-hire survey will be conducted before or concurrently with the loading operation and, if it interferes with the loading operation, the actual loss of time will be shared equally between the Owners and Charterers.
Whenever it is required, the Master will endeavour to bring the ship into an even trim, if the same is possible and circumstances allow it, in order to ensure the correct bunker soundings.
The ship's Agents are to provide the Master and/or Owners with a copy of the single issued on-hire report, as soon as the same is ready.
If be required, the Charterers and Owners have the option of appointing their own surveyors, under their time and expenses, to take fuel samples for laboratory analysis, in order to determine the quality of the delivery fuel.
2. off-hire survey.
A joint off-hire survey will be conducted by an independent surveyor formally agreed by both parties either at the port of redelivery, or, if mutually agreed, at a more convenient time/place, but always within the redelivery area and, in any case before the redelivery of the ship.
The cost of the survey will be shared equally between the Owners and Charterers.
The survey will determine the actual quantity of fuel on board at the time of redelivery and the condition of the ship for stevedoring damage. Photos will be taken of the ship's deck, hatch covers and all cargo compartments.
Whenever it is required, the Master will endeavour to bring the ship into an even trim, if the same is possible and circumstances allow it, in order to ensure the

correct bunker soundings.
Any actual loss of time due to off-hire survey will be shared equally between the Owners and Charterers.
The ship's Agents are to provide the Master and/or Owners with a copy of the single issued off-hire report, as soon as the same is ready.

5. *Cargo holds' and hatch covers' cleanliness inspection/Intermediate cargo holds' and hatch covers' cleaning.*

The Charterers have the right to carry out the holds' condition survey
. on delivery
or, at the latest, on arrival at the first loading port.
The cargo compartments must be clean, swept, washed down and dried, free of rust, free of loose rust flakes/scales and residues of previous cargo/oes, to the satisfaction of the appointed Charterers' or Shippers' inspectors or state's inspectors.
If the holds are found dirty, the ship will be placed off-hire, usually from the time of rejection until it has successfully fully passed the subsequent inspection. The Owners are to arrange for the holds' cleaning without delay, either with the Crew of the ship and provided that the same is allowed by the local regulations/labour unions, or otherwise with shore labourers and, all additional tools/ chemicals / fresh water necessary to be supplied in making the ship suitable for the intended for loading cargo will be entirely for the Owners' account. After rejection and if the holds' cleaning is not allowed at the berth, or, generally, if the berth's occupation without cargo operation is not allowed by the Port Authorities, then the Owners should additionally suffer from the charged port expenses required for the shifting of the ship to the anchorage and then back to the berth again (pilotages/tugs/ bunkers consumed etc.).

The Charterers, under the terms of the Time Charter Party, may refuse the delivery of the ship, because of the dirty holds and in the case, they are not ready on or before the cancelling date, might become entitled to cancel the Charter Party.

Under the terms of the Charter Party, the Charterers would be able to take action against the ship and to claim for damages /loss/ costs from the Owners, which may have been suffered (stevedoring stand -by, rental expenses at warehouse or for trucks, expenses for re-inspection etc.), due to the ship's un-readiness to load. In the event that certain holds that were rejected due to some minor dirty spaces could be cleaned and be ready by the crew without serious delay, then the loading operation can commence on the approved holds, provided that they will load as per loading sequence. In this case, under a provision as will be inserted in the Charter Party, the ship will be on-hire on a pro rata basis for the approved holds, in relation to the ship's total holds.

The "cargo holds' and hatch covers' cleanliness inspection" clause, which, among others, should include:
- the party that will carry out the holds' inspection and when.
- the condition of the holds on delivery.
- the consequences which may arise if the ship has not successfully passed the holds' inspection.
- who will bear the cost of the holds' cleaning and the lost time.
- whether the time will count as pro rata or not if the holds have partially passed and the ship has commenced the loading operation.

An executed clause that has been run in a Charter Party is as follows (precise copy):
Vessel on her arrival first loading port to be ready to load in all compartments in compliance with the rules of the National Cargo Bureau in USA ports or of the Port Warden in Canada ports respectively of the United States Department of Agriculture pursuant to the U.S.Grain Standards Act and/or of the Canada Department of Agriculture, as required by the appropriate authorities, if vessel loads at other than U.S. or Canadian ports, of such national and/or regulatory bodies as may be required and/or surveyors selected by Charterers, failing which she shall be off-hire prorate with actual number of holds failed and provided other holds will be loaded by shippers, otherwise vessel will remain on full off-hire, from the time of initial failure to pass until vessel passes inspections and any related and proven expenses are to be for Owners account.
Comments:
The ship must be ready for loading in all its cargo compartments, according to the local survey departments/regulations, depending on the port of loading and/ or to the Charterers' surveyors' satisfaction. The ship will be placed off-hire, from the time of its failure until it passes its re-inspection. If part of the holds is accepted, then the time will count pro rata, provided that the loading commences for the holds that have passed inspection. In any other case the ship will remain fully off-hire.

Another clause reads as follows (precise copy):
Vessel's holds/hatches, on her arrival at first port of loading to be clean-swept, washed down by freshwater and dried, free of rust, loose rust flakes/scales and residues of previous cargo/oes and in every way ready and suitable to load Charterers' intended cargo/oes..
If the vessel is rejected at loading port by Charterers'/Shippers' surveyors, vessel

to be off hire from the time of rejection until all holds/hatches are accepted by them and all extra expenses for vessel's holds cleaning to be borne by Owners.
Comments:
In case one load port apply and part of vessel's holds be accepted, and vessel start the loading operation then time shall count pro rata in relation to vessel's total holds. If more than one load port applies, then holds clean inspection shall take place only to the indented holds for loading. If part of vessel's indented holds for loading be rejected, then time shall count pro-rata in relation to vessel's total holds.
The holds' inspection will take place only at the first load port and not at the place of delivery. The ship must be presented with clean, swept, washed down by fresh water and dried holds. If it is thereafter rejected, then all cleaning expenses will be for the Owners' account and the ship will be placed fully off-hire until all holds/ hatches have been accepted. If two loading ports are to be used, then the inspection at the first load port will take place only in the holds intended for loading at this port.

Intermediate cargo holds' and hatch covers' cleaning.
If two or more loading legs are to be executed, or a short/long period is fixed, where more than one voyage will in fact be executed, an
. 'intermediate holds' cleaning' clause
must be agreed and incorporated in the Charter Party.
The majority of the Charter Parties state that the Owners, against a fee, the crew is to render customary assistance in cleaning the ship's holds between cargoes and will keep the holds clean, as if trading the ship for their own account. However, the crew are not regarded as skilled cleaning operatives and therefore there should be a limit on what cleaning can reasonably be effected by them. The removal of hard and rust/rust scale between the voyages is a major operation which cannot be performed by the crew, as such a cleaning is requiring the use of scrapping tools/rustybus machines etc. and cherry pickers for the access to the upper hopper areas/frames/corrugated bulkhead and the erection of staging. However, the crew can remove loose rust patches in accessible locations.
In fact, when intermediate hold cleaning is effected by the crew should be done without any guarantee whatsoever that the ship will pass any subsequent hold inspection and the ship/owners should not be in anyway liable if the ship's holds do not pass inspection for the next cargo.
The crew/ship/Owners, under normal circumstances should not be kept liable for rust /rust scale in ship's holds between the executed voyages, unless the Charterers have spelt out clearly in the Charter Party that the holds should remain in a specific standard cleaning for the whole duration, under the Owners'

risk/expenses. In addition, if Charterers accept the ship's holds on delivery
. without to reserve their rights for its condition,
then should be difficult for them to claim for damages, when holds are failed during the execution of the Charter.
The Charterers should be obliged to provide and pay for all chemicals, fresh water, consumable materials, tools such as water guns, pressure hoses, pumps etc. necessary to be supplied in making the ship suitable for the cargo intended for loading.
If there is corrosion/pitting in the holds, because of the Charterers' previous trade of corrosive cargoes under this Charter Party, then the lost time and extra cleaning that could be required to be performed will be entirely for their account and they should bring the ship to its initial clean condition before the redelivery takes place.
Dunnage, debris, lashing and/or other materials, which might have been stowed in the ship's holds after completion of discharge, must be removed at the Charterers risk, by shore labourers, as arranged and paid by them, at the end of each voyage and before the redelivery takes place.

The "intermediate cargo holds' and hatch covers' cleaning" clause, which, among others, should include:
-who will clean the holds between the voyages (crew or shore labourers).
-consequences if the holds eventually fail.
-the compensation of the crew for performing the holds' cleaning instead of the labourers.
-who will provide the chemicals, fresh water, consumable materials, which are considered necessary to be supplied, in making the ship's holds suitable for the cargo intended for loading.
-who will be responsible for the rust/rust scale/loose rust between the voyages.

An executed clause that has been run in a Charter Party is as follows (precise copy):
After discharging of any cargo out of vessel's first trip within this Charter Party and prior to arrival at next loadport for the successive trip, crew to perform cleaning and washing down and preparing of holds as if vessel was trading for Owners' account, provided time between discharge port and next loadport is sufficient for the crew to perform adequate cleaning/washing down. Crew to perform such cleaning/washing down always weather permitting. If shore regulations do not permit crew to carry out hold cleaning time/cost of shore labour to be for Charterers' account. In case vessel is not passed at next load port, Time charterers to arrange additional special cleaning at their time/cost. Vessel not to be placed off-hire due to any deficiency of cleanliness of holds unless rejected

due to loose rust/rust scale.
Intermediate hold cleaning USD.... per hold.
In case cargo loaded is pet coke or sulphur or salt, Charterers to pay USD ...... lumpsum for intermediate hold cleaning. All chemicals, fresh water, and consumable materials for cleaning purposes to be provided and paid by Charterers.
Comments:
The ship's crew, against a fee calculated per hold, will perform the holds' cleaning/ washing down between the voyages, always weather permitting and shore regulations allowing. In the case of rejection, the Charterers will arrange for additional special cleaning at their time/ expenses. The ship will be placed off-hire in the case that it is rejected due to loose rust/rust scales.

Another clause reads as follows (precise copy):
Vessel to remain on hire whilst cleaning/preparing cargo holds for intermediate voyages within this Charter Party and also during holds rejections periods. The Charterers have the option to request vessel's crew to clean cargo holds between voyages provided sea/weather conditions and applicable regulations permit. The Crew to perform the job to the best ability without any guarantee that the subsequent inspections will be passed. Owners /Master and/or Crew not be blamed and/or held responsible if holds are not approved at loading port and vessel always to remain on hire. However, crew to continue cleaning for cleanliness approval purposes but always taking into account above mentioned provisions. Charterers to pay Owners USD ... per hold for any cargo carried, except otherwise agreed. If required by Charterers, Master to co-operate by cleaning holds during discharge provided port Authorities permit and retain washing in ballast tanks (provided possible) if so required by port regulations.
Comments:
The ship will always remain on-hire during the cleaning of holds or during holds' rejection periods. If the Charterers request it, the crew will perform the holds' cleaning, but without any guarantee that the holds will be accepted at the intended load port. The Master's co-operation is asked, in the case that some of the empty holds are required to be cleaned during the discharging operation of the previous voyage, enabling the ship to be ready for the loading of its next cargo, as soon as possible.

A complete clause for Time Charter period, which refers to the "Cargo holds' and hatch covers' cleanliness inspection/Intermediate cargo holds' and hatch covers' cleaning' could be worded along the following lines:

CARGO HOLDS' AND HATCH COVERS' CLEANLINESS INSPECTION ON DELIVERY
It is expressly agreed that:

1.The ship on delivery:

the ship on delivery or, at the latest, on arrival or berthing at the first or sole load port will have all cargo holds and hatch covers clean:
-free of rust, loose rust flakes, rust scales, and loose scales,
-free from salt and free from residues of previous cargo/oes,
-fresh water washed and dried,
-free from insects and odours
and in all respects ready for loading the Charterers' intended cargo, according to the rules and regulations prevailing at the port of loading and to the satisfaction of the independent surveyors, as have been selected by the Charterers and/ or Shippers, under their time and expenses and/or to the satisfaction of a national cargo bureau and/or local grain surveyors' inspection.
If more than one load port is going to be used for completion of the loading, then a cargo holds' and hatch covers' cleanliness inspection will take place only in the cargo holds and hatch covers intended for loading, at each port of contact.
2. Cargo holds' and/or hatch covers' failure on delivery or at each loading port:

if, the ship fails from the cargo holds and/or hatch covers survey inspection, then the ship will be placed fully off-hire, from the time of failure until such time as all the rejected cargo holds and/or hatch covers have successfully fully passed inspection and are ready for loading.
The Owners are to arrange without delay for the cargo holds' and/or hatch covers' cleaning, at their time and expenses with the ship's Crew, if allowed by the port's regulations and/or labour unions, or, otherwise with shore labourers, in order to meet the surveyors' requirements.
The Owners are responsible for:
-the amount of the fuel consumed during the off-hire period.
-the expenses incurred for the purpose of holds' and/or hatch covers' cleaning.
-all chemicals, fresh water, consumable materials, tools such as water guns, pressure hoses, pumps, mechanical de-rusting equipment, rustybus machines etc. necessary to be arranged and supplied to the ship, in making the same suitable for the intended for loading cargo.
-any directly related expenses incurred, but supported with original vouchers, including, but not limited to, stevedores stand-by, but up to a maximum of one shift. If the Port Authorities and/or Shippers, order the ship to evacuate the berth and shift to anchorage or to change waiting place for the purpose of cargo holds' and/ or hatch covers cleaning, then all the delays incurred and extra expenses incurred, including, but not limited to, tugs and/or pilots and/or Agency fees etc., until the ship is back at the berth, or in its initial position, will be entirely for the Owners' account.

3. Commencement of the loading operation in the approved cargo holds and hatch covers at each port of loading:

if the loading operation partly commenced in the number of cargo holds and hatch covers accepted, then the time will count pro rata in relation to the ships' total cargo holds. However, if the ship commenced its full load operation which the port can provide in the approved cargo holds and hatch covers, then the time will count in full.
If the approved cargo holds and hatch covers are fully loaded or the Master stopped the loading for reasons of stability and the rest of the cargo holds and/ or hatch covers remain unacceptable, then the ship will be placed fully off-hire, from the time of the interruption, until such time as all the rejected cargo holds and/or hatch covers have successfully fully passed re-inspection and are ready for loading.
4. Master's objection to the cargo holds' and/or hatch covers' rejection at each port of loading:

if the ship's cargo holds and/or hatch covers have been partially or totally rejected and the Master has objected to the rejection, another well recognised independent surveyor then will be mutually appointed by the Charterers and Owners and his decision will be binding. The expenses incurred including the surveyor's fees will be for the account of the party in fault. Until such time as the independent surveyor's decision is given, the ship will remain on-hire.

-For Charterers, who will cancel the Charter Party due to rejection of the cargo holds/hatch covers, the undermentioned could be added:

5. Cancellation of the Charter Party:

if the ship fails its inspection and remains off-hire for 15 consecutive days or more, because of the rejection of the cargo holds and/or hatch covers on delivery or at the first load port, the Charterers have the right to reject the delivery and cancel the Charter Party, in which case:
a. a written notice will be given to the Owners.
b. the ship will be redelivered at the berth or anchorage or place, where it is performing the cargo holds' and/or hatch covers' cleaning with bunkers as onboard.
c. the Owners must refund the Charterers for any hire plus bunkers that have been paid in advance and/or for the bunkers supplied to the ship for Charterers' account, at the price of purchase, but supported with the original vouchers. All the quantity of fuel consumed from the time of delivery, until it is cancelled will

be for the Owners' account and no hire will be paid or is deemed to have earned, for the whole period.
d. the Owners will pay for all charged port expenses and/or extra expenses incurred, but supported with original vouchers, including, but not limited to, stevedore standby time, but up to a maximum of one shift.
e. the redelivery of the ship will take place, without prejudice to any other right the Charterers may have on ship.

INTERMEDIATE CARGO HOLDS'/HATCH COVERS' CLEANING.

It is expressly agreed that:
1. Cargo holds' and hatch covers' cleaning by the ship's crew:

a. upon completion of discharge of each cargo and sailing from the port, the ship's crew should render customary assistance in cleaning all cargo compartments, including hatch covers from the previous cargo and cargo residues and would endeavour to effect such cleaning 'as best as possible', in preparation for the next cargo. The cleaning operations should be performed while the ship is en route to the next loading port unless it is prevented by shore regulations and subject to adequate distance between the ports and the prevailing weather/sea conditions. In any case the ship will remain on-hire during the cargo holds and hatch covers cleaning, either at sea or in port.
b. the crew will work with the same effective manner as if the ship were trading for the Owners' account, but without any guarantee that the ship will pass the initial or any subsequent cargo holds' and/or hatch covers' inspection. The Charterers will compensate the Owners with a lumpsum amount of USD .... per cargo hold used for each time that such cleaning is performed when having last discharged grains/ agricultural products and a lumpsum amount of USD .... per cargo hold used for each time that such cleaning is performed when having discharged any other cargo except grains and/or agricultural products.
c. the ship will not be placed off-hire or held responsible due to any deficiency or subsequent deficiency of cargo holds' and/or hatch covers' cleanliness from previous cargo/es and/or for rust, rust scale, hard rust at each load port. Subject to shore regulations and crew agreement, the Charterers have the option to request from the crew to continue the cargo holds' and/or hatch covers' cleaning upon rejection, otherwise they will arrange with shore labourers for special cleaning at their time and expense.
d. the Charterers will arrange, under their time and expenses, for all chemicals, fresh water, consumable materials, tools such as water guns, pressure hoses, pumps, mechanical de-rusting equipment, rustybus machines etc. necessary to be supplied to the ship, in making the same suitable for the intended for load-

ing cargo.
e. if be requested by the Charterers, but after the Owners' and/or Master's and/or crew' consent , the crew will commence the cargo holds' and hatch covers' cleaning on the already empty cargo holds during discharge of the last cargo, provided that the Port Authorities permit it and will retain the washing liquids, slops and cargo residues in the ballast tanks (if possible), if so is required by port regulations, until the Charterers arrange, under their risk, time and expenses, for them to be discharged into shore trucks or sea barges or with any other means of disposal available in the port.
2. Charterers' obligations:

a. the ship should not be held responsible if it is rejected at the load port, due to corrosion and/or pitting and/or rust on its cargo holds and/or hatch covers, because of previous Charterers' trade with corrosive cargoes under this Charter Party. The Charterers will arrange, under their time and expenses, to make the ship suitable to load their cargo and bring the ship to its initial clean condition and before the ship's redelivery takes place.
b. any dunnage, debris, lashing materials are to be removed at the Charterers' risk by shore labourers, as arranged and paid by them, at the end of each voyage and before the redelivery takes place.
c. the Charterers will remain responsible for all costs and loss of time, including for the deviation, if it occurs, associated with the removal and disposal of dunnage and/or debris and/or lashing materials that have been stored on deck, because the free disposal and/or dumping was not allowed at the last port of contact and/or for cargo residues and/or holds' washing liquids and/or detergents and/or waste as defined by the Marpol Annex V Section 1 or other rules, relating to the disposal of such substances.
If the Charterers request the crew to perform special cleaning for loading cargoes requiring cleanliness in higher standards than for loading grain cargoes, such as alumina, sulphur etc. (even hard rust scales must be removed from the tank top) as any form of contamination would be considered a problem, the undermentioned could be agreed:
In the case of any special operations beyond customary crew's assistance is required by the Charterers, then the Owners/Master/ship and crew will not be held responsible and liable, in case the ship fails the next and/or any subsequent cargo holds' and/or hatch covers' inspection, and:
1.the Charterers will arrange, under their time and expenses, for all chemicals, fresh water, consumable materials, tools such as water guns, pressure hoses, pumps, mechanical de-rusting equipment, rustybus machines etc. and materials for the erection of stagings, necessary to be supplied to the ship, in making the same suitable for the intended for loading cargo.

2. the crew will do their utmost to have the cargo holds, including tank tops ready within the shortest time possible, but without any commitment as to the time of the cargo holds' and/or hatch covers' readiness.
3. the work will commence as soon as possible.
4. if, during the progress of work, difficulties occur or the crew is considered inadequate and/or inexperienced for such cargo holds' and hatch covers' cleaning, then the Charterers will arrange to board shore labourers to continue the work under their time, risk and expenses, but without any liability and responsibility for the work already performed by the crew.
5. the removal of residues and water liquids from the ship's holds will be arranged by the Charterers through their agents, under their time, risk and expenses.
6. the crew bonus will be paid with a lumpsum amount of USD....... directly to the Master of the ship through the Agents, before the ship's departure or directly to the Owners along with the next hire due, irrespective of the outcome (approval/failure) of the holds' inspection that will follow.
7. in the case that the ship fails the cargo holds' and/or hatch covers' inspection again, the charterers will either appoint shore labourers to continue the cleaning under their time, risk and expenses, or, after the crews' consent, will repeat the above procedure, but without any extra charge to be in effect.
8. the Charterers will supply, under their time/expenses, any other holds' cleaning tools and/or meterials, if and when it is required by the Master.

-For Charterers, who want the ship to trade between cargoes with holds clean to a high standard and/or clean for loading all the allowed cargoes under the Charter Party, at the Owners' risk and expenses, the undermentioned could be agreed:

Lack of ship's maintenance:

---

a. if at any time during this Charter, the ship's cargo holds and/or hatch covers are rejected at the loading port, because of rust and/or loose rust and/or rust scale and/or hard rust and/or lack of maintenance, the ship will be placed off-hire, from the time of rejection until such time as all the rejected cargo holds and hatch covers have successfully fully passed inspection and are ready for loading.
b. the amount of fuel consumed during the off-hire period and all expenses incurred for cleaning, in order to bring the ship to a clean standard for loading all the allowed cargoes under this Charter Party, as per Charterers request, as well as any directly related expenses incurred, but supported with original vouchers, including, but not limited to, cargo storage expenses, stevedores' stand-by, but with no more than of one stevedore shift will be entirely for the Owners' ac-

count.
c. if the Port Authorities and/or Shippers order the ship to evacuate the berth and shift to anchorage or to change waiting place for the purpose of cargo holds' and/or hatch covers cleaning, then all the delays incurred and extra expenses including tugs and/or pilots and/or Agency fees etc., until the ship is back at the berth or in its initial position will be for the Owners' account.
d. if the loading operation partly commenced in the number of cargo holds accepted, then the time will count pro rata in relation to the ships total cargo holds, but if the ship commenced its full load operation which the port can provide to the approved cargo holds and hatchcovers, then the time will count in full.
If the approved cargo holds are fully loaded or the Master stopped loading for reasons of stability and the rest of the cargo holds and/or hatch covers remain unacceptable, then the ship will be placed fully off-hire, from the time of the interruption until such time as all rejected cargo holds and/or hatch covers have successfully fully passed inspection and are ready for loading.
e. if the ship's cargo holds and/or hatch covers have been partially or totally rejected and the Master has objected to the rejection of cargo holds and/or hatch covers, because of rust and/or rust scaling and/or loose rust and/or lack of maintenance, then another well recognised independent surveyor will be mutually appointed by the Charterers and Owners and his decision will be binding. The expenses incurred including the surveyor's fees will be for the party in fault. Until such time as the independent surveyor's decision is given, the ship will remain on-hire.

6. Bunkering.

The bunkering clause for a Time Charter Party is one of the main negotiable clauses, which will be compulsorily incorporated in all issued Charter Parties, even if the Charterers do not replenish the quantity of fuel consumed for the execution of the trip.
The Charterers, at the time of the ship's delivery, will buy the amount of fuel on board the ship, enabling them to commence the Charter. Subject to agreement, the Charterers are of the obligation to redeliver the ship with about the same quantity of fuel as on the time of delivery, in which case the Owners will buy the relevant fuel.
The fuel prices are fixed at once and will remain the same for both transactions.
The main scope of this clause is to specify
. with what quantity of fuel and at what price the Owners will deliver the ship to the Charterers
and
. with what quantity of fuel and at what price the Charterers will redeliver the

ship to the Owners.
The delivery and redelivery quantity of fuel and the prices are negotiable and are fixed at once during the negotiations of this clause.
The quantity of fuel on delivery will be ascertained
. by the on-hire
and on redelivery
. by the off-hire bunker survey.
After the delivery takes place, all the subsequent supplies of fuel which will be ordered will be arranged and supplied only by the Charterers through their bunker suppliers. The Charterers remain responsible for the payment and for the quality of the ordered and supplied fuel on the ship. The fuel quality must strictly comply with
. the fuel specifications that are given and included in the ship's description by the Owners, otherwise, huge claims may arise due to delivery of improper bunkers, resulting in damage caused to the ship's engine.
From each bunkering operation
. samples will be taken from the ship's manifold, for fuel quality analysis purposes.
If the Charterers, for any reason, avoid paying for the fuel ordered by them, the suppliers, if allowed and have the right of law, might move against the ship, as a minimum action to secure the payment. Thus some Masters try to enter remarks (if allowed by the suppliers), or put a stamp on the presented for signing 'shipping order', keeping the Owners harmless for the payment of such ordered quantity of fuels.
The clause may also provide certain other requirements of the Owners or Charterers, such as the location where the ship will be supplied with bunkers etc.
For a long Charter period, the Charterers should redeliver the ship,
. with approximately the same quantity of fuel as on the time of delivery.
The payment of the value of the bunkers on board the ship at the time of delivery will usually be made together with the first 15 days hire payment. However, the Charterers have the right to deduct from the last hire payment or from the last sufficient hire payments, the value of the estimated redelivery bunkers.
For very short trips out, the ship could be delivered/redelivered
. with the quantity of fuel as on board,
i.e., without replenishment, but subject to the ship being able to provide sufficient bunkers to perform the trip and should additionally have enough bunkers, to contact the next bunkering station after redelivery. The Charterers should benefit, as they will save on the time required for replenishment, as well as, from any unexpected extra costs which may arise due to increased prices at the time of the ordering and of course from the fees of the attending Agents.
The Charterers, with their first hire payment will pay for the approximate bun-

kers that are required to be consumed for the execution of the whole trip. Upon completion, the actual bunkers consumed will be determined with the off-hire bunker survey and will be settled with the final hire statement.
Nevertheless, for such short trips, the Charterers must guarantee
. the maximum duration and remain responsible for replenishment, in the case that the duration, due to unexpected reasons (congestion/strikes on the port/ bad weather etc.), overcomes the calculated and expected maximum duration.
The Charterers, during the execution of the Charter and before its redelivery to the Owners, could ask from the Master of the ship to declare the amount of fuel necessary to be supplied, enabling the redelivery of quantities to meet the delivery quantities. A similar question could be made by the Charterers, for fuel required to execute a specific trip, enabling them to order sufficient fuel. In this case, any fuel quantity, which will be presented by the Master or Owners, should be subject:
-to the trip's final duration and destination,
and
-to the ship's actual consumption, when manoeuvring at the ports/canals/ straits and/or in confined waters,
i.e., information on any given bunker quantity to the Charterers should include the undermentioned reservation:
'The required quantity of fuel can be changed, subject to the trip's final duration and/or the ship's consumption, while it is manoeuvring at the ports and/or canals and/or straits and/or confined waters and subject to the final destination'

The "Bunkering" clause, which, among others, should include:
-with what fuel quantity of IFO&MDO or just MGO the ship will be delivered to the Charterers.
-with what fuel quantity the ship will be redelivered to the Owners.
-when the value of the fuel on delivery will be paid to the Owners.
-when the value of the fuel on redelivery will be deducted from the hire payment/s by the Charterers.
-the fixed prices which will be in effect.

An executed clause that has been run in a Charter Party is as follows (precise copy):
Bunkers on delivery to be about IFO ... mtns and about MDO .... mtns and to be redelivered with about the same quantities.
Prices both ends USD .... per metric ton for IFO and USD .... per metric ton for MDO.
Bunkers on delivery are payable together with the 1st hire payment.
Charterers have the liberty to retain funds from last sufficient hire payments in

order to cover estimated bunkers remaining on board at the time of redelivery. Owners prior to redelivery under this Charter Party, to have the privilege to supply bunkers for their account always provided that such bunkering does not interfere with the vessel's operations or result in cargo being shut out because of draft limitation. Any time lost and any expenses incurred are to be for the account of the Owners, and any time so lost is to be calculated at the Time Charter off hire of this Charter Party.

Charterers are entitled to verify the actual quantity and quality of bunkers remaining on board the vessel at any time during the currency of this time charter and the Master, Officers and Crew are to extend full cooperation and assistance in the bunker survey including but not limited acquaintanceship with deck and engine logbook entries. It is agreed that quantity of bunkers consumed to be determined by the difference between on-hire and off-hire bunker surveys with allowance for any bunkers taken en route by Owners.

The vessel will be supplied with bunkers at Singapore only at Inner Port Limits (I.P.L.) and never at Outside Port Limit (O.P.L.). Pilots required by Master of the ship will be for Charterers time/ expenses. Charterers under their time/expenses shall appoint a Surveyor who shall perform bunker survey prior and after bunkering operation in Singapore in order to determine the bunker quantities supplied to the vessel.

Comments:

The clause states with what amount of fuel and at what prices the ship will be delivered to the Charterers and with what amount of fuel and at what prices it will be redelivered to the Owners. The Charterers are entitled to deduct the value of the estimated remaining bunkers on board at the time of redelivery from the last sufficient hires.

The Owners have the privilege to bunker the ship for their own account during the period of this Charter Party, but under specific conditions.

This request could be arranged for one or more of the following reasons:

-the fuel quantity at the time of redelivery is minimal and may not be sufficient for the next intended trip.

-the quantity is not be sufficient to perform long ballast at sea or until the next available bunkering station from the point of redelivery.

- there is no available bunkering station at the redelivery point.

- the supply of bunkers could be carried out when the ship is under loading or discharging and there will be no deviation and lost time for the Owners' account.

The clause also defines the place in Singapore where the ship will be supplied and who will pay for the pilots used. The supplied fuel quantity will be determined by an appointed surveyor by the Charterers.

Another clause reads as follows (precise copy):
Vessel to be delivered with minimum 1,500 metric tons of Intermediate Fuel Oil and with minimum 80 metric tons of Marine Diesel Oil.
Bunkers on redelivery to be about 200 metric tons of Intermadiate Fuel Oil and about 40 metric tons Marine Diesel Oil.
If caused by vessel's failure to provide her agreed technical capabilities under this Charter Party increasing the normal calculated voyage time and/or consumption and Charterers have stemmed bunkers for redelivery already, Charterers are not responsible for exact replenishing of bunkers on redelivery quantity.
Prices of bunkers at both ends to be USD ....per metric ton for Intermediate Fuel Oil and USD .... for Marine Diesel Oil.Bunkers costs on delivery to be paid with initial hire.
Charterers are allowed to bunker the vessel for Charterers' account before delivery, provided it does not interfere with Owners' operations.
Owners confirm that vessel can take bunker grades in South Africa port (RMF/DMB)
Comments:
The Owners will deliver the ship with a minimum amount of fuel, otherwise they must arrange, before delivery, to supply the ship with sufficient fuel at their time/ expenses.
The redelivery bunkers are not the same as on delivery.
The lower quantity of fuel on redelivery means that:
-either the ship is heavily over-plus with fuel at the time of delivery and the lower redelivery will increase the cargo capacity, which might be required for the next employment
or
-the ship, after completion of the Charter, will proceed for dry dock/special survey commitments and the minimum amount of bunkers on board will facilitate the operations
or
-the ship will be sold for scrap.
If the ship fails to provide the agreed technical capabilities, the Charterers are not obliged to alter the already given order for the redelivery fuel quantity.
The Charterers have the privilege to supply the ship with additional bunkers for their account before the delivery takes place and obviously at the last port before the ship's redelivery from its previous trade. This operation will be made provided that it does not interfere with the ship's operation. The Owners, before they agree, must get the approval from the current Charterers (unless the ship is on a voyage basis). A provision should be made that all extra expenses which might be incurred or lost time and any potential liabilities for pollution will be for the Charterers' account. The Charterers will benefit from the time

required for the ship's additional bunkering to perform their trip and/or the cheaper prices achievable in this port.

Another clause reads as follows (precise copy):
Owners guarantee that the vessel has sufficient bunkers to perform Charterers intended voyage of about 8 days with safe margin.
Vessel to be redelivery with bunkers as onboard without replenishment.
Estimated value of bunkers to be consumed (as calculated by Owners/Charterers reasonably) to be paid with first hire payment. Actual quantities to be settled with final hire payment.
Comments:
The Charterers declare that the duration is about 8 days and the Owners will guarantee that the bunkers on board are enough to perform the trip with a safe margin. The Charterers are not of the obligation to replenish the consumed fuel. The Owners/Charterers will calculate the estimated value of the bunkers required for the trip, which will be paid together with the first hire payment.

Another clause reads as follows (precise copy):
Vessel will be delivered with about IFO 370/430 mtns high sulphur and with about IFO 190/200 mtns low sulphur. Mdo about 45/60 mtns. The value of bunkers will not be paid on ship's delivery.
Bunkers on redelivery to be about the same quantities and same qualities as on board on delivery.
Any minor difference on the bunkers value will be settled together with final settlement with exact figures, at the prices of Charterers actual purchasing price with supporting vouchers.
Comments:
The Owners present the expected bunkers on board at the time of delivery and the quality of the same (IFO with low or high sulphur). The Charterers will not pay the bunkers' value at the time of delivery. The Charterers will redeliver the ship with bunkers as on delivery and the minor differences will be settled according to the Charterers' prices.

A complete clause for Time Charter period, which refers to the "Bunkering" could be worded along the following lines:

It is expressly agreed that:
1. the ship is expected to be delivered with about VLSFO ... metric tons and about LSMGO ... metric tons.
2. the fuel, at the time of redelivery, is to be of the same quality and of about the same quantity on board, as on the time of delivery.

3. the value of the fuel at the time of delivery and redelivery will be calculated according to the prices of USD.... per metric ton for VLSFO and USD ....per metric ton for LSMGO.
4. the value of the fuel on delivery will be paid together with the first 15-day hire payment.
5. the Charterers have the right to deduct the value of the fuel on redelivery, from the last 15-day hire payment, but, if it is not sufficient, it will be extended until the Charterers deduct the full value of the fuel.
In any case, the first 15-day hire payment is not to be the last.
6. any minor differences on the value of the fuel at the time of delivery and redelivery will be settled together with the final settlement with exact figures.
7. the Owners guarantee that the ship is eligible to take bunkers in any country within the trading limits agreed under this Charter Party.
8. information on any given quantity of fuel by the Owners and/or Master to the Charterers is subject to the trip's final duration and/or the ship's consumption, while it is manoeuvring at the ports and/or canals and/or straits and/or confined waters and subject to the final destination.
9. the Charterers, before the ship is delivered to their services and at the last port of discharge of the ship's previous trade, have the option to supply the ship with fuel for their account, as long as that the same does not interfere with the full operation of the ship and is allowed (written approval) by the current Charterers and/ or after the Owners' consent. The Charterers are responsible for any loss of time which will result in the ship being placed off-hire, arising from any reason due to the Charterers' supply of fuel to the ship before its delivery to them. The Charterers will reimburse the Owners, as per the provided off-hire statement from the previous Charterers, for all time lost and fuel consumed during the off-hire period. Any expenses, losses, damages and/or consequences of any nature relating directly to the bunkering of the ship, including damages from pollution and fines imposed by any party, will be for the Charterers' account, liability, and responsibility.
10. the Owners, have the option of bunkering the ship en route to its last port of call or at the last port of call before its redelivery for their account and after the Charterers' consent, as long as that the same does not interfere with the full operation of the ship or with the effect of not loading the full intended cargo quantity. Time lost, because of the bunkering for the Owners' account should be counted as off-hire.
11. the Charterers are entitled to verify the actual quantity and quality of fuel remaining on board the ship at any time during the currency of this Charter. The Master, Officers and Crew are to extend their full cooperation and assistance during the bunkering operation. It is agreed that the quantity of fuel consumed is to be determined by the difference between the on-hire and off-hire bunker

surveys, with allowance for any fuel taken en route by the Owners.
12. the Charterers, for every bunkering operation arranged by them, under their time and expenses, will appoint a well recognised independent Surveyor, who will carry out a bunker survey before and after any bunkering operation takes place, in order to determine the quantity of fuel supplied to the ship.

-For bunkering in Singapore, the undermentioned could be added:

The bunkering operation in Singapore will take place only at the Inner Port Limits (I.P.L.) and not at the Outside Port Limit (O.P.L.). In/Out pilots that might be required by the Master of the ship will be for the Charterers' time and expenses. The Charterers, under their time and expenses, will appoint a well recognised independent Surveyor, who shall carry out a bunker survey before and after the bunkering operation in Singapore, in order to determine the quantity of fuel supplied to the ship.

-If the Owners decide to secure the replenishment of the consumed bunkers for the Charterers' account, the undermentioned could be added:

The Charterers will provide written evidence of arrangements to replenish consumable fuel en route or before the ship's arrival at the discharge port. The replenished consumable fuel will be supplied at least one day before its completion of discharge.
The Charterers' failure to arrange for the timely supply of fuel will give the Owners the right, without prejudice to any other rights they may have under this Charter Party, to suspend the performance of any or all of their obligations under this Charter Party, on immediate written notice, until the same has been proven to be arranged in writing.
Throughout the period of suspended performance, the ship will be and will remain on-hire.
The Charterers are fully liable and responsible for any expenses, losses, consequences of any nature arising from the suspension of the performance.

7. Hatch covers opening and condition test.

The ship's hatch covers, during the execution of the Charter must be
. watertight and without any leakages,
which might damage or affect the quality of the transported cargo.
However, when an expensive or sensitive cargo is intended to be loaded, such as alumina/sugar/ steels etc., the Charterers could seek additional protection, by reserving the right, before the commencement of the loading, to carry out a

hose or chalk test, to verify its water tightness and/or for an ultrasonic test on all hatch covers, to verify the thickness of the steel. If any hatch cover is found defective, the Owners must repair it immediately under their time/expenses. If the ship fails a second inspection, then the Charterers may become entitled, under the terms of the Charter Party, to cancel the Charter Party.

The 'hatch covers opening and condition test' clause, which, among others, should include:
-the Charterers' right to conduct a hose test/chalk test a/o ultrasonic test on the ship's hatch covers.
-for whose account, the expenses incurred, and lost time should be, if the ship fails its inspection.
-under whose satisfaction the repairs required should be made if the ship fails its inspection.
-whether the Charterers should maintain the fixture or cancel it, in the case the ship fails its second inspection.

An executed clause that has been run in a Charter Party is as follows (precise copy):
Opening and closing of hatches at or off each port to be done by vessel's crew if required by Charterers, provided local regulations permit at any time.
If the water tightness of ay hatch or closure appears to be deficient, Owners shall promptly take all measures to reinstate and repair the same and/or take other measures to ensure that hatches are water tight thereby ensuring that loaded cargo is kept and maintained in the prescribed manner and these measures shall be done at Owners time and expense to Charterers satisfaction.
Comments:
The opening and closing of hatches will be performed by the ship's crew, provided that the local authorities permit the same, otherwise the Charterers will arrange for shore labourers to perform the job. The ship's hatch covers are to be watertight, otherwise the Owners will bear the cost and time to reinstate and repair the same to the Charterers' satisfaction.

Another clause reads as follows (precise copy):
Owners guarantee vessel's hatch covers are watertight upon delivery of the vessel and to be maintained throughout this Charter period in such condition by the Owners and if any hatch cover found defective same to be rectified at Owners' time and expense to Charterers surveyor's satisfaction.
The Charterers have the privilege of ordering a hatch water-proof test (hose or chalk test) and/or ultrasonic testing on all hatches prior or after the delivery of the vessel and from time to time during this Charter. The time and costs of the

survey are to be for the Charterers' account, but the Master to give every facility to the Charterers and their surveyor to carry out such tests.

Comments:

The Charterers have the right, from time to time, before delivery or during the execution of the Charter, to order for a hatch water-proof test and or an ultrasonic test in order to verify the thickness of the steel on the hatch covers.

The Owners must keep the ship's hatch covers watertight throughout the Charter period, at their time and expenses.

A complete clause for Time Charter period, which refers to the "Hatch covers opening and condition test" could be worded along the following lines:

Provided that the local regulations permit it, the opening and closing of the hatch covers will be undertaken by the ship's crew, free of expenses to the Charterers and its derricks and/or booms topped without causing delays to the commencement of loading and/or discharging operation, but subject to weather and navigation permitting.

The Owners confirm that, throughout this Time Charter period, the hatch covers of the ship will be watertight and will be kept in top condition. The Charterers may also, from time to time during this Charter, or upon delivery, order through their surveyors, at berth or anchorage, an ultrasonic test and/or a hose test and/or chalk test on all hatch covers, but the Charterers' requirements will not exceed the ship's class rules. The costs and time lost for the respective survey is to be for the Charterers' account, however, the Master is to give every facility to the Charterers and their surveyors, in order to conduct such tests.

In the event that the ship fails the hatch covers inspection on delivery or during this Charter, the ship will be placed fully off-hire, from the time of failure, until such time as all rejected hatch covers have successfully fully passed inspection. The Owners will immediately arrange for shore labourers or, at their discretion, with the ship's crew, if it is permitted by the regulations and labour unions of the port, to make the respective repairs. The costs for this operation and any directly related expenses incurred, but supported with original vouchers, will be for the Owners' account.

If the Port Authorities and/or Shippers order the ship to evacuate the berth and shift to anchorage or to change waiting place for the purpose of repairs, then all the delays and extra expenses incurred, including, but not limited to, tugs and/or pilots and/or Agency fees etc., until the ship is back at the berth or in its initial position, will be for the Charterers' account.

In the case of a second rejection, the Charterers have the right of declaring the cancellation of this Charter Party for delivery or for the balance Charter period.

If the Charter Party is cancelled:

1. the Charter Party will be considered null and void and no claim will be pursued, either by the Owners nor Charterers or any other third party.
2. a written notice will be given to the Owners, while the ship remains off-hire for redelivery on dlosps of this port of inspection.
3. the ship will shift and/or sail without delay and will be redelivered with bunkers as on board.
4. the Owners must refund the Charterers for any hire not earned plus bunkers that have been paid in advance and/or for the fuel supplied to the ship for the Charterers' account, at the price of purchase, but supported with the original vouchers. No hire will be paid or is deemed to have earned from the time of the first inspection and rejection, until it is cancelled and redelivered on dlosps and all the quantity of fuel consumed during this period will be for the Owners' account.
Commencement of loading in the accepted hatch covers:
If the loading operation partly commenced in the number of hatches accepted, the ship will not be cancelled, even if it fails the re-inspection and the time will count pro rata in relation to the ships' total hatches. However, if the ship commenced its full load operation which the port can provide in the accepted hatch covers, then the time will count in full. If the holds of the hatch covers accepted for loading are fully loaded, or the Master stopped the loading for reasons of stability and the rest of the hatch covers remain unacceptable, then the ship will be placed fully off-hire, from the time of the interruption until such time as all the rejected hatch covers have successfully fully passed re-inspection/s and are ready for loading.
The Master's objection to the hatch covers' rejection:
If the ship's hatch covers have been partially or totally rejected and the Master has objected to the rejection, then another well recognised class approved independent surveyor will be mutually appointed by the Charterers and Owners and his decision will bind. The expenses incurred, including, but not limited to, possible shift, the surveyor's fees, and time lost will be for the account of the party in fault.

8. Certificates /Equipment.

Throughout the duration of the Charter, the ship must be in possession of all necessary

. valid equipment and certificates

that verify the ship's size and type, its seaworthiness, as well as its compliance with the health and safety regulations, national and international regulations and all current/local requirements at all agreed ports/regions that the ship will contact, including

. valid certificates for passing the Panama/Suez canals and several straits.
The Charterers will require from the Owners to guarantee the same, which will be incorporated in the Charter Party.
The Owners will remain responsible and liable for any delays /charges/ consequences which may arise, if the ship is unfit and not in possession of valid and up-to-date certificates, before arrival at the port/s of contact, or after inspection as will be conducted by the port Authorities and/or Port State Control inspectors. If so, the operations and sailing will not be allowed, with huge implications for the Owners.

The "Certificates /Equipment" clause, which, among others, should include:
-that, during the duration of the Charter, the ship will be in possession of all necessary equipment and certificates/specific certificates required for the trade.
-that the Owners are responsible for all delays/expenses which may arise if they are not in compliance with it.

An executed clause that has been run in a Charter Party is as follows (precise copy):
The vessel's equipment shall comply with the normal port regulations and/or requirements in effect at the port or ports of call and the canals and the countries in which the vessel will be employed. The Owners also guarantee that the vessel shall be at all times in possession of valid and up to date certificates onboard to comply with such regulations and or requirements.
If stevedores, longshoremen or other labourers are not permitted to work by reason of any failure of the Master, the Owners and/or their Agents to comply with such regulations or by reason that the vessel is not in possession of such valid and up to date certificates, then the Owners shall make immediate corrective measures. The Charterers may suspend the hire for the time lost thereby and extra expenses directly related to loading and discharging and bunkering including stevedore's standby time shall be for the Owners account.
Comments:
The Owners guarantee that the ship's equipment and certificates will comply with all regulations and /or requirements that are in effect in the countries that it will contact.
The Owners are responsible in the case of failure to comply with this clause and will bear the cost of lost time and all incurred extra expenses, including stevedores' stand-by.

Another clause reads as follow (precise copy):
Vessel to have on board valid certificates, including oil pollution certificates, necessary for the agreed trading limits and such certificates to be maintained

throughout the period of Charter. Any consequences due to vessel lacking necessary certificates, or if same should be outdated, to be for Owners' risk and expense. Vessel also to have on board valid deratisation or deratisation exemption certificate and such certificates to be maintained throughout the period of the Charter, and if fumigation is necessary, cost of same and detention to be for Owners' account except as provided in Clause 2.
Owners warrant that vessel possesses valid International Tonnage Measurement Certificate pursuant to the 1969 convention including tonnage certificate required by the Authorities in any country to which the vessel may trade.
Throughout the period of the Charter vessel shall have on board current valid Suez and Panama Canals Tonnage Certificates and will so comply with all applicable requirements, regulations, and recommendations so as to avoid any delay in transit of canals. Any uplift in tolls, fines, or other penalties due to vessel's infringement of SCA or PCA requirements or directives under this Charter Party or otherwise and all resulting consequences thereof are to be for Owners' account.
Any delay due to non-compliance or any lack of proper documentation or equipment according to vessel's specifications is to be treated as off-hire and expenses incurred thereby to be for Owners' account, unless such non-compliance is due to the default of Charterers, Sub-Charterers or their agents.
In the event that time is lost in Owners obtaining any necessary licences or permits or similar documents from any authority for Owners' own account, same will be treated as off-hire and any extra expenses, including any related to inspection of the vessel or crew by any authority, shall be for Owners' account.
Charterers to give Owners reasonable due advance notice to prepare and to process necessary documents before sending vessel to any port/country where new legislation or additional licences /permits/documents may be required. Time lost due to lack of licences/permits/documents of cargo is for Charterers' account.
Comments:
The Owners must have valid certificates and these certificates must be maintained throughout the duration of the Charter.
The ship will be placed off-hire, if there are any delays due to non-compliance or lack of proper documentation and equipment according to the ship's specifications. All the expenses that may be incurred will be for the Owners' account.
The Owners' request for reasonable advance notice to be given to the ship, before the ship reaches ports/ countries, where new legislation or additional documents may be required.

-Clause which refers to Deratting certificate.
Some Charter Parties have a separate clause, which refers to the deratting certificate. Because of the strict regulations at the port/s of contacts, the Charterers draw the Owners' attention, if the certificate is invalid or is about to expire.

An executed clause that has been run in a Charter Party is as follows (precise copy):

The vessel shall be delivered with a valid deratting or deratting exemption certificate. If such certificate does not cover the whole period of the Charter, costs of renewal and any required fumigation for obtaining of such certificate, if necessary, shall be for Owners' account.

Any detention and extra expenses incurred hereby shall be for the Owners account.

However, if because of cargo or the vessel have been sent to an infected port or ports where fumigation is ordered whilst on Charter by Charterers, cost, and time to be for Charterers account.

Comments:

A valid deratting certificate must cover the entire period of the Charter. If the certificate is going to expire before completion of the Charter, the Owners should arrange, under their expenses, for the relevant renewal.

-Clause which refers to Translated certificates.

Certain Arabian countries mainly require a number of the ship's certificates to be submitted in a language other than English, something that definitely takes time and comes at great cost. The owners must be notified well in advance of such cases. For long or even for short Time Charter period agreements, which have the intention of trading the ship through such areas, this provision is essential and must be incorporated in the Charter Party.

An executed clause that has been run in a Charter Party is as follows (precise copy):

In case a certificate has to be translated for the purpose of suiting Charterers trading a specific area/port, Owners are to be notified well in advance so as to have sufficient time to comply with such request.

Comments:

The Owners must be notified in advance, if the ship intends to contact a specific area/port, where the certificates must be translated.

-Clause which refers to the Panama and Suez certificates.

For the passage of the Panama and Suez Canal, a specific certificate must be issued and the ship must comply and be fitted with all applicable requirements to transit the same. The relevant cost for the passage will be calculated based on this specific certificate and the ship will obviously not be allowed to pass, until the same is presented or issued.

An executed clause that has been run in a Charter Party is as follows (precise copy):

Throughout the period of this Charter vessel will have on board current valid and up-to-date Panama and Suez Canal Measurement Certificates and will so comply with all applicable requirements, regulations and recommendations of the Canal Authority so as avoid any delay in transit of the Panama or Suez Canal. Any delays caused by non-compliance with the aforesaid or lack of proper documentation and/or certificates and/or equipment and fittings required to transit the Panama and Suez Canal will be considered off-hire and any expenses resulting from such delay including bunkers consumed during the period will be for Owners' account.

Comments:

Certificates and fittings of the ship for passing the Panama and Suez Canal must be valid and up to date, in order to avoid delays, otherwise all expenses incurred by such delays will be for the Owners' account and the ship will be placed 'off-hire'.

-Clause which refers to the Chinese Tonnage Certificate.

When a ship calls to China for loading or discharging, the 'Chinese Tonnage Certificate' must be issued. The certificate will be issued for the ship on behalf of the party who ordered the same and will thereafter be submitted to the port authorities on berthing.

This certificate is valid for 2 months and can be used for multiple loading/discharging purposes. If the ship is going to be redelivered in China after the completion of discharge, the Charterers, who have under their possession this certificate, could sell the same to the next Charterers at a negotiable price, always providing that the ship will remain in this country for its next trip.

An executed clause that has been run in a Charter Party is as follows (precise copy):

Chinese Tonnage Certificate to be for Charterers account.

Comments:

It is a simple clause, which states that the certificate will be for the Charterers' account.

A complete clause for Time Charter period, which refers to the "Certificates/ Equipment" could be worded along the following lines:

Owners guarantee that, throughout the duration of the Charter, the ship will be in possession of all necessary valid and up-to-date equipment and certificates, in compliance with safety and health regulations, national and international reg-

ulations and all current requirements at all ports and regions of contact. In any other case, the Owners will take immediate corrective measures and all consequences, consequential losses, detention, fines, delays, damages, costs and expenses of claims and possible shifts, will be for the Owners' account.
Time lost for the Owners to obtain any necessary permission or time lost for issuance of the missing and/or expired certificates and/or documents from any Authority and any time lost and/or costs, including those relating to the ship's inspection or crew, of any Authority and/or ship's class and/or ship's flag etc., will be for the Owners' account.
If stevedores, longshoremen and/or other labourers are not allowed to work, by reason of the Master's and/or Owners' and/or Owners' Agents failure to comply with such regulations, or because the ship is not in possession of such valid and to date certificates and the operations in the port have been interrupted or interfere with the ship's unconditional operation, then the ship will be placed off-hire for the actual time lost thereby and any directly related expenses, but supported with original vouchers, including, but not limited to, stevedores stand-by time, but up to a maximum of one shift, will be for the Owners' account.
New regulations:
The Charterers will give the Owners a reasonable advance notice to prepare and to process the necessary documents, before directing the ship to any port/country in which the new legislation or additional licences, permits, documentations are in effect. In general, throughout the duration of this Charter, when and if new regulations are imposed from an internationally recognised Maritime body within the agreed range of trade that would require additional certificates, then:
a. the certificates relating to the ship and its equipment must be obtained by the Owners' arrangements and at their expenses.
b. the certificates relating to cargoes loaded or discharged at ports of contact must be obtained by arrangements and at the expenses, time and risk of the Charterers and/or Shippers and/or Consignees and in accordance with the international regulations.
Financial responsibility:
The ship must have valid certificates of financial responsibility under U.S. government regulations, allowing it to use the navigable waters of the United States of America. The Owners also undertake to comply with all laws and regulations related to oil pollution and/or other pollution in force, in any place/region the ship can be ordered to.
Panama and Suez certificates:
Throughout the duration of this Charter, the ship will have in its possession all valid and up to date the Panama and the Suez Canal Measurement Certificates that comply with all applicable requirements, regulations and recommendations of the Canal Authorities. Any delays in transit of the Panama or Suez Ca-

nal, caused by non-compliance with that or lack of proper documentation and/or certificates and/or equipment and fittings needed to transit the Panama and Suez Canal will be considered off-hire and any costs arising from this delay, including bunkers consumed during this period, will be for the Owners' account.
Translated certificates:
If a certificate must be translated, in order to make the ship suitable to trade in a particular port/ region, the Owners must be notified well in advance, in order to have sufficient time to comply with such a request. However, the Owners and/or ship will not be held responsible, if the certificates for translation are not ready on arrival of the ship at the port, because of the Charterers' late advice. All costs for issuing the translated certificates will be for the Charterers' account. Any cargo documents that are required to be translated into another language other than English, must be arranged only by the Charterers, at their time and expense.
International Tonnage certificate:
Owners guarantee that the ship has on board 'The International Tonnage Certificate 1969' in full compliance with the requirements of IMO and the International Convention on Tonnage Measurement of Ships, 1969.
Deratting certificate:
Owners warranty that the ship will be delivered with a valid deratting or deratting exemption certificate. If the certificate does not cover the entire period of the Charter, the cost of renewal and any fumigation required to obtain this certificate will be for the Owners' account.
If any detention and extraordinary expenses arise therefore, including, but not limited to the cost to unload and reload the cargo, when fumigation takes place, it will be for the Owners' account.
Chinese certificate:
Issuance/cost of the Chinese Tonnage Certificate will be for the Charterers' account.

9. Deviation-Put back.

During the sea voyage from one port to another, it may be necessary for the ship to deviate from its normal route, for various reasons, such as

- accidents or illness of crew,
- stowaways,
- collision,
- to take spares/provisions etc.
- saving life and/or property,
- to assist or tow ship/s in distress,
- to avoid bad weather (typhoons etc.)

This clause must provide for, as possible as can be, the reason of the deviation, enabling the responsibilities to be apportioned depending on the cause of deviation.
For any reason, even for bunkering purposes, that the ship deviates from its normal route and in a laden condition (not empty), the P&I Club must be prior/immediately informed, otherwise the ship will be subject to P&I Club's rules. The P&I Club could charge certain further costs to the ship for any required additional coverage.

The "Deviation-Put back" clause, which, among others, should include:
- the reason for the deviation or why the ship is put back.
- who will be responsible for the loss of time and for the costs incurred.
- the time from which the ship will be considered off-hire and back to commission again.

An executed clause that has been run in a Charter Party is as follows (precise copy):
Should the vessel put back whilst on voyage by reason of breakdown of machinery, collision, stranding, fire or the accident of damage to the vessel, or dry docking or periodical survey, or deviate from the course of the voyage caused by sickness of or accident to the Captain, Officers, Crew or any person on board the vessel or by reason of sending stowaway or refugee, salvage, or by reason of the refusal of the Captain, Officers or Crew to do their duties without any specific and valid ground for rejection, or any Owners matters, the payment of hire shall be suspended from the time of inefficiency in port or at sea until the vessel is again efficient in the same position or regains a point of progress equivalent to that the hire ceased hereunder. Bunkers consumed while the vessel is off-hire and all extra expenses incurred during such period shall be for Owners account.
However, should the vessel alter course to avoid bad weather or be driven into port or anchorage by stress of weather reasonably for vessel's safety, the vessel shall remain on hire and all costs thereby incurred shall be for Charterers account.
Comments:
The clause states that the ship will be placed off-hire in case of a put back, or an alteration of its course for various reasons attributed to the ship. However, if the ship alters its course due to bad weather or is driven to a port for the safety of the ship, then the ship will remain on-hire and all costs will be for the Charterers' account.

Another clause reads as follows (precise copy):
In the event of actual loss of time either in port or at sea, deviation from the

course of the voyage or putting back whilst on voyage, by reason of breakdown of machinery, collision, stranding, fire or any other accident or damage to the vessel, or drydocking or periodical survey, or sickness or accident to the Master, Officers, Crew or any person on board the vessel other than persons travelling by the Charterers' request or by reason of sending stowaway or refugee, salvage including saving life, or by reason of the refusal of the Master, Officers or Crew to do their duties, or any Owners' matters, the hire shall be suspended from the time of the vessel's inefficiency in port or at sea until the time when the vessel is again efficiency in the same position or an equivalent point of progress to the place where the vessel being inefficient as above. Any direct related expenses, only in proven relation with incident caused and up to maximum first shift or present shift as the case may be, incurred including bunkers consumed during such period of suspension shall be for Owners' account.

In case of important main operational matters with the maintenance or safety of the vessel, Owners to have the right to arrange supply from port/place, with minimum possible deviation, provided Charterers will have been notified and all relevant off-hire/bunker cost for Owners' account.

Comments:

The ship is responsible for any deviation or put back whilst in voyage, by reasons attributed to the ship/Owners. Nevertheless, the clause gives the right to the ship to deviate from its route for main operational matters, particularly concerning the maintenance or safety of the ship. The ship will be placed off-hire for the time lost and the bunkers consumed will be for the Owners' account.

A complete clause for Time Charter period, which refers to the "Deviation/Put back" could be worded along the following lines:

It is expressly agreed that if there is loss of time:

1. whether in port or at sea and/or deviation from the course of the voyage or put back whilst on voyage, because of

-failure of machinery,

-collision,

-grounding,

-fire or any other accident or damage to the ship,

-dry-docking in case of emergency,

-illness or accident of the Master/Officers/Crew or any person on board except for persons travelling by request of the Charterers,

-stowaway or refugee,

-refusal of the Master/Officers/Crew to do their duties,

-any matters of the Owners,

the hire, unless caused by default of the Charterers and/or their representatives

and/or their Agents, will be suspended from the inefficiency of the ship in port or at sea, until the ship is again efficient, in the same position or at equidistance and all fuel consumed will be for the Owners' account.
The expenses incurred that are connected or not with the port expenses but proven to be directly related to the delay or the incident that caused it, as well as stevedoring expenses, but limited only to the first shift or present shift as the case may be, including bunkers that are consumed during the period of suspension, will be for the Owners' account.
2. because of the ship changing course to avoid bad weather or is driven into port or anchorage by stress of weather, which is reasonably necessary for the ship's safety, it will be for the Charterers' account. If trading in shallow harbours or in river ports or in ports with bars, any detention of the ship and/or expenses arising from such detentions, will be for the Charterers' account and the ship will remain always on-hire.
3. due to the least possible deviation to a port and/or place that is required to be made, but always en route to its destination, because of significant main operational issues related to the ship's maintenance, or safety and provided that the Charterers have been notified in advance, will be for the Owners' account. The hire will cease until the ship returns to its original position or at equidistance and the Owners will remain responsible and will cover all the expenses incurred and bunkers consumed.
4. because of the ship's deviation for the purpose of saving life and/or property and to assist or tow ship/s in distress, then such operations will not be considered as deviation during the period of this Charter Party and the ship will remain on-hire. However, all salvage attributions, which are to be paid to the ship for services provided will be shared equally with the Charterers, after appropriate deduction of the possible expenses which were incurred, including, but not limited to the Master's and Crew's shares resulting from this operation.

10. Eligibility for trading/Black list/ Boycott.

The ship could be blacklisted for various reasons by any organization or even governmental authorities before the entry in this Charter.
The Charterers through this clause will seek to protect their interests from such an event.
The Owners should be required to confirm that the ship
. is not blacklisted from the countries, which are within the allowed trade.
The blacklisted ships are to be detained at the respective port of contact, causing delays and huge expenses. For example, if a ship has called to Israel before, then it is at risk when calling to Arabian countries for any purpose. In the past, Owners have used certain standard tricks, so that their ships would not

get blacklisted, but the same are not in force anymore. Many Owners, before purchasing a ship, ask from the sellers to confirm that the ship is not blacklisted from any ports/ countries.
All lost time and expenses incurred due to blacklisting/boycotting, will be for the Owners' account.
The ship's
. flag,
. ownership,
. crew nationality,
could also be a cause for the ship to get boycotted in the ports of contact. In case the ship remains detained or is boycotted for a long time, then the Charterers could have the option, under the terms of the Charter Party, to cancel the Charter and look for substitution.
If the political situations change during the duration of the Charter and it is known that the ship will get blacklisted or be boycotted upon its contact to a port, then the Charterers must avoid ordering the ship to proceed to this port.

The "Eligibility for trading/Black list/Boycott" clause, which, among others, should include:
-that the Owners guarantee that the ship is eligible for trading within the agreed trading ranges and is not blacklisted by any organization or governmental authority.
- that the Owners will be responsible in the case that the ship is detained or is boycotted for this reason.
- that the Charterers will have the right to cancel the Charter Party in the case that the ship is arrested/detained.
- that the Charterers will not trade the ship in areas which will cause the ship to be blacklisted.

An executed clause that has been run in a Charter Party is as follows (precise copy):
The Owners guarantee that since the vessel has been under her present management she has never traded to an Israeli port and will not call at any such port prior to the Charter. To the best of Owners knowledge neither the vessel nor any other vessel owned, managed, and controlled by them is blacklisted by any Arabian country.
Comments:
Obviously, the Charterers' interests are within the Arabian countries, in which case it has become a necessity for the Owners to guarantee that the ship is not blacklisted. The Owners must also confirm, to their best knowledge that the rest of the ships which are owned by the same company or are under the same

management, are not blacklisted.

Another clause reads as follows (precise copy):
a. Owners warrant that to their knowledge the vessel has not been 'blacklisted' by governmental authorities or by any other organization such as labour union or fuel suppliers in any of the countries to which she might trade in the course of this Charter Party. Charterers warrant they will not knowingly trade the vessel to any countries which would cause the vessel to be blacklisted by such authorities or organizations.
b. Vessel to be off hire, if and for the period she is immobilized by boycott on account of non-compliance with paragraph (a) hereof. Any boycott due to vessel's flag and/ or class and/or ownership to be for Owners' time, risk, and expense. It is understood that off-hire will be calculated for the time actually lost. Charterers shall not trade the vessel to any port or place where the vessel is known to be boycotted by shore and/ or port labours and/or tugboats and/or pilots or by government and/or authority by reason of the vessel's flag/registry/class/ownership.
Comments:
The Owners guarantee that to their knowledge the ship is not blacklisted by any governmental authorities or other organizations and the Charterers from their side guarantee that they will not trade the ship in countries which will cause the ship to be blacklisted. The ship will be placed off-hire, if it is immobilized by the boycott of any Authorities or any other organization. Boycott due to ship's flag/ ownership/ class will be at the Owners' risk.
The Owners require that the Charterers do not direct the ship to ports/places where it is known that the ship will be boycotted.

A complete clause for Time Charter period, which refers to the "Eligibility for trading/ Blacklist/Boycott" could be worded along the following lines:

Owners guarantee that the ship is in all respects eligible for trading at the ports and/ or places and/or ranges, as specified in the Charter Party and complies with all International Conventions in force and meets the safety rules and regulations. The Owners further guarantee that the ship is not directly or indirectly owned, controlled by or related to any........ or........ or.......... (insert the respective countries) interests and that since the ship has been under its present management it has not been blacklisted by any governmental authority or by any other organisation, in any of the countries/ regions which are within the trading limits.
To the best of the Owners' knowledge, neither the ship, nor any other ship owned, managed or controlled by them is blacklisted by any Arab league coun-

try, as well as that the ship is not blacklisted at coal terminals in South Africa, nor by any US and/or Canadian longshoremen union.
If the ship, because of its previous trade before delivery to this Charter and/or because of its flag, ownership, class, registry and/or nationality of its Master, officers, crew and/or because of the terms and conditions of employment of the Master, officers, crew and/or due to the construction, cargo gear, fitting, other equipment and/or any other ship under the same ownership, management and control, being prevented from or is unable to perform in accordance with the terms of the Charter Party and:
a. it is boycotted, picketed, blacklisted, delayed.
b. it is not allowed to enter for loading and/or discharging and/or to sail from any port within the trading limits of this Charter Party.
c. it is rendered inoperative and/or immobilized and/or detained by the Government and/or authority and/or shore and/or port labourers and/or tugboats and/or linesmen and/or pilots and/or suppliers or the ship cannot load or discharge at the intended port/ place, the hire will cease for all actual time lost and the Owners are responsible for all the directly related expenses, but supported with original vouchers, losses, consequences, stevedores standby time, but up to a maximum of one shift. The Owners will take immediate corrective measures, but, if ordered to load and/or discharge in another place and/or port, the ship will be placed off-hire for all actual time lost and the Owners will be responsible for all the expenses, consequences, losses, consequential losses, fines, extra port dues, fuel consumed and all liabilities in respect thereof or arising therefore. Deviation incurred, in order for the ship to change crew or perform repairs or due to any other reason which is deemed necessary in order to avoid potential boycotting and/or blacklisting at the port which it will contact, all consequences, losses, consequential losses, expenses, fuel consumed, liabilities directly incurred therefrom will be for the Owners' account and the hire will cease until the return to its original position or at equidistance.
If, during the duration of this Charter, the political situation is changed in certain places within the trading areas, the Charterers guarantee that they will not trade the ship in any ports and/or places and/or regions, where the ship is known that it will be blacklisted and/or boycotted by shore and/or port labourers and/or tugboats and/ or pilots and/or by government and/or port Authorities and/or organisations by reason of the ship's flag, registry, class, ownership, manning etc.

-For Charterers, who will cancel the Charter Party, because the ship remained off-hire for more than 15 days or the ship is no longer able to trade, the undermentioned could be added:

If the ship remains for more than 15 consecutive days inoperative or immobilized or detained and has been placed off-hire or the ship is no longer able to trade at the intended place or port, as defined by the Charterers and always within the trading area, because of the above reasons, the Charterers have the right to cancel this Charter Party or the balance period.
If the Charter Party is cancelled:
1. a written notice will be given to the Owners while the ship remains so detained, for redelivery on dlosps of this port and, if there is cargo on board, the ship will be redelivered when it is practical and the ship being free of cargo.
2. the Owners must refund the Charterers for any hire plus bunkers that have been paid in advance and not earned.
3. the ship will be redelivered with bunkers as on board and any differences will be settled together with the final hire payment.
4. the Owners will pay for all extra expenses incurred, but supported with original vouchers, including, but not limited to, stevedore standby time, but up to a maximum of one shift.
5. the redelivery of the ship will take place, without prejudice to any other right the Charterers may have on the ship.
If the Charter Party is maintained:
1. the Charterers have the option to add to the maximum allowed Charter period, partly or wholly any off-hire period/s, provided that any such extension does not exceed the validity deadline of the ship's next special survey/dry dock. Any such option will be declared to the Owners a minimum of 20 days before the expiration of the minimum allowed Charter period.
2. the Charterers will trade the ship at the same rate and conditions for such extended time, as that applicable during the off-hire period.
3. the Owners will give the Charterers the full itinerary of the ship and the expected date of resumption of the service, every two days.
4. after resumption of the work, the ship will be placed at the Charterers' disposal at the same port or place where the hire was suspended, within ordinary office hours, except at sea and with quantity of fuel as when the ship was placed off-hire. The Charterers may, however, in their option, accept the ship on hire again in such a position and at such a time as the ship may again in all respects be ready to comply with the Charterers' orders and directions.

11. Capture/Arrest/Seizure/Detain.

This clause provides the responsibilities of the parties involved if the ship has been

- captured, arrested, seized, detained,

during the period of this Charter Party.

The ship can be arrested at a port/anchorage/place/yard, during the loading or discharging operation or when taking bunkers/passing its survey commitments etc. The action against the ship could be because of previous trades/cargo claims/ Charter Party disputes etc. but also because of the action against the Charterers/ Shippers/Receivers or the Carriers appearing on the Bills of Lading. In certain countries, the arrest of the ship is not allowed or is very difficult, as during the time needed to obtain the court decision to arrest the ship, the ship could in the meantime have already sailed. The claiming party will merely follow the ship's trading and when, for any purpose, it contacts a convenient place for arrest, they will immediately apply for a court decision, which could only take a few hours. The ship will only sail after the Owners have deposited a 'guarantee' at a bank.

The "Capture/Arrest/Seizure/Detain" clause, which, among others, should include:
- against whom the claim should be.
- if it is against the Owners, the period during which the ship will be placed off-hire.
- if it is against the Charterers, it should define their responsibilities.
- who will bear the extra costs etc.
- if the Charterers should have the right to cancel the Charter Party.

An executed clause that has been run in a Charter Party is as follows (precise copy):
Should the vessel be arrested/detained during the currency of this Charter Party at the suit of any party having or purporting to have a claim against or any interest in the vessel, hire under this Charter Party shall not be payable in respect of any period during which the vessel is not fully at Charterers' disposal and any consequential directly related/proven expenses shall be for Owners account, unless such arrest/detention is due to action against Charterers or sub-Charterers or their Agents or the Contractors or the cargo Shippers or Consignees, thence hire is payable and Charterers undertake the responsibility to release the vessel by taking appropriate and required measures (issuance of security/etc) as the case may be or arise.
Comments:
The ship will be placed off-hire from the time that the ship is not fully operational for the Charterers and only those expenses directly related/proven are to be for the Owners' account.
If the arrest/detention is due to action against the Charterers or sub-Charterers or their Agents or the Contractors or the cargo Shippers or Consignees, then the ship will remain on-hire and the Charterers undertake to release the ship.

Another clause reads as follows (precise copy):
Should the vessel be seized or detained by any authority during the currency of this Charter Party, the Charterers' liability for seizure or detention is ceased immediately from the time of her seizure or detention and all time lost by this reason shall be treated as off-hire until the time she resumes full service, unless such seizure or detention is occasioned by any personal act or omission or default of the Charterers or their agents or by reason of cargo carried. Any direct related expenses incurred by and/or during the above seizure or detention is to be for Owners' account, unless such seizure or detention is occasioned by any person, act, or omission or default of the Charterers or their agents, or by reason of cargo carried.
Any extra expenses incurred by and/or during above seizure or detention are to be for Owners account, unless such seizure or detention is occasioned by any personal act or omission or default of the Charterers or their agents, or by reason of cargo carried.
If the vessel is off-hire for 30 consecutive days by reason of above seizure or detention, unless such seizure or detention is occasioned by any person, act or omission or default of the Charterers or their agents or by reason of cargo carried, Charterers have the option of cancelling the balance of this Charter Party, always provided no cargo on board. In any case, should the vessel be detained, delayed or otherwise as described above, due to any fault of the Owners or their servants as outlined above, Owners will immediately take necessary action to overcome such delays.
If the ship is seized or detained by any Authority during the currency of this Charter Party, the lost time will be treated as off-hire, until the time which the ship resumes full service. All expenses will be for the Owners' account, unless such arrest or detainment is occasioned due to act or omission or default of the Charterers, in which case the ship will remain on-hire. The Owners must take all necessary actions to overcome the delays, but if the ship remains off-hire for more than 30 days, then the Charterers will have the right to cancel the balance period, but provided there is no cargo on board.

A complete clause for Time Charter period, which refers to the "Capture/Arrest/ Seizure/ Detain" could be worded along the following lines:

If the ship and/or the Master, Officers and Crew and/or any person of the Owners travelling on the ship is either captured or seized or detained or arrested by any Authority or any legal process during the period of this Charter Party, at the suit of any person having or purporting to have any claim against or interest in the ship and/or said Master, Officers and Crew and/or person of the Owners'

travelling on the ship, the hire under this Charter Party will cease for any actual period at the duration which the ship is not fully operative for the Charterers. The expenses incurred, whether they are connected or not to the port dues but are proven to be directly related to the above either capture or seizure or detention or arrest, as well as stevedoring expenses limited only to the first shift or present shift, as the case may be, including bunkers that are consumed during the period that the ship is off-hire will be for the Owners' account. The Owners will immediately take any necessary and appropriate measures in order to release the ship.
If the ship operates (load or discharge) or is allowed by the Port Authorities to operate during the above either capture or seizure or detention or arrest, the ship will remain on-hire.
If, either the arrest or detention or capture or seizure has arisen from any person or authority or legal process, due to act or omission or neglect or default or non-compliance with the provisions of the Charter Party of the Charterers and/or the Charterers' person/s travelling on the ship or sub-Charterers or their Agents or the Contractors or the cargo Shippers or Consignees or of the cargo carried, or of the name of any carriers appearing on the Bills of Lading, other than that of the actual Owners, which was inserted as per the Charterers' request, then the ship will remain on-hire and the Charterers undertake the responsibility to release the ship by taking all the necessary and appropriate measures (issuance of security/etc.), as the case may be or arise.

-For Charterers, who will cancel the Charter Party, because the ship remained off-hire for more than 15 days, the undermentioned could be added:

If the ship is off-hire for 15 consecutive days or more, because of the above either seizure or detention or capture or arrest, unless either the arrest or detention or capture or seizure has arisen from any person or any Authority or any legal process, due to act or omission or neglect or default of the Charterers and/or any person/s of the Charterers travelling on the ship or sub-Charterers or their Agents or the Contractors or the cargo Shippers or Consignees or of the cargo carried, or of the name of any carriers appearing on the Bills of Lading, other than that of the actual Owners which was inserted as per the Charterers' request, the Charterers have the right to cancel the balance period of this Charter Party.
If the Charter Party is cancelled:
1. a written notice will be given to the Owners while the ship remains so inoperative, for redelivery on dlosps of this port and if there is cargo on board the ship, it will be redelivered when it is practical and the ship being free of cargo.
2. the Owners must refund the Charterers for any hire plus bunkers that have

been paid in advance and not earned.
3. the ship will be redelivered with bunkers as on board and any differences will be settled together with the final hire payment.
4. the Owners will pay for all extra expenses incurred, but supported with original vouchers, including, but not limited to, stevedore standby time but up to a maximum of one shift.
5. the redelivery of the ship will take place, without prejudice to any other right the Charterers may have on ship.
If the Charter Party is maintained:
1. the Charterers have the option to add to the maximum allowed Charter period, any partly or wholly off-hire period/s, provided that any such extension does not exceed the validity deadline of the ship's next special survey/dry dock. Any such option will be declared to the Owners a minimum of 20 days before the expiration of the minimum allowed Charter period.
2. the Charterers will trade the ship at the same rate and conditions for such extended time, as that applicable during the off-hire period.
3. the Owners will give the Charterers the full itinerary of the ship and the expected date of resumption of the service, every two days.
4. after resumption of the work, the ship will be placed at the Charterers disposal at the same port or place where the hire was suspended, within ordinary office hours, except at sea and with quantity of fuel as when the ship was placed off-hire. The Charterers may, however, in their option, accept the ship on hire again in such a position and at such a time as the ship may again in all respects be ready to comply with the Charterers' orders and directions.

12. Quarantine/Vaccination certificate/ Free pratique.

The Master, before arrival at the port, will report to the Port Authorities through the attending Agents, that the ship complies with regulations on health, as well as that the crew is healthy without infectious diseases or symptoms of infectious diseases. The Port Authorities will allow the ship to approach the port but will board the ship before anyone else and will check that all corresponding certificates comply with the health regulations and will get Master's assurance/ confirmation that the crew is healthy, as well as comply with the vaccination and sanitary regulations at the port of contact.
The granted health clearance which will allow the commencement of the operation in the port is commonly called
. 'free pratique'.
In case of failure, the Port Authorities will put the ship in quarantine, or it will be detained, resulting in its shifting to a waiting place until it complies with the health regulations.

This clause will hold the ship responsible in the case that it fails to get the 'free pratique' and will also exclude the ship's trading in infested areas.

The "Quarantine/Vaccination certificate/Free Pratique" clause, which, among others, should include:
-that the ship is responsible in the case that the corresponding certificates do not comply with the health regulations and the crew does not hold vaccination certificates,
-that, if during the period of the Charter Party, the ship calls to infected areas, the Charterers will remain responsible in the case of failure to pass the free pratique at the next port of call.

An executed clause that has been run in a Charter Party is as follows (precise copy):
Normal quarantine time and expenses for the vessel's entering port shall be for Charterers' account, but any time of detention and expenses (except quarantine required as a result of cargo carried and ports visited) for quarantine due to pestilence, epidemics and illness of Captain, Officers and Crew shall be for Owners account.
Vessel not to be ordered to trade in infested areas or to ports where world sanitary organisations have declared contagious diseases.
Owners shall be responsible for and arrange at their own expense that the Master, officers and crew of the vessel to be vaccinated and to be in possession of valid vaccination certificates on delivery of the vessel and throughout the period of this Charter Party. Any time lost and or additional expenses incurred due to failure to provide such certificates shall be for Owners' account.
Charterers to give prior notice if vessel is calling at ports where special vaccination is required and 15 days enabling Owners to arrange for such vaccination.
Comments:
It states that the normal/customary quarantine time and expenses that are required to be spent after the ship has entered the port will be for the Charterers' account. For any other reason attributed to the ship's officers and crew or due to the Owners' failure to provide valid vaccination certificates, the same will be for the Owners' account.

Another clause reads as follows (precise copy):
Normal quarantine time and expenses to enter a port to be for Charterers' account but any time of detention and expenses for quarantine due to pestilential illness, etc of the Master, Officers or Crew to be for Owners account, unless such illness caused from areas where Charterers traded. Owners to be responsible for any delay in quarantine arising from Master, Officers or Crew having commu-

nication with shore at any infected area without the written consent or instructions of Charterers or their Agents.
Officers and crew to comply with vaccination and sanitary regulations in all ports of call and corresponding certificates to be available on board. Any detention and/or fines resulting from not having certificates on board to be for Owners account.
Owners to arrange at their expense that the Master, Officers, and crew of the vessel hold valid vaccination certificates against cholera and the yellow fever and smallpox if and where required throughout the Time Charter period.
Owners to comply for their account with all sanitary regulations at ports of call including water treatment of cholera where required.
Further, the vessel shall be in possession of valid certificate necessary to prepare radio pratique at port or ports where radio pratique is available.
Comments:
The Owners/ship remain responsible, in the case that the Master/Crew develop a pestilential illness, unless such illness is caused due to the ship's trading under the Charterers' present instructions. The ship is responsible, under the Owners' expenses, to comply with all vaccination and sanitary regulations, in all allowed ports of contact.

A complete clause for Time Charter period, which refers to the "Quarantine/Vaccination certificate/Free pratique" could be worded along the following lines:

During the currency of this Charter:
a. the Owners will be responsible and will arrange, under their own expense and time, for the Master, officers, and crew of the ship to comply with vaccination and sanitary regulations and all current requirements at all allowed as per the Charter Party ports of contact, during this Charter. The ship's Master, Officers and Crew will hold valid vaccination certificates against yellow fever, smallpox, cholera, or other necessary health certificates, where necessary and if in effect. Any actual time lost due to detention, losses, fines, additional expenses arising from non-compliance with health regulations and/or lack of vaccination certificates during the call at the ports of contact will be for the Owners' account.
b. the normal quarantine time and expenses for entering ports will be for the Charterers' account. Any actual time lost and expenses incurred for quarantine due to pestilence, epidemics and illness of the Master, Officers and Crew will be for the Owners' account. If such illness is caused by an infectious disease from ports traded under this Charter or as a result of cargo carried, which should be reasonably proven by the Owners and which will not unreasonably be rejected by the Charterers and a quarantine is required, then all delays and expense incurred, losses, damages, fines, consequences will be for the Charterers' account

and the ship will remain always on-hire.
c. the Ship is not to be ordered to trade within infested areas, or at ports where world sanitary organisations have declared contagious diseases. In the case that the ship has already commenced the loading operation and/or Bills of Lading have been signed for an infested area, the Charterers and/or Charterers' Agents then will assist and will properly direct the Master regarding the Port Authority's requirements well in advance and before the ship's arrival at the port in question. The Owners will be liable for any delays, expenses, losses, damages, fines, consequences in quarantine arising from in subject Master or any of the Officers or Crew having communication with the shore at any infected area, without the prior written consent or instructions of the Charterers or their Agents.

13. Ship's maintenance/Ladders/Lighting.

The Owners must exercise due diligence to maintain the ship's
. hull, accommodation, main deck, engine room etc. clean and in a well-painted condition, throughout the duration of the Charter.
The pilot
. ladders of the ship, including holds' ladders and accesses thereto must be kept in state of repair, in order to comply with the safety requirements at all ports of contact. In case of any refusal of the pilots or stevedores to climb/descend the stairs of the ship, then the Owners will be responsible for all the delays/consequences. Obviously and in any case of poor maintenance, the ship will be
. detained
by the port state control Authorities.
The ship must also provide lighting to facilitate the parties in loading or discharging during night work.

The "Ship maintenance/Ladders/Lighting" clause, which, among others, should include:
-that the Owners will exercise due diligence when keeping the ship in good condition.
-that the pilot and holds' ladders will be kept in a good state of repair.
-the Owners' responsibility, as a result of non-compliance.
-the Owners will provide lighting for night work.

An executed clause that has been run in a Charter Party is as follows (precise copy):
Throughout the duration of this charter, Owners to maintain the vessel's exterior hull, accommodation block, main deck, hatches and hatch covers in a clean and well-painted condition.

Vessel's accommodation and pilot ladders as well as hold ladders and accesses thereto are to be kept in good repair so as to comply with the safety requirements at all ports of call. Any delays to vessel and/or her operations as a result of non-compliance with this clause to be considered as off-hire and any expenses incurred to be for Owners' account.

Comments:

The Owners are to maintain the ship's hull/accommodation etc. in good condition and the pilot/holds' ladders are to be kept in a good state of repair so as to comply with the safety requirements.

A complete clause for Time Charter period, which refers to the "Ship's maintenance/ Ladders/Lighting" could be worded along the following lines:

Throughout the duration of this Charter, the Owners will exercise due diligence in maintaining the ship's exterior hull, engine room, accommodation block, main deck, hatches, and hatch covers in a clean and well-painted condition.
The pilot and holds' ladders and accesses thereto, are to be kept in a good state of repair, so as to comply with the safety requirements at all ports of contact.
Where the ship's accommodation ladders cannot be used due to a particular berth construction, the shore gangway rental will be for the Charterers' account.
Where a shore gangway is necessary due to the ship's accommodation ladder being out of order, rental costs, and actual time lost will be for the Owners' account.
The Owners, Master will provide lighting (as on board), to permit night work at all hatches at one and the same time, free of expenses to the Charterers.
As a result of non-compliance with this clause, the ship will be placed off-hired for all actual time lost and any directly related expenses incurred but supported with original vouchers will be for the Owners' account.

14. Ballasting / De –ballasting / Hold's flooding.

The ship, for

. stability and seaworthiness purposes

will ballast/de-ballast, during the loading/discharging operation. The Charterers will ask to have such an option and utilize the ship's full ballast at any time, either for loading and/or discharging purposes or for reaching the port restrictions (air draft or keep to a minimum the distance from the sea level to the top of hatchcoaming), even by flooding one of the cargo holds, which is known as the 'ballast hold'. The Owners should allow the same, provided that the lost time for ballasting/de-ballasting/ drying will be for the Charterers' account.
The Master has the right, for reasons of safety or navigation, to berth the ship

having water in its ballast tanks and sometimes also having the designated hold/s in full ballast.
The Charterers, before chartering the ship, should obtain all information about the de-ballasting speed the ship can provide.
Some very fast loading ports refuse to accept ships for loading which are slow at de-ballasting.
Many countries, in accordance with local regulations, require the ship, prior to entering the port of loading,
. to change ballast at a region outside the port of loading.
This way, any possible port pollution with dirty water ballasted from another area, should be avoided.

The "Ballasting/De-ballasting-Hold's flooding" clause, which, among others, should include:
-that it is the Master's right to enter the port having the ship in a ballast condition.
-that Charterers have the right to use the ship's water ballast capacity -on whose account any lost time should be.
-whether the Charterers have the right to utilize the ship's ballast hold.
-that the ship has the right to change ballast water in the open sea, according to the prevailing rules at the port of contact.

An executed clause that has been run in a Charter Party is as follows (precise copy):
Vessel to ballast/de-ballast clean water ballast tanks only, if required by Charterers or their Agents at any time during loading and/or discharging, free of expenses to Charterers but in Charterers time. All ballasting/de-ballasting shall be at the discretion of Master having due regard to stability manual and seaworthiness of the vessel.
Comments:
The Charterers have the right to ask to ballast/de-ballast the ship's ballast tanks, whenever they deem it necessary, in order to assist the loading and/or discharging operation. The Charterers' or their Agents' request is subject to ship's stability and seaworthiness.

Another clause reads as follows (precise copy):
The Charterers have the right to instruct the Master to utilize the vessel's maximum water ballast capacity and eventually to flood no4 hold only with water in order to bring down the vessel's height to get into position under loading and/or discharging appliances, however, always in conformity to free board and/or safety requirements.

Comments:
The Charterers require not only to have the right to use the ship's maximum ballast capacity, but to also flood the hold no4 (the 'ballast hold'), in order to reach the height limit (from the water line to the top of the hatchcoaming).

A complete clause for Time Charter period, which refers to the "Ballasting/ De-ballasting /Hold's flooding" could be worded along the following lines:

Whenever it is required, for reasons of safety or navigation, the ship will trade having water ballast in its ballast tanks and/or designated ballast hold/s full of water. The Master will co-operate with the attending Agents and/or Port Authorities, entering the port with the minimum possible water ballast in the ship's ballast tanks and/or in flooded hold/s. Time used for de-ballasting and time used by the crew to dry up the ballast hold, will be for the Charterers' account.
For loading and/or discharging purposes and/or because of prevailing port restrictions, the Charterers or their Agents have the right to utilize the ship's maximum water ballast capacity, including floodable hold/s, if necessary, at all ports of contact.
The ship will ballast and/or de-ballast and/or flood hold/s with clean water only, if necessary and as required by the Charterers and/or their Agents, at any time before and/or during the loading and discharging operation, at their time, but at no cost. The ship's crew will dry up the flooded hold/s as quickly as possible, always at the Charterers' time and risk, but at no cost.
All ballasting and/or de-ballasting operations will be performed at the discretion, direction, and responsibility of the Master, according to the trim, strengths, stability, seaworthiness and navigation of the ship and in conformity with the free board and safety regulations.
Whenever it is required by the regulations of the next port of contact, the ship will carry out a ballast water exchange before entering the port and/or terminal, or, as the case may be, before entering the country's territorial waters. Time lost and quantity of fuel consumed will be for the Charterers' account.
The Charterers will arrange, under their time and expenses, to load and/or discharge any solid ballast that is required and will clean the holds after discharge.

15. Double-banking/Lightering and top-off.

Many cargo ships, due to the prevailing port restrictions at the intended ports for loading or discharging, could be requested to complete

. the loading in the anchorage
where there is in fact sufficient draft to complete it, or

. to discharge part of its cargo at anchorage
until the ship reaches the available draft and berth safely.
Through this clause, the Charterers will have the right to load or discharge part or the full cargo in the anchorage by barges or alongside a mother ship, by using the ship's gear for the operation or a floating crane situated between them.
All such operations must be under the Charterers' time/risk and expense and when discharging it should be clearly stated in the Charter Party
. that the cargo's delivery over side (over the ship's rails) into the other ship/ barges etc. is to constitute right and true delivery of the cargo, under the relative Bills of Lading and the Owner/ Master/ship will have no direct or indirect responsibility or liability for any claim made by consignees or by any other party for shortage or admixture or damage or loss of cargo.
The Master must be fully convinced that the anchorage is safe and that it is common for such types of ships to load or discharge there. The Charterers must provide an adequate number of fenders and of the type designed to ensure the safety of the ship. One of the safest and most accepted is the 'Yokohama' type fender. The clause should incorporate that if adverse weather conditions prevail, the operation at the anchorage can be suspended and the mother ship/barges etc. must sail away from the ship without delay.
The most usual clause which covers the ship's double banking and is used in almost all Charter Parties, is the one called 'ship to ship transfer clause for time charter parties' as provided by Bimco. This Clause covers movements/actions which are going to be performed beyond the berth and the Owners may add any other additional conditions they deem necessary or amend some of them in order to protect their interests and the crew's/cargo's safety.
The "Double-banking/Lightering and top-off" clause, which, among others, should include: -that the Charterers have the right to conduct ship to ship cargo operations in a safe place and to load or discharge with the provided means. -that the Charterers will provide all the required adequate fendering, for the ship's safety -that the operations will be made to the Master's full satisfaction -that the additional premiums, if charged by the ship's H&M insurances, will be covered by the Charterers -that the Owners/Master/ship will not be held responsible and liable after the delivery of the cargo over the side into the lighters.

An executed clause that has been run in a Charter Party is as follows (precise copy):
It is agreed that the vessel may be lightened or partly or fully discharged into barge(s) at any safe anchorage as ordered by Charterers but always subject to Master's absolute discretion whether such lightening/discharging is safe and

feasible. The Master may, if he considers it at any time unsafe to commence or continue lightening/discharging, order the barge(s) away from his vessel to a safe distance and such barge (s) must obey such orders. Charterers shall supply adequate and proper fenders and securing equipment acceptable to the Master and shall also be liable for any damage caused to the vessel by the barge(s) during approach, securing, lying alongside, unsecuring or departure of the barge(s). Master will endeavour to substantiate such damage by holding liable in writing the party responsible. Final and sole responsibility for the positioning of fenders shall always remain with the Master of the vessel. Master shall at all times give full co-operation to Charterers and/or their agents expedite discharge and shall take all reasonable steps to protect his vessel from sustaining damage in case of any extra insurance due to such operation same to be for Charterers' account against the original vouchers.

Comments:

The Charterers have the right to lighten the ship and discharge part or the full cargo at the anchorage, as ordered by them, by using barges. The Master has the right in the case that he considers the operation unsafe, to order the barge away from the ship for safety reasons. The Charterers will supply adequate fenders and securing equipment to the Master's satisfaction. The Charterers remain responsible for damage caused to the ship by the barge.

Another clause reads as follows (precise copy):

The vessel to lie alongside barges/coasters/lighter/floating cranes at a safe dock or wharf or place (including anchorage) for loading and/or discharge of cargo and/or bunkering, if so ordered by the port authority, or when such operation is customarily carried out at the port. Such operation to be carried out always under the control of Master regarding the general safety, who may at any time order the barges/coasters/lighters away from vessel, meanwhile Charterers to take normal customary precaution for such operation. Charterers shall supply at their time and expenses fendering to the Master's satisfaction, but always within normal shipping practice. If any extra insurance applies by vessel's H&M underwriters same shall be for Charterers' account in full payable on request. Charterers will supply Yokohama fenders.

Comments:

The ship might be instructed either by the Port Authorities or by the Charterers in the case that such operations are customarily carried out at this port, to lie alongside barges/lighters etc., for loading or discharging purposes. The general safety will be under the control of the Master and any extra insurance that is in effect by the ship's H&M underwriters shall be for the Charterers' account. It also states that the Charterers must supply Yokohama fenders only.

A complete clause for Time Charter period, which refers to the "Double-banking/ Lightering and top-off" could be worded along the following lines:

Charterers have the right, during the currency of the Charter party, to load from and/or discharge into lighters and/or barges and/or alongside seagoing/mother ships, at a recognized safe lightering place, where such operations normally take place for similar sized ships, inside or off a port, where the ship can safely proceed, lie and depart, always afloat, but always subject to the Master's entire satisfaction and approval. Additional premiums, if charged by the ship's H&M insurances for the execution of such operations, will be covered by the Charterers, once the Owners have submitted the original vouchers, which are acceptable even by fax or e-mail.
Charterers will give the Owners clear notice of their intention to perform such an operation, at least 3 working days prior to commencement, stating:
-the cargo quantity to be loaded or discharged,
-the name and full characteristics of the other ship and/or barges and/or lighters,
-the location of the operation which will take place,
-the type and number of fenders.
When the ship is discharging cargo over side (over the ship's rails) into the other ship and/or barges and/or lighters will constitute right and true delivery of the cargo, under the relative Bills of Lading and the Owners, Master, ship will have no direct or indirect responsibility or liability for any claim made by Consignees or by any other party for shortage or admixture or damage or loss of cargo. The Charterers will advise of the ultimate destination.
The bimco's current 'ship to ship transfer clause for time charter parties', is deemed to be a part of and incorporated in this Charter Party.

16. Bills of Lading/Letter of Indemnity (L.O.I.).

Many definitions have been given for the bill of lading and the importance of the same. One thing is for certain, that if the Master accepts/signs and releases wrong/improper bills of lading, he could put the ship/ownership in major problems, with enormous consequences.

The Bills of Lading will be mainly signed as
. 'freight payable as per Charter Party',
or as
. 'freight prepaid' or 'freight paid'.

The Bills of Lading have the following key characteristics, which exist for both on

Voyage and Time Charter parties:

A.

The Bill of Lading is a legal document of title that acts as proof of the transportation contract between the Carriers and the Shippers.

It is a document which includes the four following key elements of information:

- *the description of the type/grade of the cargo loaded.*
- *the date of shipment (i.e., the date of completion of the loading operation).*
- *by whom the cargo was loaded and to whom the cargo will be delivered.*
- *the conditions under which the transportation will be executed.*

By the Bill of Lading, a contract of carriage is created, and the parties are identified, i.e., Shippers, Receivers, Carriers etc. and the carriage is identified, i.e., the load and discharge port.

When signed, it should be the link between the Owners and the Buyers/Consignees of the cargo, who are to receive their cargo upon the ship's arrival at the discharge port and the presentation of one originally issued Bill of Lading.

B.

There are two basic types:

1. Straight Bill of Lading:

It is a non-negotiable document, which clearly states that the cargo is consigned to a named Consignee/ assigned to a specific party (as stated in the Letter of Credit. The Bills of Lading will only be endorsed by this named Consignee and the Shippers do not have the right to alter the Bill of Lading.

2. 'To Order' Bill of Lading:

It is a negotiable document, where the title to the cargo is conferred to the 'order' of the Shipper or to the order of a bank (it is usually the party named in the Letter Of Credit) and can be transferred by one party to another by endorsement, in order to assign a title to the next party, i.e. the goods are consigned to a non-designated party.

C.

It will be issued by the Shippers or by the ship's Agents on behalf of the Carriers only after the completion of the loading of the cargo and completion of the customs formalities. After signing, the Carrier, who will appear on the Bill of Lading, acknowledges receipt of the cargo, which will remain under his custody until the time of delivery to the named Consignees and the ship is allowed to sail from the port towards its final destination. However, the Carrier does not have to be the head Owner of the ship, but even the Charterers or the Freight Forwarders or any other party that actually assumes the responsibility for the carriage, even for matters that they have no direct control over (e.g., navigation, safety etc.). Obviously, there should be no objection from the P&I Clubs about the name of the carrier, but Carriers unknown to the Owners, who are not covered by a reliable P&I Club coverage, or who have a bad reputation in the mar-

ket,
. can spell big troubles for the ship.
D.
Regardless of who the Carriers are, the Master is the only person
. who will sign the Bill.
or authorize another party to sign, but always on his behalf.
E.
The Bill of Lading is frequently issued in a set of
. three originals in standardized forms.
In the port of discharge, the party who firstly presents
. an endorsed original Bill of Lading is entitled to receive the cargo from the ship.
The other two issued originals will become invalid for the receipt of the cargo.
F.
The Bill of Lading will be issued in
. strict conformity with the Mate's Receipts, with regard to the quantity, description and condition of the cargo, as well as the date of shipment.
The Master has no obligation to agree to sign Bills containing an inaccurate or incorrect description of goods.
He should also strictly refuse to sign any
. ante-/post-dated bill of lading.
After the signing, all three original Bills of Lading (a full set) will be returned to the Shippers/Agents. The Master will keep a non-negotiable copy on board.
G.
If the ship has sailed before the Bill is issued and the Charter Party so expressly states it, the Master will allow another party involved (usually the Charterers' or ship's Agents or the ship's P&I Club's representatives or as should be agreed during the negotiations) to sign the Bill, but always on his behalf. He will hand over to the authorized party
. a Letter of Authorization,
which will contain the full and proper instructions (as per Master's wording) and the conditions under which they will sign the Bill of Lading. The Letter of Authorization must be fully accepted and signed by the authorized party.
Many Charter Parties include the term that the Bill of Lading, before being signed and released by the authorized party, will be faxed/e-mailed directly to the Owners' office for approval.
H.
The most usable type of Bill of Lading contains the following standard 'boxes':
-the Shippers' name.
-the Consignees' name.
-the notify party.
-the ship's name.

-the load port.
-the discharge port.
-the apparent condition of the goods.
-the cargo description and the loaded weight.
-the place of issuance and the date of shipment.
-the number of original Bills of Lading issued.
-the Condition of Carriage.

I.

Normal practice is for the ship to discharge the cargo and deliver it to the person entitled, who before the commencement of the discharging operation, will first hand over to the Master an endorsed original Bill of Lading, as issued at load port.

Any deviation from this practice

. may be subject to the P&I Club's rules,

mainly due to the imminent risk of the cargo being delivered to the wrong people (wrongful delivery of goods) or in general that they are conducting an inappropriate delivery of goods.

If the Consignees are unable to hand over the original Bill of Lading to the Master, thus enabling them to take their cargo, they will present to the Master of the ship

. a Letter of Indemnity (a guarantee letter abbreviated as 'l.o.i.'),

which covers the Owners from the consequences of such a delivery of goods. Nevertheless, if there is no such provision/option in the Charter Party and it is not customary in the port of contact for the ship to discharge with a l.o.i., it is at the commercial decision of the Owners/Master, whether to accept or to reject such a way of delivery of goods.

Potential complications and delays at the discharge port, due to the Owners' refusal to discharge without presentation of the original Bill of Lading, will lead the Charterers to agree to have the option to discharge without the presentation of the original Bill of Lading, from the time of commencement of the negotiations. They will ask for themselves or the Consignees to have the option to provide, before the discharging commencement, a letter of indemnity duly signed by the providing party. Many times, the fixture

. is subject to acceptance of the same by the Owners.

The l.o.i. will usually be issued in a

. standardized form,

as drafted by the International Group of P&I Clubs for the Owners'/ operators' facilitation, which gives to Owners the right to appeal against the party issuing the same. Even if those standard forms are drafted by the P&I Clubs, the ship will remain subject to P&I Club's insurance, in case of wrong delivery of the cargo and in case the l.o.i. is found to be inadequate, the risk and loss will rest with the Owners. It is important that the issued l.o.i. should also be counter-signed by a first-class bank, which, however, almost never happens and in most of the times it is rejected by the Charterers. When a bank joins the l.o.i., it will insist on a monetary limit, which in any case must not be less than 200 percent of the market value of the cargo, at the time of cargo delivery. The bank will pay any amount up to the limit of its liability, provided the party that signed the l.o.i. fails to provide bail or other security.

A l.o.i. may also be issued for delivery of cargo in a port other than that stated on the Bill of Lading or for signing a 'clean on board' when the cargo is not in apparent good order and condition.

In general, Owners, before accepting an l.o.i., must obtain advice from the ship's P&I Club, as in fact for ANY acceptance of a l.o.i. might make the ship fall outside the P&I Club's insurance. The Charterers' reliability/credibility for accepting an l.o.i. must be to the Owners' full satisfaction.

J.

Sometimes the Shippers/Charterers will ask the Master to keep on board one original Bill of Lading from the total issued.

In this case, it is a rule that all originally issued Bills of Lading

. must be properly marked,

that an original Bill of Lading is kept on board and that the cargo delivery can be properly carried out with instructions from the Shippers/Charterers. However, this method of issuance is not recommended by the P&I Clubs and the ship may be subject to rules. The Master will deliver the original Bill of Lading relating to the intended Consignees for their endorsement. The Owners, in the case that the Bill is issued 'to order', have a risk of unauthorized cargo delivery, as other Consignees could claim for delivery of cargo against presentation of one original Bill of Lading (from the other two originally issued Bills of Lading that have circulated) that has gone through the regular bank channels.

K.

Some types of Bills of Lading omit some important terms, such as the Paramount clause, the Arbitration clause etc. Care should be taken when negotiating the terms of the Charter Party and the Owners' protective clauses should be clearly agreed upon. The Owners must ensure that they are not exposed to risks that exceed the terms of the Charter Party and there should be a provision in the Bills explicitly obliging the Charterers/Shippers

. to incorporate all the terms, conditions, liberties, warranties and exceptions of the Voyage Charter Party, including the lien and arbitration clause in the Bills.

That will allow the Owners to defend against Bill of Lading claims based on the Charter Party clauses and to be able to seek an indemnity under the Charter Party terms. It will also give notice to any innocent third-party holder of the Bills that the Bills

. are subject to the Charter Party terms and conditions.

The lien of the cargo against the Receivers for unpaid due freight/demurrage/hires from the Charterers could be subject to the Owners proving that the Bill of Lading incorporates a lien clause that has the same effect as under the lien clause of the governing Charter Party, therefore it making necessary the incorporation of the Charter Party terms and conditions into the issued Bill of Lading.

When Congen form Bills are to be used, if they are of the 1994 form, they will include on the reverse side a provision expressly incorporating the law and jurisdiction clause of the Charter Party, of which the date appears on the front of the Bills. As a matter of English Law, this is normally sufficient for its terms and conditions to be in effect to the Bills. The provisions of the Hague/Hague-Visby rules and any amendments thereto (the Paramount clause general), the New Both-to-Blame Collision clause, the New Jason clause, the General average clause and the War risk clause are to be incorporated into all issued Bills of Lading.

L.

Under a single Bill of Lading, the delivery of the cargo can only be made to one consignee. However, in bulk trades, the cargo might be requested to be split between several Receivers. If such a request is made by the Charterers, then the Owners should take care to ensure that they have followed the correct procedures.

The 1st full set (all 3/3) of the original Bills of Lading must be delivered to the Owners' hands before the Charterers issue the new Bills of Lading. The Charterers must not have the right to change the total quantity of cargo/cargo description and the date of shipment. The New Bills will conform to all terms and conditions and exceptions of the originally issued in load port Bill/s of Lading and will not prejudice the ship-Owners' rights. Once the replacement Bills of Lading have been issued, then the cargo that is stated in that particular Bill of Lading should only be delivered to the Receivers named in the new Bill of Lading.

There should be also a risk of one or more of the Receivers receiving more or less cargo than they must receive. A provision in the Charter Party should state that in the case of more than one Bill of Lading have been issued to various Consignees, the Owners/Master and ship will not be responsible for delivery of the cargo to each Bill of Lading holder, however and without prejudice to the Charter Party Interclub Agreement clause, the Owners remain responsible for the total intaken cargo quantity loaded and discharged.

The Charterers may request 'Delivery Orders' to be issued instead of splitting the Bills of Lading.

The cargo interests will trade the cargo during the carriage and will deliver it without requiring Bills of Lading. The same precautions should also be in effect for delivery of the cargo with 'Delivery Orders'.

M.

If the Bill of Lading is issued to the 'order of a bank', the bank must counter sign the l.o.i. for discharging without presentation of the original Bills of Lading or to confirm directly to the Owners their acceptance/permission to discharge the cargo, i.e., to confirm delivery instructions, even though the Charterers have performed this exchange with a l.o.i. As per the issued Bill of Lading, the bank is the holder of the Bills of Lading in question. The bank that made the payment in respect to the goods holds the Bills of Lading as security, pending payment in turn upon acceptance of the Bills and shipping documents under a Letter of Credit issued by the Receivers' bank. If the Receivers' bank rejects the documents, they will not make the payment and if the cargo has been discharged/ delivered, then the bank might have a claim against the ship/Owners, as the holder of the Bills of Lading.

If issued 'to order of shippers', then the Shippers endorse the Bill of Lading and order as to whom the cargo should be delivered.

N.

The issued original Bills of Lading must contain the correct cargo quantity, which has been loaded on board the ship.

At the load port, the Shippers

. are obliged to supply the loading figures and to warrant those figures.

The same should be presented either by a carried out

. draft survey or by the shore scale measurements.

Problems may arise at the load port, in the case that the shippers' figures exceed the ship's figures or the Receivers' figures are less than the ship's figures (Bill of Lading figures) at the discharge port.

In the port of shipment and if there is a difference on the loaded cargo quantity beyond what is considered normal and/or customary (the trade allowance is 0,5 percent on the loaded cargo quantity) and the Shippers insist that the Bill of Lading will be issued in accordance with their findings (basically that happens when Shippers supply shore scale measurements), but the Master or his representative

. has been able to determine the cargo quantity on board

and there are reasonable grounds for disputing the accuracy of the shore figures, i.e., the draft survey was carried out in a calm sea, the ship was not rolling and pitching, there were no strong currents etc. and in general the draft survey was carried out in perfect conditions, under the Hague and Hague-Visby Rules Article III Rule 3

. he is not bound to accept the supplied figures to be inserted as intaken cargo quantity.

The ship should strictly follow the P&I Club's recommendations and at least the

original Bill of Lading should be claused (marked) with
. the ship's figures.
As an alternative, some reservations could be made on the Bill of Lading such as 'weight unknown' or 'quantity and quality unknown', said to be' etc. or to accept a Letter of Indemnity by the Shippers or Charterers, but in fact the Owners should be aware that such reservations might not have effect in many jurisdictions and/ or the ship could be subject to the P&I Club's rules, as it has accepted incorrect figures.

In general, the Owners and Master take considerable risks, which might lead to the ship falling outside the P&I Club's Insurance, if accepting:

1.
a l.o.i. due to non-presentation of original Bills of Lading. The Owners are of the risk of wrongful delivery of cargo or, in general, that they are conducting an inappropriate delivery of goods.

2.
a l.o.i. for change of destination, without collection of the full set as issued in the load port. The Owners are of the risk that an original Bill of Lading is presented for the cargo in the port as stated in the original Bill of Lading.

3.
a l.o.i. for 'clean on board' for cargo which appears not to be in sound condition. The Master of the ship must only sign Bills of Lading which accurately describe the condition of the cargo, or, in any other case, the Owners are exposed to receive huge claims.

4.
to sign ante-/post-dated Bills of Lading.
. *Never do it.*
The date of shipment shown in the Bill of Lading will determine the value of the cargo in the sales contract. Ante-/post-dated Bills of Lading will affect the cargo value and the Owners could be held fully liable, with huge consequences.
The correct date of shipment is the date of completion of loading or otherwise, the date of completion of each parcel, if so required by the Charterers. The Bill of Lading can be issued before the full completion of loading, provided that the Bill of Lading reflects the cargo loaded when issued.

5.
to sign a Bill of Lading containing an inaccurate or incorrect description of the cargo and cargo loaded figures. Reservations, which could be made on the Bill of Lading such as 'weight unknown' or 'quantity and quality unknown' or 'said to be' etc. or accepting a Letter of Indemnity by the Shippers or Charterers, but such reservations might not have effect in many jurisdictions and/or the ship could be subject to the P&I Club's rules, as having accepted an inaccurate or in-

correct cargo description and the cargo loaded figures.

6.

to sign a Bill of Lading incorporating the Cargo Value. The P&I Clubs reject ad valorem Bills of Lading.

7.

to sign a Bill of Lading not adhering to the originally issued Mate's Receipts with regard to the quantity, description, condition of the cargo, as well as the date of shipment.

The Charter Party should explicitly state that the original Bills of Lading will be issued in strict compliance with the Mate's receipts or Tally clerks.

8.

not to sign identical Bills of Lading which exist for the shipment, i.e., if all three (3/3) original Bills of Lading (the full set), as issued for the cargo on board the ship, have not been issued identically and have not been marked as 'original'.

9.

to sign outside of the designated box or place for signature in the Bill of Lading, which could be considered as a specific acceptance of the incorporation in the place of signature.

10.

to carry in the ship's bag one original Bill of Lading from the total issued. The Owners, in the case that the Bill is issued 'to order', have a risk of unauthorized cargo delivery, as other consignees could claim for delivery of the cargo against the presentation of one original Bill of Lading (from the other two issued and circulated original Bills of Lading) that has gone through the regular bank channels.

11.

to sign a Bill of Lading for deck cargo. The Owners, when loading deck cargo, except on certain specific circumstances (containers/logs etc.), might be in breach of the contract of carriage and the ship may fall outside of the P&I Club's rules.

The Master should avoid signing Bills of Lading (otherwise they must keep reservations until they are satisfied), which incorporate:

1.

A Charter Party date other than the Charter Party made by the head Owners.

If the Owners fix with the Operators who appeared in the market as Charterers, then, after completion of loading, the Agents might present a Bill of Lading incorporating the date of the head Charter Party, which was made between the Charterers (Operators) and the head Charterers.

However, the Owners, either on Voyage or on Time Charter, are not party to the Sub-Charter Party and ignorant of its terms.

The Owners must ensure that they are not exposed to risks that exceed the terms of their Charter Party and there should be a provision explicitly obliging the Charterers/Shippers to incorporate all the terms, conditions, liberties, warranties, and exceptions of the Charter Party in the Bills made between Owners and Operators, including the lien and arbitration clause.

In the case that the Bills are issued without the incorporation of the Charter Party made by the head Owners, then that would be a breach of the Charter Party terms and the Owners can hold the Charterers fully liable for any consequences and losses/damages flowing from that and any Bills will be issued under protest. The Owners are, in the circumstances at least, entitled to see a full copy of the Voyage Charter Party made between the Charterers and head Charterers, as this is to be incorporated into the Bill and will form the terms of the contract of carriage. The Owners must ensure that the carriage is covered and that the contract is governed and construed in accordance with the English Law unless otherwise stated in the Charter Party and the disputes will be referred to arbitration in London unless otherwise stated in the Charter Party.

2.

Letter of Credit (L/C) numbers or references to Letter of Credit on the face of the Bill of Lading / ad valorem Bill of Lading.

The question remains whether a reference of Letter of Credit number in a Bill of Lading, would have the effect of making the BL an ad valorem bill, and could prejudice Owners' interests and P&I Club's cover.

This is by reason of the fact that inserting the L/C number on the face of the B/L, could be interpreted in a number of jurisdictions as incorporating the terms of the Letter of Credit, including the agreed values of the cargo therein and thus not being able to limit Owners' liability under the Hague -Visby Rules, should the occasion arise.

Some P+I Clubs recommend that the mere reference to a Letter of Credit on the face of the B/L does not make it an ad valorem B/L, and a problem should arise, only in cases where the value of the cargo has been specifically declared and inserted in the Bill of Lading.

In general, it would be always preferable to avoid the L/C reference numbers in the description of the goods.

In all cases, if the Owners have commercially decided to accept it and accommodate the Charterers, then a reservation should be taken in order to minimize the consequences. This reservation, along with the L/C number or the references, should preferably go in the top right of the Bill (just under 'reference no' of the 'Congenbill') and not in the box referring to the cargo description.

The reservation could be worded as follows (both are a precise copy from executed Bills of Lading):

'For information only and inserted at the request of the shippers/consignees/ endorsees/holders/receivers of/under this bill of lading for their convenience and administrative purposes to identify this shipment only and is not a decla-

ration of value. The value of the cargo and the terms of any L/C are unknown to the carrier/ ship/master /owners of the ship. This is not an ad valorem bill of lading.'
or:
'The letter of Credit details noted above are for reference purposes only. The letter of Credit itself and its terms and conditions are not expressly or impliedly incorporated or deemed to be incorporated into or from part of this Bill of Lading'.
When the value of the cargo is inserted, the reservation could be worded as follows (precise copy from a Bill of Lading):
'For shipper's information and reference purposes only. No representation as to the value of the cargo carried under this Bill of Lading is made by the carrier and this is not an ad valorem Bill of Lading'.
3.
As 'carrier' other than the actual Owners.
Carriers unknown to the Owners, who are not covered by a reliable P&I Club coverage, or have a bad reputation in the market, can give rise to huge consequences against the ship and the ship may be arrested at the port by third parties, who have claimed against the named carriers appearing in the original Bills of Lading for various reasons.

The 'Bills of Lading' clause, among others, should include:
-that it will be issued in strict compliance with the originally issued Mate's Receipt.
-that, before the authorised party signs/releases the Bills of Lading, the same should be previously faxed to the Owners' office for written approval.
-that excluded Hamburg rules are to be in effect on the originally issued Bills of Lading.
-the type of Bills of Lading which are to be excluded for issuance.
-if more than one B/L is issued, who will be responsible for the distribution of the cargo to the various Consignees.
-that the Bills of Lading will incorporate the terms and conditions of the Time Charter Party.
-that the ship will discharge the cargo on presentation of the original Bills of Lading.

An executed clause that has been run in a Charter Party is as follows (precise copy):
Bs/l shall be issued in strict conformity with Mate's Receipts. In case Charterers/ Charterers agents or Sub-Charterers sign bs/l not in strict conformity with Mate's Receipts or bs/l be marked with Charter Party date other than the actual

one between present Owners and Charterers, then the Charterers agree to undertake any and all responsibilities of whatsoever nature and howsoever caused as well as Charterers agree to indemnify the Owners for any costs/ arrests/ detention/ legal implications and/or liabilities arising out of issuance of bs/l not in strict conformity with Mate's Receipts or not adhering same. Prior signing/releasing bs/l shall forward by fax to Owners office directly a draft copy of bs/l for their perusal.

Comments:

The Charterers have the option of issuing Bills of Lading, which do not adhere to the issued Mate's Receipts or even to insert a Charter Party date other than the actual one between the Owners and Charterers. Obviously, the Time Charterers have fixed the biz on a voyage basis with certain other head Charterers and this date of Charter will be requested to be inserted. By such way of agreement, the Owners will get involved in unknown terms, which have been agreed between Charterers and head Charterers. The Charterers undertake any and all responsibilities and it is to the Owners' commercial decision whether to accept such a clause or not. The Owners should at least request to receive a copy of the issued Voyage Charter Party and verify that the contract is covered, and it is governed and construed in accordance with English Law and disputes will be referred to arbitration in London.

Another clause reads as follows (precise copy):

Congen Bill 1994 edition to be used for this Charter Party.

During the currency of this Charter Party Master to sign Bills of Lading for cargo loaded on board in strict conformity with Mate's Receipts without prejudice to this Charter Party which should be presented to the Master for signature prior sailing from port of loading. Charterers have the option to present 'Freight prepaid' Bills of Lading without prejudice to this Charter Party and charterers to indemnify Owners for any consequences arising out of such signing of Bills of Lading. All Bills of Lading to be expressively governed by English Law.

In case for no time for Master to sign Bills of Lading at loading port, the Master will authorise Charterers or Sub- Charterers or their agents to sign Bills of Lading on behalf of Master strictly in accordance with Mate's Receipts and Tally Clerks receipts without prejudice to this Charter Party. Charterers shall indemnify the Owners against all consequences or liabilities arising from Charterers or their agents signing Bills of Lading.

Neither the Charterers nor their agents shall permit the issue of any Bills of Lading, waybill or other documents evidencing a contract of carriage (whether or not signed on behalf of the Owners or on Charterers behalf or on behalf of any Sub-Charterers) incorporating , where not compulsorily applicable , the Hamburg Rules or any other legislation imposing liabilities in excess of Hague or

Hague/ Visby rules, the Charterers shall indemnify the Owners against any liability, loss or damage which may result from any breach of the foregoing provision of this clause.
Comments:
Only the 'Congen' Bill edition 1994 will be presented and will be issued in strict conformity with the Mate's receipts. The Charterers have the option of marking the Bills as 'freight prepaid'. The Master will authorise the Charterers or Sub-Charterers or agents to sign the same on his behalf, provided that it does not incorporate the Hamburg rules (in places not compulsory in effect) or any other legislation imposing liabilities in excess of the Hague/Hague-Visby rules.

An executed clause that has been run in a Charter Party is as follows (precise copy):
When and if required, one original Bill of Lading being carried onboard against which the cargo may properly be released against instructions received from Shippers/Charterers.
All original Bills of Lading issued hereunder to be marked 'one original Bill of Lading retained on board against which Bill, delivery of cargo may be properly made on instructions from Shippers/Charterers.
Comments:
If it is required, an original Bill of Lading will be kept in the ship's bag for the delivery of the cargo in accordance with the instructions from the Shippers/Charterers.

A complete clause for Time Charter period, which refers to the "Bills of Lading/ Letter of Indemnity (L.O.I.)" could be worded along the following lines:

Cargo discharging on presentation of the original Bills of Lading:
During the duration of this Charter Party, the Master will sign the Bills of Lading for the cargo shipped, in strict compliance with the Mate's Receipts and/or Tally Clerks' receipts and without prejudice to this Charter Party, before sailing from the port of loading. All Bills of Lading are expressly governed by English Law.
If there is no time for the Master to sign the Bills of Lading at the port of loading, the Master will authorise the Charterers or Sub- Charterers or their Agents to sign the Bills of Lading on his behalf, strictly in accordance with the originally issued Mate's Receipts and/or Tally Clerks' receipts and without prejudice to this Charter Party. The Charterers will indemnify the Owners against all consequences and/ or liabilities arising from the Charterers or Sub-Charterers or their Agents signing the Bills of Lading on the Master's behalf. Authorised Bills of Lading, before being signed and released, will be faxed, or e-mailed directly to the Owners' office for approval. The Owners will comment on the presented Bills of

Lading, as soon as possible, after the draft is faxed or e-mailed to their office, always provided that the draft will be presented during the Owners' working days and office hours.
Neither the Charterers nor their Agents will authorise the issue of Bills of Lading or other documents evidencing the contract of carriage (even signed on behalf of the Owners or Charterers or on behalf of each Sub-Charterers) incorporating, where it is not compulsorily in effect, the Hamburg Rules or any other legislation that impose obligations in excess of the Hague or Hague/Visby rules. The Charterers will indemnify the Owners for any liability, loss or damage which may arise from any breach of this provision of this clause.
The cargo loaded will be delivered and released on presentation of the original Bills of Lading.
No Through, Way, Transhipment, Transit, Liner Bill/s of Lading will be presented and issued.
In the case of more than one Bills of Lading having been issued to various Consignees, the Owners, Master, ship will not be held responsible for the cargo delivery to each Bill of Lading holder, but without prejudice to the terms, conditions and exceptions of this Charter Party, including, but not limted to Interclub Agreement clause, the Owners remain responsible for the total intaken cargo quantity.
All Bills of Lading, which have been issued under this Charter Party or Sub-Charter Parties will incorporate all the terms, conditions, warranties, liberties and exceptions of the present Time Charter Party, including the lien and arbitration clause and the same will appear on the front side of the issued Bill/s of Lading.
The provisions of the Bimco's current:
- Paramount clause general and, where applicable, the Carriage of Goods by Sea Act of the United States (COGSA) and the Carriage of Goods by Water Act of Canada (COGWA).
-New Both-to-Blame Collision clause.
-New Jason clause.
-General average clause.
-War risks clause (Conwartime 2013)
are deemed to be a part of and incorporated in all Bills of Lading issued under this Charter Party and all sub-Charter Parties.

-When the Charterers present 'freight prepaid' or 'freight paid' Bills of Lading, the following wording could be agreed:

If the Charterers and/or sub-Charterers present original Bills of Lading marked 'freight prepaid' or 'freight paid', then the Charterers guarantee that they remain responsible for complying with their obligations and for any consequenc-

es, delays, losses arising out of marking the original Bills of Lading as 'freight pre-paid' or 'freight paid'. The Charterers undertake the responsibility to defend and release the ship, at the suit of any person or any authority or any legal process having or purporting to have any claim against the ship, by taking necessary and appropriate measures (issuance of security etc.) as the case may be or arise. The ship will remain, always and under any circumstances, on-hire.

-For Owners, who agree to reissue and/or split the Bills of Lading and/or issue delivery orders and agree to change the consignees' name/discharge port (the Owners should seek legal advice before agreeing), they may add to the above:

Charterers have the option to reissue the load port Bills of Lading and/or to split and/or issue Delivery Orders at the office of the Owners' Agents, at the cost of the Charterers, in exchange for the Charterers' l.o.i. issued according to the Owners' wording. The original l.o.i. dully issued in the Charterers' letterhead and signed by a director and the first full set (all 3/3) original Bills of Lading are to be delivered to the Owners' hands, before issuing the new Bills of Lading or Delivery Orders.
The signed l.o.i. and the draft copies of the newly issued second set and/or split Bills of Lading and/or Delivery Orders will be pre-approved by the Owners.
The second set of the reissued and/or split Bills of Lading and/or Delivery Orders will conform to all terms and conditions and exceptions of originally issued at the load port Bill/s of Lading and will not prejudice the Owners' rights. The Charterers do not have the right to change the total quantity of cargo, description of cargo, date of shipment and the Shippers' name. However, the Charterers have the right to change the consignees' name and/or discharge port. The Master's remarks, if incorporated into the old set, will be also added into the new set. Any cost, expenses, courier expenses, Agency fees will be paid directly from the Charterers to the Owner's, Owners' Agents, Owners' representatives, before signing the new issued Bills of Lading.

-Or, for split Bills of Lading and/or Delivery Orders that will be issued identically with the first set, the clause could be amended as follows:

Charterers have the option to split and/or issue Delivery Orders at the office of the Owners' agents, at the cost of the Charterers, in exchange for the Charterers' l.o.i. issued according to the Owners' wording. The original l.o.i. dully issued in the Charterers' letterhead and signed by a director and the first full set (all 3/3) original Bills of Lading are to be delivered to the Owners' hands, before issuing the new Bills of Lading and/or Deliver orders.
The split Bills of Lading and/or Delivery Orders will be issued identically with the

first set and the Charterers do not have the right to change the total quantity of cargo, cargo description and date of shipment and will conform to all terms and conditions and exceptions of originally issued at load port Bill/s of Lading and will not prejudice the Owners' rights.
The Master's remarks, if incorporated into the old set, will also be added into the new set.
The signed l.o.i. and the draft copies of the new set of split Bills of Lading and/or Delivery Orders will be pre-approved by the Owners.
Any cost, expenses, courier expenses, agency fees will be paid directly from the Charterers to the Owner's, Owners' Agents, Owners' representatives, before signing the new issued Bills of Lading and/or Delivery orders.
Neither the ship, nor the Owners will be held responsible for the separation of the undivided bulk at the port of delivery.

-For Owners, who agree to discharge without presentation of original Bills of Lading in exchange for the Charterers' l.o.i. signed only by them, the clause could be worded as follows:

Throughout the duration of this Charter Party, the Master will sign the Bills of Lading for the cargo shipped, in strict compliance with the Mate's and Tally Clerks' receipts and without prejudice to this Charter Party, before sailing from the port of loading. All Bills of Lading are expressly governed by English Law.
If there is no time for the Master to sign the Bills of Lading at the port of loading, the Master will authorise the Charterers or Sub-Charterers or their Agents to sign the Bills of Lading on his behalf, strictly in accordance with the originally issued Mate's and Tally Clerks' receipts and without prejudice to this Charter Party. The Charterers will indemnify the Owners against all consequences or liabilities arising from the Charterers or Sub-Charterers or their Agents signing the Bills of Lading on the Master's behalf. Authorised Bills of Lading, before being signed and released, will be faxed, or e-mailed directly to the Owners' office for approval. The Owners will comment on the presented Bills of Lading, as soon as possible, after the draft is faxed or e-mailed to their office, always provided that the draft will be presented during the Owners' working days and office hours.
Neither the Charterers nor their Agents or any other authorised party, will authorise the issue of Bills of Lading or other documents evidencing the contract of carriage (even signed on behalf of the Owners or Charterers or on behalf of each of the Sub-Charterers) incorporating, where it is not compulsorily in effect, the Hamburg Rules, or any other legislation that impose obligations in excess of the Hague or Hague/Visby rules. The Charterers will indemnify the Owners for any liability, loss or damage which may arise from any breach of this provision of this clause.

No Through, Way, Transhipment, Transit, Liner Bill/s of Lading will be issued. In the case of more than one Bills of Lading having been issued to various Consignees, the Owners, Master, ship will not be kept responsible for the cargo's delivery to each Bill of Lading holder, but without prejudice to the terms, conditions, and exceptions of this Charter Party, including, but not limited to Inter-club Agreement clause, the Owners remain responsible for the total intaken cargo quantity.

All Bills of Lading which have been issued under this Charter Party or Sub-Charter Parties will incorporate all the terms, conditions, warranties, liberties and exceptions of the present Time Charter Party, including the lien and arbitration clause and the same will appear on the front side of the issued Bill/s of Lading. The provisions of the Bimco's current:

- Paramount clause general and, where applicable, the Carriage of Goods by Sea Act of the United States (COGSA) and the Carriage of Goods by Water Act of Canada (COGWA)

-New Both-to-Blame Collision clause

-New Jason clause

-General average clause

-War risks clause (Conwartime 2013)

are deemed to be a part of and incorporated in all Bills of Lading issued under this Charter Party and all sub-Charter Parties.

The Charterers will make every endeavour to ensure that the original Bill of Lading will be available at the discharge port on or before the ship's arrival and the same handed over to the Master before the commencement of the discharging operation. However, if the original Bills of Lading are not available, then the Owners/Master will allow the commencement of discharge of the cargo without delay, in exchange for the Charterers' Letter of Indemnity , as issued on the Owners' P&I Club's wording, signed by the Charterers only. The L.O.I. will be issued in the Charterers' letterhead and will be signed by an office director. The original L.O.I. will be delivered to the Master's hands or to the Owners' premises, in all cases before the commencement of the discharging operation.

The Charterers remain responsible for the collection of the original Bills of Lading, regardless of whether the ship has discharged with an issued l.o.i.

The original Bills of Lading will be forwarded to the Owners as soon as they become available. On receipt of the original Bills of Lading, the Owners are to return the original l.o.i. to the Charterers, as soon as possible.

The Charterers will clearly stipulate in their issued l.o.i., the full name, full style, and address of the company to whom the cargo is to be delivered. The consignees are to try to deliver to the ship's Master a statement on their letterhead that they are taking delivery of the cargo.

17. Seaworthy and safe stress, trim condition.

Improper stowage of the cargo in the holds or part discharge of cargo, could affect the safety of the ship and make the ship unseaworthy.
The ship must always be left
. in a seaworthy stress and trim at all times, when shifting between berths or navigating between ports.
The common wording incorporated in the Charter Parties, which states that 'the ship is to be left in a seaworthy condition between berth/ports', is actually not sufficient and does not cover the Owners' interests.
In this clause,
. the party on whose account the cost and lost time for returning the ship to a seaworthy condition, should be clarified.

The "Seaworthy and safe Stress, Trim condition" clause, which, among others, should include:
-that the ship will trade between berths/ports in a seaworthy stress and trim.
-that the lost time plus expenses for returning the ship to a seaworthy condition, will be for the Charterers' account.

An executed clause that has been run in a Charter Party is as follows (precise copy):
Vessel always to be left in seaworthy stress/trim condition to Master's satisfaction between loading and discharging port/s- berth/s.
Comments:
The ship will trade in a seaworthy stress/trip condition between ports and berths.

Another clause reads as follows (precise copy):
Vessel to be always left in a seaworthy trim to the Master's satisfaction during her sailing and/or shifting between all berth/s, anchorage/s, and wharf/s.
Owners guarantee vessel can sail between loading also between discharging ports with 2-3 empty/slack holds, however final sailing configuration always subject to Master's stowage plan in conformity with vessel's loading manual and vessel always to be left in seaworthy trim between berths/ports
Final sailing configuration always subject to Master's stowage plan in conformity with vessel's loading manual and vessel always to be left in seaworthy trim between berths.
Comments:
The ship will always be left in a seaworthy trim to the Master's satisfaction, when trading between berths/anchorages/wharfs.

The Owners have also guaranteed that the ship can sail having 2/3 of its holds empty or slack but the sailing is subject to the Master's stowage plan, which is in conformity with the ship's loading manual.

A complete clause for Time Charter period, which refers to the "Seaworthy and safe stress, trim condition" could be worded along the following lines:

The final sailing configuration is always subject to the Master's stowage plan and satisfaction, which is in conformity with the ship's loading manual.
The ship should always be left in a seaworthy and safe stress, trim condition and with cargo on board safely stowed, during its sailing and/or its lightering and/or its shifting between all ports, berths, anchorages, places and wharfs, to the Master's entire satisfaction.
Upon completion of the loading and/or discharging and/or lightering operation at the first or subsequent port and/or berth, anchorage, wharf, or at any time it is requested by any party (Master, Port Authorities etc.) to bring the ship in a seaworthy and safe stress, trim condition, then all the expenses and/or extra port expenses incurred will be entirely for the Charterers' account and the ship will remain always on-hire, even during the ballasting and de-ballasting of the ship's tanks and until the Master is fully satisfied and the ship is ready to sail or shift.

18. Protective/Standard clauses.

The protective clause/s are certain
. non-negotiable clause/s,
which is part of and incorporated into the Charter Parties and usually in all originally issued Bills of Lading. Thus, this clause should be negotiated together, or when taking into account the Bill of Lading clause.
The General Paramount clause, the Both to Blame Collision Clause, the New Jason Clause, and the General Average clause (application of the York-Antwerp Rules 2004 is recommended to be excluded by the Owners for any presented Bill of Lading and agreement on the terms of the Charter Party), are the standard incorporated protective clauses (incorporated as conditions of carriage), in the most frequently used Bill of Lading called 'Congenbill'.
The General Paramount clause is the most important provision, as it ensures that cargo claims will be subject to the provisions of the Hague-Visby Rules.
When the ship trades to or from USA ports, the Bills of Lading will be subject to the provisions of the Carriage of Goods by Sea Act of the United States (COGSA), and when trading to or from Canada, it will be subject to the provisions of the Carriage of Goods by Water Act of Canada (COGWA)

Bimco has published a number of certain standard clauses, which assist the contracting parties in concluding an agreement, as these clauses are usually acceptable by all.
The Owners and Charterers may adopt and rely on some of them, depending on the trade, as well as also incorporate the same into the Charter Party.

A complete clause for Time Charter period, which refers to the "Protective/ Standard clauses" could be worded along the following lines:

The current Bimco's clauses undermentioned, are deemed to be a part of and incorporated in this Charter Party and all Sub-Charter Parties

....

....

....

-In the absence of a 'Bill of Lading' clause, the undermentioned could be added:

Neither the Charterers nor their Agents will authorise the issue of Bills of Lading or other documents evidencing the contract of carriage (even signed on behalf of the Owners or Charterers or on behalf of each of the Sub-Charterers) incorporating, where it is not compulsorily in effect, the Hamburg Rules or any other legislation that impose obligations in excess of the Hague or Hague/Visby rules. The Charterers will indemnify the Owners for any liability, loss or damages which may arise from any breach of this provision of this clause.
No through, way, transhipment, transit, liner Bill/s of Lading will be issued.
In the case of more than one Bills of Lading having been issued to various Consignees, the Owners, Master, ship will not be responsible for the cargo's delivery to each Bill of Lading holder, but without prejudice to the terms, conditions and exceptions of this Charter Party, including, but not limited to Interclub Agreement clause, the Owners remain responsible for the total intaken cargo quantity.
All Bills of Lading which have been issued under this Charter Party or Sub-Charter Parties will incorporate all the terms, conditions, warranties, liberties and exceptions of the present Time Charter Party, including the lien and arbitration clause and the same will appear on the front side of the issued Bill/s of Lading.
The provisions of the Bimco's current:

----

are deemed to be a part of and incorporated in all Bills of Lading issued under this Charter Party and all sub-Charter Parties.

19. Cargo separation in the same hold/Different Shippers for similar cargo.

This clause becomes essential, when it is the Charterers' intention to involve the loading of two or more different grades of cargo or different types of cargoes . in the same hold, with or without artificial separation.
In the absence of such a clause in the Charter Party, the Owners should ask the ship's P&I Club to board the ship and attend the loading operation.
In general, the artificial separation of cargo is difficult and often costly to arrange. The Owners should have strong reservations as to whether, amongst other things, the method of loading and packaging/stowing/separation will be sufficient to prevent
. commingling, admixture, or the puncturing of the cargo by the grabs during discharge.
Additionally, the risks
. of contamination of the cargoes are imminent resulting in damage to the cargo, i.e., one grade loaded in the bottom of the hold and covered with plastic dunnage and/or tarpaulin, which could make the cargo sweat, with the slightest humidity making, for example, the grains develop mould.
Whilst Shippers/Charterers may be possibly able to construct a cargo worthy barrier, thus keeping the cargoes separate and breathable, concerns should remain, however, on whether or not it will work and does not cause either the cargo to sweat or any other problems.
The Owners must be aware that even if an artificial separation is made and the Bills of Lading have been claused (marked) or a Charterers' Letter Of Indemnity has been presented, they cannot avoid the liabilities of carrying/separating cargo with such methods and the ship may be subject to the ship's P&I Club coverage. In respect with issuing an l.o.i. and in the event that the l.o.i. is found to be inadequate, the risk and any loss will rest with the Owners. In any case, the l.o.i. must clearly state that the Charterers agree to undertake any and all responsibilities arising from separating the cargo artificially in whatever manner and with whatever method and they also undertake the arrangements for the loading and separation at their own time/risk and cost. Furthermore, it must state that they agree to indemnify the Owners for any and all costs and/or liabilities arising out of carrying cargo with this method.
In the case that such a method of loading is inevitable and the fixture for various reasons must be solved and provided that the Owners are confident with the Charterers' financial standing and/or the intended shipment is a regular shipment for them that have always been completed without problems, then there should obviously be a little risk. However, the Owners must take care and be

covered by clausing (marking) the Mate's Receipts & original Bills of Lading, always in accordance with the recommendations of the ship's P&I Club.

The "Cargo separation in the same hold/Different Shippers for similar cargo" clause, which, among others, should include:
a. whether the Charterers can load more than one cargo/grade into the same hold.
b. if it happens, who will be liable and responsible.
c. who will supply and discharge the separation materials.
d. how the Owners will be protected with the issued documents (Mate's Receipts/Bills of Lading).

An executed clause that has been run in a Charter Party is as follows (precise copy):
Any cargo separation other than by vessel's natural separations to be for Charterers' time/risk and expense. Removal of separation materials to be for Charterers account.
Comments:
A very short clause which does not cover the Owners' interests, as it only states that the Charterers are responsible for the separation.

Another clause reads as follows (precise copy):
In case any artificial separation is required in addition of vessel's natural compartments in order to load more than one cargo type/quality/ grade in the same hold, same to be entirely at Charterers' risk/responsibility/expenses/time. Charterers also to be solely responsible for providing and installing necessary means of separation. Charterers to indemnify and/or hold Owners harmless from any liabilities, loss or damage which may arise from mixture and/or contamination and/or co-mingling of cargoes being loaded in the same hold howsoever caused whether this is due by stevedores' negligence or otherwise. All Mate's Receipts and original Bills of Lading shall be marked accordingly. Vessel will always trade in seaworthy trim condition between berth/s, port/s.
Comments:
The Charterers are entirely responsible in the case that any artificial reparation is required, in order to load more than one cargo type/quality etc. into the same hold. It also states that all Mate's Receipts and original Bills of Lading will be marked accordingly.

A complete clause for Time Charter period, which refers to the "Cargo separation in the same hold/Different Shippers for similar cargo" could be worded along the following lines:

-For Owners, who do not accept to load different grades/qualities/types of cargo in the same hold, it could be worded as follows:

The Charterers will provide only one grade, quality, type of cargo for loading, otherwise, and in the case of more than one cargo grade, quality, type, the separation will be made according to the ship's natural compartments.

-For Owners, who will accept to load different types, grades, and qualities of cargo in the same hold by making an artificial separation, it could be worded as follows:

If an artificial separation is required by the Charterers, in addition to the ship's natural compartments, then it is expressly agreed that:
1. Charterers guarantee that it will be made and the cargoes for separation will be loaded, in a maximum of 2 holds at their entire time, risk, responsibility and liability.
2. Charterers guarantee that under their own time, risk and expenses will supply, erect, and install all the necessary and required means of separation, as well as dismantle, remove and discharge them at the last discharge port and will rectify, repair any damage caused to the ship and/or conversions made in the ship's holds and restore the ship to its initial condition always before the completion of discharge and/or the ship's sailing from the port. The ship will remain always on-hire.
3. the separation will be made under the supervision and the full satisfaction of the Master and/ or the ship's P&I Club's surveyor, who will be appointed by the Owners, at the Charterers' cost. The Charterers are to reimburse the Owners for the P&I Club's Surveyor's fees, once the Owners have submitted the original vouchers, even by e-mail or fax.
4. all the presented Mate's Receipts and original Bill/s of Lading will be marked strictly in accordance with the ship's P&I Club's recommendations.
5. the ship will always trade in a seaworthy and safe stress, trim condition, between berth/s, port/s, under the Charterers' time and expenses and the ship will remain on-hire during the time lost, in order the Master to bring the ship in a seaworthy and safe stress, trim condition, even during ballasting and de-ballasting the ship's ballast tanks.
6. Charterers guarantee that will hold the Owners harmless and will keep them fully indemnified against all claims whatsoever (inclusive of any interest and costs which may be incurred for defending any such claims or recovering any such loss), which may arise and be made by any party against them, due to making separation in the ship's cargo holds. Furthermore, all risks and all expenses incurred, and/ or liabilities and responsibilities and/or loss, damages and/or pos-

sible fines and/ or third parties' claims, which may arise from mixture of goods and/or contamination and/or co-mingling and/or cargo damage, as a result of the loading of more than one cargo type, grade, quality in the same hold, howsoever caused, whether this is due to stevedores' negligence or otherwise, and/ or consequential losses will be for the Charterers' account.
If the ship is detained by any party and generally its sailing is not allowed from the port, because of mixture of goods and/or contamination and/or co-mingling and/ or cargo damage, as a result of the loading of more than one cargo type, grade, quality in the same hold, the Charterers then undertake the responsibility to release the ship, by taking all the necessary and appropriate measures (issuance of security, etc.), as the case may be, or arise and the ship will remain always on-hire until its redelivery, in accordance with the Charter Party terms.

-For Owners who will accept to load similar cargo from different shippers into the same holds with one issued Bill of Lading, it could read as follows:

If a cargo hold is loaded by different shippers with or without artificial separation, the Charterers guarantee that the cargo will be of the same grade, type and quality and all cargo documents, including the Mate's Receipts and Bills of Lading, will have exactly the same cargo description. The Charterers should indemnify the Owners against all claims whatsoever, which may arise and be made against them and/ or hold the Owners harmless from any liabilities, loss, damages or consequences, which may arise from mixture and/or contamination and/or co-mingling of cargoes, because of being loaded from different Shippers into the same hold.

20. Stevedore damage.

If a stevedoring damage is caused to the ship's holds/hatches/fittings etc, this clause clarifies:
. what measures the Master should take.
. how/when the concerned parties must be notified.
. for whose account, the resulting delays in sailing will be.
. who will pay for the damage occurred and when.
This essential clause is incorporated in all Charter Parties, whether the ship is fixed on Time Charter or on a Voyage basis.
On Voyage Charter, for the majority of the Charter Parties, the damage should be settled directly between the Owners and the stevedores, while on Time Charter the Charterers will be the party liable for returning the ship
. to such a condition, as on the time of delivery.
If the ship intends to load heavy and/or long pieces of cargo or wood or steel

in general, including pig iron/scrap or to discharge with very large grabs, but its hatches that are not as wide, then it is possible for the ship to sustain heavy stevedore damage, resulting in the ship being considered inoperative and un-seaworthy. Since the damages are not covered under the ship's Hull and Machinery insurance, the stevedoring clause becomes an important clause, and the Owners must take extra care and agree to a full and comprehensive clause.
Generally, the damage mainly occurs during the loading or discharging with the use of shore grabs or when using payloaders/bulldozers in the holds of the ship for the purpose of stowing or removing the cargo from the sides of the ship and carrying it to the places easily accessible to the grabs.
Damage could be occurred to the holds' ladders, frames, hoppers, tank top, hatchcovers etc.
Some of the stevedoring companies will try to minimize their responsibilities or to even deny that damage occurred during their work. The provisions of the stevedoring clause in conjunction with the on-hire survey, where photos of the condition of the ship on delivery should be taken, will protect the ship and will share the responsibilities accordingly.

The stevedore damage should be categorised as follows:
- slight damage which can be repaired on the spot by stevedores, without causing any delays to the ship's sailing
- damage not affecting the seaworthiness and/or cargoworthiness of the ship and which can be repaired by the crew at a later stage or together with the ship's next dry-docking. The damage will be estimated and be paid by the stevedores prior to sailing or otherwise the Charterers will be liable for the payment when they are repaired.
- damage affecting the ship's class/seaworthiness/cargoworthiness, and which must be repaired prior to sailing from the port and, in any case, prior to loading the next cargo.

The "Stevedore damage" clause, which, among others, should include:
-which measures the Master of the ship should take in case of damage occurred.
-which parties must be notified and when.
-provision for hidden damage.
-provision for the damage affecting the seaworthiness of the ship, operation etc. -provision for small damage which can be repaired at a later stage.
-who will be responsible for lost time.
-who will pay the expenses for repairing the damage.

An executed clause that has been run in a Charter Party is as follows (precise copy):

Should any damage be caused to the vessel or her fittings by Charterers or their servants, the Master is to endeavour to obtain the written acknowledgement of the responsible party, have a survey made to define and estimate the damage together with vessel's Agents. Agents will report the event to the Charterers in English language without delay, but latest prior to vessel's sailing from the respective port, however, any hidden damages to be reported as soon as discovered, but latest on sailing from the port where damage discovered. Failing the aforementioned, Charterers not to be responsible in any way for such damage and/or loss of time thus occasioned. Proven damages, if any, not affecting seaworthiness/cargo worthiness and proper workability of the vessel to be repaired by Owners after this Charter Party at Charterers' expenses. Damages affecting vessel's seaworthiness/cargo worthiness or its immediate workability to be repaired at the port of occurrence at Charterers' time and expense to class satisfaction.

Comments:

This clause obliges the Master of the ship to try to obtain written acknowledgement from the party responsible for the damage suffered and to arrange for a survey to determine and assess the damage. The damage affecting the seaworthiness of the ship/cargo worthiness and ship's workability will be repaired at the port of occurrence, at the Charterers' time/expense and to the class's satisfaction.

Another clause reads as follows (precise copy):

Stevedores to be appointed and paid by the Charterers but to work under the supervision of the Master. Should any damage be caused to the vessel or her fittings by the Stevedores, the Master shall try to arrange for Stevedores to repair such damage and try to settle the matter directly with them.

The Charterers shall not be responsible for any damage caused by Stevedores to the vessel unless the Master notifies the Charterers or their Agents of such damage within 24 hours from occurrence, except for hidden damages which is to be reported as soon as practical after discovery but not later than completion of discharge of each particular voyage. The Master shall also endeavour to obtain written acknowledgement of the damage and liability from the concerned Stevedores on occurrence.

The Charterers shall have the liberty to redeliver the vessel without repairing the damage for which the Charterers are responsible, as long as the damages do not affect the vessel's seaworthiness, but the Charterers undertake to reimburse costs of repair against the production of repair bills by repairers or dockyard unless otherwise agreed, but time used for repairs not to count as hire. Any stevedore damages affecting the ship's seaworthiness as ascertained by vessel's class surveyor to be repaired by the Charterers at their risk, expense, and time prior

to proceeding to sea.
Comments:
Upon damage occurred, the Master will endeavour to obtain written acknowledgement by the stevedores and he is entitled to arrange the required repairs directly with them.
Nevertheless, the Charterers or their agents must be notified within 24 hours from occurrence, otherwise the Charterers are released from their responsibilities.
There is a provision for hidden damage and for damage not affecting the ship's seaworthiness, when the Charterers will reimburse for the costs of repairs, but not for the time. Damage affecting the seaworthiness will be repaired prior to the ship's sailing.

A complete clause for Time Charter period, which refers to the "Stevedore damage", could be worded along the following lines:

The Stevedoring company will be appointed and paid by the Charterers, but will work under the supervision of the Master, if it is possible.
The Charterers will be liable and responsible for any damage caused by the stevedores to the cargo or ship or its fittings during the period of this Charter, including hidden damage, which should be reported as soon as detected, but in any case, before the end of this Charter.
If the ship suffers stevedore damage, the Master of the ship:
a. will report to the Charterers or their Agents in writing, within 24 hours of occurrence or from the time of discovery, but in any case, before the ship sails from the port.
b. will hold the stevedores or the respective parties responsible and liable for any damage, in writing with copies sent to the Charterers or their Agents. He will work with the Charterers or their Agents and will make every effort to reach a written acknowledgement from the parties responsible for causing the damage, unless the damage has in the meantime been repaired. If the Master is unable to obtain acceptance of liability and responsibility from the party who caused the damage, he is not allowed to sail from the port before coming into contact with the Charterers, allowing them to sort the matters out.
c. will try to get the stevedores to repair the damage and try to settle the matter directly with them in the first stage.
d. will enter such damage in the ship's log books.
For stevedoring damage which:
1. affects the effectiveness of the ship and the safe operation and/or safety of the crew and/or affect the trading capabilities and/or seaworthiness and/or cargo worthiness and/or is subject to classification requirements, then such dam-

age, as will be ascertained by the ship's Class surveyor, who will be appointed by the Owners at the Charterers' cost, is to be repaired by the Charterers at their own risk, time and expenses, to the Master's and/or Class surveyors' entire satisfaction. The Charterers are to reimburse the Owners for the Class Surveyor's fees, once the Owners have submitted the invoice, which is acceptable, even by fax or e-mail.

The damage will be repaired at the place of occurrence, if it is possible and there are adequate facilities for permanent repairs, or otherwise, temporary repairs are to be made on the spot or as when detected, at the Charterers' time, risk and expenses and to the Master's and/or Class surveyor's satisfaction.

The ship will sail or shift to the nearest shipyard, as chosen by the Charterers at their time and entire cost, in order to complete the repairs. The damage will be repaired before the ship executes the next trip, or, if it is redelivered to the Owners, the redelivery will take place on dlosps of the repair yard, or, at the Owners' option, at equidistance from the Time Charter Party point of redelivery and the ship will remain always on-hire.

2. does not affect the effectiveness of the ship and the safe operation and/or the safety of the crew and/or the trading capabilities and/or the seaworthiness and/or cargo worthiness and/or is subject to classification requirements and cannot be repaired by the stevedores, the Owners then undertake to accept the redelivery of the ship or to sail, in order to execute the next voyage, in accordance with the terms of the Charter Party.

However, the said damage is to be assessed before the ship's departure from the discharge port and the costs to be agreed upon between the Master, Owners, Charterers and the respective survey company and to be paid together with the Charterers' next hire statement. The survey company will be mutually appointed by the Owners and Charterers, if considered necessary, at the Charterers' and Owners' equal expenses, but at Charterers' time and will promptly inform the Charterers and Owners with the damage report, which will be issued before the ship sails, with a copy of all correspondence exchanged in this connection.

Owners have the option of having the damage repaired at their next convenient time or at the ship's next Dry-Docking.

In the case that the Master, Owners, Charterers cannot reach an agreement on the cost with the respective survey company, the Charterers will then undertake to reimburse for the actual cost of the repairs, but not for the time lost, upon presentation of the original bills from the repairers or dockyard, that are consistent with the surveyors' report on the extent of such damage.

3. is small and has been repaired by the crew, the Charterers will reimburse the Owners for the cost on request, but not for the time lost and upon presentation of the Owners' repair statement, which is acceptable, even by fax or e-mail.

However, and in any case:

- the Charterers are responsible for any directly related expenses resulting therefrom and incurred thereby, including, but not limited to the time lost for restoring the ship to its initial condition and before executing its next voyage or the redelivery takes place to the Master's and Class surveyor's supervision and satisfaction and any detention through any of the above causes will be for the Charterers' account and the ship will remain always on-hire.
- the Charterers must pay compensation for any damage caused and/or restore the ship to its initial condition under their own time and expense, whether or not any payment has been made by the respective parties and/or stevedores to them.
- the Charterers will not to be liable for damage, when the Master has accepted a statement in writing from the stevedores and/or the respective parties that the operations were carried out without any damage, or that the repairs were made to his satisfaction.

21. Ship in ballast condition.

When the ship completes its discharging operation, the Charterers (if it is still under the orders and directions of the Charterers regarding employment) or the Owners could order it to sail to another place at a distance to the last port.
The ship will ballast sea water and will fill the tanks designated for ballast,
. for safe passage from one place to another.
Most ships will also ballast the ballast hold (located in the middle of the ship), as designated by its class.
It is in fact a clause stating that the ship can proceed in a ballast condition (empty of cargo) from one place to another without using any
. solid ballast.

The 'Ship in ballast condition' clause, which, among others, should include:
-that the Charterers have the right to direct the ship from one place to another in a ballast condition.
-if the ship requires solid ballast, it will be for the Owners' account.

A complete clause for Time Charter period, which refers to the "Ship in ballast condition', could be worded along the following lines:

Owners guarantee that the ship is able to sail in accordance with the Charterers' request and directions in a safe ballast condition, from one place to another, without using any solid ballast.
If any solid ballast is required to be loaded, the expenses, extra expenses, risk, and time lost will be for the Owners' account.

22. Bottom fouling.

When the ship trades especially
. within warm tropical waters and remains in port for a certain period, then the bottom/propeller of the ship may become heavily fouled
. with marine growth, thereby significantly reducing the speed of its sailing and increasing its consumption at sea.
In most of the Charter Parties, since the ship was complying with the Charterers' instructions to lie for a long period of time in such waters or in an idle condition, the obligations lie upon the Charterers to return the ship's hull to its original condition and redeliver the ship in a good order and condition with the costs of doing so and lost time to be entirely for their account.
However, in the absence of such a clause in the Charter Party, then it falls within the Owners' general obligations to maintain the ship in an efficient condition, including keeping the hull clean. They will not have the right to pass the costs of hull cleaning on to the Charterers and they will additionally bear the consequences of the ship's underperformance.
It is a basic clause when negotiating a Charter Period or even single trips and especially when it is known that the ship will trade within tropical waters or areas, where port congestion is common and foreseeable.
The ship's areas which are affected by the marine growth are the flat bottom, vertical sides, bilge keels, propeller, and rudder blade. The cleaning will be performed by organized diving companies, with equipment/tools to remove the marine growth, which are in all cases to the Master's approval, in order to avoid damage to the paint of the ship's hull and damage to its propeller.
The most commonly used tools are the hand-scrapers or the more effective the airbrushes.
After the cleaning, the marine growth will have been removed, but with the possible consequence
. of the removal/destruction of the ship's anti-fouling bottom protection paint, as had been applied during its last dry-dock.
In this case, the bottom will become
. sensitive to continuous fouling from marine growth and could direct the ship to dry-docking earlier than planned.
The Owners must take care through the terms of the Charter Party and be well protected when the ship is trading within Tropical areas or not and in accordance with directions from the Time Charterers.
Sometimes, when it is possible to do so, the Charterers or Owners can instruct the ship to frequently spin around or change position in the anchorage, as a minimum action to prevent bottom fouling.

Tropical waters increase the ship's possibility of being fouled by marine growth, therefore, the number of days after which the ship might have been fouled by marine growth should be separately agreed upon in the Charter Party when trading within tropical or non-tropical ports. If the ship is fouled by marine growth the Charterers are to employ a well organised/class approved diving company, under their own time and expenses, in order to determine the extent of any bottom fouling and should it be extensively fouled by marine growth, such as to affect the ship's Charter Party performance, then the Charterers should arrange for an underwater cleaning at the place of occurrence, by an organized diving company with effective equipment/tools in order to remove the marine growth. However, if there are no facilities and/or well organised/class approved diving companies with effective equipment/tools, then the ship should be directed, always on-hire, to a mutually appointed convenient place/port, after the sailing of the ship from the port, in order for the cleaning to be carried out and provided there is no cargo on board the ship. Alternatively, and if the ship is on long period, then can be cleaned at the next available port/place.
In all cases and until a satisfactory cleaning has been conducted, the ship/Master/ Owners will not be held responsible for any reduction in the ship's declared Charter Party speed and/or an increased fuel consumption.
After the cleaning of the ship's hull,
. the Charter Party should clearly state that the Charterers are to remain responsible for any further bottom cleaning that may be required, that will be arranged and carried out immediately under their own time and expenses, regardless of the time the ship stayed in the ports, and/or anchorages and/or waiting places.

The "Bottom fouling" clause, which, among others, should include:
-who will bear the cost/lost time for cleaning if the ship becomes fouled.
-if there are no facilities for the bottom cleaning at the place of occurrence, then the ship will not be responsible for the reduction in speed and for the increased consumption during its trip to the discharge port.
-who will appoint the diving company.

An executed clause that has been run in a Charter Party is as follows (precise copy):
If vessel's bottom becomes fouled during the currency of this Charter Party because she has stayed idle for more than 30 days in tropical waters, relevant underwater cleaning by divers to be arranged and time/costs to be for Charterers' account.
Comments:
This clause states that if the ship remains in an idle condition for more than 30 days in tropical waters, then the cost/time for the underwater cleaning will be

for Charterers' account.

Another clause reads as follows (precise copy):
In case of vessel's bottom fouling due to Charterers ordering the vessel to stay in port for a period exceeding 20 days, Owners will carry out bottom cleaning at said port or next available port where there are such facilities, expenses, and time to be equally shared. Charterers will not claim on vessel's performance until such cleaning has been carried out.
Comments:
If the ship remains in port (in tropical areas or not) for a period exceeding 20 days, the Owners will arrange and will carry out the bottom cleaning. Costs and lost time will be equally shared between the Owners and Charterers.
No claim for underperformance can be presented, until such bottom cleaning has been carried out.

A complete clause for Time Charter period, which refers to the "Bottom fouling" could be worded along the following lines:

If the ship is directed by the Charterers to remain in worldwide ports and/or berths and/or in anchorages and/or in waiting places for more than.... consecutive days, or more than.... consecutive days within tropical and/or seasonal tropical waters around the world and the Master of the ship suspects that the flat bottom, including, but not limited to, the vertical sides, bilge keels, propeller, and rudder blade, may have become fouled by marine growth, then the Charterers are to employ under their time and expense, a well organised and class approved diving company to determine the extent of any bottom fouling.
Should it be extensively fouled by marine growth, such as to affect the ship's Charter Party performance, then the Charterers will arrange for an immediate underwater cleaning at the place of occurrence, by an organized diving company with effective equipment and tools to remove the marine growth. However, if there are no facilities and/or well organised and class approved diving companies with effective equipment and tools, then the ship will be directed always on-hire, to a mutually appointed convenient place or port, after the ship's sailing from the port of occurrence, in order to carry out the cleaning and provided no cargo is on board the ship. Alternatively, the ship will be cleaned at the next available port or place.
The cleaning will take place at the Charterers' time and expenses, to the Owners' and ship's class surveyor's entire satisfaction, who will be appointed by the Owners, at the Charterers' cost. The Charterers are to reimburse the Owners for the Class Surveyor's fees, once the Owners have submitted the invoice, which is acceptable, even by fax or e-mail.

The efficient equipment and tools, which will be in use to remove the marine growth are subject to the Master's acceptance, in order to avoid paint damage to the ship's hull and damage to its propeller.
Until satisfactory cleaning, or the ship's planned Dry-Dock, which will be arranged within this Time Charter period, the ship, Master and Owners will not be held responsible for any reduction in the ship's Charter Party speed and/or increased consumption. If the cleaning was carried out during the Dry-Dock, the expenses remain for the Charterers' account, but no time will be calculated for their account.
If the cleaning was carried out at the place of occurrence or at a mutually appointed convenient place and/or port, by a well organised and class approved diving company, the Charterers remain responsible and ensure that if any further bottom cleaning is required, that will be arranged and carried out immediately, under their time and expense, regardless of the time the ship stayed in the ports and/or anchorages and/or waiting places. If the bottom is eventually cleaned during the passing of its Dry-Dock, the extent of periods exists as above. In any case, the ship will be redelivered to the Owners with a clean bottom and free of any marine growth. If a prolonged stay is anticipated in an anchorage of a port of contact, the Master has the right, after having obtained relevant permissions from the Port Authorities and/or terminal and/or Shippers and/or Receivers, to spin around or change position or perform short voyages of about 6 hours every 7/10 days at a full speed, at the Charterers' time, risk and expenses, as a minimum action to prevent bottom fouling. Results are without any guarantee and without any prejudice to the Owners' rights under this clause.

23. Grabs/Bulldozers/Payloaders.

The use of grabs is a mean of loading or discharging, while Bulldozers/Payloaders assist in discharging the cargo.
Ports designated for bulk cargoes such as coal, ores, pig iron, grains etc., can provide grabs for the loading and/or for discharging operation. The most suitable type of grabs, depending on the grade of cargo, will be accommodated
. either by shore means/floating cranes
. or by the ship's provided gear.
Mainly for gearless ships, the Owners are to confirm that the ship is suitable for normal grab discharge and that the cargo is not to be loaded in places
. inaccessible to grabs (deep tanks etc.).
Loading with grabs is much easier compared to discharging. The cargo could be spread into the holds from high above or even from almost the hatch opening during the loading operation. However, difficulties could arise at the discharge port, where grabs must touch the bottom for the completion of discharge or if

unreasonably large grabs are used, incurring damage to the ship's structure, giving rise to disputes, as to whether the ship is fully suitable for grab discharge or not.

Bulldozers/payloaders could be used for stowing purposes at loadports or for removing the cargo from the sides of the ship and carrying it to the places easily accessible to the grabs at the discharge port.

When bulldozers should be used, the Owners should require the same to be with rubber wheels and their weight, including cargo weight, not to exceed the tank top strength. However, when steel, scrap, pig iron cargo, etc, is to be loaded, caterpillar tractors should be used.

The "Grabs/Bulldozers/Payloaders" clause, which, among others, should include:

-if it will discharge with normal types and sizes of grabs.

-if cargo has been loaded in places inaccessible for grabs, for whose account and time it will be.

-if bulldozers are equipped with rubber wheels for certain specific kind of cargoes.

-that the weight of the bulldozers plus cargo is not to exceed the maximum allowance regarding tank top strength.

An executed clause that has been run in a Charter Party is as follows (precise copy):

The vessel is suitable for grab discharge and no cargo to be loaded in places inaccessible to grabs or in deep tanks. Charterer to have the privilege of using bulldozers in vessel's holds. However, unit weight of bulldozers will not exceed vessel's tank top strengths permitted and her operation always under Master's satisfaction.

Comments:

The discharging operation will take place with shore grabs and the cargo must be loaded only in places which are accessible to grabs. The Charterers are entitled to use bulldozers in the holds, in order to remove the cargo from the ship's sides. The weight of the bulldozers will not exceed the ship's tank top strength.

Another clause reads as follows (precise copy):

Charterers /Shippers/Receivers have the option to load and/or discharge per magnets and/or grabs and/or buckets. Charterers/Shippers/Receivers option to attach magnets/grabs/buckets to ship's gear. Owners allowing the use of forklifts/dozers in loading and discharging provided same does not exceed vessel's tank top strength. Owners allowing that grabs, magnets, dozers and equipment to be carried by the ship between the ports if more than one loading and/or dis-

charging port and from loading to discharge port always subject to Master's discretion and any equipment carried to be loaded, lashed and secured by Charterers at their risk and expense to Master's satisfaction and provided same allowed by shore and custom regulations.

Comments:

This is a special clause, which requires the Owners to accept grabs, magnets, dozers and equipment used in the loading port, to also be carried by the ship for the use at discharge port/s. The same should be loaded lashed and secured by the Charterers at their risk and expense, to the Master's satisfaction and provided that the same is permitted by shore and customs regulations.

The Owners should seek advice from the P&I Club, on whether it is allowed and under what conditions.

A complete clause for Time Charter period, which refers to the "Grabs/Bulldozers/ Payloaders" could be worded along the following lines:

-For ships discharging with grabs fitted on its gear and Bulldozers/Payloaders will assist in discharging the cargo:

Charterers may fit suitable grabs on the ship's cranes, in accordance with the class of the ship and remove the same after the loading and/or discharging operation is completed. Risk, expenses, and time used to fit, un-fit and restore the grabs, is to be for the Charterers' account. The loading and/or discharging operation with grabs is to be entirely for the Charterers' risk and expenses, who are also responsible for any damage caused to the ship and/or cargo and/or third party and the ship will remain always on-hire. The ship will not be considered off-hire in the case of improper grabs fitted on the ship's cranes. No cargo will be loaded in places inaccessible to grab discharge, such as deep tanks etc. The ship's holds sides and/or frames are not to be considered as an inaccessible location.

The Charterers will only use rubber-wheeled bulldozers, payloaders and/or tractors in the ship's holds and/or similar equipment, as deemed necessary by the port's customs. The unit weight of bulldozers and/or tractors and/or similar equipment, plus cargo weight, will not exceed the ship's tank top strength and its operation will be under the Master's supervision and satisfaction.

-For gearless ships, for grains or other bulk cargoes:

The ship is suitable for a normal grab discharge operation, with normal size/ type grabs and as far as a standard bulk carrier/ singledecker can be. No cargo will be loaded in places inaccessible to grab discharge, such as deep tanks etc. The ship's holds sides and/or frames are not to be considered as an inaccessible locations. The Charterers will only use rubber-wheeled bulldozers and/or tractors in the ship's holds and/or similar equipment, as deemed necessary by the port's customs. The unit weight of bulldozers and/or tractors and/or similar equipment, plus cargo weight, will not exceed the ship's tank top strength and its operation will be under the Master's supervision and satisfaction.

24. Logs/Reports/Charterers' trading orders.

The Master must keep on board the ship
. a full and correct deck and engine log in the English language for all the ship's activities. The Master in the ship's deck log and the Chief Engineer in the ship's engine log, must insert every 3 hours all the events that take place on the ship's deck and engine room, including the prevailing weather at that time, the ship's speed, consumption etc.
Some Charterers provide their own type of forms for the execution of the voyage, which must be completed by the Master and Chief Engineer in conformity with the Charterers' instructions. The same will be delivered to the Charterers or to the Charterers' Agents, after completion of each voyage.
In case of disagreement between the Owners and Charterers on the ship's executed performance, the Charterers may request
. a true copy of the ship's logs.
Depending on the agreed Ocean Route clause, the ship's logs might
. have equal value with other data if the case is referred to arbitration.
The Master must keep the Charterers informed of the ship's daily speed/consumption/prevailing weather conditions, or otherwise as per the Charterers' instructions.
The Charterers must furnish the Master with clear instructions of the intended voyage and of sailing directions, in the English language. The Master is to receive clarifications upon and when it is requested. There is a provision in the NYPE 46 Charter Party (clause 11) which refers to the Charterers' employment instructions and of the Master's obligation to keep a full and correct log, but some Charterers or Owners prefer to have a separate clause.

The "Logs/Reports/Charterers' trading orders" clause, which, among others, should include:
-that the Charterers should provide clear trading instructions in English and should immediately respond to the Master's request for clarifications.
-that the Master will keep a full and correct log, in the English language.
-that a copy of the logs should be ready for dispatch to the Charterers upon request, after completion of each voyage.

-that the Master should complete the Charterers' forms if so required.

An executed clause that has been run in a Charter Party is as follows (precise copy):
Charterers shall furnish the Master from time to time with all requisite instructions and sailing directions in writing or by telegram and the Master shall keep a full and correct deck and engine log in English of the voyage or voyages showing inter alia the course of the vessel and distance run and the consumption of fuel oil. A true copy of which is to be sent to Charterers from each port of call on the voyage and immediately after completion of the voyage, together with any other information which the Master deems necessary.
Comments:
Master to co-operate in giving any information Charterers may require.
It states that the Charterers will provide instructions to the Master, who he is obliged to keep a correct deck and engine log in English. The deck logbook, among others, should include the distance run and the ship's consumption. A true copy of the logs should be delivered to the Charterers after completion of each voyage.

A complete clause for Time Charter period, which refers to the "Logs/Reports/ Charterers' trading orders" could be worded along the following lines:

Charterers will furnish the Master on fixing and from time to time and for each voyage separately, with all requisite clear voyage instructions and sailing directions and cargo information, only in writing, by e-mail or fax and only in the English language.
Charterers will reply, as soon as possible, to any of the Master's requests for clarification on the given written instructions, or clarifications on the new and/or revised instructions, at the time of occurrence, otherwise, the Master, ship should not be kept responsible for any delays and expenses, which may arise, due to the Charterers' failure to immediately respond. Master will report daily to the Charterers as per the given instructions and will keep a full and correct deck and engine log in the English language of the voyage, or voyages, to the Charterers' forms or not, showing, among others, the course of the ship, the running distance, the fuel consumption, the prevailing weather conditions etc. A true copy of which, at the Charterers' request, will be sent to them or will be delivered to their attending Agents or their representatives, by the end of each voyage, along with harbour reports, off-hire reports, stevedoring damage reports or any other reports, which the Master considers necessary to be sent.
Master is to co-operate in providing all information that the Charterers may require, at any time.

25. Mobile crane/Evacuators/Equipment and fittings on deck.

The Charterers would need the option to
. fit/weld any extra equipment and fittings for loading or discharging and/ or securing the cargo on deck.
For discharging purposes, the Charterers should also request to have the right
. to place mobile cranes and/or evacuators on the ship's deck.
The Charterers should be informed of the allowable weight per square meter of available spaces on the deck, in order to provide the most functional Mobile cranes/evacuators. In fact, the discharging operation with evacuators is more acceptable by the Owners, than the discharging with Mobile cranes. Nevertheless, it cannot be avoided when this method is mandatory at the intended port of discharge. The Owners should take all the necessary precautions, in order to avoid possible damage to the ship and the Charterers must be kept fully responsible for any damage occurred.

The "Mobile crane/Evacuators/Equipment and fittings on deck" clause, which, among others, should include:
-the Charterers' right to place equipment/fittings/Mobile Cranes and/or evacuators on the ship's deck.
-the Charterers' responsibility to restore the ship, after completion of discharge, to its original condition.
-whether a class surveyor will attend the operation/works.

An executed clause that has been run in a Charter Party is as follows (precise copy):
Charterers have the liberty to place mobile cranes on deck/hatch covers, provided suitable protection is made if necessary, and maximum allowable weights are not exceeded according to Master. In case of any damages same to be repaired at Charterers' time/expenses.
Comments:
It is a short agreement, formally incorporated in the Charter Parties and on trips where it is known that the ship will not discharge with mobile cranes.

Another clause reads as follows (precise copy):
Charterers have the liberty to place mobile cranes/evacuators on deck, but only in areas where there are sufficient deck spaces for grab and cargo, to facilitate discharge of the cargo. All costs, time and risk to be for Charterers account and sufficient dunnage (if required by class or Master) to be placed underneath the cranes to spread the weight which in any case is not exceed permissible deck

strength, should any cutting or welding or reinforcement be necessary on vessel's hatches and hatch covers to accommodate the placement of such cranes on deck, the risk/ expense/time of such works to be for Charterers account and such work always to be carried out subject to the classification society surveyor's approval, Charterers to be fully responsible for any and all damages, time, expenses and costs (including but not limited to all burn areas of paint on deck and underneath which to be reconditioned to its original state prior to redelivery) and that all works to be under Master's /Officers' supervision and to the classification society's and Master's satisfaction. In case of shifting/sailing with such equipment on board lashing always to be to Master's satisfaction and at Charterers' cost, time and risk. If required by Charterers vessel to supply power but only as available on board. Any extra bunkers consumed therefrom shall be for Charterers account.

Comments:

The Charterers have the right to place Mobile cranes/evacuators on the ship's deck for discharging purposes. The Charterers will provide sufficient dunnage and all projects undertaken to accommodate the placement of these cranes on deck. All works will be performed at the Charterers' risk/expense/time and subject to the approval of the class surveyor. The Charterers are fully responsible for any damages which may be incurred.

A complete clause for Time Charter period, which refers to the "Mobile crane/ Evacuators/ Equipment and fittings on deck" could be worded along the following lines:

Charterers have the right to fit and weld padeyes on deck and/or fit, install and weld any additional equipment and fittings and/or placing mobile cranes and/or evacuators on deck, in order to perform the discharging operation.

All these operations will take place provided that:

1. an independently recognized surveyor will be mutually appointed, but at the Charterers' time and cost, to take photos and make a report on the condition of the ship's deck, hatch covers, rails etc., before the fitting and welding of padeyes and/or the fitting and welding of any additional equipment and before the commencement of the work for the placement of the mobile crane/s and/or evacuators. After completion of the operations and the removal, discharge of any padeyes and/or any additional equipment and fittings and/or removal of the mobile crane/s and/or evacuators and when all repairs have successfully been performed, the surveyor will board the ship again and take photos and make a report, ensuring that the ship has been returned to its original condition.

2. the ship's Class Surveyor, as will be appointed by the Owners, will attend the ship's full operation in the port and all works which will be carried out, from the time of berthing until sailing. The Charterers are to reimburse the Owners for the Class Surveyor's fees, once the Owners have submitted the invoice, which is acceptable, even by fax or e-mail. The original should be followed if it is requested by the Charterers.

3. the overall operations, installations, fittings, welding, placement of any additional equipment and fittings and/or placing mobile crane/s and/or evacuators, as well as, their lashing, securing and unlashing, dismantling, removal and discharge upon completion of the operations from the ship's deck, will always be carried out at the Charterers' time, risk and expenses, by shore labourers and to the supervision and full satisfaction of the Master and/or the ship's Class surveyor.

4. the weight of any additional equipment and fittings, as well as that of the mobile crane and/or evacuators plus cargo weight, should not exceed the deck's allowable weight per square meter. The Charterers will provide and install the required dunnage underneath, so as to evenly distribute the weight, to the supervision and full satisfaction of the Master and/or the ship's Class surveyor.

5. time, costs and risks for any cutting and/or welding and/or strengthening that is needed to be performed, in order to strengthen the ship's hull for accommodating the placement and/or is needed to be performed, in order to secure the additional equipment and fittings and/or the Mobile cranes and/or evacuators on deck and/ or modification of the hatch covers that have been made, will be for the Charterers' account and all works will be carried out by the Charterers' trained and skilled shore labourers, always under the Master's and/or Class surveyor's supervision and to their entire satisfaction at all stages.

6. if the ship shifts to another berth, or as will be directed by the port Authorities, the Charterers then will provide and install all the required lashing and securing materials, at their own expenses. The shifting will be performed at the Charterers' risk and the ship will remain always on-hire.

7. after completion of the discharging operations and the dismantling, removal and discharge of the additional equipment and fittings, as well as that of Mobile crane/s and/or evacuators, the hatch covers are to return to their original condition to the Master's and/or Class surveyor's entire satisfaction, at the Charterers' time and expenses. A hose test should be carried out at the Charterers' time and expenses, to ensure that all hatch covers are watertight, and their integrity is verified. Any degradation of the ship's paint work (on hatch covers and/or coamings and/or deck and/or tanks etc.) must be returned to its original condition, with such paints, as are in accordance with the Owners' original specification.

8. Charterers guarantee that they will remain liable and responsible for any consequences, loss, damages, costs, expenses, fines, Owners may suffer and will indemnify the Owners against third party claims (inclusive of any interest and costs which may be incurred for defending any such claims or recovering any such loss),, including, but not limited to, cargo claims and oil/air pollution arising from placing any additional equipment and fittings and/or evacuators and/or mobile cranes on the ship's deck. If the ship is detained by any party and generally its sailing is not allowed from the port, the Charterers then undertake the responsibility to release the ship, by taking all the necessary and appropriate measures (issuance of security, etc.), as the case may be or arise and any expenses incurred will be for the Charterers' account. The ship will remain always on-hire.

9. the Charterers will be fully responsible for damage caused to the ship's deck and/or hatch covers and/or rails and/or hydraulic motors etc.

For damage which:

a. affects the effectiveness of the ship and the safe operation and/or safety of the crew and/or affect the trading capabilities and/or seaworthiness and/or cargo worthiness and/or is subject to classification requirements, then such damage as will be ascertained by the ship's Class surveyor, who will be appointed by the Owners at the Charterers' cost, is to be repaired by the Charterers at their own time, expenses and risk, to the Master's and/or Class surveyors' entire satisfaction. The Charterers are to reimburse the Owners for the Class Surveyor's fees, once the Owners have submitted the invoice, which is acceptable even by fax or e-mail. The original should be followed if it is requested by the Charterers. The damage will be repaired at the place of occurrence, if it is possible and there are adequate facilities for permanent repairs, or otherwise, temporary repairs are to be made on the spot or as when detected, at the Charterers' time/risk and expenses and to the Master's and/or Class surveyor's satisfaction.

The ship will sail or shift to the nearest shipyard, as chosen by the Charterers, at their time and cost, in order to complete the repairs. The damage will be repaired before the ship executes the next trip, or, if it is redelivered to the Owners, the redelivery will take place on dlosps of the repair yard, or, at the Owners' option, at equidistance from the Time Charter Party point of redelivery and the ship will remain always on-hire.

b. does not affect the effectiveness of the ship and the safe operation and/or the safety of the crew and/or the trading capabilities and/or the seaworthiness and/or cargo worthiness and/or is subject to classification requirements and cannot be repaired by the stevedores, the Owners then undertake to accept the redelivery of the ship or to sail, in order to execute the next voyage, in accordance with the terms of the Charter Party.

However, the said damage is to be assessed before the ship's departure from the discharge port and the costs to be agreed upon between the Master, Owners, Charterers and the respective survey company and to be paid together with the Charterers' next hire statement. The survey company will be mutually appoint-

ed by the Owners and Charterers, if considered necessary, at the Charterers' and Owners' equal expenses, but at Charterers' time and will promptly inform the Charterers and Owners with the damage report, which will be issued before the ship sails, with a copy of all correspondence exchanged in this connection.
Owners have the option of having the damage repaired at their next convenient time or at the ship's next Dry-Docking.
In the case that the Master, Owners and Charterers cannot reach an agreement on the cost with the respective survey company, the Charterers will then undertake to reimburse for the actual cost of the repairs, but not for the time lost, upon presentation of the original bills from the repairers or dockyard, that are consistent with the surveyors' report on the extent of such damage.
c. is small and has been repaired by the crew, the Charterers will reimburse the Owners for the cost on request, but not for the time lost and upon presentation of the Owners' repair statement, which is acceptable, even by fax or e-mail.
However, and in any case:
- any directly related expenses resulting therefrom and incurred thereby, including time lost for restoring the ship to its initial condition and before executing its next voyage or the redelivery takes place to the Master's and/or Class surveyor's supervision and satisfaction and any detention through any of the above causes will be for the Charterers' account and the ship will remain always on-hire.
- the Charterers must pay compensation for any damage caused to the ship and/or restore the ship to its initial condition under their own time and expense, whether or not any payment has been made by the respective parties and/or stevedores to them.
- the Charterers will not to be held liable for damage when the Master has accepted a statement in writing from the stevedores and/or the respective parties that the operations were carried out without any damage, or that the repairs were made to his satisfaction.

26. Fresh water.

The cost and supply of the required fresh water for cleaning of the holds between two trips should be agreed during the negotiations of the clause of
. 'intermediate holds' cleaning'.
In the past, the general cargo ships used to stay in the ports of discharge for a very long time. If they were discharging in the anchorage, the stevedores stayed on board the ship and they were used the ship's water for drinking/washing themselves/cooking etc.
Therefore, it was necessary for a separate clause to be incorporated into the Charter Party, stating:
. whose account should the water consumed by the stevedores would be for.

In our days, the same is limited, but, in any case, the clause could still exist, even if only for the water consumed for cargo holds' cleaning.

The "Fresh water" clause, which, among others, should include:
-on whose account the domestic consumption should be.
-on whose account the fresh water used for stevedores and/or for holds cleaning should be.

An executed clause that has been run in a Charter Party is as follows (precise copy):
Fresh water to be supplied and paid for by Charterers for stevedores at both ends.
Comments:
The fresh water consumed for stevedoring purposes, should be supplied, and paid by the Charterers.

Another clause reads as follows (precise copy):
Fresh water consumed under this Charter for the purpose of drinking, washing of Master, Officers and Crew onboard is to be for Owners' account. When trading, limits the vessel's ability to provide fresh water through its desalination plant, the extra expense for replenishment of fresh water for Charterers' account.
All fresh water consumed in washing down of holds/decks to be for Charterers' account and Charterers are to pay Owners accordingly to Master bill at supplying price or at Charterers proposal to arrange the consumed quantity at their expenses Master consider necessary and convenient.
Owners will arrange for the supply of fresh water at the best possible price. Charterers are to pay to Owners the cost according to presented bill at supplying price or at Charterers option to arrange the required quantity at their expenses. In case there is no available fresh water at the port of required, the Owners will not be held responsible and Charterers or Owners will arrange for the supply at the first convenient place/time.
When Charterers trade the ship to areas which limits the ship's ability to provide fresh water through its desalination plant, the extra expense and time lost for replenishment of fresh water to be for the Charterers' account.
Comments:
The fresh water consumed for domestic purposes will be for the Owners' account, but the water used for washing down the holds/decks etc. will be for the Charterers' account. The Owners will arrange for the supply of the required fresh water and the Charterers will reimburse the Owners according to the presented bill.

A complete clause for Time Charter period, which refers to the "Fresh water" could be worded along the following lines:

All fresh water consumed during the period of this Charter for domestic use (drinking, washing etc. for the needs of the crew, ship's maintenance etc.) is for the Owners' account. For all other purposes, such as fresh water needed for the holds' cleaning or for the stevedores' use etc., it is for the Charterers' account.
The supply of fresh water that is required for the Charterers' account will be arranged by the Owners/Master through the Charterers' Agents at the ports of call, at the best possible price and at the Charterers' time. The Charterers are to pay the Owners at cost and on presentation of the vouchers, which will be accepted even by fax or, if it is requested by the Owners, the cost could be included in the ship's general port disbursements and will be paid to the attending Agents directly by the Charterers.
Owners and/or ship will not be held liable if there is no fresh water available at the port of contact, as has been directed by the Charterers. The ship will be supplied with fresh water under the Charterers' time and expense, at the first convenient place and time.
Owners guarantee that the ship's desalination plant will operate normally throughout the duration of this Charter. Fresh water consumed, due to improper operation of the ship's desalination plant will be for the Owners' account, unless its improper operation was caused due to the Charterers' trade, which affected its good working condition (trading within tropical waters or in rivers etc.).

27. ITF.

The 'International Transport Workers' Federation' (ITF) is a union which
- protects and assists seafarers worldwide and represents their interests and their rights.

The ITF covers all
- crew nationalities, irrespective of the ship's flag/ownership.

This clause will request from the Owners to trade the ship under the terms and conditions of the ITF or according to the terms which are in any case acceptable to the ITF. Additionally, the clause stipulates and attributes the responsibilities in the case that the ship is inoperative or delays from its sailing due to an 'ITF problem'.

The "ITF" clause, which, among others, should include:
-that the ship's crew are employed under the terms/conditions of the ITF or equivalent terms/ conditions acceptable to them
-that in case of stoppages/boycotting etc. due to an ITF problem, the ship will be placed off hire and the expenses incurred will be for the Owners' account.

An executed clause that has been run in a Charter Party is as follows (precise copy):

Owners warrant that the vessel's crew are always employed under terms/conditions acceptable to ITF or its affiliates. Any cost or loss of time due to Owners failure to comply with the above to be for Owners' account.

Comments:

A clause holding the Owners responsible for any cost or for any lost time that may arise, in the case that the ship's crew are not employed under the terms/ conditions which are accepted by the ITF.

Another clause reads as follows (precise copy):

The Owners of the vessel guarantee that the minimum terms and conditions of employment of the crew will be, prior to the presentation of the vessel for loading, and will remain for the period of this Charter Party, covered by an ITF agreement or equivalent or an acceptable bonafide trade union agreement, failing which vessel to be off-hire for any time lost and any extra expenses such as stevedore standby time, port expenses, barge demurrage and the like to be for Owners' account.

Comments:

The Owners guarantee that the ship's crew are employed with the minimum terms and conditions of the ITF agreement or equivalent. The ship will be placed off hire for any lost time and all extra expenses incurred will be for the Owners' account.

A complete clause for Time Charter period, which refers to the "ITF" could be worded along the following lines:

Owners guarantee that on the ship's delivery and during the duration of this Charter, the crew will be employed and covered with the minimum terms and conditions of an ITF agreement or with an equivalent agreement or with a bona fide trade union agreement, but which must always be accepted by the I.T.F. However, in the event of a boycott and/or strike and/or stoppage and/or refusal of work of shore labourers and/or labour unions and/or pilots and/or tugs against the ship and/or if all ship's activities in general have stopped and it is inoperative due to the crew's employment terms and conditions or actions of the ITF against the ship, then the ship will be placed off hire from the time of stoppage until the time that all works resumed and the ship is again in full operational condition. All directly related proven costs supported with the original vouchers, including, but not limited to, stevedore stand-by, but up to a maximum of one shift and/or additional port expenses incurred directly resulting therefrom, will be for the Owners' account.

-For Charterers, who will cancel the Charter Party, because the ship remained off-hire for more than 15 days, the undermentioned could be added:

If the ship is delayed and remains inoperative and has been placed off-hire for more than 15 consecutive days, then the Charterers will declare whether they will maintain or cancel the fixture for the remaining balance period.
If the Charter Party is cancelled:
1. a written notice will be given to the Owners while the ship remains so detained, for redelivery on dlosps of this port and if there is cargo on board, the ship will be redelivered when it is practical and the ship being free of cargo.
2. the Owners must refund the Charterers for any hire plus bunkers that have been paid in advance and not earned.
3. the ship will be redelivered with bunkers as on board and any differences will be settled together with the final hire payment.
4. the Owners will pay for all extra expenses incurred, but supported with original vouchers, including, but not limited to, stevedore standby time, but up to a maximum of one shift.
5. the redelivery of the ship will take place without prejudice to any other right the Charterers may have on the ship.
If the Charter Party is maintained:
1. Charterers have the option to add to the maximum allowed Charter period any off-hire period/s, partly or wholly, provided that any such extension does not exceed the validity deadline of the ship's next special survey/dry dock. Any such option will be declared to the Owners a minimum of 20 days before the expiration of the minimum allowed Charter period.
2. the Charterers will trade the ship at the same rate and conditions for such extended time, as that applicable during the off-hire period.
3. the Owners will give the Charterers the full itinerary of the ship and the expected date of resumption of the service, every two days.
4. after resumption of the work, the ship will be placed at the Charterers disposal at the same port or place where the hire was suspended, within ordinary office hours, except at sea and with quantity of fuel as when the ship was placed off-hire. The Charterers may, however, in their option, accept the ship on hire again in such a position and at such a time as the ship may again in all respects be ready to comply with the Charterers' orders and directions.

28. Taxes-Dues.

All taxes/dues, either levied or calculated on cargo or on ship or on voyage freight must be charged for the Charterers' account. Nevertheless, in some

countries there are certain taxes/ commissions levied on the ship's daily hire rate or there are even taxes levied on hire remittance. For example, when hire is remitted from Brazil to a company with their domicile in a tax haven, there is a 25 percent tax on the remitted amount, which will actually kill everything. The Owners should take care to ensure through this clause that the hire will be credited to their bank account,
. net of any taxes/bank taxes/commissions etc.

The "Taxes-Dues" clause, which, among others, should include:
-that all taxes/dues on cargo/ship/hires/remittances are for the Charterers' account.
-that taxes/dues imposed by the ship's flag/Owners' domicile will be for the Owners' account.

An executed clause that has been run in a Charter Party is as follows (precise copy):
Any dues and /or taxes on cargo and/or hire to be for Charterers' account, however, any taxes on hire levied by country of vessel's flag to be for Owners account.
Comments:
A clause which states that the taxes/dues imposed on cargo/hire are to be for the Charterers' account, but the taxes on hire levied by the country of the ship's flag are for the Owners' account.

Another clause reads as follows (precise copy):
Any dues/taxes on cargo and/or freight or hire earned under this Charter Party or its sub-hires /headhires and or sub-freights/headfreights and/or cargo to be for Charterers' account. Any extra insurance on cargo to be for Charterers account excluding taxes levied by the country of the flag of the vessel on the Owners.
Comments:
An improved clause, which covers the ship's interests.

A complete clause for Time Charter period, which refers to the "Taxes-Dues" could be worded along the following lines:

All local, national, and international taxes, dues, charges, extra insurance on the ship and/or flag and/or ownership and/or age and/or cargo and/or charter hire and/or its sub-hires, head-hires and/or freight and/or sub-freights, head-freights and/or cargo, arising out of carried or ports visited under this Charter Party, will be for the Charterers' account. On every occasion, the hires are payable to the

Owners' nominated bank, net of any taxes, bank taxes, commissions etc.
All the loading and discharging operations will be performed free of expense to the Owners.
All dues, duties, charges and/or taxes on crew and/or stores and/or income taxes on Time Charter hire, levied by the state of the ship's flag and/or the Owners' domicile will be for the Owners' account.

29. Weather routing.

The ship's performance during the execution of the Charter is one of the most important points for the Charterers and it is their success when making the
. voyage calculation for their fixed Voyage Charter.
The ship's actual speed/consumption and the validity of the agreed relevant clause could be a base for disputes/claims arising between the Owners and Charterers.
The method, which will be used and will be acceptable by both parties, to assess the ship's actual performance across the oceans during the currency of the Charter will be agreed when discussing this clause. In fact, the Charterers should not only rely on the route which the Master will select and on evidence of weather conditions taken from the ship's deck logs. They will seek to protect their interests and ensure the fastest, safest, and most economical route, by appointing a recognised
. International Weather Bureau (Weather Routing Company or 'Ocean Route' as it is commonly called)
that will inform the Master of the expected winds, waves and prevailing currents on its route and recommend to (not impose on) him the fastest and safest route, from the point where the ship will commence the voyage up to its final destination.
The Master will choose the best route for his ship and will report the ship's daily noon position to them and will inform them (and document in the ship's log book) of any changes in his route which he deems necessary for the safety of the crew/ cargo, as well as, the reason for it.
At the end of the voyage at sea, the 'Ocean Route' will assess
. how fast and with what consumption the ship performed the trip from pilot to pilot.
Because of this clause's validity, the ship's declared speed/consumption must be described on an
. "about" basis.
The Owners should be protected if the ship fails to maintain its description, as it provides
. a negative speed variation of half a knot and a 5 percent increase in the de-

clared fuel consumption.
It is commonly accepted that the weather and the sea waves affect the performance of the ship, when
. the Beaufort Force 4 and Douglas Sea State 3 are exceeded and the same should be clearly incorporated in the Charter Party.
When there are large discrepancies regarding the prevailing weather conditions between the log books and the weather reports that are issued by the Ocean Route, the Charter Party should spell out whose report will be binding for both parties, otherwise and upon referral to Arbitration, both should have the same value.

The "Weather routing", which, among others, should include:
- the weather and sea waves which will affect the ship's performance.
-whether an ocean route will be appointed and who will pay them.
-whether the ship's log books will be taken into consideration for the calculation of the ship's performance and if they will be equally counted with the other data in arbitration, if be referred.

Given the importance of this clause, the contracting parties, in the event of a disagreement and in order to save the fixture, will discuss various ways to bridge the difference.

Some of them from executed clauses are (precise copy):
1. Within the context of this Charter Party 'good weather conditions' are to be taken as wind speed not exceeding but including Beaufort Force 4/Douglas Sea State 3. Evidence of weather condition to be taken from vessel's deck logs and independent weather reports. In the event of a consistent discrepancy between deck logs and the independent weather report which arranged and paid for by Charterers, then Owners to appoint for their account a further major International Weather routing company and in case there is still a consistent discrepancy between all data and unable to settle amicably, then all data to have equal value in arbitration.
Charterers do not have the right to withhold any amount on account of a speed or performance claim for which there is no mutual agreement or arbitration award.
Comments:
In case of consistent discrepancies between the ship's deck logs and the Charterers' Ocean Route's report, the Owners have the right to appoint for their account a further major Ocean Route company and in the case that they are unable to settle amicably, then all data will have equal value in arbitration.

2. The Charterers and the Owners may supply an independent weather bureau advice to the Master, during voyage specified by the Charterers. The Master at his sole discretion shall comply with the reporting procedure of the routing service selected by the Charterers. Evidence of weather conditions shall be taken from the Owners and Charterers independent weather bureau reports. In the event of a consistent discrepancy between the Owners and Charterers' independent weather bureau reports, the average of the two independent weather bureau final reports shall be taken as ruling. In case both sides Charterers and Owners decide to proceed to Arbitration then all data including Vessel's logs to have equal value in Arbitration.

Comments:

Both Owners and Charterers will appoint their Ocean Route company and, in case of discrepancies, the average of the two final reports will be taken as ruling.

3. The service of Ocean routes or equivalent, which company to be subject to Owners' approval which not to be unreasonably withheld may be engaged at Charterers' expense to recommend optimum route and advise Master about weather en route. Master shall be free to follow Ocean routes or equivalent, which company to be subject to Owners' prior approval, recommendations, and direction to route course and/or whether or not in accordance with the safety of the vessel and crew which are left entirely to his discretion. Charterers shall not be responsible in any way for Ocean routes information, directions and recommendations which are provided as assistance only and entirely without prejudice. The Master shall prepare a deck log abstract for each sea passage showing daily noon position and all other relevant information. In case of consistent discrepancies between the deck logs and Ocean Routes' report, the Ocean Routes' report to be taken as ruling.

Comments:

The Ocean Route Company will be appointed by the Charterers, but it is subject to the Owners' approval. In case of consistent discrepancies between the deck log and the Ocean Route's report, the Ocean Route's report will be taken as ruling.

4. The Charterers have the option of providing the vessel with Ocean Routes Weather Routing Service with all cost in this respect for the account of the Charterers. The Master at his reasonable discretion may not follow the suggested route, in which case he is to detail in the log book the reason for departing from same and advise the Charterers accordingly. The Charterers are to instruct the Master to keep Ocean Routes fully advised of the vessel's actual time of departure daily position and actual time of arrival.

Evidence of weather conditions to be taken from vessel's deck logs and inde-

pendent weather bureau reports. In the event of a consistent discrepancy between the deck logs and independent weather bureau report, then the matter to be referred to arbitration.
Comments:
Evidence of weather conditions will be taken from the ship's deck logs and the Ocean Route as appointed by the Charterers. In case of consistent discrepancies, the matter will be referred to arbitration.

5. Throughout the currency of this Charter Party, Owners and Charterers have the option to employ their own independent weather routing service monitor weather conditions experienced during sea passages and assess the vessel's performance. Each party to pay their own costs for such services.
The Master will comply with Charterers' independent weather routing services reporting procedures at all times and endeavour to follow Charterers' (or their routing services) recommended route. However, the final choice of selecting the safest route will be the Master's decision, providing it is justified, in which case he has to inform Charterers as soon as possible.
In case of discrepancies between the two, a third mutually agreed independent weather bureau to be appointed the findings of which to be binding by both parties and expense of above bureau to be equally shared.
Comments:
The Owners and Charterers will appoint their own Ocean Route company under their cost. In case of consistent discrepancies between the two reports, a third mutually agreed Ocean Route company will be appointed with the cost to be shared equally and their findings are to bind both parties.

A complete clause for Time Charter period, which refers to the "Weather routing" could be worded along the following lines:

Throughout the currency of this Charter Party, 'good weather conditions' are to be taken as wind speed not exceeding, but including, Beaufort Force 4 and sea state Douglas 3.
Throughout the currency of this Charter Party, the Charterers, at their cost, may appoint an established and recognised independent weather bureau company to monitor the ship and report and recommend to the Master the safest and fastest route, from the point of commencement of the voyage to the final destination and of the expected winds, waves and prevailing currents on its route.
The Master is to report all the information required daily and to comply with the reporting procedures and recommendations, provided that it is safe for the ship, cargo and crew, at his sole discretion.
The Master will keep a deck log abstract (in Charterers' forms if so be required),

for each sea passage, showing the daily noon position and all other relevant information and he is to detail in the log book the reason of any change in the ship's route, as recommended by the weather routing company or otherwise. After completion of each voyage, a signed true copy of the deck log will be delivered to the Charterers' Agents or Charterers' representatives in the port.
The Charterers will not be held responsible for the Ocean routes information, directions and recommendations, which are provided as assistance only and entirely without prejudice.
The calculation of the ship's performance will be made in accordance with English Law and from pilot station to pilot station.
The ship's reduced speed or delay or any deviation made will not be taken into consideration and will not be considered as underperformance, if made due:
-to Charterers' instructions for their commercial reasons (subject to the Owners' consent).
-to trading in narrow or restricted waters with regard to the ship's size/draft.
-to the safety of the ship/crew/cargo, due to weather distress.
-to save lives/property or assist a ship in distress.
-to the ship's hull bottom becoming fouled by marine growth, but without prejudice to the 'bottom fouling clause'.
-to a distance between the ports of less than 10 hrs.
-to navigation which is not in an even keel condition.
-to poor visibility (fog etc.).
-when trading coastally and within 15 nautical miles from the coast.
-to the Charterers' supply of fuels not being in compliance with the Owners' provided fuel specifications or, generally, for any other reason, which can affect the ship's performance and is beyond the Owners', ship's or Master's fault and/or defect.
Evidence of the ship's actual speed and consumption and of the prevailing weather conditions at sea during the currency of this Charter Party, will be taken from the ship's deck, engine logs and the independent weather bureau, as appointed and paid by the Charterers.
In the event of a consistent discrepancy between the ship's logs and the independent weather bureau's reports and, if the Owners and Charterers fail to settle amicably, the Owners then have the right to appoint for their account a further major independent weather bureau company. In the case that there is still a consistent discrepancy between all data and they have failed to settle amicably, then all data, including the ship's logs, is to have equal value in Arbitration.
The Charterers do not have the right to withhold any amount on account of a speed or performance claim, for which there is no mutual agreement or arbitration award.

30. Gypsy moth.

Another specific clause for the ships that would trade within the allowed region/ ports, which will be affected by the Asian Gypsy Moth.
The areas and periods when Moth flights occur are listed from time to time
. by the Animal and Plant Health Inspection Service (Aphis) of the United States dept. of agriculture, or jointly with the Canadian Food Inspection Agency (CFID). Asian Gypsy Moths are forest pests which can be found and carried on the ship's superstructure or even in any other place on the ship and could infest the U.S. and Canadian forests.
The circular recommends that ships calling this area should take several measures in order to reduce the risk of infestation, otherwise, and when calling U.S and Canadian ports, they may cause
. significant delays or even miss the cargo charter.
When the ships trade within the designated areas which change year by year (but mainly include North China, Korea, Japan and Far Eastern Russia) and during Moth flight periods (mainly from June until September), prior to departure from the discharge port they must obtain
. a Phytosanitary certificate or other approved certificate/s that the ship is free of Asian Gypsy Moths.
Australia, Canada, New Zealand, Chile and Argentina (recently joined) have imposed AGM regulations and have procedures in place to prevent the risk of infestation.
Nevertheless, some Charterers will only seek to fix ships which have not called at all to the areas or have not called within a certain period of time.

The "Gypsy moth " clause, which, among others, should include:
-whether the ship has contacted areas affected by the Asian Gypsy Moth.
-if it has indeed contacted them, who will be responsible in the case that the ship is rejected by the Authorities at the port of loading.
-who will pay the cost for the issue of a pre-departure certification, in the case that the ship contacted affected areas during the currency of this Charter.
-that the Charterers have the right to cancel the Charter Party if the ship remained off-hire for a certain period.

An executed clause that has been run in a Charter Party is as follows (precise copy):
Vessel not to has called at Russian/Soviet Pacific ports in the last 24 months and is to the best of Owners knowledge free of any Asian gypsy moth egg or larvae or any form of gypsy moth life and is not on U.S.A./Canadian blacklist.
Comments:

The Owners are to confirm that the ship has not contacted Russian/Pacific ports during the last 24 months.

Another clause reads as follows (precise copy):
Owners herewith confirm that the vessel has not called at Russian Far East ports ranging from Posyet to Olga Bay, including Vladivostock, Nakhodka and Vostochniy, during the months of June through September and that there is no danger of the vessel being rejected entry and/or being delayed by the U.S./Canadian authorities. However, if the vessel has called Russian Far East ports as aforementioned, vessel is to present certification from the State Plant Quarantine Service of Russia stating that the vessel is free of Asian Gypsy moth. In case certification is not provided and/or vessel is rejected by U.S./Canadian authorities, then Charterers have the option to either cancel the Charter Party or to instruct the vessel to clean and re-present with all time/expenses until the vessel is accepted being for the Owners' account.
Comments:
The Owners are to confirm that the ship has not contacted the area during the specific time of period, otherwise they must present the relevant issued certificate confirming that the ship is free of Asian Gypsy Moths.

A complete clause for Time Charter period, which refers to the "Gypsy moth" could be worded along the following lines:

Owners confirm that the ship has not contacted the designated area affected by the Asian Gypsy Moth and when a seasonal risk exists, as listed from time to time by the 'Aphis' of the United States dept of agriculture or jointly with the Canadian Food Inspection Agency. However, if the ship has contacted the designated area affected by the Asian Gypsy Moth, during the designated Asian Gypsy Moth flight periods, the ship is to present on request and before loading, a pre-departure certification from the affected areas, stating that the ship is free of Asian Gypsy Moth life stages.
If the ship fails its initial inspection and after the carried out re-inspection the ship is accepted, then the actual time lost, quantity of fuel consumed, directly related expenses incurred and possible fines imposed, which will be supported with original vouchers and all consequences arising therefore, will be for the Owners' account.
However, if the respective certification is not provided and/or is not approved and/ or the ship is rejected by the U.S. and/or Canadian and/or any other Authorities, or if after Charterers' instruction the ship to be cleaned and readied for re-inspection, the ship remained off-hire for 15 consecutive days or more, then the Charterers have the right to reject the delivery and to cancel the Charter Par-

ty, in which case:
1. a written notice will be given to the Owners.
2. the ship will be redelivered at the berth or anchorage or place, where the inspection carried out or it is performing the cleaning, with bunkers as onboard.
3. the Owners must refund the Charterers for any hire plus bunkers that have been paid in advance and/or for the bunkers supplied to the ship for Charterers' account, at the price of purchase, but supported with the original vouchers. All the quantity of fuel consumed from the time of delivery until it is cancelled will be for the Owners' account and no hire will be paid, or is deemed to have earned, for the whole period.
4. the Owners will pay for all port expenses and extra expenses incurred, but supported with original vouchers, including, but not limited to, stevedore standby time, but up to a maximum of one shift.
5. the redelivery of the ship will take place, without prejudice to any other right the Charterers may have on ship.
If, during the currency of this Charter, the ship arrives for loading or discharging purposes within the season of the Gypsy Moth in the area of high risk, as designated by the CFID/ AMPHIS, then the Charterers, under their time and expenses and before sailing from the port, will arrange to appoint competent Authorities, fully recognized by the CFID/APHIS, to inspect the ship and issue a pre-departure certification, stating that the ship is free of the Asian Gypsy Moth life stages and eggs. All time lost, expenses incurred until the ship is free from the Asian Gypsy Moths and the pre-departure certification is issued, will be for the Charterers' account. The Charterers, during the duration of this Charter Party, will remain responsible for any rejection by the U.S. and/or Canadian Authorities and/or from any other Authorities, because of their instructions the ship to contact within the season of the Gypsy Moth in the area of high risk, as designated by the CFID/AMPHIS and the ship under any and all circumstances will remain on-hire.

31. Port State Control inspection.

=Ship's inspection at the ports of contact for operation:
During the operations in the port, the ship may be subject to inspection by the inspectors
. of the Port State Control (abbreviated as 'PSC').
The PSC are regional organizations covering the entire coastal world as follows:
-The Europe and the north Atlantic is covered by the Paris MOU.
-The Asia and the Pacific Ocean from the Tokyo MOU.
-The Caribbean Sea from the Caribbean MOU.
-The Mediterranean Sea from the Mediterranean MOU.

-The Indian ocean from the Indian Ocean MOU.
-The Latin America from the Acuerdo De Vina Del Mar.
-The West and Central Africa from the Abuja MOU.
-The Black Sea from the Black Sea MOU.
-The Arab States of the Gulf from the GCC MOU.

The PSC inspectors will check/inspect/ensure:
-the ship's compliance with the International conventions (MLC- Maritime Labour Convention / SOLAS -Safety of Life at Sea / STCW – International Convention on Standards of Training).
- the ship's condition (if there is lack of maintenance and if affecting its seaworthiness).
- the ship's documents, (if they are valid and up to date. All ship's certificates according to SOLAS must be originals with certain exceptions).
- the crew's certifications for inadequate certifications (safe manning with qualified officers and crew).
- if the equipment is working properly.
- the operation of lifeboats (life-saving equipment).
- the fire safety on board.
- the cleanliness in the engine room etc.
and generally, if the ISM procedures are followed by the Master/officers/crew, as well as, ensuring whether the ship meets and maintains the safety and the security standards and in an operational condition in compliance with the international maritime conventions/law.
The well-trained officers and crew and their familiarization with the company's safety management system (SMS) is of the obligation of the Master/company. Officers'/crew's failure to comply with the plans for shipboard operations will result in possible ISM failure, with huge consequences to the ship.
If the ship fails to pass the PSC inspection, it may be detained in the port and its sailing prohibited. The ship, depending on the nature of the deficiency/ies, must comply with any repairs and/or recommendations made by the Authorities, rectifying them, as the case may be, before departure or until the ship reaches the next port or within a certain time.
The most important codes to action taken are:
- Code 15: rectify deficiencies by next port.
- Code 16: rectify deficiencies within 14 days.
- Code 50: rectify deficiencies within 30 days.
- Code 17: rectify deficiencies prior to departure.
- Code 30: detained ship.
If the ship was detained (code 30), it means that the deficiencies were serious, and the ship was considered by the PSC inspectors as unsafe to trade and sail

from the port. The ship will remain
. detained,
until all the deficiencies are rectified, either at the port of occurrence, or otherwise, and, if due to lack of repair yards or competent shore equipment etc., the ship under special conditions and provided it has no cargo on-board, may be granted permission to sail to a nearby port/yard, which can provide all the required means to rectify the deficiencies.
The ships under its flag, are categorised as being in a
. black or grey or white list.
Obviously, the ships which are black, or grey listed constitute a 'red flag' for the inspectors.
It is a fact that a ship is profitable for the Owners only when it is working properly and without delays, hence the ship's Management should take all the necessary measures, to ensure that the ship's hull condition, maintenance, equipment and crew to be to the PSC inspectors' full satisfaction, with adequate manning and in such an operational condition, as to be always in compliance with the international maritime conventions. Normally, even if the ship is detained, it is not usually for a long time and the loading or discharging operation/program is not disrupted.

The "Port State Control inspection" clause, which, among others should include:
-that the ship trades to the Port State Control inspector's full satisfaction.
-that the Owners are responsible due to detentions or deficiencies imposed on the ship, resulting in delays/expenses etc.
-that the Charterers have the right to cancel the Charter Party if the ship remained off-hire for a certain period.

An executed clause that has been run in a Charter Party is as follows (precise copy):
Owners to be fully responsible for any directly related and proven costs and directly related consequences as a result of a detention for vessel by Port State Control, provided loading or discharging operations are disrupted.
Comments:
In case the loading or discharging operation is disrupted due to the ship's detention by the PSC, then the Owners are responsible for all costs and for the directly related consequences which could arise.

Another clause reads as follows (precise copy):
In case the vessel delays to sail from the final discharging port latest by 10 or more days after completion of discharge, due to Port State Control and/or ITF action against the vessel which is solely attributable to the vessel/Owners then

Charterers have the option to cancel this Charter Party.
Comments:
This clause is in effect for Time Charter periods or for Charters with more than one loading leg. In the case that the ship's sailing is delayed by 10 or more days due to PSC action, then the Charterers have the option to cancel the balance period. However, this clause provides the Owners with a certain time to rectify all deficiencies implied by the PSC and to ready the ship for trading again.

A complete clause for Time Charter period, which refers to the "Port State Control inspection', could be worded along the following lines:

The Owners guarantee that the ship will be kept in a well-maintained condition, throughout the duration of this Charter and the ship, crew will comply and maintain its safety and security, in accordance with the international maritime conventions and law, to the Port State Control inspector's full satisfaction.
If the ship fails the inspection, the Owners/Master will take all the appropriate measures to comply with the required repairs or recommendations made by the Authorities. However, if the loading or the discharging operation is interfered with or interrupted as a result of detention and/or deficiencies of the ship from the Port State Control, the Owners remain fully responsible for any directly related and proven costs and extra expenses incurred and for consequences arising therefrom. The ship will remain off-hire with the quantity of fuel consumed to be for the Owners' account during this period and until such time as the Authorities have approved the repairs and the ship is returned to its full operational condition.

-For Charterers, who will cancel the Charter Party, because the ship remained off-hire for more than 15, the undermentioned could be added:

If the ship is delayed its sailing from the port or generally remained off-hire for more than 15 consecutive days, due to Port State Control action against it, which is due solely to the ship's lack of maintenance and/or the Owners' or crew's fault or omission, and/or because the safety and security is not in compliance with the international maritime conventions and law, then the Charterers have the right to cancel the balance period of the Charter Party.
If the Charter Party is cancelled:
1. a written notice will be given to the Owners while the ship remains so detained, for redelivery on dlosps of this port and if there is cargo on board, the ship will be redelivered when it is practical and the ship being free of cargo.
2. the Owners must refund the Charterers for any hire plus bunkers have been paid in advance and not earned.

3. the ship will be redelivered with bunkers as on board and any differences will be settled together with the final hire payment.
4. the Owners will pay for all extra expenses incurred, but supported with original vouchers, including, but not limited to, stevedore standby time, but up to a maximum of one shift.
5. the redelivery of the ship will take place without prejudice to any other right the Charterers may have on the ship.
If the Charter Party is maintained:
1. the Charterers have the option to add to the maximum allowed Charter period any off-hire period/s, partly or wholly, provided that any such extension does not exceed the validity deadline of the ship's next special survey/dry dock. Any such option will be declared to the Owners a minimum of 20 days before the expiration of the minimum allowed Charter period.
2. the Charterers will trade the ship at the same rate and conditions for such extended time, as that applicable during the off-hire period.
3. the Owners will give the Charterers the full itinerary of the ship and the expected date of resumption of the service, every two days.
4. after resumption of the work, the ship will be placed at the Charterers' disposal at the same port or place where the hire was suspended, within ordinary office hours, except at sea and with quantity of fuel as when the ship was placed off-hire. The Charterers may, however, in their option, accept the ship on hire again in such a position and at such a time as the ship may again in all respects be ready to comply with the Charterers' orders and directions.

32. Cargo exclusions.

This main clause provides the grade/type of cargoes which are
. not allowed for carriage,
as well as certain allowed cargoes, which should be loaded
. under specific conditions.
If the Charterers present an excluded grade/type of cargo for loading, then the Master of the ship has the right to refuse the commencement of the loading operation and instead to request an allowable cargo to be presented. The Owners will face the same problems, in the case that they refuse to load cargoes which are allowed by the Charter Party, but which have just been omitted/forgotten to be excluded during the negotiations.

It is a valuable clause when negotiating long period charters, which, together with the 'Trading exclusions' clause,
. have serious implications on the amount of hire which the ship will be fixed on, thus, it is almost always discussed during the negotiations of main terms.

In fact, the proposed 'cargo exclusions' from the Owners and/or due to the Charterers' intended cargoes for loading (dirty/dangerous etc.), might even prevent the commencement of the negotiations.
For single trips, the intended cargo/es can be declared by the Charterers from the commencement of the negotiations or even from the moment when the biz was quoted in the market, thus enabling both parties to avoid arguments and long negotiations as to the cargoes allowable for carriage.
When this clause is negotiated, the Charterers will obviously try to obtain

- the maximum possible number of allowable cargoes,

enabling them to be more flexible/play better in the market and minimize their risk of losing money when their own cargoes are not available.
The Owners must clearly state the grade/type of cargoes which the ship is unable to load, either due to her class or due to her certificates or because they are dirty or dangerous cargoes or even because of the Owners' previous bad experience with this cargo.
For cargoes that are dirty or could expose the ship to danger or could cause damage to the holds etc., the owners can set a limit on the number of voyages allowed for transport of such cargoes and place specific conditions for carriage, which are considered necessary for the protection of the ship/crew.
All cargoes should be loaded/stowed/trimmed/discharged by the Charterers at their own time/risk/ expense and always in accordance with IMO/ IMSBC Code/ international regulations/ recommendations and local regulations, but to the extent that local regulations are more onerous/restrictive than those under IMO/ IMSBC Code and/or regulations/recommendations of the countries through whose waters the ship will be navigated.
Solid bulk cargoes (solid are all cargoes except liquid or gas) and especially dangerous solid bulk cargoes will comply with the

- International Maritime Solid Bulk Cargoes Code (IMSBC Code).

All cargoes listed in the IMSBC Code have an official name (Bulk Cargo Shipping Name-BCSN). All the issued sea transport documentations for the cargo must be identified by the BCSN sign.

The dry bulk cargoes are classified by the IMSBC Code into three groups:
Group A:
Solid bulk cargoes which may liquefy during the sea voyage, resulting in cargo shift and possible capsizing of the ship, are listed in this group. Such cargoes are nickel ore/mill scale/iron ore fines/pyrites etc. Special care must be taken by the Master/Owners prior to commencement of the loading, in the areas where the cargo is exposed to rain without proper protection. The moisture content of the cargo must be measured/determined by the recognised international methods, which in any case must not be in excess of their Transportable Moisture Limit

(TML). However, dry cargoes are not liable to liquefy.

Group B:

All solid bulk cargoes, which may pose a chemical hazard and could expose the ship and crew to danger, are listed in this group. These cargoes are dangerous, because when they come into contact with air or by interaction with water, they can come to be self-heating or emit flammable gases in dangerous quantities during the voyage or turn into radioactive materials or even contain toxic substances etc.

Dangerous goods are classified according to their particular characteristics (4.1/ 4.2/ 4.3/ 5.1/ 6.1/ 7/8/9) while certain others when carried in bulk are characterised as 'Materials Hazardous only in bulk' (MHB). Shipments will only be executed on ships that can provide the Appendix 'B' certificate as will be issued by the ship's class and can provide a CO2 firefighting system in their cargo holds.

Group C:

All solid bulk cargoes, which are simply not liable to liquefy and do not pose chemical hazards are listed in this group.

However, certain mineral cargoes are listed in both Group A and Group B, i.e., pose a chemical hazard and may liquefy during the voyage.

For all cargoes, whether they have been classified as dangerous or not, the Shippers are obliged before loading to provide to the Master of the ship a cargo declaration, as outlined in the IMSBC and the BCSN, as well a valid certificate with the information on the physical and chemical properties and the required conditions for the carriage.

The Charterers, during the execution of the Charter, could ask from the Master of the ship to present a stowage plan of the cargo, which is intended to be loaded. The same should be given as a 'preliminary stowage plan' and should state that:

- the cargo quantity, which will be loaded, and the holds' breakdown is subject to

. the actual stowage factor of cargo in the ship's holds.

. the ship's actual trim/stability/strength/stresses (hogging/sagging bending moments/ shear forces etc.).

. the available draft/salinity of the port.

. load line zones limit.

For cargo that is in the form of bags/pallets etc. the wording could be amended accordingly.

The "cargo exclusions" clause could be agreed with the three following paragraphs:

A. generally includes the cargoes which are not allowed for loading and under what terms they can be loaded etc.

B. includes cargoes which are excluded from the trade.
C. includes cargoes, which are allowed for loading, but under certain specific conditions.

PARA A.
For ships which have in their possession the appendix 'B' certificate, the undermentioned clause may be amended accordingly.
PARA A:
The ship is to be employed in lawful trades for the carriage of lawful merchandise only.
Any cargo of which the nature could affect the safety of ship, crew, other cargo loaded on board and all dangerous, inflammable, injurious, hazardous and corrosive cargoes, explosive of any kind, combustible, toxic, as listed in the latest IMDG CODE and/or any subsequent modifications and amendments thereof and all cargoes listed under Group 'B' of the IMSBC Code and/or cargoes requiring electrical or mechanical ventilation and/or Co2 and/or insulation between the engine room and the last hold, are all to be excluded from loading.
All cargoes are to be loaded, stowed, trimmed, and discharged by the Charterers at their time, risk and expenses, always in accordance with the IMO/IMSBC Code and/or international regulations and recommendations and/or local regulations and to the extent that local regulations are more onerous and restrictive than those under the IMO/IMSBC Code and/or regulations and recommendations of the countries through whose waters the ship will navigate and/or call and U.N./ US. sanctions/ approvals.
For all cargoes, whether they have been classified as dangerous or not, the Shippers before loading, are to provide the Master of the ship with a cargo declaration as outlined in the IMSBC and the BCSN, as well a valid certificate with the information of their physical and chemical properties and the required conditions for their carriage. The Master is entitled to refuse the commencement of loading if such certificates are not provided.
For cargoes, which are not listed in the IMSBC Code, the shippers will provide a certificate with the characteristics and properties of the cargo, in order to enable parties concerned to assess the acceptability of the cargo for safe shipment, handling and carriage.
For all solid bulk cargoes belonging to Group A that can liquefy during the voyage, the latest 'Bimco Solid Bulk Cargoes That Can Liquefy Clause for Charter Parties" is to apply. The Charterers are to pay for all expenses, consequences, and time lost, due to and or in connection with the carriage of the Charterers' intended carriage of such commodities or goods.
California/Vertical block stowage is not permitted.
The Shippers and/or stevedores will load and/or discharge strictly as per the

agreed ship's loading sequence, stowage plan, method of cargo operations and as per the provisions of SOLAS chapter VI, regulation 7.3 (loading, unloading and stowage of solid bulk cargoes) and the Code of Practice for the Safe Loading and Unloading of Bulk Carriers (BLU Code). In the case that the Shippers and/or stevedores do not follow the agreed stowage plan, loading sequence, method of cargo operations, then the Master has the right to stop the loading and/or discharging, and the Charterers, with immediate effect will give proper instructions to the parties concerned. Under any and all circumstances, the ship will remain always on-hire and any expenses incurred therefrom (possible shifting etc.) will be for the Charterers' account.
All dirty cargoes which are not grains and/or agricultural products, but are allowed under this Charter Party, can be transported to a maximum of two consecutive trips and not to be the last two cargoes before the redelivery of the ship.
For all cargoes loaded in wet condition (cargo of sulphur, coal etc. irrigated with freshwater), the pumped-out quantity of water during the sea voyage from the load port towards discharge port, as will be ascertained by the ship's logs should be regarded as a cargo and the Receivers and/or any other third party will not be entitled to submit a claim for shortage of this quantity.
All stowage plans provided by the Master or Owners of the ship, are subject to the actual cargo stowage in the ship's holds and/or the ship's actual trim, stability, strength and the port's final available draft and salinity.

PARA B.
It is a list of certain cargoes, which could be specifically excluded from loading in the Charter Party. The majority of them have already been excluded from the foregoing clause, as they belong to group B of the IMSBC Code. The Brokers/ Owners can choose which of them are to be stipulated in the Charter Party:
PARA B:
Without prejudice to the generality of the foregoing and those cargoes under the ship's class appendix certificates which are allowed for loading, the following cargoes are in addition specifically excluded:
Asphalt – asphalt hides- asbestos in any form - aggregates- alumina – anthracite-
ammunitions and war materials of any kind - arms - ammonium nitrate- aluminium dross - aluminium silicon - aluminium silicon powder - ammonium sulphate - aluminium ashes - aluminium nitrate - aluminium residues - aluminium smelting by-products - ammonium chloride - aluminium ferrosilicon briquettes - aluminium ferrosilicon powder – ashes- ammonium sulphate - ammonium nitrate - acids
black powder - borax in bulk or bagged - baler/twine- blasting caps - bitumen or any of its products - barium nitrates - bleaching powder - bombs - bones and

meat bones meal - barium nitrates – bauxite
chloride of lime - chlorinated lime - chrome cake- clay - coal - cocoa - coffee - coke - cotton - cotton seeds - cyanide ore pellets - detonators - detonator caps - calcium carbide-coils-
cuttings - caustic soda - caustic potash - calcium chloride - calcium nitrate - containers - copra - canola - canary seeds - calcium oxychloride - caporit - cement - cement clinker - calcium hypo chloride (including all cargoes/products that are/ contain, or are related to calcium hypochlorite) - concentrates - charcoal whether in gunny bags or not- carbon black asphalt - calcined pyrites - castor beans - charcoal briquettes - carbon black -copra and/or its products including raw copra - Chilean natural nitrate - Chilean potassic mixture - creosoted goods - calcined pyrites – corrosives – cars - car vehicles
direct reduced iron - dynamite - detonator caps - devco coal - explosives of any kind, including blasting and detonator caps - expellers of any kind
expellers of all kind
ferrosilicon - Ferro manganese - fishmeal - ferrous metal boring - ferrophosphates - ferro silicon, including silicon manganese - fluorspar - ferrous metal boring - fertilizers - harmful and corrosive ferrophosphorus - ferrophosphorus briquettes - ferous metal borings - fireworks - fire briquettes - hot briquette iron, grain expellers - garbage of any kind - granite blocks and other stone blocks - gypsum
iron oxide spent - iron pellets - iron sponge- iron briquette - incendiaries - iron oxide
hypo chloride - harmful and corrosive fertilisers - hot briquetted iron (h.b.i.) - heavy lifts with weight exceeding the ship's strength - hides of any kind - hypo chloride solutions - heavy hardwood logs (softwood and pine logs are allowed)
livestock/animals - logs of any kind - lead nitrate/concentrates - lime - lime chloride - limestones – locomotives
magnesia unslakes - magnesium nitrate - metal sulphide concentrates - mobile homes - motor spirits - macoya pellets-manioc and/or manioc pellets - milled rice - metal sulphide-mineral sands - mill scale - mammalian meat and bone - meat bone and meal – millscale - mahogany logs - motor blocks
naphtha - nuclear products - nuclear fuel or substances or radioactive materials of any kind and /or their wastes- nuclear isotopes - nefeline - nefeline cyanide - nefeline cyanide ore pellets- nefeline syenite - nigerseeds - nytro glycenne - nickel ore
oilseed cakes and expellers - overweight cargo (which exceeds the ship's tank top strength limitation) oily cargoes - organic peroxides - oilcakes
petroleum and its products - pyrites - pig iron - prefabricated houses & mobile buildings - pyrites ore - pitch prill - pitch in bulk or drums - potassium nitrate - prilled tar - prilled coal tar, pencil pitch (in bulk/drums) – peatmoss - pesticides

- petroleum and its liquid derivatives/products - pond coal - petroleum coke calcined and/or uncalcined and/or green delayed - pollard pellets - potassium chloride - pulplogs/pulpwoods of any kind - pond coal - palm kernels
quicklime - quebracho extracts
resin - rice bulk a/o bagged - rockets - railway wagons - radio isotopes - radioactive cargo and/or nuclear materials and waste materials or its products
sunflower seeds/expellers or their by-products - salt - sludge ore - sponge iron/ pre - reduced iron - reduced sponge - sawdust - scrap, including motor blocks and metal turnings/borings /cuttings/shavings - silicon metal - silicon carbide - sulphur - sodium nitrate - sulphur - seed cakes in any form, including palm kernels - silicon - silicon manganese - sludge ores - slurry - soda ash - sulphate in bulk - silicon manganese - sulphate - steel coils-slabs - saltpetre - sand - sawdust - sodium nitrate - salt cakes - slag (blast furnace slag ok provided harmless/ non-dangerous) - soda as - sun pellet seeds
turpentine - tar - tnt - tobacco - tankage - tankage fertilisers - turnings - toxic cargoes
vanadium ore
unprotected pipes
yellow phosphorous
wet hides - wheat flour bulk - woodchips or shavings - woodpulp pellets - wheat bran - war and war like materials
yachts,
zinc ashes/dross/residues/skimming/concentrates.

PARA C.
It contains cargoes which are allowed for carriage, but under specific conditions. When the ship will load corrosive or messy or oily cargoes such as sulphur/coal/ petcoke/HBI etc., the Owners could request that the holds be applied with RBM Holdblock, or similar product, for the protection of the metals. RBM is more effective than lime washing, and it is much more preferred by the majority of the Owners.
PARA C.
For ships intending to load bulk sulphur or coal or petroleum coke cargo, the term could be worded along the following lines. It could be also used for Salt cargo and the 'RBM Holdblock' could be substituted, if the parties involved wish, with 'limewashing'or with 'slip-coat' in places which is acceptable.

Sulphur Clause:
Notwithstanding the cargo exclusions as above, the Charterers have the right, during the currency of this Charter, to load up to a maximum of two cargoes of bulk sulphur, whether loaded as full or part cargo, provided it does not belong to

the group 'b' of the IMSBC.
or
Petroleum coke Clause:
Notwithstanding the cargo exclusions as above, the Charterers have the right, during the currency of this Charter, to load up to a maximum of two cargoes of bulk petroleum coke, whether loaded as full or part cargo, provided it does not belong to the group b of the IMSBC Code and it is limited to the non-hazardous, non-oily, non-dangerous, green delayed type and/or calcined type and metallurgical type. (some of the undermentioned items must be amended, in order to comply with the loading of Petroleum coke cargo)
or
Salt Clause
Notwithstanding the cargo exclusions as above, the Charterers have the right, during the currency of this Charter, to load up to a maximum of two cargoes of bulk salt, whether loaded as full or part cargo, provided it does not belong to the group 'b' of the IMSBC (some of the undermentioned items must be amended, in order to comply with the loading of salt cargo)
The Charterers guarantee that:
1. the ship's P&I Club's approved independent surveyor, as will be appointed by the Owners, but at Charterers' cost, will attend, supervise, and approve the ship's full operation in the port of loading and discharging and all works which will be carried out, from the time of berthing until sailing. The whole process of loading and/ or discharging and/or supplying materials etc. should be performed to his and/or Master's entire satisfaction. The Charterers are to reimburse the Owners for the P+I Club's Surveyor's fees, once the Owners have submitted the invoice, which is acceptable even by fax or e-mail. The original should be followed if it is requested by the Charterers.
2. they will not drill holes and or make any opening on any hatch covers for loading, stowing, trimming, and discharging purposes.
3. after the Master's consent and subject to the ship's stress, they will use the minimum possible holds for loading their cargo
4. a. At Load port:
- the loading will commence when the Charterers have applied the holds with RBM holdblock or similar product at their time, cost, and risk. The applying of the RBM holdblock or similar product is to be performed by shore labourers, at the Charterers' time, risk and expenses, to the Master's and Owners' P+I Club's surveyors' supervision and satisfaction.
- if the local regulations permit it and it is approved by the Master and crew, the Charterers may use the ship's crew to apply the RBM holdblock or similar product in the ship's holds, against paying USD...... per hold, besides normal intermediate holds' cleaning.

- the related applying materials, protective package for RBM holdblock or similar product, which are required to be supplied, will be arranged, and supplied by the Charterers at their time and cost.
b. At discharge port:
- the RBM holdblock or similar product will be removed after discharge with RBM holdwash or similar product, as arranged and supplied by the Charterers, at their time, risk and cost to the Master's and Owners' P+I Club's surveyors' supervision and satisfaction. The applying of RBM holdwash or similar product and the holds' cleaning is to be performed by shore labourers at the Charterers time, risk and expenses.
- the related applying materials, protective package and any additional materials needed for holds' cleaning according to the Master's request will be arranged and supplied by the Charterers at their time and cost.
- if the local regulations permit it and it is approved by the Master and crew, the Charterers may use the ship's crew to perform the applying of RBM holdwash or similar product and the holds' cleaning, against paying USD...... per hold, besides normal intermediate holds' cleaning. However, the risk for the holds passing their cleaning inspection at the next load port will remain with the Charterers, who, if rejected, will arrange for shore labourers to complete the cleaning operation under their time and cost. Slops, washing liquids etc will be removed and pumped out from the ship's holds at the Charterers' time, risk and cost, at facilities or other means of disposal that accept slops,washing liquids etc and before the ship's sailing from the port. The ship's pipeline will not be allowed for the pumping out of bilge water with residues of the sulphur (or other) cargo.
5. the cargo will be loaded, stowed, trimmed, and discharged by the Charterers, at their time, risk and expenses always in accordance with the IMO/ IMSBC Code and/or international regulations and recommendations and/or local regulations and to the extent that local regulations are more onerous/restrictive than those under the IMO/IMSBC Code and/or regulations and recommendations of the countries through whose waters the ship will navigate and/or call and U.N./US. sanctions/ approvals.
6. the Charterers will supply the ship at the load port, at their time and cost, with sufficient freshwater, as will be required by the ship's Master, for sulphur irrigation during loading and voyage and for cleaning and washing down after completion of discharge.
7. such cargo is not to be the last cargo before redelivery.
8. if any corrosion has been caused to the ship's holds, the Charterers then are to immediately arrange with shore labourers, at their time and expenses, to remove the same by chipping or shot blasting or with any other method acceptable by the Master and/or independent surveyor.
9. any directly related expenses resulting therefore and incurred thereby, includ-

ing restoring the ship to its initial condition and any detention through any of the above causes will be for the Charterers' account and the ship will remain always on-hire.

For ships intending to load bulk cement or clinker cargo from holes cut on hatch covers, the term could be worded along the following lines:

Cement/Clinker Clause:
Notwithstanding the cargo exclusions as above, the Charterers have the right, during the currency of this Charter, to load up to a maximum of two cargoes of bulk Clinker and/or Cement, whether loaded as full or part cargo.
The Charterers have the right to cut cement holes on hatch covers, in order to perform the loading operation with closed hatch covers.
The Charterers guarantee that:
1. the loading and discharging operations will fully comply with the local and/or international regulations and/or environmental restrictions. The Charterers are liable and responsible for fines, extra expenses, time lost and any consequences that may arise, due to non-compliance with the aforesaid regulations and restrictions.
2. the ship's class approved independent surveyor, as will be appointed by the Owners, but at Charterers' cost, will attend and supervise and provide a written approval of the cutting of holes and the restoration of the hatch covers. The Charterers are to reimburse the Owners for the class surveyor's fees, once the Owners have submitted the invoice, which is acceptable even by fax or e-mail. The original should be followed if it is requested by the Charterers.
3. after the Master's consent and subject to ship's stress, they will use the minimum possible holds for loading their cargo.
4. they will supply, at their time and expenses, proper and sufficient covering materials, such as tarpaulin, plastic sheets to protect the deck area, accommodation, and equipment from the dust ingress.
5. the cement holes will be opened at a suitable place as per the class surveyor's recommendation on the hatch covers and they will be restored to their original condition, only by an experienced shore side workshop and/or stevedores. All operations and all materials required to be fitted for welding in order to reinstate the hatch covers back to their original condition, including, but not limited to, chill plates or similar materials, if required by the Master and/or class surveyor, will be performed at the Charterers' time, risk and costs to the Master's and class surveyor's full satisfaction.
6. after completion of discharging and of restoration, it is the Master's right to ask for a hose test to be carried out to verify that the hatch covers have been restored to their original condition. A hose test will be carried out at the Charter-

ers' time, risk, and expenses.
7. after completion of discharge, the Charterers will supply, under their time and cost, cleaning materials, including submersible pumps, as will be required by the Master of the ship, including fresh water for the cleaning of the cargo hold and the deck. The Cargo holds and the deck, before being cleaned and washed down with water, will be swept well, and all cargo residues, including cargo solidification on the cargo holds and on deck, will be removed from all compartments to the Master's full satisfaction, at the Charterers' time and cost.
Slops and washing liquids etc. will be removed and pumped out from the ship's holds at the Charterers' time, risk and cost, at facilities or other means of disposal that accept slops, washing liquids etc. and prior to the ship's sailing from the port. The ship's pipelines will not be allowed for the pumping out of bilge water with residues of clinker and cement cargo.
8. the holds' and deck cleaning is to be performed by shore labourers at the Charterers' time, risk and cost, under the Master's supervision and satisfaction. If local regulations permit it and it is approved by the Master and crew, the Charterers may use the ship's crew to perform the holds' and deck's cleaning, against paying USD...... per hold, besides normal intermediate holds' cleaning.
However, the risk for the holds passing cleaning inspection at the next load port will remain with the Charterers, who, if rejected, will arrange for shore labourers to complete the cleaning operation under their time and cost.
9. such cargo is not to be the last cargo before redelivery.
10. any directly related expenses resulting therefore and incurred thereby, including restoring the ship to its initial condition and any detention through any of the above causes will be for the Charterers' account and the ship will remain always on-hire.

For ships intending to load pig iron or scrap cargo, the term could be worded along the following lines:

Pig Iron Clause:
Notwithstanding the cargo exclusions as above, the Charterers have the right, during the currency of this Charter, to load up to a maximum of two cargoes of bulk pig iron, whether loaded as full or part cargo.
or
Scrap cargo Clause:
Notwithstanding the cargo exclusions as above, the Charterers have the right, during the currency of this Charter, to load up to a maximum of two cargoes of bulk scrap HMS 1and 2 and/or shredded scrap, but always excluding motor blocks and/ or turnings and/or metal borings and/or cuttings and/or oily and/or radioactive, whether loaded as full or as part cargo.

The Charterers guarantee that:
1. after the Master's consent and subject to ship's stress, they will use the minimum possible holds for loading the cargo
2. the ship's P+I Club's approved surveyor, as will be appointed by the Owners, but at Charterers' cost, will attend and supervise and approve the ship's full operation in the port of loading and discharging and all works which will be carried out, from the time of berthing until sailing. The whole process of loading, discharging, supplying materials etc. should be performed to his and/or Master's entire satisfaction. The surveyor will carry out a cargo hold, hatch covers and deck condition survey at Charterers' time, before commencement of the loading, as well as, after completion of the discharge. The survey reports, together with the photos taken, will be sent to the Owners or Master and to the Charterers' office
The Charterers are to reimburse the Owners for the P+I Club' surveyor's fees, once the Owners have submitted the invoice, which is acceptable even by fax or e-mail. The original should be followed if it is requested by the Charterers.
3. the appointed surveyor will carry out a pre-loading survey of the cargo, under the Charterers' time and expenses. Any rejected cargo quantity will be substituted with sound cargo. The original Mate's receipts and Bill/s of Lading will be claused strictly in accordance with the P+I Club's surveyor's recommendations, in order to reflect the apparent order and condition of the goods at the time of loading.
4. they will arrange and supply on board, at their time and cost, dunnage and/or any other materials, which are necessary and required by the Master and/or surveyor, in order to provide safe protection of the cargo holds, tank top, deck from damage, when loading such a cargo.
5. that the loading of the first layer of pig iron will be lowered and released, as close as, possible to the tank top and is not to be dumped or dropped during loading, so as to provide a cushioned flooring for the balance of the cargo, under the Master's and/or surveyor's full satisfaction.
6. the cargo will be loaded up to the ship's allowed tank top strength, according to the ship's stress and to the Master's and/or surveyor' entire satisfaction.
7. it is the Master's right to stop the loading procedure at any time, if damage is caused during the loading or if the loading procedure is not as per the Master's and/or surveyor's recommendations and directions. The Charterers should immediately take care and will additionally provide, under their cost, proper materials in order to protect the ship's tanktop, parts, places, frames and deck. The ship will remain always on-hire.
8. for any damage incurred to parts, places, frames in the ship's cargo holds, hatch covers and deck, including the Australian hold ladders, whether they affect the seaworthiness and/or cargo worthiness of the ship or not, will be re-

paired at the port of occurrence and the ship will generally be restored to its initial condition, before its sailing from the port. If there are not any adequate facilities for permanent repairs, temporary repairs then are to be made on the spot or as when detected, at the Charterers' time, risk, and expenses to the Master's and/or surveyor's entire satisfaction. The ship will sail or shift to the nearest shipyard, as chosen by the Charterers, to complete the repairs to the Master's and/or surveyor's entire satisfaction.
The shipyard's expenses, port expenses and other expenses incurred, including but not limited to, in and out tugs, pilots, agency fees etc. will be entirely for the Charterers' account and the ship will remain always on-hire.

or, if the ship executes a single trip, the item no 8 could be worded as follows:

8. for any damage caused to parts, places, frames in the ship's cargo holds, hatch covers and deck, including the Australian hold ladders, whether affecting the seaworthiness and/or cargo worthiness of the ship or not, will be repaired at the port of occurrence and the ship will generally be restored to its initial condition, before its sailing from the port. If there are not any adequate facilities for permanent repairs, then temporary repairs are to be done on the spot or as when detected, at the Charterers' time, risk and expenses, to the Master's and/or surveyor's entire satisfaction. The ship will sail or shift to the nearest shipyard, as chosen by the Charterers, to complete the repairs. The ship will remain on-hire until redelivery takes place on dlosps, or, at the Owners' option, at equidistance from the Charter Party's point of redelivery, with the damage repaired to the Master's and/or class surveyor's entire satisfaction.
The shipyard's expenses, port expenses and other expenses incurred, including but not limited to, in and out tugs, pilots, agency fees etc. will be entirely for the Charterers' account and the ship will remain always on-hire.
9. in the case that magnetic equipment is used for loading or discharging, the Charterers will be responsible for the adjustment of the ship's magnetic compass, if it becomes necessary and time lost and expenses incurred will be for their account.
10. any directly related expenses resulting therefore and incurred thereby, including restoring the ship to its initial condition and before executing the next voyage, to the Master's and/or class surveyor's supervision and satisfaction, and any detention through any of the above causes to be for the Charterers' account and the ship will remain always and under all circumstances on-hire.
11. in the event of the shifting of cargo en route to the discharging port, which might affect the safety of the ship, the Owners have the right to call to the nearest port for any necessary trimming, under the Charterers' arrangements, time, cost and risk. The ship will remain always on hire.

When loading pig iron cargo, the Owners may wish to be more specific on the loading procedure by adding:

The Shippers and/or Charterers are to provide wood protection for the tank top of each hold, at their cost and time. Such protection is to consist of 5 wooden pallets of 2mx2m each, placed in a square (4mx4m) with the 5th pallet on top in the centre. The cargo is to fall over the pile to cushion the impact. For ships intending to load bulk rice the term could be worded along the following lines:
Bulk Rice Clause:
Notwithstanding the cargo exclusions as above, the Charterers have the right, during the currency of this Charter, to load up to a maximum of two times and up to a maximum of two holds with bulk rice cargo. If bulk rice is to be loaded as first cargo, the Owners are not to be kept responsible for the cargo holds and/or hatch covers condition, however, they guarantee that the holds are suitable for grain cargo. The Owners will appoint an independent surveyor at the Charterers' time and expenses to inspect the intended holds and hath covers for loading bulk rice and to determine if they are suitable for grain cargo. If the ship fails its inspection, the ship will be placed off-hire from the time of rejection, until passes the re-inspection, which rearranged and paid by the Owners. If the ship's holds and hatch covers are suitable for grain cargo, the Charterers then remain responsible for the ship's subsequent inspections for loading bulk rice cargo. If the ship fails its inspection for loading bulk rice, the Charterers will arrange for shore labourers at their time, risk and expenses for cleaning the holds and/or hatch covers for bulk rice cargo and will provide and pay for all chemicals, fresh water, consumable materials, tools such as water guns, pressure hoses, pumps etc., are necessary to be supplied, in order to make the ship suitable for bulk rice cargo. The ship will remain always on-hire and the Charterers will load up to maximum hold's capacity.

For ships intending to load steel or even steel slab cargoes (steel slabs are considered as semi-finished and low-value cargo and actually no special pre-loading survey is required), the term could be worded along the following lines:

Steel/Steel slab Cargo Clause:
Notwithstanding the cargo exclusions as above, the Charterers have the right, during the currency of this Charter, to load up to a maximum of two cargoes of steel and/or steel slabs, whether loaded as full or part cargo.
The Charterers guarantee that:
1. they will give the Owners no less than 72 hours' notice, Saturdays, Sundays and Holidays excluded, that their intention is to load any steel and/or steel slab

cargo under this Charter.
2. the ship's P+I Club's approved surveyor, as will be appointed by the Owners, but at Charterers' cost, will attend, supervise, and approve the ship's full operation in the port of loading and discharging and all works which will be carried out, from the time of berthing until sailing. The whole process of loading, discharging and of the supplying materials etc. which were arranged and supplied by the Charterers and/or Shippers, under their time and expenses, that are necessary and required for the cargo lashing and securing, should be performed to surveyor's and/or Master's entire satisfaction. The surveyor will carry out a cargo hold, hatch covers and deck condition survey, at Charterers' time, before commencement of the loading, as well as, after completion of the discharge. The survey reports, together with the photos taken, will be sent to the Owners or Master and to the Charterers' office. The Charterers are to reimburse the Owners for the P+I Club' surveyor's fees, once the Owners have submitted the invoice, which is acceptable even by fax or e-mail. The original should be followed if it is requested by the Charterers.
3. the appointed surveyor will carry out a pre-loading survey of the cargo, under the Charterers' time and expenses. Any rejected cargo quantity will be substituted with sound cargo. The original Mate's receipts and Bill/s of Lading will be claused strictly in accordance with the P+I Club's surveyor's recommendations, in order to reflect the apparent order and condition of the
goods at the time of loading and
4. the cargo will be loaded up to the ship's allowed tank top strength and according to ship's stress and to the Master's and/or surveyor's entire satisfaction.
5. will supply, under their time, expenses and risk, sufficient dunnage and/or any other securing materials with due regard to the safety and seaworthiness, as will be required by the Master and/or surveyor and will be responsible for the chocking up, welding, lashing, securing, stowage, tallying of the cargo to the surveyor's and Master's entire satisfaction. Charterers guarantee and will remain responsible for providing the Master of the ship, under their time and expenses and before the ship's sailing from the port, with an original phytosanitary certificate issued by the plant protection organization of the port/s of loading to the plant protection organization of the discharge port/s, certifying that the dunnages and/or other materials have been inspected according to appropriate procedures and fully comply with the regulations and is fully accepted by the plant protection organization of the importing country.
The unlashing, unsecuring, as well as the removal, discharge of all dunnage and/or other materials used for the ship's safety, will be performed under the Charterers' time, risk and expenses and before the ship executes the next voyage or the redelivery takes place. The ship will not sail from the port, unless all such materials have been removed and discharged from the ship and any delays, expens-

es, consequences and/or possible fines imposed by the Port Authorities, arising from the Charterers' inappropriate discharge of the materials or inappropriate delivery ashore, will be for the Charterers' account.
6. any directly related expenses resulting therefore and incurred thereby, including expenses to restore the ship to its initial condition and to paint all the affected areas to the Master's and/or surveyor's supervision and satisfaction and any detention through any of the above causes is to be for the Charterers' account and the ship will remain always and under all circumstances on- hire.
7. California block loading is not allowed under this Charter Party.
8. in the event of cargo shifting en route to the discharging port, which might affect the safety of the ship, the Owners have the right to call to the nearest port for any necessary trimming, under the Charterers' arrangements, time, cost and risk. The ship will remain always on-hire.

For ships intending to load grains/grain products/agricultural products, the term could be worded along the following lines:

Owners guarantee that the ship is suitable for carrying grain and/or grain products and/or agricultural products in all cargo holds without requiring any grain fittings and/or bagging, strapping, securing, but always based on a full cargo loading. The ship has on board a valid grain loading booklet in accordance with the Solas 1974 regulations and the IMCO resolution a-264(vii) as adopted in 1974 and any subsequent amendments thereto. Additionally, the ship has on board an approved table of heeling moments for 'filled holds' and 'untrimmed ends' in accordance with IMCO BC XIX/INF .4.
The grain and/or grain products and/or agricultural products are harmless, non-oily and they are not listed under the Group 'b' of the IMSBC Code. The Owners, in any case and always provided that no Appendix B certificate is required, are to allow the carriage of rape seed meal and/or pellets, sunflower seed meal and/or pellets and soya bean meal and/or pellets, provided that all these cargoes are of solvent extracted grade and that they contain no more than 4pct oil and 15pct oil moisture combined for which the observations under IMO regulations do not apply, as agreed by the Maritime Safety Committee's Seventy Second Session (17 to 26 May 2000). If it is required by the Owners and/or Owners' P&I Club, the Charterers and/or Shippers will provide, before loading, an analysis certificate of the above mentioned cargoes, as issued from a competent authority, confirming that the cargo is harmless, under their time/expenses.
If the ship loads soya bean meal from Argentina and/or Brazil, the Charterers will arrange, under their time and expenses, a preloading inspection to verify if the cargo is Salmonella contaminated. In any case the Ship and Owners will be kept fully harmless, if the cargo is found to be salmonella contaminated at the

discharge port/s and the ship will remain always on-hire. For ships intending to load oil cakes/seed cakes and palm kernel, the term could be worded along the following lines:
Oil cakes/seed cakes listed under the Group 'C' of the IMSBC Code, are permitted cargoes, but always limited to mechanically expelled oilcakes, seedcakes and always excluding raw copra, raw palm, and sunflower seed expellers, as well as solvent extracted.
All cargoes will be loaded and discharged as per the local and/or latest IMO regulations and UN/ U.S. sanctions and approvals. If the cargo is Palm kernels expellers, it will be mechanically expelled, containing no more than 10pct oil, no more than 10pct moisture and no more than 20pct oil and moisture combined.
Charterers and/or Shippers will provide, before loading, an analysis certificate, as issued from a competent authority, confirming the cargo's actual water and moisture content and that the cargo can be safely carried in accordance with the IMSBC Code. For ships intending to load iron ore, the term could be worded along the following lines:
The ship will load one grade iron ore, always excluding DRI and its products and/ or DRIP and/or HBI, as per Bimco's latest 'solid bulk cargoes that can liquefy clause for Charter Parties' and as per the ship's P&I Club's surveyor's instructions, as will be appointed by the Owners, but at the Charterers' cost.
The ship will load and discharge, strictly as per the provisions of the SOLAS chapter VI, regulation 7.3 (loading, unloading and stowage of solid bulk cargoes) and the Code of Practice for the Safe Loading and Unloading of Bulk Carriers (BLU Code). If the Shippers/Receivers breach these provisions, then the Master has the right to stop loading and the Charterers will give proper instructions with immediate effect. Any directly related expenses resulting therefore and incurred thereby and any detention through any of above causes, is to be for the Charterers account and the ship will remain always on-hire.

33. Trading exclusions- NAABSA clause.

This clause, together with the 'cargo exclusions' clause
. has serious implications on the amount of hire
which the ship will be fixed on, thus it is almost always discussed during the negotiations of the Main Terms.
The Owners' proposal for the allowed trading is directly linked with
. the ship's flag,
as if the ship is hoisting a convention flag, it could be unworkable for the Charterers' intended trade. In fact, these ships will avoid trading in countries, where the inspection from the ITF is strict, while some ships will even avoid trading in countries which may lead

. to a subsequent boycott by other countries.
Countries on which there is an embargo or economic sanctions from International Organizations and/ or areas where there is risk of piracy or war/war like areas or are not ISPS compliant ports could be excluded from the Owners.
Through this clause, the ship will be protected if navigating in narrow waters and if the ship is allowed to trade outside of the I.N.L. and under what conditions.
The Charterers, during the negotiations will try to obtain the maximum possible allowed trade, so that they are more flexible in the open market, if and when it is required.
The Owners must clarify in which countries the ship can trade and under what conditions.

The "trading exclusions" clause could be agreed with the three following paragraphs:
A. generally includes in which countries they are not allowed to trade and additionally that the trade will be consistent with the embargoes or sanctions imposed by International organizations and/or are banned from the country of the ship's registry and/or the country of the Owners' incorporation.
B. includes specific countries, which are excluded from the trade.
C. includes specific countries, which are allowed for trading, but under certain specific conditions.

PARA A.
For ships willing to trade in war areas or in areas non-compliant with the ISPS regulations or affected by the gypsy moth etc. the undermentioned clause may be amended accordingly
PARA A.
The ship will be under the orders and directions of the Charterers, as regards employment and agency, for a worldwide lawful trade and for the carriage of lawful merchandise only. The ship will trade between safe port/s, safe berth/s, safe anchorage/s, where it may safely lie always afloat during the duration of the whole charter, always within the International Navigating Limits.
The ship is strictly prohibited and will not be directed to navigate, contact and neither any Bill of Lading will be issued for countries, regions, places, ports, berths, anchorages:
- that are not allowed by the country of the ship's registry,
- that are not allowed by the Owners' domicile,
- that are outside of the International Navigating Limits,
- which may lead to a subsequent boycott by other countries,
-that are areas or ports infested, where the world sanitary organisations have

declared contagious diseases,
-where there are prevailing narrow waters and/or shallow harbours or tidal and/ or ports in rivers or with bars,
-that commercially restrict the ship's qualification in the USA/Australia/ Canada and enforce costs in accordance with the laws of the countries respectively,
- that are subject to UN Security Council Resolutions and/or embargoes or sanctions of NATO, EU, USA or any other International Organization or ship's flag state,
- that are ice blocked or threatened with ice or ice bound ports. The ship should not be required to force ice or to follow icebreakers. The Master has the right at any time to sail from the anchorage, berth port, place, waiting place to a convenient open place, if he considers it dangerous to remain there, due to ice and according to weather reports and advice of the Port Authorities for the ship. The ship will always remain on-hire and the Charterers will instruct on whether the ship will wait for weather improvement or will sail to another safer port, place in order to complete the loading or discharging operation.
- that are at war or are war-like or are piracy zones or to trade in territorial waters of any of the countries or places that are described in the current War Trading Warranties, as adopted by the ship's underwriters,
- that it is known that the ship will be boycotted by shore and/or port labourers and/or tugboats and/or pilots and/or by government and/or Authority by reason of the ship's flag, registry, class, ownership,
- that are a non ISPS compliant port/s,
- that are affected during the high risk season of the Asian Gypsy Moth In the event of new rules, regulations, or laws enacted in the countries, ports, ranges allowed for trading under this Charter, or in the country of flag or registry, that significantly affect the trading allowance, or incurring substantial risks or expenses to the Owners, then the Owners and Charterers will discuss in a spirit of good faith and attempt to mutually reach a satisfactory solution, which will comply to new requirements, including changing the flag or registry of the ship. If the political situation changes in any of the excluded countries, areas, regions, during the currency of this Charter, to the extent that the Owners are able to verify that there is no risk to the ship and/or crew, the Owners' approval for trading should not be unreasonably withheld. Additional premium charged by the Owners' insurances and/or the crew's bonus and/or extra costs incurred, will be for the Charterers' account.
Without prejudice to war risks or piracy clauses, if the political situation changes in any of the allowed countries, areas, regions during the currency of this Charter, to the extent that the Charterers are able to verify that there is a risk to the ship and/ or crew, the Charterers' consent to exclude these countries, areas, re-

gions from trading should not be unreasonably withheld.
Charterers are not allowed to consecutively employ the ship for more than 30 days in any trade/s, where steaming at the full Charter Party speed between ports is less than 7 days.

PARA B.
Specifically, the following countries/ports/places/areas/passages are excluded from the trading:
……
…….

Comments:
Owners can insert specific countries/ports/places/areas/passages, which are excluded from the trade.

PARA C
When the ship trades in countries where there is a risk for the cargo, ship, crew and/or additional regulations are in effect, additional protection then should be required, which could be covered by this clause.
PARA C:
-For ships intending to trade in Arabian countries with translated certificates the term could be worded along the following lines:

If a certificate must be translated to make the Charterers able to trade in a particular port, region, the Owners must be notified well in advance, in order to have sufficient time to comply with such a request. However, the Owners and ship will not be held responsible, if the certificates for translation are not ready on arrival of the ship at the port, because of the Charterers' late advice. All costs of issuing the translated certificates will be for the Charterers' account. Any cargo documents that are required to be translated into another language other than English, must be arranged only by the Charterers, at their time and expense.

-For ships intending to load in Brazil and, due to very fast loading of mineral cargoes, or, generally, cargoes with a stowage factor above 01 ton/01 cbm, (iron ore, bauxite, manganese and phosphate) for ships aged equal or more than 18 years old, a condition survey is compulsory prior to commencement of loading (DPC Survey). This survey is valid for one year.
Also, the ship will be subject to a port control inspection upon arrival and prior to the commencement of loading.

Loading in Brazil:
The DPC survey in Brazil is for the Charterers' time and costs.

-For ships intending to trade in the Amazon River, Brazil, the term could be worded along the following lines:

The ship is allowed to trade in the Amazon River, but up to and including .... port and provided it is customary for ships of similar dimensions and size to navigate there. The Charterers will employ pilots at their cost in accordance with the Master's request, from the first outer sea boy to the port, berth and from the port, berth to the last outer sea boy.
Time lost for exchange of ballast water in the river or before entering the river, as per the local regulations, which will be made at the utmost dispatch will be for the Charterers' account. All Additional Premiums charged from the Owners' insurances will be for the Charterers' account, which are payable on request and the supported vouchers are acceptable, even by fax or e-mail.

-For ships intending to trade in shallow harbors/rivers etc., the term could be worded along the following lines:

In the event that the ship is navigating as per the Charterers' employment instructions, in shallow harbours or rivers or ports with bars, any detention of the ship and/or expenses resulting from such detention will be for the Charterers' account and the ship will always remain on-hire.

-For ships intending to trade between China and Taiwan, the term could be worded along the following lines:

The ship is not allowed to sail and/or trade directly between P.R. China and Taiwan and vice versa. If Taiwan is the last port of call, then the Charterers, before the redelivery takes place, will clear the ship up under their time and expenses, at a port other than Taiwan, enabling the ship to trade in China.

-For ships intending to trade in ports, where the ship can safely aground, the term could be worded along the following lines:

NAABSA Clause:
Charterers have the right to load or discharge the ship at any ports Up River, but not north of San Lorenzo timbues area and Nueva Palmira or in Brazilian port/s not north of Vitoria port (Vitoria to be excluded), provided that it is customary for ships of similar dimensions and size to navigate there, where they can safe-

ly lie aground on the soft muddy sea-bed, but always at the Charterers' risk and subject to the permissible and prevailing draft declared by the Port Authorities and/ or pilots.
If the ship lies aground and the Master suspects that the hull is damaged, then the Charterers are to employ, under their time and expenses, an inspection of the ship's flat bottom, including, but not limited to, the vertical sides, bilge keels, propeller, and rudder blade, by a well-organized and ship's class approved diving company, to determine the extent of any damage. The inspection will be carried out under the appointed by the Owners' ship's class surveyor attendance, but at the Charterers' cost.
Should the ship's bottom and underwater parts, propeller, rudder be extensively damaged, affecting the ship's seaworthiness and class, the ship will be directed immediately and always on-hire to a mutually appointed convenient place or port or yard to carry out the repairs at the Charterers' time and cost and to the class surveyor's entire satisfaction. The Shipyard's expenses and port expenses, including in and out tugs, pilots, agency fees etc., will be entirely for the Charterers' account.
Time lost for unloading the cargo at the load port, if it is required, and any expenses incurred will be for the Charterers' account.
Damage not affecting the ship's seaworthiness and class will be repaired at the Charterers' cost, but not for the time lost, at the next scheduled dry docking, normal wear and tear excepted.
If, after the ship's underwater inspection, no damage is found, either minor or not, then all the directly related expenses incurred such as the diving company's expenses, ship's class surveyor's fees etc. will be shared equally between the Owners and Charterers, but the ship will remain always on-hire.

-NAABSA procedure in a place other than at ECSA:

In the case that the ship is directed for an NAABSA procedure at any place other than those aforementioned, then after completion of each loading and/or discharging operation or at the nearest next available port, place, the Charterers, under their time and expenses, will appoint a diving inspection, in order to certify whether any damage is caused due to the ship's running aground.

-For ships intending to trade in the U.S.A., the term could be worded along the following lines:

The U.S.A. is allowed for trading, provided that, if the ship on arrival is considered by the Port Authorities as an HIV (High Interest vessel), then all time lost and expenses incurred due to inspection, are to be for the Charterers' account.

If the ship fails its inspection, then the ship will be placed off-hire from the time of rejection, until it is fully approved, and any directly related expenses incurred, but supported with original vouchers, will be for the Owners' account.
Notwithstanding anything else contained in this Charter Party, all costs or expenses arising out of or related to security regulations or measures required by any U.S. Authority, including but not limited to, Security Guards, launch services, tug escorts, port security fees or taxes and inspections, including 'High Interest Vessel', will be for the Charterers' account.
Watchmen employed by the ship to safeguard the cargo, or port Authorities' compulsory watchmen will be for the Charterers' account.
The crew's watchmen in the U.S.A. or in U.S.A. controlled areas, for not having granted shore visas, will be for the Charterers' account. The Charterers are to inform the Owners on issuing USA crew entry visas, well before the ship's calling to the U.S.A. and the Owners are to best endeavor for such insurance. Nevertheless, if the issue of this visa becomes impossible, the compulsory crew watchmen will be for the Charterers' account.
For USA ports the following current bimco's clauses are to apply:
....
....
....

-For ships intending to pass through the Magellan straits:

The ship is not allowed to pass through Cape Horn, but the Charterers have the right to instruct the ship to pass through the Magellan straits provided:
1. it will only pass during the summertime from the beginning of November to the end of March,
2. the Charterers will pay for all compulsory and/or customary and/or recommended pilotages and/or tugs, as will be required by the Master of the ship,
3. any additional premium charged by the ship's insurances, will be covered by the Charterers, which are payable on request and the supported vouchers, are acceptable even by fax or e-mail.

-For ships intending to pass through the Cape Horn:

Charterers, during the currency of this Charter Party, have the right to pass through Cape Horn provided that:
1. they will pass only during the summertime from the beginning of November to the end of March,
2. they will pay for all compulsory and/or customary and/or recommended pilotages and/or tugs, as will be required by the Master of the ship,

3. a weather routing company, as will be appointed and paid by the Charterers, will monitor the ship and report, recommend to the Master on the expected weather conditions and on the safest route,
4. the voyage will be from the Pacific (Far East) to the Atlantic and they will navigate according to the currents' direction,
5. any additional premium charged by the ship's insurances, will be covered by the Charterers, which are payable on request and the supported vouchers, are acceptable even by fax or e-mail,
6. the Master, at his sole discretion, in the case that weather conditions are extremely bad and he considers it unsafe to proceed, has the right to slow down or to stop at a safe anchorage and wait for the weather's improvement. The ship will remain always on-hire.

-For ships intending to trade in the USG Mississippi River:

The ship will not trade above the port of Baton Rouge, Louisiana in the Mississippi river.

-For ships intending to trade in the Orinoco River:

The ship is allowed to trade in the Orinoco River, but up to and including the......... port, but provided that it is customary for ships of similar dimensions and size to navigate there. The Charterers will employ pilots at their cost in accordance with the Master's request, from the first outer sea boy to the port and berth and from the port and berth to the last outer sea boy.
Time lost for ballast water exchange in the river or before entering the river, as per the local regulations, which will be made at the utmost dispatch will be for the Charterers' account. All Additional Premiums charged from the Owners' insurances will be for the Charterers' account, which are payable on request and the supported vouchers are acceptable even by fax or e-mail.

-For ships intending to trade in countries, where the political situation is unstable:

The ship is allowed to trade in .... /......../ ..../ ...., but provided that the Charterers will remain 100pct responsible and liable for all and any cargo claims made by any third party, arising as a result of actual or alleged cargo shortages and/or cargo damage, as well as on the quality of the cargo, unless proven caused by default of the ship.
In connection with the above, the Charterers guarantee that:
a. they will deal and settle the claims directly with the cargo claimants and here-

by agree to indemnify the Owners, servants and agents and hold all of them harmless, in respect of any liability, loss, damages or expenses of whatsoever nature of which the Owners may sustain, by reason of any claim, however so arisen.
b. they will remain responsible and liable for any delays, losses, expenses, charges, additional port expenses and agency fees, possible fines and penalties or sanctions imposed by the action of the local Authorities, Consignees or Agents or by any other party having interest on the cargo.
c. they are responsible for negligent, wrongful and/or improper acts of pilots, immigration port or other local officials, which may result in the arrest, detention or delay of the ship.
d. they will remain responsible for accidental, unwitting, or unknown acts of the Master, officers or crew, which may contravene customs, immigration, or other port regulations in such countries.
e. if the ship and/or any other ship and/or property under the same beneficial Ownership and/or under associated Ownership, management and/or control, is detained or arrested by any party, the Charterers will provide on demand such bail or other security, as may be required to prevent such arrest or detention, and/or to secure the release of the ship and/or property and will indemnify the Owners and/or the Owners of such ship and/or property in respect of any loss, damages or expenses incurred by such arrest or detention, whether or not the same may be justified.
f. they have no right to claim for any contribution from the Owners and the NYPE Inter-club agreement clause will not be in effect when the ship contacts these countries.
g. the ship will remain always and under any and all circumstances on-hire.

-For ships intending to trade in Alaska:

Alaska is not allowed, but the Red Dog port is allowed for trade from the end of April to the end of August, but always subject to the Owners' and ship's H&M insurances prior approval. All the Additional Premiums charged from the Owners' insurances, will be for the Charterers' account, which are payable on request and the supported vouchers, are acceptable even by fax or e-mail.

For ships intending to trade to a non ISPS compliant ports:

During the currency of this Charter Party, the Charterers are entitled to trade the ship to a non ISPS compliant port/s subject to the Owners' prior approval that they will check the safety of the port/s. However, the Charterers remain fully responsible and liable whether the ship increases the level or not, for any and all

consequences of any nature, delays, expenses, extra expenses for the crew and insurers, arising out or incurred, because of the ship's trading and/or because the ship traded to a non ISPS compliant port/s or areas. The Charterers guarantee that the ship's last 5 ports before its redelivery will imperatively be an ISPS compliant ports and the Owners have the right to refuse to proceed to a non ISPS compliant port, in the case that the Charterers do not satisfy them, according to their presented redelivery schedule.
The Bimco's ' ISPS/MTSA Clause for Time Charter Parties 2005' to be in effect or any subsequent amendment thereto.

-For ships intending to trade in areas affected by the Gypsy Moth:

If, during the currency of this Charter, the ship arrives for loading and/or discharging purposes within the season of the Gypsy Moth in the area of high risk as designated by the CFID/ AMPHIS, then the Charterers, under their time and expenses and before sailing from the port, will arrange to appoint competent Authorities, fully recognized by the CFID/APHIS, to inspect the ship and issue a pre-departure certification, stating that the ship is free of Asian Gypsy Moth life stages and eggs. All time lost, expenses incurred until the ship is free from Asian Gypsy Moths and the pre-departure certification is issued, will be for the Charterers' account. The Charterers, during the duration of this Charter Party, will remain responsible for any rejection by the U.S. and/ or Canadian Authorities and/ or from any other Authorities, because of their instructions the ship to contact within the season of the Gypsy Moth in the area of high risk, as designated by the CFID/AMPHIS and the ship under any and all circumstances will remain on-hire.

-For ships intending to break the International Navigating Limits:

Charterers are to have the option to break the International Navigating Limits, but only after the Owners' and ship's H&M insurances written consent. The Charterers are responsible and liable and will pay for:
1. all the Additional Premiums charged for the Owners' insurances, which in any case will not exceed the Lloyds of London insurance rate. The Additional Premiums are payable on request and the supported vouchers, are acceptable even by fax or e-mail.
2. any deductible amount made by the Owners' Underwriters, in relation to any damage whatsoever arising from breaking the International Navigating Limits, including, but not limited to ice damage.
3. any kind of damage to the ship caused, because of breaking the International Navigating Limits. All repairs which affect the ship's class and/or seaworthiness

and/or cargoworthiness will be made good at the port of occurrence or at a mutually agreed place, port, yard, always at the Charterers' time, expenses and to the appointed by the Owners and paid by the Charterers ship's class surveyor's entire satisfaction. Minor damages will be repaired at the Charterers cost, but not for the time lost, at the ship's next routine dry-docking.
4. all the expenses, dues, charges, fees incurred and/or losses/damages the Owners may suffer of any nature, for breaking the International Navigating Limits.
Furthermore, it is expressly agreed that:
a. the Charterers guarantee that the ship is suitable for loading or discharging in such ports, berths, anchorages, places and all the required certificates and fittings and/or Port Authorities' acceptances for the ship to proceed in such ports, berths, anchorages, places will be for their time, risk and expenses.
b. the Charterers guarantee that the ship will not be obliged to force ice, however, touching ice does not mean forcing ice.
c. the ship will follow the ice-breaker/s in an ice convoy as per normal practice. Tugboats and/or ice-breakers used while the ship is in the ice convoy or, if the ship cannot follow the convoy, then the individual tug boat/ice-breaker assistance, as required by the ship's Master, will be for the Charterers' account.
d. the Master has the right at any time to sail from the anchorage, berth, port, place, waiting place to a convenient open place, if he considers it dangerous to remain there, due to ice and according to weather reports and the advice of Port Authorities for the ship. The ship will remain always on-hire and the Charterers will instruct the ship on whether it will wait for weather improvement or sail to another safer port, place in order to complete the loading or discharging operation.
e. the ship will remain on-hire for time lost, due to non-ability to move into the ice due to tugboat and/or ice-breaker unavailability and/or delay and/or the ship's inability to follow the ice-breaker, because of ice conditions.
f. the ship will remain on-hire at all times and at any circumstances, when breaking the International Navigating Limits, unless delays incurred due to Owners' or Master's fault.

34. NYPE Inter-club agreement.

This clause stipulates that the liabilities for cargo claims to third parties will be apportioned between the Owners and Time Charterers in accordance with the rules
. of the Inter-Club New York Produce Exchange Agreement.
It should be considered as a top clause, which could be in effect for all modes when fixing on Time Charter, either for a single trip/short or for long periods.

When an Inter-Club Agreement (abbreviated ICA) clause is incorporated in the Charter Party, then the same

. binds the contractual parties and settles the liability claims depending on the cause of the cargo claim, so that extra costs and lengthy litigations of the involved P&I Clubs can be avoided.

One of the most important terms of the ICA, are the liabilities for cargo shortages and the cargo handling. Particularly, in the case of

. shortages, the Owners are entitled to recover 50 percent of the paid amount from the Charterers' /Charterers' P&I Club. Shortages or damages due to cargo handling are 100 percent for the Charterers' account.

The "ICA" clause, which, among others, should include:

-how the cargo claims will be settled.

An executed clause that has been run in a Charter Party is as follows (precise copy):

Any liability to third parties for loss of or damage to cargo including reasonable survey costs and other reasonable costs incurred to negotiate and settle such liabilities, shall be apportioned between the Owners/Time Charterers in accordance with the New York Produce Interclub Agreement of September 1, 1996 and any amendments thereto promulgated by the International Group of P and I Clubs.

Charterers/Owners shall be discharged and released from all liabilities in respect of any claim or claims which Owners/Charterers respectively may have under this Charter Party and such claim shall be totally extinguished, unless such claims have been notified in detail to Charterers in writing accompanied by all available supporting documents (whether relating to liability or quantum or both) within 12 months from completion of discharge of the appropriate cargo under this Charter Party.

Comments:

The clause states that the liabilities to third parties, including the survey cost and other costs incurred for negotiating and settling such liabilities will be apportioned between the Owners and the Charterers. It also states that, unless the claim has been notified in detail in writing within 12 months from completion of discharge, then such claim will be totally abated.

A complete clause for Time Charter period, which refers to the "NYPE Inter-club agreement" could be worded along the following lines:

It is expressly agreed that:

1. the liabilities for cargo claims to third parties will be apportioned between the

Owners and Time Charterers, in accordance with the Inter-Club New York Produce Exchange Agreement as agreed on September 2011 by the International Group of P&I Clubs and any subsequent amendments thereto.
2. for cargo claims which fall under the agreement and both parties the Owners and Charterers are involved, then:
a. the cargo claim will be notified by either the Owners or the Charterers to the other party in writing and in detail accompanied by all available supporting documents, as soon as possible, after receiving notification of such claims.
b. no party is authorised to proceed with any cargo claim settlement or grant time extension unless the written approval from the other party has been obtained.
c. the survey costs and other necessary costs incurred to negotiate and settle such liabilities will be equally shared between the Owners and Charterers, supported with original vouchers.
d. neither party will between themselves refer to the one-year time limit, as a defence.
3. for cargo claims for which the Owners or Charterers remain wholly responsible, in accordance with the Inter-Club New York Produce Exchange Agreement, as agreed on September 2011by the International Group of P&I Clubs and any subsequent amendments thereto, then any time lost and extra expenses incurred in that respect, will be in full for the party in fault.

35. Charterers'/Owners' Agents.

The attending Agents at each port of contact will be
. appointed and be paid by the Charterers.
Nevertheless, the Owners reserve the right to appoint their
. protective Agents
to follow their own matters and/or to protect their interests, during the ship's stay in the port. For major ship husbandry matters, such as dry-docking, ship accidents etc., the Charterers should request to not be involved through their Agents. The Owners will appoint their own Agents and will be responsible for the payment of all the expenses incurred. However, under normal circumstances and for a regular trip, the Charterers' Agents could provide their services for the Owners' normal and/or minor husbandry matters, without charging any additional fees.

The "Charterers'/Owners' Agents" clause, which, among others, should include: -who will appoint and pay for the attending Agents at all ports.
-whether the Owners can use the Agents for their husbandry matters, without

paying additional agency fees.
-if and when the Owners have the right to appoint their own protective Agents.

An executed clause that has been run in a Charter Party is as follows (precise copy):
Charterers will have their agents to attend the vessel's normal matters without charging agency fees. However, should there be an extraordinary matter such as attending to crew's hospitalisation or survey work for Owners account, Owners shall employ Charterers' agents as Owners' husbandry agents and pay agency as per local recognised tariff or Owners shall appoint their own husbandry agents at their expenses.
Comments:
The Charterers' Agents will attend to the ship's normal matters, as the same are covered with the already charged customary agency fees.
All extraordinary matters, which might be requested from the Owners, will be paid to the Agents as per tariff, otherwise the Owners will appoint their own protective Agents to follow them.

Another clause reads as follows (precise copy):
Whilst on this charter Charterers' agents will attend Owners minor husbandry matters such as postage, supply of fresh water, cash to Master, doctors etc without paying agency fees.
Owners will provide advance founds for the actual expenses to agents otherwise same will be deducted from hire.
For major ship's husbandry matters such as dry-docking, vessel's accident, hospitalisation, repatriation of crew, repairs, supply of the vessel's stores and provisions etc. then agency fees/attendance fees will be negotiated directly with agents.
For these major ship's husbandry matters Owners will provide advance founds to agents for the expenses involved and for the agency fee/agency.
Should Charterers agents disagree then Owners have the right to appoint their own agents to attend to such matters and they will be remaining responsible for funding the agents accordingly.
Comments:
The Owners may use the Charterers' Agents for their minor husbandry matters without paying agency fees. The actual expenses incurred will be paid in advance to the Agents from the Owners, otherwise, the Charterers are entitled to deduct the same from the next due hire.
For all major ship husbandry matters, agency/attendance fees will be negotiated directly with the Agents, otherwise, the Owners have the right to appoint their own Agents to attend their matters.

A complete clause for Time Charter period, which refers to the 'Charterers'/ Owners' Agents" could be worded along the following lines:

The Agents at all ports of contact, except for contacts made for the Owners' matters, will be selected, and be paid by the Charterers.
It is expressly agreed that:
a. the Charterers' Agents will attend to the Owners' minor and customary husbandry matters, such as postage, supply of fresh water, cash to the Master, medical treatments, doctors, shore passes, supply of the ship's stores and provisions, ordering small repairs etc. without the Owners paying additional agency fees.
b. for any other actual extra expenses incurred for the shore Owners' account, the Owners will provide advance funds to shore agents, but, in any case, before the ship's sailing from the port. If be requested by the Owners and after the Charterers' consent and approval, such funds (which are approved and agreed by the Charterers' Agents) may be covered by the Charterers and be deducted from the next due hire, provided that the last hire payment has not yet been effected.
c. any time lost, due to the Owners' late settlement of any accounts with the Agents, will be considered as off-hire and the Owners will remain responsible for all the additional port and extra expenses which may be incurred.
d. for major ship's husbandry matters such as dry-docking, ship accidents, attending to crew hospitalisation, changes of major parts of the crew, large repairs, survey work made for the Owners' account, in order to monitor the loading or the discharging operation, or in order to protect the Owners' interests in the ports etc., the Owners have the right to appoint their own selected Agents to attend to such matters and will remain responsible for funding them timely.
The Owners may also use the Charterers' Agents if they agree to do so, provided that the Agents will directly send a message to the Charterers that they will keep them harmless from the entire Owners' expenses incurred.
For these major ship husbandry matters, the Owners should provide advance funds to the Agents for the expenses incurred and for the additional agency fees incurred, otherwise, any time lost due to settlement of any accounts with the Agents, is to be considered as off-hire and the Owners will be responsible for all consequences, including, but not limited to, additional port and extra expenses which might be incurred.

36. Grace period for late and/or incorrect payment / Anti-technical.

Grace period for late and/or incorrect payment.
The Charterers will be protected through this clause if they are running out of a

punctual and regular hire payment.
This clause will grant them
. additional time (an extension)
to remit the due hire, which has been delayed being credited to the Owners' bank account, either by their oversight/negligence, or by a bank error outside of the Charterers' fault or otherwise.
It removes the right to protect their interests from the Charterers and actually
. retracts the agreement that the hire is always payable in advance.
However, the Owners would be in a position and would have the right to
. withdraw the ship or suspend its performance (subject to the agreement)
if the Charterers have additionally breached this clause and have not fulfilled their obligation during the grace period agreed on.
For long Time Charter periods, it is considered
. a standard clause for the Charterers and the granted period does not usually count during bank holidays/holidays/weekends.
For short trips it
. should be avoided,
as it could cause the ship completing the discharge without the payment having been made.
It should also be in effect in the case that the Charterers have not remitted the correct hire and/or have made deductions in breach of Charter Party terms, in which case extra time should be required in order to rectify their fault.
The grace period clause for the payment
. of last hire/s should be explicitly excluded.

The "grace period for late and/or incorrect payment/ Anti-technical" clause, which, among others, should include:
-the allowable time for the Charterers to remit the unpaid hire.
-whether the holidays/weekends are excluded within the grace period.
-that the clause will not be in effect when the ship is towards its last discharge port.

An executed clause that has been run in a Charter Party is as follows (precise copy):
When hire is due and not received by the Owners, before exercising the option to withdraw the vessel from Charter Party, Owners will give the Charterers three banking days' notice to Charterers and will not withdraw the vessel if the hire is received within these three banking days.
Grace period not to apply at the port of redelivery when the hire becomes due and is to be paid on a day-by day basis or so.
Comments:

The clause allows the Charterers to remit the hire within three banking days from the day for which the hire was due. The grace period will not be in effect at the port of redelivery and the Charterers should take care for the timely remittance of the due hire.

Another clause reads as follows (precise copy):
Referring to lines 60 and 61, where there is any failure to make 'punctual and regular payment', due to oversight or negligence or error or omission of Charterers bankers or agents or otherwise, for any reason where there is absence of intention or failure to make payment as set out, Charterers shall be given by Owners two banking/ working days' notice to rectify the failure where so rectified, the payment shall stand as punctual and regular payment.
Evidence of the timely remittance of the hire payments by the bankers designated by the Charterers fulfils the Charterers obligations to pay hire, whether the Owners are so notified by the bankers or not.
Comments:
The lines 60/61 of the NYPE 46 Charter Party give the Owners the right to withdraw the ship from the Charterers' service, in the case that the Charterers have failed on the punctual and regular payment of the hire. In the case that the payment is delayed and in the case of which there is absence of intention, the Charterers will be given two banking days' notice to rectify the payment. Obviously, the grace period will not be in effect when the Charterers have delayed the payment on purpose.
If the payment is made within the two allowed banking days, then the payment will stand as punctual and regular, whether the Owners have been notified by the bank or not.

A complete clause for Time Charter period, which refers to the "Grace period for late and/or incorrect payment/Anti-technical" could be worded along the following lines:

If the hire has not been paid in time and in full with no intention, but it is caused due to a technical banking failure or the Charterers' oversight or negligence or error or omission, in accordance with the terms and conditions of this Charter Party, the Owners are to put the Charterers in two London banking days written notice that unless payment is made in full as per the Charter Party terms, then the Owners have the right to exercise all of their rights under this Charter Party, in order to protect their interests.
If the payment is rectified timely, it will stand as a regular and punctual payment and no interests will be paid.
The grace period clause will not be in effect, when the Charterers exercise this

clause for more than two times during the currency of this Charter Party, or when the ship is towards its last discharge port or the redelivery area and the hire becomes due and is to be paid on a day-by day basis, or for such length as the time necessary to complete the Charter period.

The Owners can combine the 'Grace period for late and/or incorrect payment' with an Anti-technical (Suspension) clause. The suspension clause could grant a further extension for Charterers to remit the due hire, as they will receive some additional notices to comply with their obligations, failing which the Owners may suspend the performance of any or all of their obligations, under this Charter Party in port or at sea, until time as the payment due is received by the Owners.

Alternatively, the grace period clause may be omitted, and the Anti-technical clause could be agreed directly.

Anti-technical clause:

The main body of the NYPE 46 Charter Party incorporates a provision that, in the case of non-payment or late payment of hire, the Owners have the right to lien the cargo, freights etc., as well as to have the right to withdraw the ship from the Charterers' services. On these respective lines, the Owners may negotiate and insert certain additions or amendments necessary to protect their interests. In the case that there is no provision, then the Owners should seek legal advice for a complete

. lien/withdrawal clause which could be added in the Charter Party accordingly. Generally, if the hire failed to be timely paid, the Owners should have the right, under the terms of the Charter Party, to withdraw the ship from the Charterers' services, as well as to have the right to lien the cargo and/or freights or sub-freights/demurrages etc. due to the Charterers from the sub-Charter Parties. The Owners could have the right to refuse to release the cargo to the consignees and to keep it at their possession either on board the ship, or, after discharge, at the port's custody, until the Charterers fulfil their obligation, or otherwise, until they sell the cargo to others. The lien and the withdrawal will be directed from the Owners' law firm, as it is subject to the way in which the originally signed Bill of Lading was marked.

The provision

. for suspension

of the ship's performance of any or all of their obligations, provides the Owners with the privilege to exert pressure on the situation in order to get the due hires, without exercising their right to withdraw the ship. The suspension clause could be made by the Owners' law firm or their legal department.

Bimco has published a non payment of hire clause for Time Charter, which provides the Owners with the right to suspend the performance until payment

of due hire/s is made. The ship will remain on-hire and the suspension will be made without prejudice to any rights that the Owners may have under the Charter Party.
The notice for the ship's suspension of performance must be given after
. the grace period of the respective clause has expired,
provided that such a clause is also incorporated in the Charter Party.

37. Deductions.

It stipulates whether the Charterers have the right to make deductions from the hire payments, the kind of deductions and, sometimes, the maximum amount of the same.
Under the terms of the Charter Party, the hire is due until the last day of the Charter, which is always payable in advance. The Charterers will look to cover themselves through deductions, for any
. overpayment,
which could had been incurred due to the ship's underperformance or overconsumption or for possible fines, which were imposed to the ship or even to cover any minor expenses charged for the Owners' account in the port/s and basically paid by them together with the ship's general port expenses.
There should also be a provision in this clause, which refers to the deduction of the value of the fuel at the time of redelivery, if no reference was made in the 'bunkering' clause.
The Charterers' rights for deductions must be clearly written, for disputes to be avoided.

The "deductions" clause, which, among others, should include:
-if and what amounts the Charterers are entitled to deduct from the due hire.

An executed clause that has been run in a Charter Party is as follows (precise copy):
Charterers have the right to withhold from the Charter hire during the currency of this Charter, provided they have priory obtained Owners consent which not to be unreasonably rejected by Owners, such amounts due to them for off-hire and Owners' disbursements or any expenditure is incurred by Charterers on behalf of the Owners, but Charterers shall produce supporting evidence for such deductions. Charterers also have the right to withhold from the Charter hire Owners' estimated advances and disbursements including any fine and any other amount for Owners' account, but maximum USD .... per port, also the estimated value of bunkers on redelivery from the last most suitable hire payments.
However, final accounting shall be arranged by Charterers as promptly as possi-

ble and Charterers shall produce supporting evidence for such deductions.
Charterers are in no way entitled to withhold money on account of a speed or performance claim for which there is no mutual agreement or arbitration award.
Comments:
From the last 'most suitable hire payments', the Charterers have the right to deduct from the due hires such amounts as to cover the ship's undisputed off-hires, the Owners disbursements and any other expenses incurred by the Charterers on the Owners' behalf. Additionally, they will deduct the estimated value of the bunkers on redelivery.

Another clause reads as follows (precise copy):
Deductions of any amount from hire or any other amounts due to Owners, under this Charter are only permitted subject to mutual agreement of Charterers and Owners.
Charterers may deduct only value of bunkers from last hire payment/s as well as Owners' share on/off hire surveys. No other deductions on account of Owners' disbursements or fines or off hire are permitted during the currency of this Charter.
Comments:
It states that the Charterers have the right to deduct from the due hires only the value of bunkers and the Owners share the on and off hire surveys. No other deductions are allowed without the Owners' consent.

A complete clause for Time Charter period, which refers to the "Deductions" could be worded along the following lines:

The Owners will pay directly to parties concerned, through their Agents or otherwise, for all the expenses incurred for their account, unless otherwise stated in this Charter Party.
Nevertheless, the Charterers have the right to withhold from the last hire payment and, if the last hire is insufficient, it will be extended until deductions of sufficient funds are enough to cover:
a. the estimated value of the bunkers remaining on board at the time of redelivery.
b. the domestic fuel consumption allowance.
c. the estimated minor Owners' disbursements, but with a maximum of USD 500 per port, irrespective of whether the Owners have paid their expenses for their husbandry matters directly to the attending Agents. The actual cost of the Owners' expenses will be settled together with the final hire statement, which will be supported with original vouchers.
d. the advances made to the ship's Owners and/or Master, but only after the

Owners' written consent, including a commission of 2,5pct thereon.
e. the payment of any amounts, which may reasonably be held to be for the Owners' account, either due to violation of the Charter Party terms or for any other expenses incurred and paid by the Charterers, due to the ship's fault, including, but not limited to, possible port fines and/or additional port expenses, due to rejection of cargo holds and/or cargo gear breakdown and/or, generally, any off-hires because of its inability to be wholly and operationally at the Charterers' disposal.
f. the undisputed off-hires, due to the ship's underperformance and overconsumption, during the execution of this Charter Party.

38. Last voyage.

An argument between Owners and Charterers over the actual time of redelivery of the ship under the Charter Party terms is something heard more than once. The last voyage is the point of focus and as to whether the Charterers are entitled to fix the ship to a biz, where it is sure or about to be, that will
. run out of the maximum agreed period.
However, when the market has a downward direction and the hire is much higher than the prevailing one, nobody from the Owners' office should complain.
The provision of this clause will clarify the situation, if and when it arises.
For Time Charter, either for short or long periods, such a provision should be incorporated in the Charter Party, for the avoidance of disputes and arguments between the contracted parties.

The "Last voyage" clause, which, among others, should include:
-that the Charterers will redeliver the ship within the allowed period.
-that, if the ship has remained longer at the Charterers' services, the hire will be adjusted if the market goes up.

An executed clause that has been run in a Charter Party is as follows (precise copy):
Should the vessel on her last voyage under this Charter Party be delayed and the maximum Charter period thereby exceeded, the Charterers shall have the use of the vessel to enable them to complete the voyage. Hire for such excess period to continue at the Charter Party rate, provided the delay is not caused by reasons which were reasonably foreseeable at the time the Charterers gave orders to the last voyage.
Comments:
In the case that the ship's contractual redelivery time has been delayed, the Charterers may have the use of the ship until completion of the fixed voyage at

the same rate, provided that the delay was not foreseeable at the time that the Charterers fixed the ship.

Another clause reads as follows (precise copy):
The Charterers warrant that they will not order the vessel to commence a voyage (including any preceding ballast voyage) which cannot reasonably be expected to be completed in time to allow redelivery of the vessel within the period agreed in this Charter Party. In any event, in case the freight market for the vessel (or similar vessels) should then be higher than the rate of hire agreed under this Charter Party, the Charterers shall pay the Owners the rate of hire then obtainable in the market (instead of the hire agreed under this Charter Party) for the number of days by which the period agreed is exceeded.
Comments:
The Charterers do not have the right to fix the ship for its next trip which will prevent the redelivery of the ship in time, as per the Charter Party terms. In the event that the ship is delayed, and the market rates are higher, then the Charterers will adjust the hire according to the prevailing market rates.

A complete clause for Time Charter period, which refers to the "Last voyage" could be worded along the following lines:

Charterers guarantee that they will not fix and order the ship to commence a voyage, which cannot reasonably be expected to be completed in time and allow the redelivery of the ship to take place within the period agreed in this Charter Party. Unless the Charterers provide a satisfactory voyage plan, the Master is entitled to refuse the commencement of the voyage and request for an alternative voyage, which allows the redelivery of the ship within the period agreed in this Charter Party. If the ship has delayed and runs out of the maximum allowed period, due to unforeseen circumstances, the Charterers will continue to have the use of the ship, enabling them to complete the voyage. For any period in excess of the agreed period under the terms of the Charter Party and provided that the market has risen or has had an upward trend since the ship's delivery to the Charterers' services, the hire rate will mutually be adjusted according to the current prevailing market rate ships of similar type and size, at the area of trade. In any case, the hire will not be below that of the contractual hire rate.

-As an alternative, the last four lines could be amended as follows, irrespective of the market's tendency:

For any period in excess of the agreed period under the terms of the Charter Party, the hire rate will be increased by 50pct on top of the agreed Charter Party hire rate.

39. Gangway watchmen.

The cost of ordering 'Gangway Watchmen' will be apportioned in this clause. The employment of shore gangway watchmen will be ordered for one or more of the following reasons:
- because of the prevailing local regulations, for foreign flag ships, where the employment of shore gangway watchmen is compulsory.
- for the defence against stowaways, smugglers, or thefts, as ordered by the ship's Master in order to assist and/or the protect ship's crew.
- because of the cargo on board the ship, as ordered by the Charterers, in order to watch and secure the discharging of the cargo.
- because of the nationality of the crew, as ordered by Port Authorities.

Whilst the shore security guards are employed for any of the above reasons, in order to assist the ship's crew, their presence on board does not relieve the ship's crew from their
. ISPS responsibilities.
The negligence of their obligations could potentially lead to heavy fines from the Port Authorities or Port State Control surveyors.

The "Gangway watchmen" clause, which, among others, should include:
-for whose account will the ordering of gangway watchmen be.

An executed clause that has been run in a Charter Party is as follows (precise copy):
Watchmen, if any, to be for the account of the party ordering same. If compulsory, always to be for Charterers account.
Comments:
The clause states that the party ordering the watchmen for their various/own reasons, will bear the costs incurred, but, when it is compulsory, then the costs will be for the Charterers' account.

Another clause reads as follows (precise copy):
Expenses for gangway watchmen, if ordered by the vessel to be for Owners' account, but if ordered by Charterers or required by port Authorities/Regulations, except in the case such watchmen are necessitated due to nationality or conduct of any ship's members, such expenses shall be for Charterers account.
Comments:
The clause states that in the case that the Gangway Watchmen were required

by the Charterers or Port Authorities/regulations and they are in effect compulsory as per the local regulations, then they will be for the Charterers' account. However, if they were ordered by the ship's Master or employed due to the ship's nationality or conduct of any of the ship's members, then those expenses will be for the Owners' account.

A complete clause for Time Charter period, which refers to the "Gangway watchmen" could be worded along the following lines:

It is expressly agreed that the costs incurred for the employment of shore gangway watchmen and ordered:
- by the ship, for the ship's and/or crew property and/or protection etc. will be for the Owners' account.
- by the Charterers for the cargo and/or the ship, will be for the Charterers' account.
- as per the local regulation, which is in effect compulsory and/or were ordered by the Port Authorities for any reason, such costs will be for the Charterers' account, except in the case that such watchmen are necessitated, due to nationality or conduct of any of the ship's members, in which case they will be for the Owners' account.

40. War and extra war risk insurance.

The Charter period can be terminated in the case that certain major countries, as defined during negotiations, or in the area where the ship is scheduled to trade, are engaged in a war.
A war between major countries could affect the global economy, with the consequent cancellation of many contracts bringing the Charterers who were speculating in the market with their fixed ships, in a weak position to find suitable cargoes and to fulfil their contractual obligations.
For ships which have been fixed to perform within a standard trade between certain countries which have no direct effect of the war, they could definitely try to maintain the fixture.

An executed clause that has been run in a Charter Party is as follows (precise copy):
In the event of any of the following countries, or the country of vessel's flag ,Japan, U.S.A., U.K., Germany, Italy, France, China, Russia, and Australia becoming engaged as an actual participant in a war or warlike operation with one or more of the other named countries and the war or warlike operation shall last continuously for a period of at least six months and at that time be of an unpredicta-

ble duration, Charterers and/or Owners shall have the right to cancel this Charter Party without liability. If the Charter Party is cancelled, such cancellation shall take place at the port of destination after discharge of any cargo on board. It is understood that a war or warlike operation means direct war or warlike operation between these countries and does not include local hostilities or civil war where any of the above countries supports opposing side.

Comments:

The clause states the countries between which the war could terminate the contract. If the war continues for a period of at least six months, the ship can be redelivered to the Owners without liability at the port of discharge and after having discharged the cargo. Local hostilities or civil war cannot terminate the Charter Party.

Another clause reads as follows (precise copy):

If war breaks out between any two or more of the following countries:

United Kingdom, U.S.A., Russia, People's Republic of China, France, Norway, Japan, directly affecting the performance of this Charter, both Owners and Charterers shall have the option of cancelling this Charter without liability whereupon Charterers shall redeliver the vessel to Owners. If she has cargo on board after discharge thereof at destination or, if debarred from reaching or entering it, at a near, open and safe port as directed by Charterers, or if she has no cargo on board, at a port which she stays or if at sea at a near and safe port as directed by the Charterers. When the port where the vessel stays could not be regarded as a safe port because of the major war, then the vessel not to be redelivered until she enters into the safe zone outside the port. In all cases, hire shall be paid until vessel's redelivery dolsps.

Comments:

The Charter Party will be cancelled if one of the contractual parties declares the termination due to war between the defined countries. It also states where and when the ship will be redelivered to the Owners.

War Risk Insurance

The ship's Hull&Machinery insurance excludes any cover in case of damage or losses caused due to wars/civil wars/revolutions/riots/terrorist attacks/piracy etc. The Owners/Managers should arrange for the ship's coverage of all these risks through the recognised

. War Risk insurance companies.

A basic war risk annual premium is payable to them, provided that the ship does not trade within high risk areas. However, if the Charterers are entitled from the Charter Party to order the ship to trade at ports/places/countries/zones or areas, where there is a war/war like operations/piracy etc. or are generally consid-

ered high risk areas, as listed from time to time by the War Risk insurance companies, then extra cover is required, and certain additional premiums (extra war risk insurance) are in effect and will be imposed on the ship. The Owners will pay for those additional premiums (under the terms of the Charter Party the Charterers will reimburse the Owners for the cost in full), which vary depending on the specific area/country in which the ship is trading and of the prevailing situation at the time of call.
A 48-hour notice must be given to insurances prior the ship's entry in these high-risk areas with the coverage renewed every 7 days.

The "war and extra war risk insurance" clause, which, among others, should include:
-the countries between which the war can terminate the Charter Party.
-when the Owners or Charterers have the right to declare the termination of the contract.
-where the ship will be redelivered in case of war.
-that the basic war risk insurance premiums are for the Owners' account.
-if the ship trades in war risk areas, who will cover the cost.
-when the premium is payable to the Owners.

An executed clause that has been run in a Charter Party is as follows (precise copy):
Basic war risk insurance premium for world-wide trading shall be for Owners' account but additional premia –if any- for hull and machinery, Officers/crew and official crew war bonus as well as blocking and trapping and loss of hire due to the vessel's trading to a restricted area, will be for Charterers account. Payment of such premia to be effected by Charterers upon receipt of relevant debit notes issued by Owners' insurance brokers. The order of Owners' war risks underwriters should be followed. Conwar clause to be part of this Charter Party. wording as attached.
Comments:
The basic war risk insurance is for the Owners' account. All additional premiums for H&M, crew, crew war bonuses etc. are for the Charterers' account, payable on presentation of relevant debit notes issued by the Owners' insurance brokers.

Another clause reads as follows (precise copy):
If some of the intended trade areas are subject to extra insurance Charterers shall reimburse Owners for any such additional insurance Hull and Machinery premia, not exceeding the Lloyd's of London market.
The amount of such premia is payable by Charterers to Owners, separately from

hire payments, immediately after receipt of Owners, invoice on fax enclosing quotation received from the underwriters. Owners shall send the original of such documents/underwriters invoice to Charterers once in hand and any difference to be adjusted accordingly. Charterers will be given notice of such expenses as soon as practicable (it is understood that customary practice nowadays is that the underwriters give 7 days coverage firm quote 48 hours before vessel's arrival in extra war risk insurance area and after 7 days it must be renewed should the vessel not have left the war risk insurance applicable area by then.) It is agreed by Charterers that the Charterers shall be liable to pay the extra war risk insurance amount to Owners to such time until vessel leaves the extra war risk insurance applicable area, even if she is redelivered within the area, basis a duration calculation by redelivery surveyor.
Charterers failure in due payment of extra war risk insurance shall be deemed equivalent to failure in payment of hire.
Comments:
The Charterers will cover the additional war risk insurance premium for Hull and Machinery provided that the total cost does not exceed the Lloyd's London market. The Charterers should be given notice of such expenses and the cost is payable upon receipt of the Owners' invoice even by fax. The original invoice will be sent to the Charterers at a later stage and if there are any differences, they will be adjusted accordingly. The payment of the premium is equivalent to payment of hire.

A complete clause for Time Charter period, which refers to the "War and extra war risk insurance" could be worded along the following lines:

If war breaks out between two or more of the following countries:
...., ..., ..., ..., ..., ..., directly affecting the performance of this Charter and/or the Charterers' or Owners' ability to perform their obligations under this Charter Party, both parties, after having given at least 10 days written notice, will have the right to cancel this Charter, without liability to the other party.
If the ship is empty of cargo, it will be redelivered on dlosps at a safe port, place or at sea. If the ship has cargo on board, it will be redelivered on dlosps at the port, after completion of discharge. If the country of discharge has been engaged in a war, the Charterers have the right to change destination and discharge in a safe port, place, where the ship will be redelivered on dlosps.
The ship will proceed to load, navigate and discharge always without prejudice to the Charter Party war risk and piracy clause.
In all cases, the ship will be redelivered with quantity of fuel as on delivery and the Charterers are obliged to route the ship to a safe place for bunkering purposes, in which case the redelivery will take place on dlosps after the supply of

the bunkers.
It is understood that a war or act of war means an immediate operation of war or armed conflicts between these countries and does not include local hostilities or civil war, where any of these countries support opposing sides. The Owners or Charterers will not unreasonably take advantage of this clause, in case of a limited local conflict.
The basic War Risk insurance premium for the ship's world-wide trading will be for the Owners' account. All additional premiums, for the ship's Hull and Machinery, Officers, crew and official crew war bonus, blocking and trapping and loss of hire, because of the ship's trading to an area designated by the insurance companies as restricted or high risk, will be entirely for the Charterers' account.
The ship's trading to restricted or high-risk areas is without prejudice to Charter Party terms and conditions and signed Bills of Lading and always after the Owners' consent. The orders and recommendations of the Owners' war risk underwriters will always be followed.
Additional premiums are payable by the Charterers to the Owners on request or with the next due hire, on receipt of the relevant vouchers, as issued by the Owners' insurance brokers, which are acceptable even by fax or e-mail to the Charterers. The original vouchers will be sent at a later stage and, if there are differences, they should be adjusted accordingly.
The extra war risk premium will not exceed the Lloyd's London rate.
After the Charterers' payment of all additional premiums for the ship's trading to restricted areas, the Owners/Owners' insurers do not have the right to make any claim against the Charterers for loss, damage or delay relating to or resulting from any war risk, including blocking and trapping. The ship will be placed off-hire, if it is blocked or trapped or detained, because of these war risks.
For redelivery purposes, the ship will remain covered by the Charterers until it sails to the open sea and outside of the restricted area, even if the ship is actually redelivered on dlosps, unless it is delayed for their own reasons or they intend to load from the same port or from a port/s within the restricted area.
Charterers' failure of payment of extra war risk insurance will be deemed equivalent to failure of payment of hire.

41. Requisition.

It is a non-top clause. However, and due to the politically unstable situation worldwide, it should be included in all Charter Parties and be in use if and when necessary.
During the execution of the Charter Party and because of
. war or for any other emergency reason,
the ship could be requisitioned by the Government of its flag.

The Owners
. will lose control of the ship for a long or short time,
in which case there should be a provision in the Charter Party, entitling the Charterers to cancel the balance period or the conditions under which they will maintain the same.
In fact, the requisition will place the Owners and Charterers in an unpleasant situation and both will have a serious impact on their investments/program and development of their companies.

The "requisition" clause, which, among others, should include:
-that the ship will be placed off-hire during the period of requisition.
-whether the Charterers will have the right to cancel the balance period and when such a right should be declared.
-with what quantity of fuel, the ship will be redelivered to the Owners or with what quantity of fuel at the time of which the ship will again be in commission.

An executed clause that has been run in a Charter Party is as follows (precise copy):
Should the vessel be requisitioned by the Government of the vessel's flag during the period of the Charter, the vessel shall be deemed to be off-hire during the period of such requisition and any hire paid by the said government in respect of such requisition period shall be retained by Owners. In the event of such requisition, Charterers shall have the option to cancel the balance period of this Charter.
Comments:
In case the ship is requisitioned by the Government of the ship's flag, the Charterers have the option to cancel the balance period of the Charter.

Another clause reads as follows (precise copy):
Should the vessel be requisitioned by any Government or Governmental authority during the currency of this Charter Party, vessel shall be put off-hire hereunder for the period of such requisition and any hire or other compensation receivable for such requisition shall be for Owners account. If the period of requisition is, or is estimated to be, more than 15 consecutive days, Charterers have the option of cancelling this Charter Party.
Comments:
It declares that the Charterers have the option to cancel the Charter Party if the requisition is or estimated to be more than 15 days. It additionally declares that any hire or other compensation received for requisition will be for the Owners' account.

A complete clause for Time Charter period, which refers to the "Requisition" could be worded along the following lines:

If the ship is requisitioned by any Government and/or Governmental Authority, during the currency of this Charter Party, it is expressly agreed that:
1. the ship will be placed off-hire for the whole period of such requisition.
2. any hire or other compensation receivable for such requisition from any party will be for the Owners' benefit.
3. if the period of requisition is, or estimated to be, more than 15 consecutive days, the Charterers have the right to cancel the balance period of this Charter Party.
Such an option will be declarable as soon as possible and before the 15 days have expired.
4. the ship will be redelivered with fuel on board as on the day of requisition. Differences on the bunker's value and/or hire paid in advance and not earned will be settled together with the final hire statement.
If the Charter Party is maintained:
a. the Charterers will accept the ship's position as a point of delivery back to them and the ship will be on-hire again, upon the Master's tendering of the relevant delivery notice
b. the Master of the ship will endeavour to tender 5/4/3/2/1 delivery notices to the Charterers, stating the estimated date of the ship's readiness and place.
c. the ship will be delivered back to the Charterers with quantity of fuel as on the day of requisition or with enough quantity of fuel, until the next available bunkering station. The deviation time and the fuel consumed, enabling the ship to be supplied with fuel, in order to meet the quantity of fuel as on the date of requisition, will be for the Owners' account.
d. the ship will continue to serve the Charterers for the balance period at the same hire rate.
e. any time for which the ship is off-hire during its requisition will be added to the maximum allowed period of this Charter, provided that any such extension does not exceed the validity deadline of the ship's next Special Survey, Dry Dock, Intermediate Survey.

42. Oil pollution.

Along with other marine liability risks, the ship is insured for oil and other types of pollution risks. The
. supply of fuel to the ships constitutes the majority of pollution incidents.
If the ship pollutes the pier/sea etc. the consequences
. are severe

and the ship could be arrested by the Port Authorities and sustain imposed fines and clean-up/damage costs etc.
An acceptable form of security from the Port Authorities could be provided by the Protection and Indemnity Associations (P&I Clubs). The International Group of P&I Clubs in consultation with BIMCO has drafted a recommended clause for inclusion in the Charter Parties, which addresses the specific situation of criminal fines and civil penalties.
Some countries require additional insurance for possible pollution by ships when they call to their territorial waters for any purpose. All ships over 300 gross of any size are not allowed to operate in U.S. waters, unless they comply with the COFR (certificate of financial responsibility) regulations. This certificate covers the clean-up and damage costs to the limits of liability required by the Oil Pollution Act.

The "Oil pollution" clause, which, among others, should include:
- the Owners' insurance obligations which should be covered on the ship's delivery to the Charterers.
-liabilities for the contractual parties if pollution is caused.

An executed clause that has been run in a Charter Party is as follows (precise copy):
Vessel must comply with all requirements for calling .... including Oil Pollution conditions.
Charterers shall bear no responsibility for all consequences (including fines if any imposed to Charterers) or oil or other pollution damage and any time lost due to pollution of oil or its consequences shall be deemed off-hire unless Owners prove that the oil pollution is caused by Charterers representatives/Charterers agents/ Charterers servants.
Comments:
The ship must comply with all requirements for calling the intended countries, including oil pollution conditions.
Unless the oil pollution was caused by the Charterers' servants/Agents all other damages/time lost will be for the Owners' account.

Another clause reads as follows (precise copy):
a. If the Owners are required to establish or maintain financial security or responsibility in respect of oil or other pollution damage to enable vessel lawfully to enter, remain in or leave any port, place, territorial or contiguous waters, of any country or state in performance of this Charter Party, the Owners shall make all arrangements by bond or otherwise as may be necessitated to satisfy such requirements at the Owners' sole expense,

b. The Charterers shall not be responsible for any consequences (including loss of time) as a result of oil or other pollution damage or failure or inability of the Owners to make provisions as specified in Paragraph a. Any loss of time incurred thereby shall be considered off-hire.
c. The Owners shall indemnify and hold the Charterers harmless against all consequences (including fines if any imposed on the Charterers) of oil or other pollution damage except in case such pollution is caused by the Charterers and any failure or inability of the Owners to do so as provided for in the preceding paragraph a. particular reference is made to a Certificate of Financial Responsibility in compliance with requirements of the U.S. Water Quality Improvement Act of 1970 and amendments thereto including OPA 90.
Comments:
The Owners are responsible for the ship being in all respects covered for oil or other pollution damage to enable it to lawfully enter/remain/leave any port within the Charter Party trading limits.

A complete clause for Time Charter period, which refers to the "Oil pollution" could be worded along the following lines:

Owners guarantee that, during the currency of this Time Charter Party, the ship will comply with any international and/or government and/or state and/or provincial regulations, for oil or other type of pollution damage, risk and provide, at their expense, and maintain the respective certificates of financial security or responsibility in connection with oil or other pollution damage, risk, enabling the ship to lawfully trade within, to and from ports, ranges, areas specified in this Charter Party, including, but not limited to, trading in the U.S.A.
Actual time lost, cost, expenses, fines, consequences, which may arise due to the Owners' failure to comply with these provisions, will be for the Owners' account. The 'International Group of P&I Clubs' oil pollution indemnity clause for penalties and fines' will be in effect in this Charter Party.

43. Marine insurance / Return of premium.

The ship's Marine Insurance includes three main types of coverage:
-H&M insurance (hull and machinery).
-P&I Club insurance (protection and indemnity).
-War risk insurance.

The H&M insurance will cover the ship from
. marine perils/its total loss (actual or constructive) and for damage incurred to the ship's structure and machinery and those that are connected therewith,

either by collision or by any other way that the ship can be damaged such as storms etc. and will contribute to the general average and pay for the salvage expenses.
If the ship remains in port for more than 30 days, the Charterers, through this clause, will ask to benefit from any returns of insurance premiums by the ship's H&M underwriters. This is subject to such a provision in the Owners' contract with their underwriters, but, due to certain additional fees charged, the majority of the Owners avoid such agreements.
The P&I Club (Protection & Indemnity) provides coverage to their members for a wide range of liabilities to third parties, such as
. cargo claim, oil pollution, wreck removal, injury, or sickness of crew etc.
The liabilities for cargo claim to third parties will be apportioned between the Owners and the Time Charterers in accordance with the Inter-Club New York Produce Exchange Agreement clause, thus both Owners and Charterers must guarantee in this clause that they are fully covered by a P&I Club's rules, preferably belonging to an International Group of P&I Clubs. The International Group of P&I Clubs with its thirteen members (Groups) covers more than the 90 percent of the world's trading ships.
The thirteen Groups are:
-American Steamship Owners Mutual Protection and Indemnity Association, Inc. -Assuranceforeningen Skuld.
-The Britannia Steam Ship Insurance Association Limited.
-The Japan Ship Owners' Mutual Protection & Indemnity Association.
-The London Steam-Ship Owners' Mutual Insurance Association Limited.
-The Swedish Club.
-The Steamship Mutual Underwriting Association (Bermuda) Limited.
-The North of England Protecting & Indemnity Association Limited.
-The Standard Club Ltd.
-Gard P&I (Bermuda) Ltd.
-The Ship Owners' Mutual Protection & Indemnity Association (Luxembourg).
-United Kingdom Mutual Steam Ship Assurance Association (Bermuda) Limited.
-The West of England Ship Owners Mutual Insurance Association (Luxembourg).

The "Marine insurance/Return of premium" clause, which among others, should include:
-the Owners' guarantee that the ship is and will remain covered for H&M and P&I Club risks.
-the Charterers' guarantee that they are covered for P&I Club risks.
-if Charterers will have the benefit of any return of insurance premiums.

An executed clause that has been run in a Charter Party is as follows (precise

copy):
Notwithstanding anything contained in this Charter Party to the contrary, it is expressly agreed that the Owners remain responsible for all personal injury and claims which might be due to their fault or fault of the vessel to the extent of a full ship-owners' P and I Club cover and the Owners guarantee to maintain such a P and I cover for the currency of this Charter Party.
Charterers shall have the benefit of any return insurance premium receivable by Owners from their Underwriters by reason of the vessel being in port for a minimum period of 30 days if on full time for this period or pro rata for the time actually on hire.
Comments:
The Owners guarantee that the ship will be covered by the P&I Club insurance during the currency of this Charter Party. Some Charterers, before the fixture is concluded, request from the ship's P&I Club to confirm directly to them that the ship is fully covered for the duration of the trip and that fees due to them have been paid by the Owners.
The Charterers will have the benefit of any return of insurance premiums if the ship has been in port for more than 30 days.

A complete clause for Time Charter period, which refers to the "Marine insurance-Return of premium" could be worded along the following lines:

Owners guarantee that, the ship is and will remain entered and covered for marine risks and liabilities to third parties, during the whole duration of this Charter.
The Owners guarantee that the ship is fully covered by a P&I Club, which is a member of the International Group of P&I Clubs for cargo claims, oil pollution risks and wreck removal risks and will remain so throughout the duration of this Charter.
Charterers guarantee that, they are and will remain covered by a P&I Club which is a member of the International Group of P&I Clubs, throughout the duration of this Charter.
Notwithstanding anything contained in this Charter Party to the contrary, it is expressly agreed that the Owners remain responsible for all personal injuries and claims, which may be due to their fault and/or the ship's fault, to the extent of a full Owners' P&I Club coverage.
The Owners guarantee that the ship is fully covered for Hull and Machinery risks and entered with Messrs. ...... for a value of USD...... Charterers will have the benefit of any return of insurance premiums receivable by the Owners from their Underwriters, as and when received from the Underwriters, because the ship is in port for a minimum period of 30 days, if on full time for this period or pro rata, for the actual time on- hire.

44. Exports/Imports/Government approval of cargo.

Export and import licences/permits to the ports of contact should be entirely for the Charterers' risks and costs.

The ship should

. not be involved in this issue and the fixture should not be subject to import and export licences/ permits, which might give rise to delays/extra expenses.

The "Exports-Imports-Government approval of cargo" clause, which, among others, should include:

-that the export and import licences and Government permits/approvals will always be for the Charterers' risk/expense and time to grant the same.

-that the Charter Party will not be cancelled if the licences have not been granted.

A complete clause for Time Charter period, which refers to the "Exports /Imports/ Government approval of cargo" could be worded along the following lines:

Export and Import licences and Government permits and/or approvals of cargo, as well as trade and taxation and levies whatsoever applicable for these purposes, will always be for the Charterers' risk, expenses, and time to grant the same. The Owners and/or ship will not be asked to be involved and delays and/or expenses incurred therefrom, will be entirely for the Charterers' account. The ship will remain on-hire at all times and at any circumstances and the Charter Party will not be cancelled if the import or export licences or the Government permits and approvals be withdrawn for any reason and the Charterers will look for alternatives, but always in compliance with the Charter Party terms.

45. Fumigation.

Fumigation work could be ordered because of

. the illness of the ship's crew,

. the ship's previous trading,

. the cargo holds'/hatches' rejection, due to live insects on/in them,

. the Charterers' cargo loaded or intended to be loaded.

All fumigations should be performed as per IMO regulations/recommendations and the Owners/ Master should take all the required precautions for the ship's crew safety. The Port Authorities could request, during the fumigation, that the

ship's crew be away from the ship, in which case all lodging expenses should be agreed to be for the Charterers' account and the ship to remain on-hire.
In general, and as per P&I Club recommendations, the fumigators should provide written documentation in respect of the following:
-pre-fumigation inspection certificate.
-Standard safety recommendations for ships with fumigated grain cargoes.
-Manufacturers' information or safety data sheet.
-First aid and medical treatment instructions.
-Fumigation certificate.
-Fumigation plan.
-Statement of vessel suitability for fumigation and fumigant application compliance.
-Detailed Precautions and procedures during voyage and period of time during which the ventilation of holds is not allowed and the period of time for which the holds must be open, enabling the discharging operation to commence, for Crew's/Stevedores' safety, as well as all information required for the Crew safety during the voyage.
-Precautions and procedures during discharge.
The Owners should officially seek the ship's P&I Club's recommendations concerning the fumigation procedures or the way in which to clause the Bill of Lading, in the case that the cargo appears not to be in a sound condition, after the fumigation has been performed.

The "Fumigation" clause, which, among others, should include:
-for whose account, the ordered fumigation should be.
-the conditions under which the fumigation will be carried out.
-for whose account, the lost time/expenses will be.
-that the fumigation Authorities/fumigation company should provide full and detailed instructions in writing for the sea voyage to the Master as well as when the holds must be open, enabling the discharging operation to commence, for the Crew's/Stevedores' safety.

An executed clause that has been run in a Charter Party is as follows (precise copy):
Charterers have the liberty to effect the fumigation of cargo on board the vessel at their risk and expenses. In this respect, the Master to render his utmost assistance and co-operation together with crew including loading and victualling for crew's staying ashore while such fumigation is being carried out. Al expenses related to fumigation including accommodation and victualling of crew ashore, if required, to be for Charterers' account.
Comments:

The Charterers have the right to carry out fumigation on board the ship at their risk/expenses. The Master is required to render his utmost assistance and co-operation together with the crew. The Charterers will be responsible for all lodging expenses incurred.

Another clause reads as follows (precise copy):
If during the on-hire survey upon delivery, dead or live insects are discovered in vessel's cargo compartment, the costs for fumigation at port of delivery, vessel's detention, subsequent survey and issuing of fumigation certificate are for the Owners account. Otherwise, the vessel to be considered/accepted as fumigation-clean and if within the duration of Time Charter the vessel has to be fumigated on account of cargo loaded and ports visited under instructions of Charterers (including as per local rules/regulations at discharging ports in respect of origin of cargo to be discharged) the cost of fumigation and lodging expenses to be for Charterers account.
If cargo is fumigated after loading on request of Charterers and/or Shippers and/ or Receivers and vessel requested by fumigation company in writing stating number of days and/or time to elapse not to ventilate and not to enter fumigated cargo holds during sea voyage, then Owners not to be responsible for any cargo damage whatsoever as a result of fumigation of cargo and/or holds.
Comments:
The ship is responsible for fumigation ordered due to live insects discovered in the ship's cargo compartments during the on-hire survey. The ship will thereafter be considered fumigation-free and all expenses/lost time for subsequent fumigations ordered due to the ship's trading or due to previous cargoes loaded will be for the Charterers' account. The clause also states that in the case that the cargo is fumigated, the fumigation company will deliver written instructions to the Master of the ship for the number of days for which the ventilation of holds is not allowed and for the number of days for which the holds must be open before the commencement of the discharging operation.

A complete clause for Time Charter period, which refers to the "Fumigation" could be worded along the following lines:

Fumigation ordered because of and directly related to:
-live or dead insects found in the ship's holds or compartments at the time of delivery or at first port of loading after delivery or during the holds' inspection after delivery -Officers'/Crew's illness at the time of delivery -the ship's previous trade before the delivery
will be entirely for the Owners' account. The Owners remain responsible for all costs and extra expenses, including, but not limited to, port expenses, shifts,

lodging expenses, delays, detention, which may arise. The ship will be placed off-hire, from the time when the fumigation is ordered, until the fumigation certificate is granted, fully certifying that the ship is again in full operational condition and free to load its intended cargo.
However, for all other ordered fumigations or subsequent ordered fumigations by the Shippers and/or Charterers and/or by any other party and for any reason within the duration of this Charter, whether ordered or not, due and are related or not to: - live or dead insects found in the ship's holds or compartments -intended cargo to be loaded or already loaded -the ports that the ship visited under the instructions of the Charterers, with consequent illness of the Officers, Crew or not will be entirely for the Charterers' account. The Charterers remain responsible for all costs, extra expenses, delays, risks, detention, which may arise.
The ship will always remain on-hire and, if the Port Authorities or the Master deem it necessary for the Officers and crew to stay ashore, then all lodging expenses, including, but not limited to, expenses for transportation, hotel and meals will be entirely for the Charterers' account.
All fumigations will be performed as per IMO regulations and recommendations and under the Master's supervision with respect to crew safety.
If the cargo is fumigated after the completion of loading, then the fumigation Authorities and/or fumigation company will provide in writing:
- full and detailed instructions for the sea voyage, to the Master's satisfaction,
- it will state the period of time that the ventilation of holds is not allowed,
-when the holds must be opened, enabling the discharging operation to commence, for the Crew's and Stevedores' safety
and
- all other information required for the Crew's safety during the sea voyage and during the discharging operation.
The Master will record in the ship's log books all the operations and in the case that the cargo that was loaded appears not to be in a sound condition and has been damaged or contaminated after the performed fumigation, then the Bills of Lading will be issued strictly in accordance with the ship's P&I Club's recommendations.
The Charterers will be liable and responsible for any cargo damage and/or impact on the cargo quality whatsoever, as a result of fumigation of the cargo and/or cargo holds and will hold the Owners harmless and keep them indemnified against all claims whatsoever, which may arise and be made by any party against them.
If the ship is detained by any party and generally its sailing is not allowed from the port, because of any cargo damage and/or impact on the cargo quality whatsoever, as a result of fumigation of the cargo and/or cargo holds, the Char-

terers undertake the responsibility to release the ship, by taking all the necessary and appropriate measures (issuance of security, etc.), as the case may be or arise. The ship will always remain on-hire.

46. Hire/Hire payment/ Owners' bank details.

If the specific clause in the Main Body of the Charter Party which refers
. to the hire/hire payment/Owners' bankers etc. is deleted, then this clause should stipulate all the agreements made in this respect.
The fixed amount as 'hire rate' which is agreed upon during the negotiations of the Main Terms, should include the crew's/officers' overtime (abbreviated as 'inclot'), i.e., the Charterers should not be obliged to pay any additional/extra amount when the ship's officers/crew render their services out of their daily maximum normal time of work.
This clause should also include when the payment of the agreed
. ballast bonus is due for payment (it is usually payable together with the payment of the first 15-day hire).
For short/long periods, the standard procedure is for the hire to be due every 15 days in 'advance'.
Nevertheless, there might be an agreement for the first 15 day hire to be paid within 2/3 banking days after the ship's delivery, as the Charterers prefer to effect the payment after
. the ship's actual delivery to them and sometimes after the on-hire survey takes place.
However, for all the subsequent due (15 day) hires, the Charterers' obligation is to make the remittances before the running 15 days have been expired, i.e.
. in advance and before the period of the next 15 days commences, even if it exists during weekends/holidays, in which case the hire should be remitted on a working day before the commencement of a weekend/holiday.
For very short trips, the first hire payment could be adjusted according to the expected maximum duration of the trip, i.e., if the expected period is 8 or even 20 days, then the hire could be paid for the entire expected period and if any balance period remains, it could be payable day by day or so.
For the last hire, which is much less than 15 days for payment, a day-by-day payment could be agreed or every two/three days or so.

The Owners' bank details must be given in full and must include the correspondent banks etc. The Charterers must protect their interests and declare that, unless full details are given, they will bear no responsibility for any payment delays.

The "Hire/Hire payment/Owners' bank details" clause, which, among others, should include:
-the amount of daily hire or pro rata and what services it includes.
-when the ballast bonus should be payable.
-when the first hire/subsequent hires and the last hire should be due for payment.
-the Owners' bank details.
- for whose account, the bank charges should be.

An executed clause that has been run in a Charter Party is as follows (precise copy):
Owners' bankers as follows:
Notwithstanding anything contained herein to the contrary, if at any time during the currency of this charter party hire shall become due on or during a Saturday, Sunday or a national holiday or outside normal office hours, or at any time which for reasons beyond their reasonable control prevent charterers from effecting payment of hire on the due date, payment of hire may be made on the next banking day immediately following the date on which hire becomes due.
Comments:
The Owners' bank details are given, and it additionally states when the hire will be paid, if it is due during weekends or holidays.

Another clause reads as follows (precise copy):
Hire is to be USD....... per day or pro rata including overtime payable every 15 days in advance.
Owners to provide by fax invoice on their headed paper to the Charterers covering each Charterers' remittance but such fax invoices not to be considered a condition required for hire or other payments and on no occasion are such payments to be delayed by charterers on this account when they are due.
First hire and agreed estimated consumable bunkers for the whole voyage to be paid to Owners' account within 2 banking days after delivery.
Should the vessel be on her voyage towards port of redelivery at a time a payment of hire becomes due, said payment shall be made for such length of time as the Charterers or their agents estimate as the time necessary to complete the voyage, always based on agents discharging/completion prospects less disbursements arranged by Charterers for Owners account and other deductions allowed under this Charter, and when the vessel is redelivered to Owners any difference shall be refunded by Owners or shall be paid by the Charterers as the case may require. Final hire payment will be made according to actual time of redelivery.
Owners bank account as follows: ....

Comments:
The agreed amount of hire stated in this clause and when the hire shall become due. Additionally, the Charterers request the presentation of the Owners' invoice that will cover each of their remittance, which is to be issued in their headed paper and be faxed to them. However, it specifies that the payment of hire is not subject to invoice.
The first hire and the agreed consumable bunkers are payable within 2 banking days after the ship's delivery.
The last payment of hire will be made for such length of time as the Charterers and/or their Agents may agree upon as the estimated time which is required for completing the final voyage. The Owners do not have the right to object, if they foresee through their sources, that the ship will delay and that the Charterers' expectations are inaccurate. From the last hire payment, the Charterers are entitled to deduct any of the Owners' disbursements arranged and paid by the Charterers on the Owners' behalf or for any deductions which are entitled through the terms of the Charter Party.

A complete clause for Time Charter period, which refers to the "Hire/Hire payment/ Owners' bank details" could be worded along the following lines:

It is expressly agreed:
1. hire USD.... per day/pro rata, including Master's, Officers', Crew's overtime and lubricants. The hire is payable to the Owners' nominated bank account in United States currency, every 15 days, always in advance, except for the first hire payment of 15 days, which is payable within .... London banking days, after the ship's delivery to the Charterers' full services.
2. that the amount of hires will be remitted to the Owners' bank account, net of tax and bank charges, which will be for the Charterers' account. The hire/s is to be remitted less any deductions agreed and allowed under this Charter Party.
3. that if the hire becomes due during weekends or Holidays, the payment of hire will be arranged and be made by the Charterers, during a banking day before the commencement of the weekend or holidays.
4. that if the ship is towards the redelivery port and the hire payment becomes due, said payment will be made for such length of time as the Charterers and Owners will mutually agree, according to the information given by the attending Agents for completion of discharge. However, if the ship further delays, the hire will be payable every two days or day by day, as it becomes due, always in advance.
5. that the payment of hires and all amounts due to the Owners under this Charter Party, to be made to the Owners' bank account as follows:
....

....
....

-If a ballast bonus is agreed the payment could be worded as follows:

Gross (or net of commissions as should be agreed) *Ballast Bonus of USD ...., is payable together with the first 15-day hire payment, net of tax and bank charges.*

47. Final settlement.

In most of the Charter Parties this clause should be omitted,
. as reasonable time should be given to the Charterers to present their final hire settlement (abbreviated as 'fhs') with the possible due balance hire to the Owners.
The Charterers usually avoid paying hire for the last day/s (in breach of Charter Party terms) and withhold this amount in order to cover any unexpected expenses, which have arisen during the Charter or may arise for the Owners' account or even to cover the ship's possible underperformance etc.

The "Final settlement" clause, which, among others, should include:
-when the Charterers' preliminary final hire statement should be presented for the Owners' consideration.
-when the Charterers' final hire statement should be presented for the Owners' confirmation.

A complete clause for Time Charter period, which refers to the "Final settlement" could be worded along the following lines:

Charterers are to present their preliminary final hire statement within 5 working days after the ship's redelivery for the Owners' perusal, comments and any agreed amount or undisputed amount will be refunded within 2 banking days by the Owners or will be paid by the Charterers, as the case may require. Charterers are to produce their final hire statement within one month after the ship's redelivery and provided that all original vouchers of the Owners' possible port expenses have been collected and submitted for the Owners' approval.

48. Delays/Off-hire.

The ship, during the execution of the Charter, might be delayed for various reasons, attributed or not to the ship's or Owners' or even due to the Charterers' or their Agents' fault/default. Obviously, if it is due to the Owners' or the ship's fault/ default/defect, the ship will be placed

. off-hire and the Charterers should have the right, under the terms of the Charter Party, to add to the end of the allowed Charter period any lost time or, if it is heavily delayed, to even

. cancel the balance period.

For any other delays which have not be caused due to the ship's or Owners' fault/ default, the Charter Party should state that the ship will remain on-hire and no additional period should be added to the end of the Charter, without the Owners' prior consent.

Delays could be caused due to technical reasons, collision, grounding, gear breakdown, pollution, delays to disembark stowaways, crew illness, crew strikes, etc.

The "Delays/Off -Hire" clause, which, among others, should include:

-for how many days the ship may delay for serious reasons without being cancelled.

-when the Charterers have the right to declare the cancelling of the Charter.

An executed clause that has been run in a Charter Party is as follows (precise copy):

If the vessel will be off-hire for more than 30 days continuously, the Charterers have the option of cancelling this Charter provided no cargo on board. The Charterers have the option of adding the times lost by vessel being put off-hire to Charter period.

Comments:

A simple clause which states that in the case the ship is off-hire for more than 30 consecutive days, the Charterers have the option to cancel the balance period provided that there is no cargo on board. The period for which the ship was off-hire may be added to the Charter period.

Another clause reads as follows (precise copy):

In the event of any delay to the vessel arising from one or more of the following reasons:

1. strike of the crew and/or deficiency of officers/crew or stores,

2. deviation from lading invalid crew and/or stowaway and from salvage,

3. breakdown partial or otherwise, damage to hull, machinery, equipment,

4. detention by average incidents to ship/cargo including collisions/stranding,

5. repairs/dry docking or other necessary measures to maintain the efficiency of the vessel,

6. fail lure to be in possession of valid certificates or other documentations for

the vessel/officers/crew,
7. breakdown partial or otherwise, or non-conformity or disablement at one or any/all winches/cranes or any other part of vessel's equipment for loading/discharging cargo preventing full operation of vessel,
8. insufficient power to operate all winches/cranes at the same time if required,
9. cargo residues,
10. failure to pass free pratique at loadport due to crew members and/or rust scale and/or loose rust in holds, accommodation any other structures and provisions etc ...,
11. in the event of detention of proved oil pollution, unless pollution arose by reason of Charterers' or their servants negligence,
the vessel to be off-hire and/all-time thereby lost and Owners to pay all extra expenses incurred together with the cost of any fuel/gas/diesel oil consumed in consequence thereof, including but not limited to stevedore standby time, if any, and which shall all be deducted from the hire.
Comments:
The specific reasons for which the ship could be placed off-hire have been stated, as well as that all expenses incurred, including fuel consumed, will be for the Owners' account.

A complete clause for Time Charter period, which refers to the "Delays/Off – Hire" could be worded along the following lines:

Because of one or more of the undermentioned reasons, and provided that it has not been caused due to the Charterers' act or omission or neglect or fault or default and/or the Charterers' Agents or Servants and/or the Charterers' person/s travelling on the ship and/or sub-Charterers and their Agents or Servants and/or the Contractors and/or the cargo Shippers or Consignees and/or of the cargo carried and ports contacted under this Charter Party, the ship is delayed, then it will be placed off-hire for any time lost and will be held responsible for the fuel consumed and for any directly related expenses incurred, which will be supported with original vouchers and consequences arising therefore.
If the ship:
-is no longer able to trade at the intended places, ports as defined by the Charterers and allowed by the Charter Party,
-is delayed and remains inoperative and has been placed off-hire for more than 15 consecutive days, unless otherwise stated in the Charter Party for each specific case, for one or more of the following reasons:
a- it is failed to be in possession of all necessary valid and up-to-date equipment and certificates in compliance with health and safety regulations, national and international regulations and all current and/or local requirements, including

valid certificates for passing the Panama/Suez canals and several straits.
b- the Master, Officers and Crew and/or any Owners' person travelling on the ship is captured, seized, detained, or arrested by any Authority or any legal process, during the period of this Charter Party, at the suit of any person having or purporting to have any claim against or interest in the ship and/or said Master, Officers and Crew and/or the person travelling on the ship.
c- because of its previous trade and/or because of its flag, ownership, class, registry, nationality and/or the terms and conditions under which the officers and crew are employed and/or to the construction, cargo gear, fitting, other equipment and/or any other ship under the same ownership and/or management and/ or control:
- it is boycotted, picketed, blacklisted,
-it is not allowed to enter, load and discharge or leave any port within the trading limits of this Charter Party,
-it is rendered inoperative or immobilized or detained by Governments and/or Authorities and/or shore and/or port labourers and/or tugboats and/or linesmen and/or pilots and/or suppliers and/or the ship cannot load and/or discharge at the intended place, port and is ordered to load or discharge in another place, port,
d- it is inoperative from default and/or strike and/or deficiency of the Master, Officers, crew or stores,
e- it is fired and/or breakdown damage to hull and/or machinery and equipment and/or stranding and/or grounding and/or collision,
f- it is under repairs or other necessary measures, in order to maintain its efficiency,
g- it is inoperative, due to winches' and/or cranes' partial or other breakdown, and/or due to non-conformity and/or disablement of one or any or all winches and/or cranes and/or any other part of the ship's equipment, which prevents the ship's full operation and/or there is insufficient power to operate all winches and/ or cranes at the same time, if it is required,
h- it is failed to pass free pratique due to pestilence, epidemics and illness of the Captain, Officers and Crew,
i- it is detained by the port state control,
j- it is detained due to proven oil pollution,
k- it is detained due to desertion of crew or all crew not being on board when the ship is ready to sail,
l- it is unable to pass the initial holds' and/or hatch covers' inspection due to rust scales and/or loose rust in its holds,
m- it is underperforming and generally does not comply with the Charter Party description, or, for any other reason, including deviation, the ship is not to the Charterers' full service and instructions,

the Charterers then will declare whether they will maintain or cancel the fixture for the remaining balance period.
If the ship will be cancelled (unless otherwise stated in the Charter Party for each specific case):
1. a written notice will be given to the Owners while the ship remains so detained, for redelivery on dlosps of this port and if cargo is on board, the ship will be redelivered when it is practical and the ship being free of cargo.
2. the Owners must refund the Charterers for any hire plus bunkers that have been paid in advance and not earned.
3. the ship will be redelivered with quantity of fuel as on board and the differences will be settled together with the final hire payment.
4. the Owners will pay for all extra expenses incurred, but supported with original vouchers, including, but not limited to, stevedore standby time, but up to a maximum of one shift.
5. the redelivery of the ship will take place without prejudice to any other right the Charterers may have on ship.

If the Charterers maintain the fixture (unless otherwise stated in the Charter Party for each specific case):
1. the Charterers have the option to add to the maximum allowed Charter period, any off-hire period/s partly or wholly, provided that any such extension does not exceed the validity deadline of the ship's next special survey/dry dock. Any such option will be declared to the Owners a minimum of 20 days before the expiration of the minimum allowed Charter period.
2. the Charterers will trade the ship at the same rate and conditions for such extended time, as that applicable during the off-hire period.
3. the Owners will give the Charterers the full itinerary of the ship and the expected date of resumption of the service, every two days.
4. then, after resumption of the work, the ship will be placed at the Charterers disposal at the same port or place where the hire was suspended, within ordinary office hours, except at sea and with quantity of fuel as when the ship was placed off-hire. The Charterers may, however, in their option, accept the ship on hire again in such a position and at such a time as the ship may again in all respects be ready to comply with the Charterers' orders and directions.
This clause does not apply when the ship passes its normal survey commitments (special survey/Dry Dock etc.)

49. Periodical Dry Dock/Special Survey/Intermediate Survey.

The Owners should avoid fixing the ship for a long period, when a periodical Dry Dock or Special Survey or Intermediate survey will be conducted during the

Charter, without
. first agreeing on this particular clause.
The ship will be placed
. off-hire for a reasonable period, in order to enable the ship to meet its class commitments.
The area where the ship will perform its works should be clearly stated in this clause (Far East or within the Mediterranean or the Continent etc.) and from when it will be placed off-hire and when it will be placed in commission again.
The Owners should take note of when the surveys are due and agree that the Charterers should make every effort to bring the ship as close as possible to the area where it will pass its surveys. However, Charter Parties for single trips or short period Charters should explicitly state that the ship will not pass a Dry Dock or any other class commitments, during the execution of the Charter, except in case of emergency.
The Charterers, particularly when fixing ships, in order to cover their contractual obligations (not as much in effect when only fixing to speculate on the open market) should learn from the outset of negotiations when the dry dock/special surveys are due, as they must secure the continued smooth execution of the Charter and not have to search for a substitution ship later.

The "Periodical Dry Dock/Special Survey/Intermediate Survey" clause, which, among others, should include:
-that the Owners have the right to temporary withdraw the ship from the Charterers' services, in order to pass its regular survey.
-when and where the Owners intend to pass the surveys.
-the period during which the ship will be placed off-hire.
-the quantity of fuel at the time of resumption of services.

An executed clause that has been run in a Charter Party is as follows (precise copy):
Charterers to place vessel within Singapore/Japan range during summer 2014 to allow Owners to perform Dry Dock works and/or Special Survey and/or Intermediate Survey (Expected in PRC). Vessel is to go off-hire from DOP/Passing within the aforementioned range and to return on-hire after Dry Dock/Special Survey/ Intermediate survey in same or equidistance position in Charterers option as she was when the vessel went off-hire. All bunkers used whilst off-hire to remain for Owners account and be replenished prior the vessel be in commission. For any time the vessel is off-hire during the Charter period to be added to the period of this Charter provided that any such extension does not exceed the validity deadline of the vessel's next Special Survey/Dry Dock/Intermediate Survey.
Comments:

The clause states that the Charterers are obliged to bring the ship to a specific range within a specific time in order for the ship to pass its surveys. It additionally states the period during which the ship will be placed off-hire and the Owners' obligation to replenish the bunkers consumed. Any time for which the ship is off-hire will be added to the period of this Charter.

Another clause reads as follows (precise copy):
The Owners have the right of withdrawing the vessel, subject to mutual agreement between Owners and Charterers as to time and place, from the Charterers' service for a reasonable period to Dry-Dock vessel or to execute annual or maintenance repairs and/or survey.
However, Owners to notify Charterers of their intention to perform such periodical surveys with six months notice, unless is emergency and obtain Charterers' consent as to time and place. Charterers' consent shall not be withheld unreasonably.
Charterers will assist Owners in order to position the Vessel as close as possible to the Dry-Docking position. Charterers have the option of adding any time the vessel is off-hire to the Charter period. All bunkers consumed whilst vessel is off-hire to be for Owners account and vessel to be replenished and return on hire with about the same quantities of IFO and MDO on board as when she was placed off-hire.
Comments:
The Owners and Charterers will mutually agree as to the time and place where the ship will pass its surveys. The Owners must give notice to the Charterers about six months earlier, advising them of their intention and obtain the Charterers' consent as to the time and place. The Charterers will assist the Owners by fixing the ship as close as possible to the Dry Docking position. It is to the Charterers' option whether the off-hire periods will be added to the Charter period.

A complete clause for Time Charter period, which refers to the "Periodical Dry Dock/ Special Survey/Intermediate Survey" could be worded along the following lines:

Owners have the right, during the currency of this Charter and as should be required by the ship's class, to temporally withdraw the ship from the Charterers' services, for a Dry Dock and/or Special Survey and/or Intermediate Survey commitments.
Charterers agree to place the ship off-hire within the trading area, for a Dry Dock and Special survey during .... (month and year) to.... (month and year) and for an Intermediate Survey and Dry Dock during .... (month and year) to.......(month

and year).
The exact time and place where the ship will pass its surveys will be entirely at the Owners' option. However, the Charterers will be informed a minimum of four months earlier of the Owner's intention, which in any case should be before they fix the ship for its the next trip to the performing one, enabling them to bring the ship as close as possible to the area where the ship will pass its surveys. The Owners will inform the Charterers of the estimated time required to conclude the ship's commitments and will fully update them every 5 days on the ship's schedule and the anticipated date of its return to the Charterers' services. Also, they will tender 5/4/3/2/1 days definite notice thereto. The ship will be placed off-hire, as from the date and time it is placed under the Owners' control on dropping last outwards sea pilot station of the ship's last discharge port, before passing its surveys and until the date and time it is placed again in commission and under the Charterers' control, at the Charterers' option:
a. at the place where the ship was placed off hire,
b. at equidistance to the place where the ship was placed off-hire,
c. at the place and at such time as the ship may again be in all respects ready to comply with the Charterers' orders and directions,
d. on arrival first sea pilot station at a port of which the distance is less than the distance between where the ship was placed off-hire and the Owners' survey port and/or yard.
All fuel consumed while the ship is off hire will be for the Owners' account. The ship will be placed on-hire again with quantity of fuel as on the time in which it was placed off-hire or with enough fuel until the Charterers' next bunkering station, where the Owners will replenish the fuel consumed for their account.
Any time for which the ship is off hire while passing its surveys will be added to the maximum allowed period of this Charter, provided that any such extension does not exceed the validity deadline of the ship's next Special Survey, Dry Dock, Intermediate Survey.

-Or, in the case that the added period should be as an option:

The Charterers have the option of adding any time which the ship was off hire during passing its surveys to the maximum allowed period.

-Survey/Dry Dock in case of emergency:
This clause does not allow the ship to pass its routing survey commitment during the currency of the Charter, except in case of emergency. Additionally, the conditions under which the ship will be temporarily withdrawn from the Charterers' services should be stated in this clause.

The "Survey/Dry-Dock in case of emergency" clause, which, among others, should include:
-that the Owners do not have the right to program for a Dry Dock and/or Special Survey and /or Intermediate Survey during the currency of this Charter.
-the conditions under which the ship will be withdrawn from the Charterers' services.

An executed clause that has been run in a Charter Party is as follows (precise copy):
No routine Dry Docking during the currency of this Charter except in case of emergency.
In cases of unforeseen emergencies occurring, no such notice needs to be given to Charterers, but if required by Charterers, Owners to provide Charterers promptly with all required documentary evidence pertaining to the emergency occurrence, including but not limited to Classification society or their agents' reports, certificates, correspondence, repair yard documentation and/or invoices.
Comments:
The Owners do not have the right to pass the routine Dry Dock commitments during the execution of this Charter except in case of emergency. However, the Owners must provide the Charterers with evidence pertaining to the emergency occurrence.

Another clause reads as follows (precise copy):
No Dry Docking during the currency of this Charter Party except in case of emergency.
In case of emergency, Charterers will bring the vessel within the trading range for dry docking to a yard at their choice.
Vessel shall be off-hire from time of dropping last outwards sea pilot on vessel's last discharge port before Dry Docking or from time of arrival at or off the Dry Docking port or from arrival at Dry Dock Yard if vessel is loading, discharging or bunkering at that port.
Vessel to be placed back in Charterers' service at an equivalent point as when vessel was off-hire for Dry Dock.
Charterers have the option to add any off-hire period including Dry Dock period to the hire period.
Comments:
In case of emergency the Charterers will choose the yard for passing its Dry-Dock survey. Obviously, the Charterers will bring the ship to a yard with a very good track record in order for the works to be completed as soon as possible. It states as from when the ship will be placed off-hire and that the Charterers have the option of adding any time to the Charter for which the ship was run-

ning off-hire.

A complete clause for Time Charter period, which refers to the "Survey/Dry-Dock in case of emergency" could be worded along the following lines:

No routine Dry-Docking and/or Special Survey and/or Intermediate surveys will be performed during the currency of this Charter.
A Dry-Docking survey is allowed only in the case of emergency.
In this case, the Owners, at their choice, will bring the ship for dry docking to a yard located within the trading range, evidencing a good track record.
The ship will be placed off-hire, as from the date and time it is placed under the Owners' control on dropping last outwards sea pilot station of ship's last discharge port, before passing its surveys and until the date and time it is placed again in commission and under the Charterers' control, at the charterers' option:
a. at the place where the ship was placed off-hire.
b. at equidistance to the place where the ship was placed off-hire.
c. at the place and at such time as the ship may again in all respects be ready to comply with the Charterers' orders and directions.
d. on arrival first sea pilot station at a port of which the distance is less than the distance between which the ship was placed off-hire and the Owners' survey port and/or yard.
Owners are to advise the Charterers of the estimated time required to conclude the ship's survey/repairs and to keep them fully updated every day on the ship's schedule and anticipated date of return to the Charterers' services. Also, to tender 2/1 days definite notice thereto. All fuel consumed, while the ship is off hire, will be for the Owners' account. The ship will be placed on-hire again with quantity of fuel as on the time in which it was placed off-hire or with enough fuel until the Charterers' next bunkering station, where the Owners will replenish the fuel consumed for their account.
Any time for which the ship is off hire while passing its surveys will be added to the maximum allowed period of this Charter, provided that any such extension does not exceed the validity deadline of the ship's next Special Survey, Dry Dock, Intermediate Survey.
If any Dry-Docking or any other survey and/or repairs are required, which have resulted due to act or omission or neglect or default of the Charterers and/or their Agents or servants and/or their representatives or sub-Charterers, the Charterers will bring the ship to a yard at their choice evidencing a good track record and will bear the cost of the fuel consumed and all the yard's expenses incurred which are payable directly by them to the yard and to any other parties involved. The ship will remain on-hire under any and all circumstances.
All the surveys and repairs will be under the ship's class surveyor's supervision

and approval, as will be appointed by the Owners, whose fees are payable by the Charterers to Owners on request and on presentation of the supporting original documents.

50. Draft survey/ Plans.

At the load port, the Shippers
. are obliged to supply the loading figures and to guarantee those figures.
The same should be presented either by a carried out
. draft survey
or by the shore scale measurements.
At the discharge port, the Receivers are to determine the discharged cargo quantity
. by a carried out draft survey
or by shore measurements or even by weighting the trucks at the customs house.
Problems could arise at the load port, in the case that the shippers' figures exceed the ship's figures or the Receivers' figures are less than the ship's figures at the discharge port and both are beyond a normal and/or customary difference on the loaded and/or discharged cargo quantity.
At the load port, reservations which could be made on the Bill of Lading such as 'weight unknown' or 'quantity and quality unknown', 'said to be' etc. or to accept a Letter of Indemnity by the Shippers or Charterers, might not be enough to cover the Owners and the ship could be subject to the P&I Club rules.
However, if the cargo quantity is only going to be determined by a draft survey at both ends, it is common and for the avoidance of any arguments, for the Charterers'/Shippers' draft surveyors, to jointly conduct the draft survey with the Master/Chief Officer of the ship and/or with the Owners' appointed draft surveyors.
At the discharge port, the Owners are not obliged to accept the shore measurements or to be involved in the weighting of trucks. In all cases the Master should carry out his own draft survey in all ports of call and the ship's P&I Club's assistance and/or recommendations should be requested, especially in places and Charter Parties where the draft survey is not to be taken into consideration for the determination of the cargo quantity.
The ship must be in possession of all of the ship's plans, enabling the surveyors to carry out a correct draft survey. The Owners/Master should take care, before the ship's approach to the ports, that the plimsoll marks and draft marks on the port and starboard sides, bow, amid ship and stern are clearly cut and marked on the shell platting and are always readable.
For an accurate draft survey, the ship's tanks must be empty or fully pressed,

otherwise the Master should indicate their remaining ballast to the surveyors. During the time of the carried out draft survey, the bunkering operation, or the ballasting/ de-ballasting of the ship's tanks, or the general shifting from one tank to the other of liquids is not allowed.
The
. wrong/improper measurement at the load port could result in the Owners sustaining heavy delays/expenses, due to short or surplus cargo delivery at the port of discharge.
The ship must be covered by a first-class P&I Club, which should be in position to issue a letter of undertaking upon request, in order to release the ship from the detention of an alleged cargo claim (most of the times it is for cargo shortage), as might be presented at the discharge port. In case of shortages and in accordance with the Inter-Club Agreement clause, the Owners are entitled to recover 50 percent of the paid amount from the Charterers.

The "Draft survey/Plans" clause, which, among others, should include:
-that the loaded and discharged cargo quantity will be determined by a draft survey.
-that all related plans to perform the draft survey are available on board the vessel.
-that all draft marks on the ship's sides will be clearly readable.
-that, during the draft survey, liquids will not be allowed to be shifted.

An executed clause that has been run in a Charter Party is as follows (precise copy):
As cargo quantities will be determined by a joint draft survey, vessel must be in a condition to perform such surveys. Hence at load port and discharge port, Charterers' appointed surveyor/s is to be allowed full access to the vessel for the purpose of the draft survey. The vessel is to furnish a calibration scale for all tanks, including fore/aft peaks and double bottom tanks and deep tanks. Plimsoll marks and draft marks on port and starboard side, bow, amid ship and stern to be clearly cut and marked on shell platting. Vessel to furnish capacity plan and displacement scale, hydrostatic curves and trimming table.
At no time during draft survey (initial/intermediate or final) liquids should be shifted, taken in or pumped out without prior advice to Charterers. Master/Chief Officer to follow first and final draft surveys/soundings at each load port/load berth and to sign such reports/findings when presented.
Comments:
The laden and discharged cargo quantities will be determined by a draft survey. All plans must be available to the surveyors and all draft marks must be clearly cut and marked on shell platting. During the performed draft survey, the shifting of liquid that may influence the cargo measurement is not allowed. The

Master/ Chief officer must jointly sign the draft survey report, which declares the Owners' acceptance of the loaded or discharged cargo quantities.

Another clause reads as follows (precise copy):
Charterers and Owners to perform a joint draft survey at load and discharge port. Owners are not to be responsible for Silo figures and Charterers to indemnify Owners from these Silo scale and Charterers to remain solely responsible for any cargo shortage.
Comments:
The loaded and discharged quantities will be determined by a joint draft survey. In the case that silo figures were presented and there are discrepancies with the draft survey findings, then the Charterers will indemnify the Owners and will remain responsible for any cargo shortage.

A complete clause for Time Charter period, which refers to the "Draft survey/ Plans" could be worded along the following lines:

Loaded and discharged cargo quantities will be ascertained by a joint draft survey, which will be held between the Owners, Charterers and Shippers at the load port and between the Owners, Charterers and Receivers at the discharge port. Each party will appoint and pay their surveyors.
It is the Owners' option to appoint the ship's Master or Chief Officer as their surveyor.
The appointed surveyors are to be allowed to have full access to the ship for the purpose of the draft survey.
Owners guarantee that the ship will, throughout the duration of the Charter Party have on board and make available to the Charterers or their Agents or to the appointed surveyors, the capacity plan, the deadweight and displacement scales, the hydrostatic curves and tables of displacement, the calibration scales for all ship's tanks and trimming correction tables, as well all sounding tables and all sounding tubes which will be properly maintained, in good condition and free from impediments. Plimsoll marks amidships and draft marks on the port and starboard side, bow, amid ship and stern are always to be clearly cut/ marked on shell platting and readable.
During the carrying out of the draft survey (initial, intermediate, and final) no liquids are allowed to be shifted, taken in or pumped out without the Charterers' and surveyors' prior consent. If it is required, the Master is to forward to the Charterers, upon arrival at the loading and discharging port and before the commencement of the loading or discharging operation, a certificate indicating all remaining ballast in the ship's tanks.
One draft survey report will be issued, which will be signed by all the surveyors

involved.
At the load and discharge port, the shore scale figures are not to be taken into consideration and the ship will load and discharge according to the findings of the carried out draft survey. The Owners, Master, ship will not be held responsible for any discrepancies which may arise between the draft survey and the shore figures and it is explicitly agreed that the Bill of Lading will be issued according to the draft survey findings. If the ship is detained by any party and generally its sailing is not allowed from the port, because of any discrepancies between the draft survey and the shore figures, the Charterers then undertake the responsibility to release the ship, by taking all the necessary and appropriate measures (issuance of security, etc.), as the case may be, or arise and the ship will remain always on-hire.

-When the Bills of Lading will be issued as per the shore figures (Shippers' figures), the clause could be worded as follows:

In the case that the Shippers' figures at the load port exceed the ship's figures beyond a normal and/or customary difference on the loaded cargo quantity, and the Shippers insist that the original Mate's Receipts and the Bill of Lading will be issued in accordance with their findings, then the Master has the right to clause the Mate's Receipts and the Bill of Lading with the ship's figure and/or to strictly follow the ship's P&I Club's recommendations.

51. Lay -Up condition.

This clause is in effect for
. long Time Charter periods,
which allows the Charterers to lay up the ship or not.
The Charterers, for commercial reasons or in order to create a lack of tonnages in the area or due to a collapsed market, should seek to have the option through this clause to lay-up the ship. For short periods there is no reason for the Owners to accept such a clause, as the lay-up is not so easy to be placed and may involve risk to the machine when it is in an out of order condition for a long time. The Owners should not accept that the ship be redelivered to them in a lay-up condition or immediately after reactivation.

The "Lay-up condition" clause, which, among others, should include:
-whether the Charterers are allowed to lay-up the ship.
-if they are allowed, the conditions under which the ship will be placed on lay-up.
- for whose account should the cost/time/expenses associated with placing the ship in lay-up will be.

An executed clause that has been run in a Charter Party is as follows (precise copy):
Should Charterers wish to lay up the vessel for any period of time under this Charter Party, Charterers and Owners to discuss same and endeavour to reach a reasonable agreement mutually acceptable to both parties.
Comments:
The Charterers have the right to lay up the ship for any period of time that they wish, and it leaves all discussions open for a mutual agreement on if and when it is required. It is a general clause which could be agreed between parties having a good and long relationship.

Another clause reads as follows (precise copy):
The Charterers shall have the right to order the laying-up of the vessel at any time and for any period of time at a safe berth/anchorage and in such a manner as mutually agreed upon and acceptable to the vessel's hull Underwriters and in the event of any such lay-up, the Owners shall promptly take steps to effect all the economies savings in operating costs including insurance which may be possible and give prompt credit to the Charterers in respect of all such economies savings giving due regard for any activation expenses that Owners may incur. At the request of the Charterers, the Owners shall at any time provide an estimate of the economies which would be possible in the event of the laying-up of the vessel. Should the Charterers, having exercised the option granted hereunder, desire the vessel again be put into service, Owners shall, upon receipt of written notice from Charterers to such effect, immediately take steps to restore the vessel to service as promptly as possible. The option granted to Charterers hereunder may be exercised one or more times during the currency of this Charter Party. Hire shall continue to be paid, net of any economies in operational costs and insurance referred to in paragraph on above, throughout the period of lay-up. All costs associated with placing the vessel in lay-up during lay-up period and on reactivation are to be in Charterers' time and expense. It is understood that hire over lay-up period will anyway be mutually discussed and agreed.
Comments:
The Charterers have the right to lay-up the ship at any time and for any period. The Owners will take steps to make savings in operating costs, including insurances. The new hire which will be in effect over the lay-up period will mutually be agreed on.

A complete clause for Time Charter period, which refers to the "Lay-Up condition" could be worded along the following lines:

Charterers, after having given sufficient notice to the Owners, which should be no less than 20 days, will have the right to lay-up the ship one or more times, at any time during the period of this Charter and for any period of time, at an acceptable by the Owners and the ship's hull underwriters safe berth or safe anchorage or safe place. The lay-up procedure and performance will mutually be agreed upon and be fully acceptable by the ship's hull Underwriters.

The Owners will present as soon as possible to the Charterers a statement with all possible estimated economic savings in operating costs, including the signing off of crew and insurance coverage (but not on the insured ship's values).

Charterers will accept for their entire account and will be responsible for all costs, time and expenses associated:

a. with placing the ship in a lay-up condition, including the crew's expenses for repatriation (transportation, tickets etc.), indemnities payable to the crew and other expenses and/or port expenses incurred, including, but not limited to, lunch, tugs, pilots, Agency fees etc. during the lay-up period.

b. with placing the ship on reactivation upon being requested, including the crew's expenses for embarkation (transportation, tickets etc.). The Charterers will provide no less than 20 days written notice to the Owners of their intention to reactivate the ship. The Owners will immediately take steps to restore the ship to the Charterers' service, as promptly as possible and within the 20 days period but without any guarantee and responsibility to the Owners for the readiness of the ship.

c. with the cleaning of the ship's hull and/or bottom including, but not limited to, vertical sides, bilge keels, propeller, rudder blade from the marine growth, before returning it again to the Charterers' services and to the Master's satisfaction.

d. with minor or serious damage and/or the supply of stores and/or repairs required to be made to the ship's hull or engine which might be incurred during the time of lay-up until reactivation and are not covered by the ship's insurances.

The hire, during the lay-up period, will be mutually adjusted and agreed between the Owners and Charterers, taking into consideration only the respective savings, as may be actually achieved. The lay-up hire will commence counting from the time of lay-up and crew's disembarkation from the ship, until the time of its sailing from the lay-up position and return to the Charterers' services. The hire will continue to be paid throughout the period of lay-up, as per the agreed payment clause.

During the execution of this Charter, any period of time waiting for the Charterers' orders will not be considered or declared as a lay-up period.

The ship will not be placed in a lay-up condition during the last three months of this Charter and if it is already in lay-up, the ship will be reactivated as soon as possible.

52. Sales -Substitution.

Such a clause should facilitate
. the Owners, in the case that they decide, for commercial or other own reasons, to sell or to substitute the performing ship, in which case it specifies the conditions under which they could make such a move.
Usually, such an option can be made at any time during the execution of the Charter.
The new Ownership/Management or the presented characteristics of the substitute ship should be previously approved by the Charterers.
However, the Charterers, before agreeing, must take into consideration all additional risks arising from such an agreement.

The "Sales-Substitution" clause, which, among others, should include:
-whether the Owners have the right to sell/substitute the ship and when/under which conditions.

An executed clause that has been run in a Charter Party is as follows (precise copy):
Owners guarantee that the vessel will not be sold during the Charter period without Charterers' prior consent, which not to be unreasonably withheld.
Comments:
The Owners do not have the option to sell the ship prior to obtaining the Charterers' consent, who, in any case should not object for an unreasonable reason.

Another clause reads as follows (precise copy):
The owners to have the option of selling this vessel during the currency of this Charter Party. No changes as such in vessel's name or owning company/etc. to be made whilst vessel performs a trip where documents such as Bills of Lading etc. are already being used showing vessel's current name/ management/ ownership.
Such option to be communicated to Charterers in due course enabling Charterers to approve the new management/ownership. Such approval shall not be unreasonable withheld. Takeover by vessel's new management/ownership to take place at a port convenient for such purpose. The vessel to be off-hire during takeover and all bunkers consumed to be for present/next Owners' account unless replenish in the meantime.
The Charterers to be given 5/3 days approximate and 2/1 days definite notice of re-entering the service.
Eventual change of ships shall not change any Charterers' rights or Owners obligations under this Charter Party.

Comments:
The Owners have the right to sell the ship during the period of this Charter Party, but the ownership cannot be changed during the execution of a voyage. The Charterers must be timely informed about the Owners' intention to sell the ship and the new ownership must be previously approved by them. Certain notices must be given before the ship is again ready for trade.

Another clause reads as follows (precise copy):
Owners have the right to sell the vessel at any time during the period of this Charter Party in which case Charterers to have the option to either to accept new Owners, their management, possible new registration and flag of the vessel and the Charter Party then to remain in full force with the new Owners, or instead to require Owners to substitute the vessel with a vessel of same size/type and with at least the same cargo lift, vessel's gear capacity, same age or newer. Vessel's speed and vessel's consumption to be at least as good as that of the m/v ...... Owners are to give Charterers a minimum 30 days advance notice of such intention. The vessel to be delivered to Charterers at a port/place acceptable by Charterers or at a distance from termisa no greater than that of the port/place where m/v... is handed back to the Owners. The substitute vessel to have similar quantities of bunkers on board on delivery as the m/v... (but always sufficient to reach nearest main bunkering port) as at time of withdrawal. Any adjustment to be at Charter Party bunker prices. Such vessel substitution always to be subject to Charterers final approval.
Comments:
The Owners have the right to sell the ship, in which case the Charterers should confirm the sale and accept the new management/flag etc., otherwise they can request from the Owners to substitute it with a sister or better ship. The Owners must tender at least a 30-day advance notice of their intention.

A complete clause for Time Charter period, which refers to the "Sales –Substitution" could be worded along the following lines:

Provided that the Charterers have not nominated the ship under their contractual voyage/s and the Charterers can provide their consent at the time of request, then the Owners have the option:
1. to sell the ship at any time during the currency of this Charter Party, provided: -they will tender a minimum of 45-day advance notices of their intention to the Charterers.
-the new Ownership, Management and the possible changing of flag and/or nationality of crew etc. will be previously approved by the Charterers, however, such approval is not to be unreasonably withheld.

-the Charter Party made with the old Owners will be fully acceptable by the new Ownership, Management, who will countersign the same before the ship's delivery to the Charterers' services, under the new Ownership, Management.
-the ship will be sold at the place, area, port, where the ship last discharged at berth or on dlosps at the Owners' option. The ship will be placed off-hire from the time of completion of discharge or on dropping last outward sea pilot, until the ship is again fully ready and operational to countersign the Charterers' services and at the same position. All the expenses incurred, or even paid by the Charterers on the Owners' behalf, will be for the new or previous Owners' account and as the case may be. The ship will not change Ownership, Management when it is fixed for its next employment or before the end of the running voyage.
-the new Ownership, Management guarantee that the ship will be ready to be delivered to the Charterers' services again within 20 days from the time when the ship was placed off-hire for the purpose of changing the Ownership, Management. The new Ownership, Management will tender 7/5 approximate and 3/2/1 definite notice for delivery of the ship back to the Charterers' services.
-the ship will be delivered back to the Charterers with about the same quantity of fuel/grades as on the time it was placed off-hire. Any minor differences on the bunkers' value will be settled together with the next hire payment at the Charter Party's prices.
-all delays incurred will be added or not, at the Charterers' option, to the maximum period allowed under this Charter.
2. to substitute the ship, in the case that the Charterers are unable to approve the new Ownership, Management, or for any other reason, provided:
- the Charterers can grant their consent at the time of request.
- the Owners will tender a minimum of 20-day advance notices of their intention.
- the substitute ship will meet the Charterers' needs and/or contractual commitments and is previously fully approved by them.
- the hire will be mutually adjusted based on the differing ship's specifications.
- the substitute ship will be delivered at the same place, port or to an area fully acceptable by the Charterers and at the same dates. If the substitution cannot be arranged at the same place, port and at the same time, then it can take place at an area, place and at a time fully acceptable to the Charterers. The financial differences and/or impact of the different delivery place and/or dates, whether they are to the Charterers' or to the Owners' favour, will be mutually agreed and the same will be settled with the next hire payment.
- the substitute will be delivered with the minimum quantity of fuel, grades and prices as will be mutually agreed between the Owners and Charterers.
- the Owners will tender 10/5 approximate and 3/2/1 definite notice for delivery of the substitute ship to the Charterers' services.

- all delays incurred will be added or not, at the Charterers' option, to the maximum period allowed under this Charter. If the Charterers are unable to approve the new Ownership, Management and/or the Owners cannot provide the right substitution to the Charterers' entire satisfaction, the Owners and Charterers then can come to an agreement of cancellation of this Charter Party. The Owners will compensate the Charterers with a mutually agreed cancellation fee, according to the prevailing market.
In any and all cases, and if there is a disagreement in finding a solution between the Charterers and Owners for the sale and/or for substitution, then the ship will remain at the Charterers' services as initially fixed and the Charter Party will continue to be in force, for the execution of the balance period and no sale or substitution or cancellation will be made.
If the ship is lost, the Charter Party will be terminated.

53. Communications/Victualing/Entertainment (C/V/E).

The monthly or pro rata
. lumpsum amount,
which will be paid to the Owners by the Charterers, in lieu of reimbursement for expenses incurred on behalf of the Charterers for communication, victualing, and entertainment, during the ship's stay in the ports or at sea, will be stipulated in this clause.
It lies within the Master's duties to tender, either by e-mail or cable or by fax, the number of notices as will be required by the Charterers and to make all the necessary phone calls to the respective parties. He is to serve drinks/cigarettes, if deemed necessary, to the pilots/customs officers/foremen etc. and to victual the Charterers' representatives or to victual any important persons, as might be requested by the Charterers. All these costs are included in the agreed monthly amount, which covers the period of a month or pro rata.

The agreement should be made based on one of the following three ways:
a. with a monthly or prorate for any part of the monthly coverage, which is widely used in the Charter Parties and is acceptable by both Charterers and Owners. The resulting monthly equal amount will be payable to the Owners together with the 15 days hire payment
b. with a mutually agreed lumpsum amount for the entire duration of the Charter.
It is mainly applicable when the expected duration is less than one month, and it is usually payable at the end of the Charter.
c. payable at cost, in which case the Master should provide a statement of all expenses incurred on the Charterers' behalf. It should obviously be paid at the

end of the Charter.

The "c/v/e" clause, which, among others, should include:
-the monthly or prorate lumpsum amount.
-when this amount will be paid.

An executed clause that has been run in a Charter Party is as follows (precise copy):
Charterers to pay Owners a lumpsum USD .... per month or prorate in lieu of reimbursement of communication, victualling and entertainment expenses incurred on Charterer's behalf.
Comments:
A clause stating the monthly or pro rata amount due to the Owners for expenses incurred on the Charterers' behalf.

Another clause reads as follows (precise copy):
Charterers will remit to Owners with each hire payment a lump sum of USD .... per month or pro rata for cable/victualling/entertainment for the whole Time Charter period. This payment includes cost of incidentals such as cigarettes, drinks and petty expenses incurred by Master/Officers of the vessel on Charterers' behalf and the cost of radio telegrams, telexes, fax, and telephone communications made by the Master/Officers to the Charterers or their agents or servants in direct performance of this Charter Party.
Comments:
A clause, which specifically states which expenses are for the Charterers' account.

A complete clause for Time Charter period, which refers to the "Communications/ Victualing/ Entertainment" could be worded along the following lines:

Charterers, for the whole duration of this Charter period will compensate the Owners with a lump sum amount of USD.... per month or pro rata for any part of the month, for communication, victuals, entertainment.
This payment includes incidental expenses made on the Charterers' behalf, such as cigarettes, drinks and petty expenses incurred by the Master, Officers to the Charterers' representatives and/or pilots and/or custom officers and/or foremen etc. It also includes the cost of radio telegrams, telexes, fax and telephone communications made by the Master, Officers to the Charterers or their Agents or servants, in direct performance of this Charter Party. The monthly equal amount is payable to the Owners together with the 15 days hire payment.
Meals to tally clerks and/or stevedores and/or foremen etc. as well other ex-

penses, which have been previously approved by the Charterers and incurred by the Master, Officers on the Charterers' behalf will be paid at cost and on presentation even by fax or e-mail of the supporting vouchers and/or the Master's statement of expenses.
or
C/V/E are payable by the Charterers at cost and in accordance with the presented even by fax bills or the Master's statement of expenses incurred on the Charterers' behalf.

54. Charterers' house flag/Markings/Colours.

For their own commercial reasons, the Charterers may request to hoist their house flag and paint their markings on the funnel and/or the ship's sides. The Charterers are obliged to restore the ship's condition, under their time and expense.

The "Charterers' house flag/Markings/Colours" clause, which, among others, should include:
-whether the Charterers have the right to hoist their house flag and/or paint the ship's sides/hull.
-that, if the Charterers hoist their flag and/or paint the ship's sides/hull, the lost time and costs should be for their account.
-that for the ship's restoration, the quality and type of colours that will be used should be at the Owners' choice.

An executed clause that has been run in a Charter Party is as follows (precise copy):
Charterers to have the privilege of painting the funnel and/or the ship's sides to Charterers colours/names at their expense and in their time. On redelivery of the vessel Charterers are to repaint the funnel and/or ship's sides to Owners colours at Charterers time and expense. The ship also to run Charterers house flag during the Charter Period, if so required.
Comments:
The Charterers have the option to paint the funnel and/or the ship's sides with their colours/names, but they must restore the same under their time/expenses.

A complete clause for Time Charter period, which refers to the "Charterers' house Flag/ Markings/Colours" could be worded along the following lines:

Charterers have the option of painting their markings on the funnel and/or the ship's sides, including painting the ship's hull, with the Charterers colours and/

or names at their expenses and time. Before the ship's redelivery, the Charterers are to restore and repaint the funnel and/or the ship's sides and/or hull to the Owners' markings and colours at the Charterers' time and expense. The quality and type of colours for repainting and/or restoring will be at the Owners' choice, which will be agreed before the Charterers exercise their right to paint the ship with their colours. The ship will hoist the Charterers' house flag during the Charter Period, if so be required.

-For Owners who do not agree to run the Charterers colours/markings, but it could be discussed at the time of request, the clause could be worded as follows:
The Charterers do not have the right to paint their markings or paint the ship's sides/hull with their colours without the Owners' consent.

55. Slow/Fast steaming.

This clause gives the right to the Charterers to instruct the ship to reduce speed and delay the arrival at the port of load or discharge. Some of the reasons could be for example that due to the prevailing heavy congestion at the load or discharge port, navigation of the ship with economic speed should benefit them or that the voyage freight from the head-Charterers has not yet been paid and the arrival at the discharge port is prompt or for any other of the Charterers' commercial reasons.
The navigation with reduced speed towards the load port is not expected to cause problems as the ship should be empty of cargo, however, when it is in a loaded condition, the violation of the Owners' obligation to proceed with the utmost dispatch to the discharge port, could give rise to claims against them. In fact, the 'main' and 'eco' speed in the Charter Party is simply a description and it is not linked to the separate obligations arising from the signing of the Bills of Lading.
A reduced speed increases the Owners'
. liabilities towards third parties.
Nowadays, because of the high cost of fuel supply, the majority of the Charterers require from the Owners, along with the main ship speed, to also declare its 'eco' speed/consumption and to guarantee the same. This can assist them in their voyage calculations and so a good 'eco' speed/ consumption could bring the ship on top of the other proposed ships for the particular biz and might have the 'first refusal'.
On the other hand, because it is possible for another ship to arrive a little earlier at the port and occupy the berth intended for loading or discharging, resulting in the ship facing berthing delays, the Charterers could also instruct the Master

to increase the speed to the maximum allowed, with the increased consumption, however, to be covered by them.
Bimco has published a relevant clause called 'slow steaming clause for time charter parties' to assist the parties involved on reaching an agreement.

The "Slow/Fast steaming" clause, which, among others, should include:
-whether the Charterers have the right to instruct the ship to reduce or fast steam.
-under which conditions they will reduce/fast steam.
- the Owners' coverage in the case of reduction/fast steaming.

An executed clause that has been run in a Charter Party is as follows (precise copy):
The Master shall, if instructed by the Charterers, slow/fast steam the vessel in accordance with Charterers instructions subject to the discretion of the Master and always subject to safety of hull and machinery, Owners to supply the Charterers, details of expected fuel consumption at the reduced/increased speed or speeds required.
Comments:
The Charterers have the right to instruct the Master of the ship to slow/fast steam in accordance with the Charterers' needs, but subject to the safety of the ship's hull and machinery.

Another clause reads as follows (precise copy):
Charterers shall be entitled from time to time to instruct the vessel to proceed at reduced speed, always in line with manufacturer's instructions for economic or other reasons, at the speeds/consumptions given in good faith but without guarantee. Charterers shall always indemnify Owners and hold them harmless against all consequences or liabilities towards third parties resulting from such instructions. Further, the provisions regarding performance shall cease to apply throughout the currency of such instructions.
Comments:
The Charterers may instruct the ship at any time to proceed at a reduced speed. However, the Charterers remain responsible towards third parties and hold the Owners harmless for all consequences which may arise.

A complete clause for Time Charter period, which refers to the "Slow/Fast steaming" could be worded along the following lines:

Charterers, after the Owners' consent, who will take into account the safety reasons for the ship and/or cargo and the obligations towards third parties, have

the option, for their commercial reasons, to request the Master to proceed towards the load and/or discharge port, at a reduced or increased speed, always in line with the manufacturer's instructions, for economic or other reasons, at the speeds and consumptions as given by the Master in good faith, but without guarantee.
Charterers will always indemnify the Owners and hold them harmless for all consequences or liabilities towards third parties resulting from such instructions.
The Owners' responsibilities as to the Charter Party speed and consumption will cease when such orders have been given. For reduced speed, the Bimco's current 'slow steaming clause for time charter parties' will be in effect.

56. Master's and Crew's assistance/Services/Duties.

During the execution of the Charter, the Master/Officers and Crew of the ship will render their
. unconditional
but regular/customary full assistance day and night and on overtime and will cooperate with all the parties involved, as they should normally perform when trading for the Owners' account, but subject to local regulation allowances. These procedures will contribute to the smooth execution of the Charter and to a fruitful cooperation between the Owners and Charterers.
Some of the most important tasks of the crew could be provided in this clause, in order to avoid possible disputes and misunderstandings. The additional work which may be required by the Charterers and not included in the crew's customary assistance should be clearly stated in it, otherwise, it is to the Master's/crew's right to refuse to execute the same or to discuss it at the time of occurrence. However, the Charterers are to come, either directly or through their Agents, but always after the Owners' consent, in contact with the crew and agree on a compensation for their intended extra work. The Charterers are to benefit from such agreements, as the work from shore is costly and there would be loss of time. Such work could be the lashing/unlashing of the cargo on deck, the operation of the ship's cranes for loading and/or discharging purposes, the welding of padeyes on deck or in holds etc.
The ship will work day and night and on overtime, without extra charges for the Charterers, as the hire includes such work during overtime.

The "Master's and Crew's assistance/Services/Duties" clause, which, among others, should include:
-the regular work which the Master/officer/crew will perform.
-the extra work which the Master/officer/crew are allowed to perform.

An executed clause that has been run in a Charter Party is as follows (precise copy):

Master and crew shall co-operate with Charterers, Receivers or their Representatives insofar as local regulations permit to facilitate loading and discharging operation.

Before and upon arrival at a port the vessel's Officers/Crew to shape up the vessel's hatches, cranes, derricks and gangway in order to commence loading and/or discharging without any delay.

Opening/closing of all hatch covers shall be done by Officers /Crew free of cost to Charterers, if allowed by shore regulations.

Comments:

A general clause, however, does not relieve the Master's/officers'/crew from their regular duties as the ship was trading for the Owners' account.

Another clause reads as follows (precise copy):

Master to have vessel ready for immediate cargo operation on arrival at each port in accordance with Charterers' local agent's instructions as far as this is possible.

Additionally, the Owners will provide at their own risk and expense for the Officers and/or crew to carry out the following duties and any other duty as the Crew should perform when trading for Owner' account:

- *a) All opening and closing of hatches for loading and/or discharging, weather and local regulations permitting.*
- *b) Supervision in loading and/or discharging*
- *c) Watchmen in holds during loading and/or discharging as far as permitted by port regulations and as far as possible.*
- *d) Maintaining necessary power while loading and/or discharging*
- *e) Shifting ship during loading and/or discharging and boards*
- *f) Docking and undocking*
- *g) Bunkering and ballasting*
- *h) Officers and crew to shape-up hatches as much as possible, weather permitting prior to arrival at loading and/or discharging operations*
- *i) Driving vessel's cranes for loading and/o discharging operations. Charterers to make their own arrangements/agreement, in respect of payment directly with the Master.*

All above subject to local shore regulations.

The Master shall be responsible for the safe keeping of all gear, equipment and/or stores supplied by or for Charterers' account and the Master shall keep a record of all such gear, equipment and /or stores so supplied. Master to maintain same in good condition, as Owners' property but without guarantee. Such gear, equipment and/or stores to be redelivered to Charterers prior to the redelivery of the vessel to the Owners, or if required by the Charterers at any time during the Charter in like good order and condition as supplied, fair wear and tear always excepted.

Comments:

Some of the ship's major crew duties have been described in this clause. Apart from the crew's regular work, it is agreed that the crew will operate the ship's cranes for the loading and/or discharging operation. It seems that there will either be lack of competent or experienced shore drivers at the intended ports, or the Charterers are trying to save some money. The Owners must provide enough experienced drivers from the crew to work the cranes, at any time that it will be required.

A complete clause for Time Charter period, which refers to the "Master's and Crew's assistance/Services/Duties" could be worded along the following lines:

Master will render all customary assistance and work with the ship's Officers, Crew, and all nautical duties, which are necessary for the safe navigation day and night, weekends and holidays included.

The ship will work and provide, at all ports, during day and night, weekends and holidays included, efficient operation in all respects and will co-operate with all parties involved to facilitate the loading and discharging operation.

Master, Officers and Crew would normally perform as when the ship is trading for the Owners' account, free of risk and expenses to the Charterers. The Charter hire includes any of the Master's, Officers', and Crew's overtime and the following, but not limited customary duties:

a. the ship should be ready for immediate cargo operation on arrival at each port of call. The crew will open and close hatches in connection with the loading and discharging operation.

b. supervision of the loading and/or cargo stowage and discharging and everything related thereto and inserting necessary remarks in the Mate's Receipts and/or Tally clerks and/or Statement of facts.

c. the bunkering operation of the ship.

d. whenever it is required, they will ballast and/or de-ballast the ship's tanks, during the loading and discharging operation, in order to reach the port restrictions and/or the ship's stability.

e. docking, undocking, shifting between berths and/or warping alongside berths, whenever it is required by the Charterers and/or Agents and/or Port Authorities.

The above duties will be considered as a minimum and will always be subject to weather conditions and safe navigation, as well as, subject to allowances by local regulations and/or stevedores and/or labour unions. The Master, Officers,

Crew will not be requested to violate the same.

-For geared ships, the following wording could be added:

-raising and lowering and general preparation of the ship's cranes and/or gangways and/or grabs and/or grabs devices, before the arrival at each port for the commencement of loading and discharging operation without delay.
-maintaining sufficient and continuous electric power for the efficient work of the ship's equipment.

-When deck cargo is loaded, the following wording could be added:

- attending the deck cargo and its lashing while the ship is at sea.

-In the case that the Charterers have supplied gear/equipment/stores on board the ship, the following wording could be added:

-any gear, stanchions, lashing etc. of materials supplied by the Charterers are to remain at the Charterers' property and to be delivered to the Charterers' Agents at any time it is requested, but always before the ship's redelivery, at the Charterers' time, risk and expenses. The Master will keep a record of all materials so supplied and will maintain the same in good condition, but without guarantee and always subject to wear and tear and of any damage incurred, which are connected or attributed to the Charterers' trade.

-In the case that the Charterers request extra work, the following wording could be added:

- the ship's Officers, Crew, after their consent and the Owners', and/or Masters' approval, will perform extra work, if so requested by the Charterers. The Charterers will make their own arrangements and agreement, in respect of compensation, directly with the Master and/or the Officers/Crew.

57. Superficial inspection/Port Captain/Purplefinder.

Even though there is a provision for supercargo on the Main Body of the more frequently used Charter Parties, sometimes an additional clause is required to specify the conditions of visit.
The Charterers or their representatives or their surveyors
. are allowed to visit the ship during its stay in the ports.
They should be entitled to have easy access to the spaces where the Charterers

have interests and visit the engine room, the bridge, the cargo spaces etc. They should be able to verify the ship's actual speed and consumption, the bunker quantities remaining on board and the ship's general condition. They could also monitor the loading or discharging and can assist the Master whenever it is required. Unless this survey is interfering with the ship's operation in the port as per the

. ISPS/ISM regulations,

there is no other reason for the Owners to refuse such a visit.

The Master should provide to them - if available - accommodation and meals for their whole stay on board the ship. Sometimes, the Charterers could request from their surveyors to also remain on board the ship during its sea voyage for various commercial reasons.

The Charterers could also request to monitor the ship's speed and daily position at sea, by the installation and use of a purplefinder fleet management gps/tracking system, in the ship's bridge. The Owners are not obliged to accept it and they may refuse it, if such a term has not been explicitly stipulated in the Charter Party. The Master must command the ship according to his discretion for the crew's/cargo safety and not as per the Charterers' installed system. The Charterers have the opportunity to assess the ship's speed/position, through their appointed 'Ocean Route' company.

The "Superficial inspection/Port Captain/Purplefinder" clause, which, among others, should include:

-the allowance of the Charterers' representatives to visit the ship during its stay in the port/s.

-whether the Charterers' representatives can be accommodated on the ship, during its sea passage.

-what they have access to when visiting the ship.

-whether the Charterers have the right to install and use a purplefinder fleet management gps/tracking system.

An executed clause that has been run in a Charter Party is as follows (precise copy):

The Charterers and/or their Supercargoes shall have free and reasonable) access to the whole vessel including bridge, holds and engine room, and also to all vessel's tanks, including but not limited to, bunker, lubricating oil, sludge, ballast, and fresh water tanks. Whenever required, the Master must bring the vessel into even trim, if possible, to ensure correct bunker soundings. The Charterers and/or Surveyors to have free and reasonable access to the vessel's deck- and engine-log books-tank plans, calibration scales and /or other plans as requested and are allowed to make copies of the original Log Books on board or ashore.

Owners to supply accommodation/ victualling to Charterers' Supercargo/es only during stay of the vessel in load and/or discharge port in order to be able to assist/ follow best possible for loading/discharging operations.
Comments:
The Charterers' supercargoes will have full access to the ship's places, in order to have a clear and complete picture of the ship's condition/performance etc. The Master should provide any assistance and should arrange for their accommodation/victualing.

Another clause reads as follows (precise copy):
The Charterers shall have the right and privilege of having their representatives visit the vessel while in port or at sea, provided written pre-notice is given and without interference with vessel's normal operation. Charterers' representatives shall have access to the entire vessel (excluding accommodation spaces) and the Master, Officers and Crew of the vessel shall co-operate with rendering any and all reasonable assistance that the Charterers' representative may require. Refusal by the Owners or the Master, Officers or Crew to allow such visits or to co-operate, as required by this provision, shall be considered as a breach of Charterers' orders and vessel shall be off-hire from the time of such refusal.
Comments:
The Charterers' representatives can visit the ship while in port or at sea provided a pre-notice is given to the Owners and without interfering with the ship's operation. It states that they will have access to the entire ship and the Master/ officers/crew must render any possible assistance. Any refusal of the Owners/ Master to allow such visits will be considered as breach of the Charterers' orders and the ship will be placed off-hire.

A complete clause for Time Charter period, which refers to the "Superficial inspection/Port Captain/Purplefinder ", could be worded along the following lines:

During the duration of this Charter, the Charterers and/or their Supercargoes and/ or their representatives, can visit the ship during its stay at any port of contact.
They will have free and reasonable access to the ship's:
- bridge, deck, holds and engine room.
- tanks, including, but not limited to, fuel, lubricating oil, sludge, ballast, and freshwater tanks.
- deck and engine-log books, tank plans, calibration scales
and/or other plans, as will be requested, and are allowed to make copies of the original Logbooks on board or ashore.
Master, Officers, crew will co-operate and provide all possible assistance to

them, but provided that the same does not interfere with the Master's, Officers', crew's normal work and does not violate the ISPS and/or ISM regulations. Whenever it is required, the Master will endeavour to bring the ship into an even trim, if the same is possible and circumstances allow it, in order to ensure the correct bunker soundings.
The Owners should supply accommodation - if it is available - and victualing, if so required, to the Charterers and/or their Supercargoes and/or their representatives, during their stay on the ship, in order to be able to assist/follow the best possible loading and/or discharging operation. All expenses incurred related to their visit on board the ship, are for the Charterers' account and will be paid at cost, as per the Master's relevant statement.

-For Owners who accept the use of a purplefinder fleet management gps/tracking system, the following wording could be added:

The Charterers are allowed, under their cost, expense, and time lost, to monitor the ship's speed and daily position by the use of a purplefinder fleet management gps/ tracking system.

58. Deck /Hatch covers loading.

For multipurpose ships or loggers etc., usually all available cargo spaces will be utilized by the Charterers and it is common for deck loading to be requested. Obviously, the deck cargoes must be in such a form/condition, enabling the stevedores to lash/secure them. Under the Charter Party terms, the Charterers must provide any and all required lashing/securing materials and/or any other extra fittings/ equipment for the safe stowing/carriage. The Master must supervise the loading and the loading must be performed to his absolute and full satisfaction.
The Owners are to be aware that when loading deck cargo, apart from certain specific circumstances (containers/logs etc), they are in breach of the contract of carriage and the ship may fall outside the P&I Club rules.
The clause should state whether the Charterers will have the right to load cargo, on the ship's deck and under which conditions.
Even if a Panamax ship is going to be fixed and deck cargo is impossible to be arranged, the Owners should clearly state in the Charter Party whether deck cargo is allowed or not.
When loading on deck, the Owners should seek advice from the ship's P&I Club, as the Bill of Lading must be issued strictly as per their recommendations.

The "Deck/Hatch covers loading" clause, which, among others, should include:

-whether the deck cargo (including or not the hatch covers) is allowed and under which conditions.
-who will provide the requisite materials/equipment etc. and under whose account/time.
-how the Bills of Lading will be marked.
-who will be responsible for incurred damages due to deck loading.
-the Master's right to dispute and stop the loading, if the lashing/securing is not to his satisfaction.

An executed clause that has been run in a Charter Party is as follows (precise copy):
Charterers have the option to load cargo on deck and hatch covers, if required, in accordance with normal marine practice, and the deck load will be controlled by stability, seaworthiness and strength of the vessel, but at Charterers' time, risk and expense for cargo on deck under Master's supervisions. Charterers will be responsible for time and cost for any repairs of the deck structure caused by cargo on deck. Relevant Bill of Lading to be issued as per Owners P&I Club instructions.
Comments:
Deck cargo is allowed. The cargo can also be loaded on hatch covers according to the ship's stability/seaworthiness and strength at the Charterers' time/risk/expense. The Charterers are responsible for damages incurred on deck and the Bills of Lading will be issued as per the Owners' P&I Club's recommendations.

Another clause reads as follows (precise copy):
Charterers have the option to load cargo on deck (excluding hatch covers) at Charterers' entire risk/time and expense but always in accordance with vessel's deck permissible strength and always in accordance with vessel's stability / visibility.
Deck cargo to be loaded/stowed/dunnaged/lashed/carried/secured/unlashed and discharged by shore hands at Charterers entire time, risk, and expense.
Stanchions /gear/lashing etc. materials required to properly secure the deck cargo to be supplied by Charterers at their time and expense. Any gear/stanchions/lashing etc. materials supplied by Charterers to remain Charterers' property and to be delivered (subject to fair and wear and any damages connected or attributed to the trade) to Charterers agents prior to vessel's redelivery at Charterers time and expenses.
For all deck cargo separate Bills of Lading to be issued which will be claused as follows: 'shipped/carried on deck at Charterers, Shippers and Receivers risk, expense and responsibility without any liability on the part of the vessel, or her Owners any loss, damage or delay howsoever caused.'

Charterers further to remain fully responsible for any damage/s caused to vessel's hull and/or machinery/equipment /fittings and/or other cargo as a result of the carriage of cargo on deck and if such damage affect vessel's seaworthiness, cargo worthiness or vessel's trading capabilities damage to be repaired by Charterers at their time, risk and expense at the port of occurrence and the vessel is to remain o full hire until such repairs are completed and if required passed by vessel's classification society. Any damages which do not affect vessel's seaworthiness, cargo worthiness of vessel's trading capabilities to be repaired at Owners' option either upon first opportunity or during vessel's next dry dock but in both cases Charterers to be fully responsible for the cost of the repairs against Owners providing supporting vouchers.

Comments:

The Charterers remain responsible for the loading on deck and for the damage incurred therefrom. They will supply all the materials required to secure the cargo on deck. The Bill of Lading will be marked as per the provisions of this clause.

A complete clause for Time Charter period, which refers to the "Deck/Hatch covers loading" could be worded along the following lines:

Charterers have the right to load cargo on deck and/or hatch covers at the Charterers' and/or Shippers' and/or Receivers' entire risk, responsibility, liability, time, and expenses, but always strictly in accordance with the ship's deck and/or hatch covers permissible strength and with the ship's stability and/or visibility.

It is expressly agreed that:

1. the cargo on deck and/or hatch covers will be loaded, stowed, dunnaged, lashed, carried, secured and unlashed and discharged always in accordance with the international regulations, by shore labourers, at the Charterers'and/or Shippers' and/or Receivers' entire time, risk, responsibility, liability and expenses and to the supervision and satisfaction of the Master and/or the ship's P&I Club's surveyor, who will be appointed by the Owners, at the Charterers' cost. The Charterers are to reimburse the Owners for the P&I Club's Surveyor's fees, once the Owners have submitted the original vouchers, which are acceptable even by fax/e-mail. The original should be followed if it is requested by the Charterers.

2. the Charterers and/or Shippers, at their own time, risk, expenses with shore labourers will weld padeyes and/or supply, fit and install stanchions and/or sockets, as well as will supply sufficient lashing, securing, dunnaging materials and/or any other extra fittings, equipment etc. requisite for the safe stowage, carriage of the cargo on deck and/or hatch covers, to the Master's and/or ship's P&I Club's surveyor's supervision and satisfaction.

Charterers guarantee and will remain responsible for providing the Master of

the ship, under their time and expenses and before the ship's sailing from the port, with an original phytosanitary certificate issued by the plant protection organization of the port/s of loading to the plant protection organization of the discharge port/s, certifying that the dunnages and/or other materials have been inspected according to appropriate procedures and fully comply with the regulations and is fully accepted by the plant protection organization of the importing country.
Any gear, stanchions, lashing etc. materials supplied by the Charterers, are to remain at the Charterers' property and the Master should keep a record for all materials that were supplied in this way and should be kept in good condition, but without guarantee and always subject to wear and tear and for any damage which occurred and is connected or attributed to the voyage.
At the discharge port, the welded padeyes and/or the fitted, installed stanchions and/or sockets and/or debris, dunnages and/or any other extra fittings, equipment etc, supplied to the ship and it were necessary for the safe stowage, lashing, dunnaging and carriage of the cargo, will be dismantled, removed and discharged and/or delivered in the custody of the Charterers' Agents, by shore labourers, before completion and the ship's sailing from the port and the redelivery takes place, at the Charterers' own time, risk and expenses. The ship will not sail from the port, unless all such materials have been removed and discharged from the ship and any delays, expenses, consequences and/or possible fines imposed by the Port Authorities, arising from the Charterers' inappropriate discharge of the materials or inappropriate delivery ashore, will be for the Charterers' account.
3. any directly related expenses resulting there from and incurred thereby will be for the Charterers' account and those covered by the Owners are payable on request, supported with the original vouchers, which are acceptable, even by fax or e-mail.
4. for all cargo loaded on deck and/or hatch covers, separate Bills of Lading will be issued, which will be marked strictly as per the Owners' P&I Club's recommendations.
5. Charterers guarantee that they will remain liable and responsible for any consequences, loss, damages, costs, expenses, possible fines Owners may suffer and will indemnify the Owners against third party claims (inclusive of any interest and costs which may be incurred for defending any such claims or recovering any such loss), including, but not limited to, cargo claims and oil/air pollution arising from loading cargo on deck and/or hatchcovers. If the ship is detained by any party and generally its sailing is not allowed from the port, then, the Charterers undertake the responsibility to release the ship, by taking all the necessary and appropriate measures (issuance of security, etc.), as the case may be or arise and any expenses incurred will be for the Charterers account. The ship un-

der any and all circumstances will remain on-hire.
6. the Charterers will be fully responsible for any damage caused to the ship's deck, hatch covers, rails, hydraulic motors etc.
For damage which:
a. affects the effectiveness of the ship and the safe operation and/or safety of the crew and/or affect the trading capabilities and/or seaworthiness and/or cargo worthiness and/or is subject to classification requirements, then such damage, as will be ascertained by the ship's Class surveyor, who will be appointed by the Owners at the Charterers' cost, is to be repaired by the Charterers at their own time, risk and expenses, to the Master's and/or Class surveyors' entire satisfaction. The Charterers are to reimburse the Owners for the Class Surveyor's fees, once the Owners have submitted the original vouchers, which are acceptable, even by fax or e-mail.
The damage will be repaired at the place of occurrence, if it is possible and there are adequate facilities for permanent repairs, or otherwise, temporary repairs are to be made on the spot or as when detected, at the Charterers' time, risk and expenses and to the Master's and/or Class surveyor's entire satisfaction.
The ship will sail or shift to the nearest shipyard, as chosen by the Charterers at their time, cost, in order to complete the repairs. The damage will be repaired to the Master's and/or class surveyor's entire satisfaction, before the ship executes the next trip, or, if it is redelivered to the Owners, the redelivery will take place on dropping last outward sea pilot station of the repair yard, or, at the Owners' option, at equidistance from the Time Charter Party point of redelivery and the ship under any and all circumstances will remain on-hire.
Shipyard's expenses and/or port expenses, including in and out tugs and/or pilots and/or agency fees etc. will be entirely for the Charterers' account.
b. does not affect the effectiveness of the ship and the safe operation and/or the safety of the crew and/or the trading capabilities and/or the seaworthiness and/or cargo worthiness and/or is subject to classification requirements and cannot be repaired by the stevedores, the Owners then undertake to accept the redelivery of the ship or to sail, in order to execute the next voyage, in accordance with the terms of the Time Charter Party. However, the said damage is to be assessed before the ship's departure from the discharge port and the costs to be agreed upon between the Master, Owners, Charterers, and the respective survey company and to be paid together with the Charterers' next hire statement. The survey company will be mutually appointed by the Owners and Charterers, if considered necessary, at the Charterers' and Owners' equal expenses and will promptly inform the Charterers and Owners with the damage report, which will be issued before the ship sails, with a copy of all correspondence exchanged in this connection. The ship will remain always on-hire.
Owners' have the option of having the damage repaired at their next convenient

time or at the ship's next Dry-Docking.
In the case that the Master, Owners, Charterers cannot reach an agreement on the cost with the respective survey company, the Charterers will then undertake to reimburse for the actual cost of the repairs, but not for the time lost, upon presentation of the original bills from the repairers or dockyard, that are consistent with the surveyors' report on the extent of such damage.
c. is small and has been repaired by the crew, the Charterers will reimburse the Owners for the cost on request, but not for the time lost and upon presentation of the Owners' repair statement, which is acceptable, even by fax or e-mail.
However, and in any case:
- the Charterers are responsible for any directly related expenses resulting therefrom and incurred thereby and must pay compensation for any damage caused to the ship and/or restore the ship to its initial condition under their own time, risk and expenses, whether or not any payment has been made by the respective parties and/or stevedores to them, before executing the next voyage or the redelivery takes place to the Master's and Class surveyor's supervision and satisfaction. Any detention through any of the above causes will be for the Charterers' account and the ship under any and all circumstances will remain on-hire.
- the Charterers will not to be liable for damage, when the Master has accepted a statement in writing from the stevedores and/or respective parties that the operations were carried out without any damage, or that the repairs were made to his satisfaction.
If the ship's P&I Club does not supervise the loading operation on deck, then the above term should be amended, and the following wording could be added:
If any dispute for over stowage, lashing, securing of such deck cargo arises, then an internationally recognised independent surveyor will be mutually appointed by the Charterers and Owners, in order to ascertain and confirm that such deck cargo has been stowed, lashed and secured for safe carriage at sea. The cost of the independent surveyor and any extra expenses resulting therefrom and incurred thereby and any detention and loss of time through any of the above will be for the Charterers' account.

-For large ships or for Owners who are unwilling to load on deck, the following wording could be in effect:

No deck cargo is allowed.

59. Arbitration/Mediation.

A top clause for all Charter Parties, which provides that the disputes arising from the Charter Party will be resolved

. by Arbitration.
The place of Arbitration and the prevailing law should be discussed and agreed during the negotiations of the Main Terms. The Owners' legal department and/ or their FD&D Department must be involved in the fixture and the Owners must obtain their advice during the negotiations.

The main arbitration venues are
. in London with English Law to apply
and
. in New York with U.S. Law to apply.
Bimco has added the option for two more arbitration venues
. in Singapore with Singapore or English law to apply.
. in Hong Kong with Hong Kong or English law to apply.

The most acceptable and usable clause for Arbitration is the Bimco's published Arbitration clause called the 'dispute resolution clause' either for the London venue or New York or outside of these two venues. They also provide a mediation clause.

The "Arbitration-Mediation" clause, which, among others, should include:
-that disputes should be resolved by Arbitration.
-the place of Arbitration and which law should apply.
-the number of the Arbitrators and their qualifications.
-that for small amounts in dispute, reference will be in accordance with the small claims procedure of the LMAA.

An executed clause that has been run in a Charter Party is as follows (precise copy):
Arbitration shall be held in accordance with the rules for arbitration procedures of the London Maritime Arbitrators Association and English Law to apply. The two arbitrators appointed by the Owners and the Charterers shall be commercial persons, familiar with the maritime business and the third arbitrator shall be a commercial person as well, who shall also be the chairman of the panel.
If one party fails to appoint an arbitrator, either originally or by way of substitution, for thirty clear days after the other party, having appointed his arbitrator, has served the party making default with notice to make the appointment, the party who has appointed an arbitrator may appoint that arbitrator or act as sole arbitrator in the reference, and his award shall be binding upon both parties, as if he had been appointed by consent. If the two arbitrators fail to agree upon a third arbitrator, the third arbitrator shall be appointed by the president of London Maritime Arbitrators Association.

In the event the amount of claim and counter-claim does not exceed USD .... the parties agree to refer any dispute to a sole arbitrator in accordance with the London Maritime Arbitrators Association shorten for arbitration.

A complete clause for Time Charter period, which refers to the"Arbitration/Mediation" could be worded along the following lines:

This contract is governed by and construed in accordance with English Law and any disputes will be referred to arbitration in London.
The current Bimco's "Dispute Resolution Clause" will apply.

60. Cargo gear.

There are two methods to load and/or discharge the cargo. Either the ship will be fitted with 'cargo gear' (geared ships) and the loading and/or discharging operation will take place with
. its own means
or the port will provide
. shore means/floating cranes,
in which case gearless ships are also workable to trade in this port.
If the ship is a 'geared ship', then the Charterers have the right to load and/or discharge by using its 'cargo gear' as has been described and also have at their disposal any related equipment as on board.
The 'cargo gear' clause should be incorporated into the Charter Party, in order to stipulate the events in case of breakdown or malfunction or disablement of the provided gear. For any lost time of improper working, the ship should be placed off-hire and in case of breakdown the Owners should bear the cost for hiring shore cranes.

The "Cargo gear" clause, which, among others, should include:
-that the provided gear should be in a fully efficient state throughout the currency of the Charter Party and should comply with the regulations of the countries in which the ship will be employed.
-that in the case of breakdown or malfunction of cargo gear, the ship will be placed off-hire for the lost time.
-that in the case of breakdown or malfunction of cargo gear, the Owners will be responsible for the expenses incurred, like stevedore standby/shifting from the berth to the anchorage etc.
- who will arrange and hire the shore cranes, if so, required in the case of a breakdown or malfunction of the cargo gear.
-if there should be any grace period for the Owners in order to repair the dam-

aged gear.

An executed clause that has been run in a Charter Party is as follows (precise copy):
The vessel's cargo gear and all other equipment shall comply with the regulations and/or requirements in effect at the port or ports of call, canals, and countries in which the vessel will be employed. Owners also guarantee that the vessel shall be at all times in possession of a valid and up-to-date certificate onboard to comply with such regulations and/or requirements. If stevedores, longshoremen or other labours are not permitted to work or refuse to work by reason of failure of the Captain, Owners and/or their agents to comply with such regulations or by reason that the vessel is not in possession of such valid and up-to-date certificates of efficiency, then Owners shall take immediate corrective measure. Charterers may suspend hire for time lost thereby and any extra expenses including stevedore standby charges/time waiting and/or demurrage charges for trucks, barges and/or rail wagons and/or overtime charges for Shippers, Receivers, Charterers employees/servants and all other expenses incurred shall be for Owners' account.
All crane drivers grab operations to be arranged/employed by the Charterers at their risk and expense in both load and discharge port.
All cargo handling gear and equipment including derricks/cranes/winches to be maintained in good working order by Owners. In the event of a breakdown or malfunction of derrick/s, crane/s, winch/es by reason of damage, disrepair, disablement or insufficiency in relation to the number of hatches affected, and any stevedore standby costs incurred are to be for Owners' account. Alternatively, Charterers have the option to hire shore gears, in which case Owners to pay for the relevant cost but vessel to remain o hire.
Comments:
The Owners guarantee that the ship will be in possession at all times of valid and up to date certificates, otherwise the Owners will bear the cost of the consequences.
In the event of a breakdown or malfunction of the ship's gear, the Charterers have the option to hire shore gear, but at the Owners' cost.

Another clause reads as follows (precise copy):
Owners confirm vessel's gear is in good working condition. Actual time lost on account of breakdown of the vessel's gear or other gear essential to the loading/ discharging to be deducted from hire in relation to number of workable holds. If due to vessel's gear breakdown, Charterers to engage shore crane with actual cost to be for Owners' account and vessel to be remained on hire, but always subject to Owners/Master's approval which not to be unreasonably with-

held-and if any stevedores/terminals standby times are charged or incurred thereby, such actual expenses shall be for Owners' account.

Comments:

The actual lost time due to breakdown of the ship's gear and the cost to hire shore gear will be for the Owners' account.

A complete clause for Time Charter period, which refers to the "Cargo gear" could be worded along the following lines:

Charterers have the right to load and/or discharge by using the ship's cargo gear as described and have at their disposal any related equipment as on board the ship.

Owners guarantee that the ship's cargo handling gear, including derricks and/or cranes and/or winches and/or grabs and/or equipment, have been serviced and will be maintained in good working order, throughout the duration of this Charter Party.

All the provided as per the ship's description derricks and/or cranes and/or winches, can work simultaneously and can serve all cargo holds during day and/or night and/or during weekends and holidays, as may be requested free of expenses to the Charterers. The ship's auxiliary engines and/or generators have been serviced and will be maintained in good working order, throughout the duration of this Charter Party and can provide continuous and sufficient power to the ship's cargo gear.

It is expressly agreed that:

1. Cargo gear certification:

Owners guarantee that, throughout the duration of this Charter Party, the ship will be in possession of all necessary valid and up-to-date cargo gear certification, in compliance with safety regulations, national and international regulations and all current requirements at all ports and regions of contact.

If the ship is not in possession of valid and up-to-date cargo gear certification, then the Owners will take immediate corrective measures and all consequences and all consequential losses, delays, detention and/or possible fines, damages, costs and expenses of claims and possible shifts will be entirely for the Owners' account. Expenses for Stevedores, longshoremen or other shore labourers' standby, but up to maximum of one shift, supported with the original vouchers will be for the Owners' account. Time lost for the Owners to obtain any necessary permission or time lost for issuance of the missing and/or expired certificates and/or documents and time lost, costs, including those relating to the inspection of the ship and/or ship's cargo gear, by any Authority and/or ship's class and/or ship's flag etc., will be for the Owners' account.

2. Cargo gear breakdown and the hire of shore means.

In the case of a breakdown and/or malfunction and/or disablement of the ship's full cargo gear and/or if the provided electric power to work the full cargo gear is not continuous and/or insufficient, the ship will be placed off-hire for all actual time lost. However, should a breakdown and/or malfunction and/or disablement of the ship's full cargo gear occurs and/or if the provided electric power to work the cargo gear is not continuous and/or is insufficient to a particular hatch/ es, then the ship will be placed off-hire prorate, in relation to the number of hatches that are workable.

In the case of a breakdown and/or malfunction and/or disablement of the ship's full cargo gear or to a particular hatch/es and/or if the provided electric power to work the full cargo gear or to a particular hatch/es is not continuous and/ or is insufficient, the Charterers will provide the Owners with a maximum period of 48 hours, in order to rectify the damage and/or failure etc. before ordering for shore means and/or equipment and/or shore engines, and/or for a possible shift to a berth, as will be ordered by the Charterers and/or Port Authorities, which can provide shore means and/or equipment and/or shore engines, and the time lost, as well as, bunkers consumed and possible shifting expenses incurred will be for the Owners' account. The cost of the shore means and/or equipment and/or shore engines, if they are available, will be for the Owners' account, but should be ordered after the Owners' approval of the cost. Otherwise, and in the case of a disagreement, the Owners should provide and arrange for cheaper shore means and/or equipment and/or shore engines. The ship will be placed again on full or pro rata on-hire, from the time when the ship will again be in a full or pro rata operational condition and commences the loading or discharging operation with the shore means and/or equipment and/or shore engines, as the case may be.

In all cases, any extra expenses incurred, including stevedores' standby charges, but only up-to one shift, as well as the waiting time and/or demurrage charges for trucks, barges and/or rail wagons and/or overtime charges for Shippers', Receivers', Charterers' employees, servants and all other expenses incurred, including possible shifts, whether their affect is partial or not, will be for the Owners' account, but will be supported with the original vouchers. The Owners will also be responsible for all the consequences, fines and all liabilities in respect thereof or arising therefrom.

Any actual time lost due to inefficiency in working the ship with shore means and/ or equipment and/or shore engine will be for the Owners' account.

3. Competent and experienced drivers

All cargo gear drivers, including the grab operators, will be arranged and be employed by the Charterers at their own time, risk, and expenses at all ports of contact. The Charterers guarantee that they will arrange and employ competent and well experienced cargo gear drivers. The Master has the right at any time to stop the loading or discharging operation if the drivers and/or stevedores are not competent and experienced and cause or could cause damage to the ship's gear and/ or equipment and the ship will remain always on-hire. The Charterers, with immediate effect, will substitute the drivers with competent and experienced ones. If any breakdowns and/or damage and/or stoppages are caused due to bad handling by the drivers and/or stevedores and/or other Charterers' servants, then any expenses incurred, delays, consequences will be for the Charterers' account and the ship will remain always and under all circumstances on-hire.

4. Ambient temperature

If the ambient temperature falls below 15 degrees Centigrade while the ship is operating its cargo gear, then the ship is under no circumstances to be held responsible for the general low performance of the gear or for any other delays thereof, because of the low temperature and the ship will remain always on-hire.

For ships discharging with grabs fitted on its gear and Bulldozers/Payloaders will assist in discharging the cargo, the Owners may add to the above:

Charterers may fit suitable grabs on the ship's cranes, in accordance with the class of the ship and remove the same after the loading and/or discharging operation is completed. Time, risk, and expenses incurred to fit, un-fit and restore the grabs, is to be for the Charterers' account. The loading and/or discharging operation with grabs, is to be entirely for the Charterers' risk and expenses, who are also responsible and liable for any damage caused to the ship and/or cargo and/or third party. The ship and/or Owners are not to be kept responsible, in the case of improper grabs fitted on the ship's cranes.

No cargo will be loaded in places inaccessible to grab discharge, such as deep tanks etc. However, the ship's holds sides and/or frames are not to be considered as an inaccessible location.

The Charterers will only use rubber-wheeled bulldozers and payloders and/or tractors in the ship's holds and/or similar equipment, as deemed necessary by the custom of the port. The unit weight of bulldozers and payloaders and/or tractors and/or similar equipment plus cargo weight, will not exceed the ship's tank top strength and its operation will be under the Master's supervision and satisfaction.

For gearless ships for grains or other bulk cargoes, the clause could be worded as follows:

The ship will be suitable for a normal grab discharge operation, with normal size/ type grabs and as far as a standard bulk carrier/ singledecker can be. No cargo will be loaded in places inaccessible to grab discharge, such as deep tanks etc. The ship's holds sides and/or frames are not to be considered as an inaccessible location. The Charterers will only use rubber-wheeled bulldozers and/or tractors in the ship's holds and/or similar equipment, as deemed necessary by the port's customs. The unit weight of bulldozers and payloaders and/or tractors and/or similar equipment plus cargo weight, will not exceed the ship's tank top strength and its operation will be under the Master's supervision and satisfaction.

61. Pilots/ Tugs acts.

The acts of pilots/tugs etc. should usually be agreed on when discussing the main body cl 26 of the NYPE 46 form where the Charterers keep the Owners additionally responsible for the
. 'acts of pilots and tugs'.
Nevertheless, some Charterers prefer to have a separate clause for the acts of pilots/tugs, but the Owners should, however, specifically state that
. drunk or non-competent
pilots/tugs will be rejected and substituted by others at the Charterers' time/ expenses.

A complete clause for Time Charter period, which refers to the "Pilots/Tugs acts", could be worded along the following lines:

Charterers will not be held responsible for acts of pilots, tugboats, or linesmen, who although employed by the Charterers will be deemed to be the Owners' servants and under their instructions.
It is the Master's right to refuse their services, if they are not in a normal state (drunk etc.) and/ or incompetent, in which case they will be substituted by others, at the Charterers' time and expenses.

62. Smuggling-Fines.

This clause apportions the responsibilities in the case smuggling or other infractions take place during the ship's stay at the ports of call. In the case of smuggling, heavy fines may be imposed, and the Master should previously inform and draw the attention of the crew. There should be a provision in the clause that smuggling, or any other offence made by the Charterers' supercargo or representatives etc. should be for the Charterers' account.

The "smuggling-fines" clause, which, among others, should include:
-who will be responsible and what the consequences should be in the case of

smuggling by the ship's crew.
-who will be responsible and what the consequences should be in the case of smuggling by the Charterers' representatives.

An executed clause that has been run in a Charter Party is as follows (precise copy):
Any fines imposed on the vessel, Owners, Master, Officers, or members of the Crew or on Charterers, originating from Master, Officers or Crew contravening local port and/or custom regulations, particularly as regards smuggling, to be for Owners' account and Charterers are not to be responsible for any consequences resulting from such offence.
Comments:
Any time due to above circumstances to be for Owners' account and to be deducted from hire.
The Owners are responsible for all fines which may be imposed for smuggling by the ship's crew. Any lost time will be for the Owners' account and will be deducted from the due hire.

Another clause reads as follows (precise copy):
Any delay, expenses and/or fines incurred on account of smuggling or other infraction shall be for Owners account if caused by the Officers and/or Crew or shall be for Charterers account if caused by Charterers supercargo and/or their staff and/ or agents and/or persons appointed by Charterers.
Comments:
This clause separates the responsibilities in the case that the smuggling was caused by the ship's crew or by the Charterers' representatives.

A complete clause for Time Charter period, which refers to the "Smuggling-Fines" could be worded along the following lines:

During the execution of this Charter Party, the Owners will be liable and responsible for any loss of time and/or for other directly related expenses that were caused by the ship's detention by the Authorities and/or for the possible imposed fines and/or penalties and/or security required, as a result of actions of the Master, Officers, crew, such as smuggling and/or theft and/or desertion and/or suspected intention of the same and/or for any other violation of the laws and/or contravening local port and/or custom regulations.
It is strictly forbidden for the crew, Officers or Master to have any illegal merchandise on board. Any such merchandise found on board is to be confiscated by the Master.
During the execution of this Charter Party and at all ports of contact, the Char-

terers will be liable and responsible for any loss of time and/or other directly related expenses that were caused by detention of the ship by the Authorities and/or for the imposed fines and/or penalties and/or security required, as a result of actions of the Charterers' supercargo and/or their staff and/or Agents and/or persons appointed by the Charterers, such as smuggling and/or theft and/or desertion and/or suspected intention of the same and/or for any other violation of the laws and/or contravening local port and/or custom regulations.
In the event that contraband and/or un-manifested drugs or goods are found to have been shipped as part of the goods and/or in containers on board, any fines, taxes etc. which might be imposed will be for the Charterers' account and the ship will remain on hire during any time lost as a result thereof, unless it can be established that the Master, Officers and/or Crew were involved in such action.

63. In Lieu Of Holds' Cleaning (ILOHC).

The Charterers are to redeliver the ship in such a clean condition
. as on delivery.
After completion of discharge the Charterers must arrange and employ on board the ship shore workers to clean the ship's holds' compartments, prior to its redelivery to the Owners. Obviously, that will take time and cause heavy expenses. Alternatively, they could agree with the Owners through this clause, to compensate them and redeliver the ship with holds
. as left after the completion of the discharge of the last trip,
but in any case, without cargo residues, as will be stipulated in the Charter Party. The ship's crew will clean the holds, as the ship will now trade for the Owners' account. The paid amount is negotiable and varies depending on the size of the ship, the total holds which have been used etc.

The agreement should be made based on one of the following two ways:
- on a 'lumspum' amount, which is widely used in the Charter Parties and it is acceptable by both Charterers and Owners. The agreed amount will be paid at the end of the Charter, regardless of the holds used by the Charterers.
- on a 'lumpsum' amount for each hold used by the Charterers. That is preferred by the Charterers, when they are to load the ship in alternative holds, therefore saving some money.

The inclusion or not to the agreed amount of the removal/discharge of the debris/dunnage and/or all other materials that have been used to secure/protect the cargo during the execution of the trip, is to be specified during the negotiations. Owners should take into consideration the fact that in some ports the discharge of such materials has a considerable cost while in others they are not even allowed to be discharged. The Owners should agree that the same must be discharged from the ship, before its redelivery, entirely for the Charterers' own time/ risk/expenses. The ship should not sail from the port unless all such materials have been removed/discharged from the ship and any delays/ expenses/consequences and/or fines imposed by the Port Authorities arising from the Charterers' inappropriate discharge of such materials or for their inappropriate delivery ashore must be for the Charterers' account and under the terms of the Charter Party, the ship must always remain on-hire. It should also be agreed that in the case that the materials are not allowed to be discharged at the last port of discharge and before the ship's redelivery, the ship should proceed to the most convenient anchorage/ berth of another port or country and discharge the same in barges or ashore, under the Charterers' own time/risk/ expenses. The ship should always remain on-hire until it fully discharges the particular materials.

The "in lieu of holds' cleaning" clause, which, among others, should include:
-whether the Charterers have the option to redeliver the ship with unclean holds.
- whether the removal/discharge of the debris/dunnage or other materials are included in the agreed amount.
-the amount of compensation for delivery of the ship with unclean holds and when it should be due.

An executed clause that has been run in a Charter Party is as follows (precise copy):
Charterers to have the option of redelivering the vessel against paying Owners lumpsum USD.... in lieu of hold cleaning. Debris/dunnages if any shall be removed from vessel's holds/deck prior vessel's sailing from discharge port under Charterers time/risk/expenses.
Comments:
The Charterers have the option to redeliver the ship with unclean holds by paying the Owners a lumpsum amount as will be agreed. Debris/dunnage etc. will be removed from the ship, before sailing from the discharge port, at the Charterers' time/risk/expenses.

Another clause reads as follows (precise copy):
Vessel to be redelivered to Owners with holds in same condition as on delivery, normal wear and tear excepted. Charterers shall have the option to redeliver the vessel with unclean holds paying an amount of USD .... per used hold in lieu of hold cleaning.
Comments:

The Charterers keep the option to redeliver the ship with unclean holds and will pay an amount for each hold which has been used. The agreement per hold is mainly in effect when the Charterers intend to load the ship in alternative holds.

Another clause reads as follows (precise copy):
Charterers to redeliver the vessel with unclean holds paying the Owners a lumpsum of USD .... in lieu of cleaning including removal/disposal of dunnage which to be for Owners' time and expenses, however, always understood that Charterers are to remove dunnage from holds and place same on deck at their time/ risk/ expenses.
Comments:
The removal of dunnage will be for Owners' account, provided that the Charterers will remove such dunnage from the holds on the ship's deck under their time/ risk/expenses.

A complete clause for Time Charter period, which refers to the "In Lieu Of Holds' Cleaning (ILOHC)" could be worded along the following lines:

The ship will be redelivered to the Owners with its holds in the same condition as on delivery, normal wear and tear excepted.
However, the Charterers will have the option to redeliver the ship with unclean holds, by paying a lumpsum amount of USD ...... The amount due will be paid together with the Charterers' final hire statement.
The ship, in any case will be redelivered to the Owners free of cargo residues in its cargo holds.
Debris and/or dunnage and/or lashing or any other materials that have been used to secure and/or protect the cargo during its sea voyage will be discharged from the ship before its redelivery, at the Charterers' entire time, risk and expenses. The ship will not sail from the port, unless all such materials have been removed and discharged from the ship and any delays, expenses, consequences and/or possible fines imposed by the Port Authorities, arising from the Charterers' inappropriate discharge of the materials or inappropriate delivery ashore, will be for the Charterers' account and the ship will remain always on-hire. In the case that the materials are not allowed to be discharged at the last port of discharge, then, before the ship's redelivery, the ship will proceed to the most convenient anchorage, berth of another port or country in order to discharge the same in barges or ashore, under the Charterers' time, risk and expenses. The ship will remain always on-hire, until having fully discharged the respective materials and the ship is redelivered on dropping last outward sea pilot station or at the Owners' option at equidistance from the actual point of redelivery.

64. Instructions to the Charterers' Servants-Charterers' procurement.

This clause should protect the Owners and will request from the Charterers' servants, such as Agents/bunker suppliers etc. to confirm directly to the Owners, that all of the ship's costs which are for the Charterers' account, have been paid and/ or to notify them that they will not accept any responsibility or claim, owing to delay of the Charterers' payment.
The Charterers are not allowed to order any supplies in the name of or credit of the ship/Owners.
This clause should be in effect for single trips or very short trips or when the Owners are fixing with 'names' not well known in the market.

The 'Instructions to the Charterers' Servants-Charterers' procurement' clause, which, among others, should include:
-that the Charterers will remain responsible for the payment of their Agents/ fuel suppliers and all other parties involved.
-that the Charterers shall not procure, or permit the procurance for the ship, of any supplies, necessities, or services in the name or for the ship's/Owners' credit.

A complete clause for Time Charter period, which refers to the "Instructions to the Charterers' Servants/Charterers' procurement" could be worded along the following lines:

During the duration of this Charter, the Charterers undertake to instruct their Agents and/or fuel suppliers and/or all parties involved in general, which will provide services to the ship on the Time-Charterers' behalf, to countersign the Master's presented letter that they are rendering these services on the Time Charterers' behalf only and without any liability or responsibility for the payment of their services from the ship, Master, Owners. The ship, Master and Owners will not be held responsible for the Charterers' unpaid bills. If the Master is unable to obtain such a letter, the Owners have the right to place all parties acting for the ship on the Charterers' behalf, on notice that they will not accept any responsibility or claim owing to payment delay or non-payment by the Charterers.
The ship's port disbursements and/or canals and straits costs will be paid in advance by the Charterers to the respective Agents and the Owners have the right to request from the Agents to confirm in writing that all the disbursements and/ or canal and straits' costs have been paid before the ship approaches the respective places. The Charterers will remain responsible for all delays, expenses, liabilities, consequences arising from their failure to keep the Agents paid in advance

and the ship will remain on-hire, until the Charterers fulfil their contractual obligations.
In no event are the Charterers to procure, or permit the procurance for the ship, of any supplies, necessities, or services in the name or for the ships' or Owners' credit.
When the Charterers supply the ship with fuel for the voyage or for the redelivery purposes, the Charterers will clearly instruct the suppliers to accept the Master's remark on the delivery bill that the supply of the fuel was ordered and arranged by the Time Charterers and the ship, Owners of the ship accept no liability for the cost value of the fuel supplied. The Charterers will remain fully responsible for all delays, expenses, liabilities, consequences arising from their suppliers' refusal to accept the Master's remarks on the delivery bill.

65. Winter navigation.

The ship's performance is in fact affected, when the ship navigates or is under any operation in
. low temperatures.
This clause should protect the Owners in the case that no such provision has been added to the ship's initial Time Charter description.

The "Winter navigation" clause, which, among others, should include:
-that the ship will not be held responsible due to low temperatures.

A complete clause for Time Charter period, which refers to the "Winter navigation", could be worded along the following lines:

If the ship navigates within I.N.L., where the ambient and sea water temperatures fall below 15 degrees Centigrade, the ship is under no circumstances to be held responsible for general low performance (main engine, cargo gear, auxiliary machinery) and/or for any other delays thereof, because of the low temperatures.

66. Drug and alcohol guidelines.

The shipping companies should have
. a written policy on drugs and alcohol abuse on board the ship.
The Master/Officers/crew should not be impaired by alcohol at any time and especially during the performance of their duties. Obviously, no seafarer should navigate the ship or operate any ship equipment when his perception is affected by drugs or alcohol. The Company's policy should keep alcohol distribu-

tion under control and monitor its consumption. The Charterers should be held harmless in the case that the ship has received claim from any country/party in respect to alcohol or drugs found on board the ship.

The "Drug and alcohol guidelines" clause, which, among others, should include:
-that the Owners are responsible to keep policy/guidelines on drug and alcohol on board the ship.
-that the Owners are responsible in the case that the ship is detained or for imposed fines, due to alcohol/drugs found on board the ship.
-the Charterers are responsible in the case that the ship is detained or for imposed fines, due to drugs/alcohol found on board, but on the persons representing them.

A complete clause for Time Charter period, which refers to the "Drug and alcohol guidelines" could be worded along the following lines:

Owners guarantee that there is a policy and guideline on Drug and Alcohol abuse (Policy) applicable to the ship, which meets the minimum standards of the Oil Companies International Marine Forum Guidelines for the control of Drugs and Alcohol on board dated June 1995 and any later amendments or equivalent. The Owners will exercise due diligence throughout the period of this Charter, to ensure that the policy is complied with.
Owners are the sole responsible party in the case that the ship is detained, or for any possible imposed fines to the ship's Master, Officers, Crew, due to alcohol, drugs etc. found on board the ship. The ship will be placed off-hire, from the time when the ship's full operation is prevented, until such time as the ship is returned to its full operational condition and all extra expenses incurred and/or consequences will be for the Owners' account.
Charterers are responsible in the case that drugs, alcohol etc. are found on the person/s representing them, or on their servants or on their Agents or placed on the ship without the knowledge of the Master or loaded on board together with the cargo or equipment or other property. The Charterers are responsible for all consequences and for any and all possible imposed fines, extra costs, including for the ship's detention, in which case the Charterers undertake the responsibility to release the ship by taking all the necessary and appropriate measures, as the case may be or arise, and the ship will always remain on-hire.

67. ISPS compliant ports.

The International Ship and Port Security Code (ISPS Code) adopted by the parties contracting to SOLAS in December 2002 and applied on the 1st of July 2004

as a defence due to the September 11th, 2001 attacks in New York.
The ISPS is contained in the SOLAS (Safety Of Life At Sea) and imposes the minimum measures which any ship/port must take, in order to ward off outside threats and the security level at which the ship will trade at each port of contact. As it is mandatory for all ships, there should obviously not be any Charter Party issued, which does not include a clause, which refers
. to ISPS.
Some countries, such as the U.S.A., accept ships coming from a non ISPS compliant ports only after the ship's Master has taken measures while contacting such port/s, and has increased the security level accordingly. It is possible for the U.S.A. Port Authorities to refuse the ship's entrance into their territorial waters or into the port.

The "ISPS compliant ports" clause, which, among others, should include:
-that the ship will trade only between ISPS compliant ports.
-under which conditions the ship will trade to a non ISPS compliant port/s.

An executed clause that has been run in a Charter Party is as follows (precise copy):
The vessel will trade only between ISPS compliance ports. BIMCO ISPS clause for Time Charter to apply.
Comments:
This clause specifies that the ship will trade only between ISPS compliant ports. Therefore, the Charterers do not have the right to direct the ship to a non ISPS compliant port and the Master would be entitled to refuse trading in such areas.

Another clause reads as follows (precise copy):
Charterers will endeavour to trade only between ISPS compliant ports and in case vessel trades to non ISPS compliant port/s then Charterers shall be held fully responsible for any and all consequences arising out of vessel's trading in such port/s. The vessel's last 5 ports before her redelivery shall be imperatively ISPS compliant ports.
In case vessel be instructed to trade to the areas /ports where vessel must provide Level 2 or more (always subject the area/port be considered safe) then any kind of vessel's delays/expenses/additional port and cargo expenses be incurred will be for Charterers' account in full.
Comments:
This clause is mainly in effect for long periods, where the ship will contact many ports. The Charterers have the right to trade the ship in a non ISPS compliant ports, but provided that the last 5 ports before the ship's redelivery are ISPS compliant ports.

A complete clause for Time Charter period, which refers to the "ISPS compliant ports" could be worded along the following lines:

During the currency of this Charter Party, the Charterers are entitled to trade the ship to a non ISPS compliant port/s subject to the Owners' prior approval that they will check the safety of the port/s. The Charterers remain fully responsible and liable whether the ship increases the level or not, for any and all consequences of any nature, delays, expenses, extra expenses for the crew and insurers, arising out or incurred, because of the ship's trading, or because the ship traded to a non ISPS compliant port/s or areas. The Charterers guarantee that the ship's last 5 ports before its redelivery will imperatively be ISPS compliant ports and the Owners have the right to refuse to proceed to a non ISPS compliant port in the case that the Charterers do not satisfy them, according to their presented redelivery schedule.
The Bimco's ' ISPS/MTSA Clause for Time Charter Parties 2005' to be in effect or any subsequent amendment thereto.

68. ISM.

A standard clause incorporated in all Charter Parties, requesting the Owners to guarantee that both the ship and the company will comply with the requirements of the ISM Code.
The International Safety Management Code was made mandatory for dry cargo ships from July 2002 and generally provides standards for the safety at sea and at port, in order to prevent accidents and to avoid damage to the marine environment and to the ship.

A complete clause for Time Charter period, which refers to the "ISM Clause" could be worded along the following lines:

The Bimco's current 'Standard ISM Clause for Voyage and Time Charter Parties' will be in effect.

69. Confidential fixture.

When a fixture is required by one party to be kept private and confidential (abbreviated as 'p&c') for their own or commercial reasons, then this clause will provide the same.
Certain Charter Parties also provide for penalties payable from the party which would violate this clause, but it is in fact difficult to prove which party has

breached this clause.

A complete clause for Time Charter period, which refers to the "Confidential fixture", could be worded along the following lines:

This Charter Party and the negotiations that led to this Charter Party should be kept strictly private and confidential by the Owners and Charterers and brokers involved and should not be referred to and/or discussed and/or reported anywhere.

70. Supersede.

Many times, there are contradictions or ambiguities between all legs of the agreement, i.e., Main Terms/Main Body of the printed Charter Party/Rider clauses. This clause should assist the contractual parties to specify which terms supersede others in the government Charter Party.

A complete clause for Time Charter period, which refers to the "Supersede" could be worded along the following lines:

The main terms agreed will supersede the Charter Party.
In the event of contradictions or ambiguities between the agreed rider or additional clauses and the agreed terms and conditions of the printed Charter Party form, the rider or additional clauses will supersede.

THE AUTHOR

George Tsoudis is a Greek citizen, born on the 1st of July 1959, on the island of Cephalonia that is located in the Ionian Sea. His family moved to Athens in 1965, in order to provide a better life and education for him and his brother. He received his early education at public schools in a northern suburb of Athens. After graduating from high school, he decided to follow his island's tradition and get involved with the sea. He enrolled in the most famous at that time Greek school for maritime studies, where he studied shipping. Upon graduation, he also received the State Degree for Specialization in the Shipping Sector, and then moved to London in order to complete his studies in shipping at an independent college for further education, the 'London school of Foreign Trade'. Professionally, he started out as a competitive broker in late 1983 and has since been employed in shipping agencies (ship-owners and charterers) working as a contractor but mainly as a 'home-broker', while he has also been involved in operations and claim departments for all type/size of dry cargo ships. He has worked for many years in the dry cargo shipping industry, having reached the level of operations and chartering manager. Seeing that the new blood coming in to work as brokers may need his knowledge in starting their careers, as well as to also to assist his colleagues in certain specific areas, he decided to publish his first book containing all the experience gained during the performance of his duties in the various departments of shipping agencies.
He now lives in one of Athens' southern suburbs with his lovely wife, enjoying life with his family and spending time with his two grandchildren, Taxiarches and Lydia.

Made in United States
Orlando, FL
04 November 2026

90200818R00372